# Modification of Proteins

## Food, Nutritional, and Pharmacological Aspects

# Modification of Proteins

## Food, Nutritional, and Pharmacological Aspects

**Robert E. Feeney,** EDITOR
*University of California, Davis*

**John R. Whitaker,** EDITOR
*University of California, Davis*

Based on a symposium
jointly sponsored by the Divisions of
Agricultural and Food Chemistry
and Biological Chemistry
at the 180th Meeting of the
American Chemical Society,
Las Vegas, Nevada,
August 27–28, 1980.

ADVANCES IN CHEMISTRY SERIES **198**

AMERICAN CHEMICAL SOCIETY

WASHINGTON, D. C. 1982

Library of Congress CIP Data
Modification of proteins.

(Advances in chemistry series, ISSN 0065-2393; 198)

Includes bibliographies and index.

1. Proteins—Congresses.
I. Feeney, Robert Earl, 1913- . II. Whitaker, John R. III. American Chemical Society. Division of Agricultural and Food Chemistry. IV. American Chemical Society. Division of Biological Chemisty. V. Series.

QD1.A355 no. 198 [TP453.P7] 540s [641.1'2]
82-1702 AACR2 ADCSAJ 198 1-402
ISBN 0-8412-0610-4 1982



PRINTED IN THE UNITED STATES OF AMERICA

# Advances in Chemistry Series

# FOREWORD

Advances in Chemistry Series was founded in 1949 by the American Chemical Society as an outlet for symposia and collections of data in special areas of topical interest that could not be accommodated in the Society's journals. It provides a medium for symposia that would otherwise be fragmented, their papers distributed among several journals or not published at all. Papers are reviewed critically according to ACS editorial standards and receive the careful attention and processing characteristic of ACS publications. Volumes in the Advances in Chemistry Series maintain the integrity of the symposia on which they are based; however, verbatim reproductions of previously published papers are not accepted. Papers may include reports of research as well as reviews since symposia may embrace both types of presentation.

# CONTENTS

# PREFACE

The symposium upon which this book is based was a continuation and updating of a similar one held in 1975 at the First Chemical Congress of the North American Continent in Mexico City, Mexico. The first symposium was entitled "Food Proteins: Improvement through Chemical and Enzymatic Modification." The present symposium has a greater emphasis on nutrition and adds pharmacological science. Not only is there a large increase in the amount of information available on these subjects since 1975, but the information is of higher quality and level of sophistication.

Since the last symposium there have been several changes in attitudes or directions. One of these has been the recognition that alkali treatment of proteins is a chemical procedure, and in particular, a chemical modification. Another has been the recognition that, although there are still difficult problems in assessing the safety aspects of chemical modifications of foods, both chemical and enzymatic treatments can be useful in laboratory experimentation and, perhaps, in eventual application to food production. Increase in nutritional value through modifications is now recognized.

The main purposes of this symposium were to delineate the current status in protein modification and to describe the major advances. In addition, the speakers and the subjects were selected to show how the different objectives in food, nutritional, and pharmacological sciences have many common threads, and how chemical and enzymatic modifications are used in research in all three fields. These interrelationships are reflected not only in the practical interest in some of the research results, but also in the actual research laboratories in which some of the industrial organizations have closely intertwined research programs in more than one of these fields. The laboratory of Thomas Richardson at the University of Wisconsin has used chemical modification of amino groups by reductive methylation with isotopically labeled reagents as an analytical tool to follow the highly important interactions of milk proteins during handling, processing, and storage. Another use has been the removal of unwanted constituents as shown by the laboratory of John Kinsella at Cornell University where carboxymethylacetylation of amino groups has been used to separate nucleic acids from yeast proteins.

Other uses have been to protect against modifications (the Maillard reaction) during processing.

Increases in nutritional quality have been achieved through the covalent attachment of amino acids by chemical or enzymatic modifications.

The pharmacological area, the new one introduced in this symposium, has been one of intense activity since 1975. Chemical derivatization of natural products such as endocrines has been used to control their utilization in the body. On the other hand, fundamental studies of the specific chemical derivatization of enzymes by chemicals destroying their active sites is a rapidly expanding area of research. Some of these reagents, frequently looked upon as possible perfect target-specific drugs, have been called "suicide reagents" by some workers. In the reactions of these suicide reagents, a chemical modification is actually an enzymatically driven process.

Most readers will probably agree that there have been extensive advances in the fundamental areas of chemical and enzymatic modifications. The organizers do not expect similar agreements as to the success of the application of these methods in food, nutrition, and pharmacological areas. This lack of agreement would be particularly evident for the eventual adoption of some methods for human use. Perhaps it will be evident, however, that fundamental studies will show what types of materials are required for human use and that, after proper research into the health and safety aspects, eventually some of these methods will be applied for human use.

The symposium organizers hope that our attempt to interrelate the different areas will show the common thread running through them. We apologize for any subjects that have been omitted and trust that the excuse of inadequate space for such a wide subject is understood. Any lack of coherence between the chapters is certainly not the fault of the authors but could only be laid to the efforts of the organizers to emphasize the ramifications.

ROBERT E. FEENEY
Department of Food Science and Technology
University of California
Davis, California 95616

JOHN R. WHITAKER
Department of Food Science and Technology
University of California
Davis, California 95616

December 24, 1980.

# OVERVIEWS

# 1

# Chemical Modification of Proteins: An Overview

ROBERT E. FEENEY, R. BRYAN YAMASAKI[1], and KIERAN F. GEOGHEGAN[2]

Department of Food Science and Technology, University of California, Davis, CA 95616

*Protein chemical modification is a problem-solving technique in research and technology. Modifications also occur in natural deteriorations. Generally these modifications are with the most reactive side chains and are predominantly oxidations, reductions, and nucleophilic and electrophilic substitutions. Deteriorations include peptide bond scissions, racemizations, β-eliminations, and formation of products by the reaction of proteins with added chemicals. Proteins are modified intentionally for structure–function relationship studies or for development of new and improved products. Although appearing quite varied, the techniques used in pharmacological, food and feed, or other industrial areas differ more operationally than from major differences in the levels of chemical sophistication that are used.*

Among the numerous special techniques that have been developed in recent years to assist the investigation and exploitation of protein structure, chemical modification of proteins has assumed a prominent role (*1, 2, 3, 4*). A glance at the contents page of any modern biochemical journal bears witness to this. A specialized approach to research using chemical modification of proteins has developed and is used presently in fundamental studies of the structure and function of enzymes and other proteins as well as in studies related to industrial needs, foods, nutrition,

[1] Present address: Department of Chemistry, 2145 Sheridan Road, Northwestern University, Evanston, IL 60201.
[2] Present address: Biophysics Research Laboratory, Brigham and Women's Hospital, 75 Francis Street, Boston, MA 02115.

0065-2393/82/0198-0003$10.75/0

and pharmacology. The common denominator is the effects of chemical modification of the side chains of the amino acids on the properties of the proteins.

Even before scientists recognized the distinctive chemical features of proteins, workers were using chemical modifications in such areas as leather tanning and fabric dyeing. Older procedures using organic compounds (e.g., formaldehyde) and inorganic salts (e.g., chromates) have been refined and extended on the basis of a modern understanding of how they function. For example, formaldehyde now is often replaced by glutaraldehyde, a reagent for cross-linking proteins. Specific modifications can be used to give materials with particular properties and performance. For example, the controlled scission and reoxidation of disulfide bonds is the key to the permanent press and permanent wave industries, both of which are of considerable economic importance. (One hopes that their social impact occasionally measures up to the effort and expenditure that are required of their customers!)

Not even the proteins that we eat escape chemical modification. In earlier times this normally was the result of treating food with acid or alkali, and it may be a result of the long acceptance of such procedures that their consequences in terms of alterations to protein structure have been ignored by many today. The ancient Mayan practice of treating corn and other grains with alkali preserved the food, but is was valued also for its pleasing effect on the product (*5*). In nutritional terms alkali treatment had the mixed effect of causing losses of some amino acids but increasing the digestibility, a valuable benefit to these people. Some believe that when corn first was grown in Southern Europe and became a major factor in the diet there, pellagra became endemic in some areas because the crop had been imported without the cultural habit (alkali treatment) that made it an effective staple food for the Maya.

The chemistry of the effect of formaldehyde in the tanning process probably is related closely to its action in the production of toxoids (*6*). This procedure was developed in the first part of this century mainly by trial and error and still thrives today. Toxins are treated with formaldehyde for several weeks at temperatures close to 40°C. This results in the toxin being changed in such a manner that it retains its capacity to elicit an immune response when injected into an animal or person but it lacks the capacity for the specific and damaging attack that is associated with the native toxin. Imprecise as this procedure seems today, the satisfactory results obtained through strict quality control have resulted in its continuing use and the production of many millions of doses of life-saving vaccines. However, more specific modifications should be available for such purposes from current research.

Some drugs and toxic compounds owe their activities to an ability to form covalent products with important target substances (*7, 8, 9*).

Among these targets are enzymes and other important proteins such as hemoglobin. For a number of years, the covalent modification of the α-amino groups of hemoglobin with cyanate was studied intensively as a possible means of therapy for the sickle cell trait. Interest was based on the finding that the modification displaces the oxygen association curve of hemoglobin to the left, reducing the molecule's tendency to give up oxygen and to be converted to its deoxy form (in which it is liable to the gelation that results in the sickling of the erythrocyte). Unfortunately, orally administered cyanate had forbidding side effects in volunteer patients and did not appear to reduce the incidence of the disorder's "painful crisis." (For a concise review on this matter, *see* the article by Harkness (*10*)).

It is difficult to devise a precise system by which chemical modifications of proteins can be classified. A useful division is that between (a) those that are performed intentionally for specific purposes and (b) those that can be described as deteriorative or incidental to the processing, storage, or aging of protein-containing materials.

Specific modifications of proteins result from adding a selected reagent to the pure protein or crude protein-rich material. This may be done in the course of a fundamental study in protein chemistry or as a step in the production of a bulk protein product for practical purposes. The same chemical modification can be useful in both processes. For example, enzyme chemists use charge-changing modifications to dissociate oligomeric proteins to their monomer components, while the same modifications are proposed as a means of solubilizing yeast proteins to permit their extraction for use in foods (*11*). This chapter is concerned mainly with the many types of intended modifications.

Deteriorative chemical modifications of protein structure are found in nearly all biological systems and have been discussed at length in a recent symposium (*12*). They will be described only in outline here. (For an overview, *see* the article by Feeney (*13*)).

The aim of this chapter is to describe briefly the progress and current status of the chemical modifications of proteins and how they are being applied to food, nutritional, and pharmacological sciences. In many instances, the subjects covered superficially here are covered in depth in other chapters in this volume. We hope that omissions will be understood sympathetically as due to the wide scope of the subject.

## *Chemistry of the Reactions*

**Reagents for Individual Amino Acid Side Chains.** Although there is a wide range of protein modification reactions reported in the literature, the similarities among these reactions make it possible to categorize most of them into a small number of classes. The most important classes of reactions between reagents and proteins are (a) acylations and related reactions, (b) alkylations, (c) reductions and oxidations, and (d) aromatic ring substitutions. Other important reactions are those between proteins and reagents such as cyanogen bromide, the primary products of which undergo spontaneous or induced rearrangements that lead to peptide bond cleavage. In some cases, modification of a particular side-chain group is possible with different reagents involving different classes of reaction, whereas two nonrelated side-chain groups may be modified by different reagents involving the same class of reactions.

Modifying reagents usually are directed towards side chains of a single type, but few are completely specific (*see* Table I). Because of this, no discussion of a reagent's effect on proteins is complete without attention to its minor side reactions. Examples of the major reactions for each side-chain functional group are given in Figures 1–9 along with details of the optimum conditions for each reaction. However it should be remembered that a group whose modification proceeds under conditions atypical for its class of side chain often may be especially interesting because its unusual behavior implies an unusual environment within the protein molecule.

**Sulfhydryl Groups.** Sulfhydryls are probably the most reactive side groups commonly found in proteins and, as such, lent themselves at an early stage to the development of reagents for their specific modification. Several colorimetric procedures for estimating the number of thiol groups in a protein have been based on the relative ease with which they are oxidized (*see* Figures 1a–1d). Perhaps the most popular of these methods uses 5,5′-dithiobis(2-nitrobenzoic acid) (DTNB), commonly called Ellman's reagent (*14*). The strongly colored thionitrobenzoate anion released as each protein thiol group reacts (*see* Figure 1a) can be determined from its absorption at 412 nm ($\epsilon = 1.36 \times 10^4$ $M^{-1}$ $cm^{-1}$ at pH 8). Ellman's reagent also can be used to quantitate hydrogen sulfide in solution, but 2 mol of thionitrobenzoate anion and 1 mol of elemental sulfur are formed from 1 mol of hydrogen sulfide and 1 mol of DTNB (*15*).

Wilson et al. (*16*) has presented a spectrophotometric method whereby a protein thiol group that reacts slowly with Ellman's reagent may be studied using a variety of nonchromogenic disulfides as intermediates. In this method, a protein thiol is allowed to react with a small molecule disulfide, RSSR, in the presence of DTNB. If the rate of the reaction of the generated RSH with DTNB is fast (*17, 18*) relative to the rate in which RSH is generated, then the rate of increase in absorption at 412 nm would be the sum of the rates of reaction of the protein thiol with DTNB and RSSR. The characteristics of the environment of the protein thiol and the shape and charge of the low-molecular-weight disulfide strongly affect the capacity of the low-molecular-weight disulfide to serve

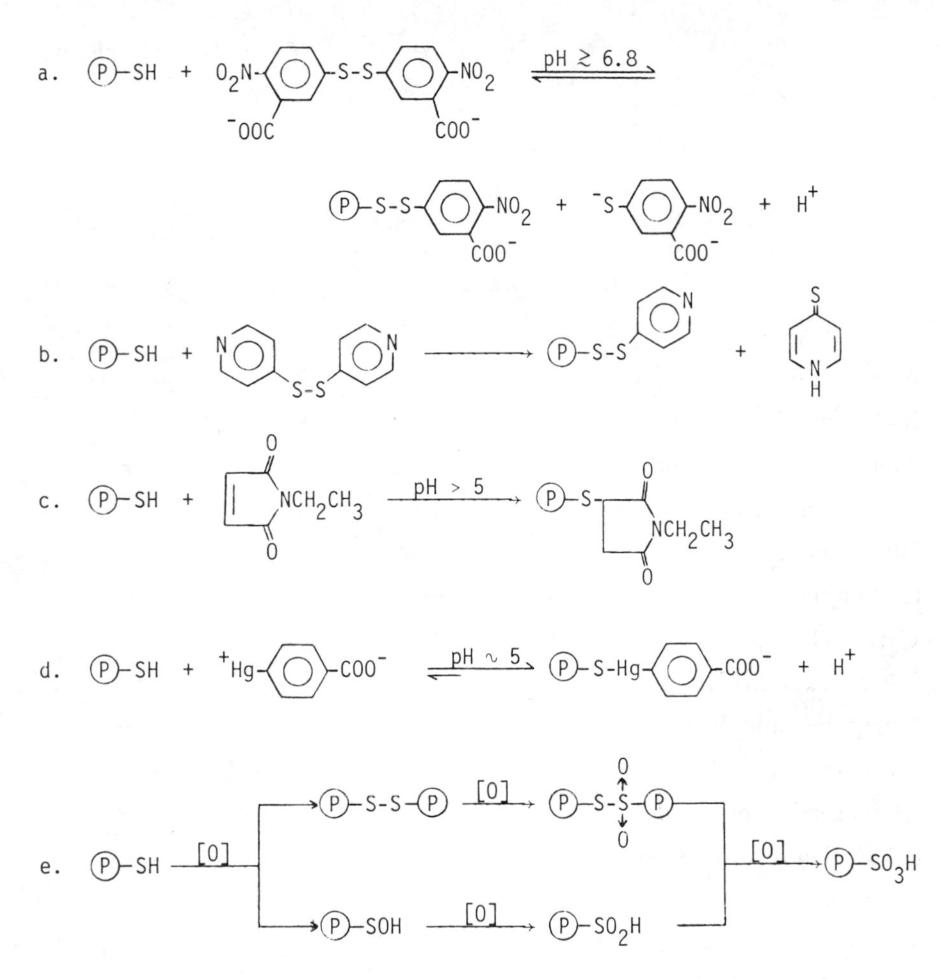


*Figure 1. Modifications of sulfhydryl groups: (a) DTNB or Ellman's reagent; (b) 4,4′-dithiodipyridine; (c) NEM; (d) PMB; (e) hydrogen peroxide oxidations.*

as an intermediate. Aminlari et al. (*19*) applied this technique on penalbumin and bovine serum albumin (BSA) using cystamine as the intermediary RSSR (*see* Figure 10).

Hydrogen peroxide commonly is used to oxidize sulfhydryls in proteins to disulfides or, on more extensive treatment, to sulfonic acids (*1*). However, under certain conditions reacting hydrogen peroxide with protein sulfhydryls may lead to the formation of sulfenic acids (*21*). The inactivation of highly purified papain by stoichiometric amounts of hydrogen peroxide appears to be due almost exclusively to the formation of papain sulfenic acid (*22*). The formation and reactions of sulfenic acids in proteins have been reviewed elsewhere by Allison (*21*).

**Table I. Side-chain**

| *Reagent*[c] | $-NH_2$ | $-SH$ | —C₆H₄—OH |
|---|---|---|---|
| Acetic anhydride (5-1, A-1) | +++ | +++[d] | +++[e] |
| *N*-Acetylimidazole (5-1, A-6) | ±± | +++[d] | +++[e] |
| Aldehyde/$NaBH_4$ (6-8, A-1) | +++ | − | − |
| *N*-Bromosuccinimide (8-9, A-8) | − | +++ | ++ |
| Butanedione | ± | − | − |
| Carbonyl compound/amine borane | +++ | − | − |
| Carbonyl compound/$NaBH_3CN$ | +++ | − | − |
| *N*-Carboxyanhydrides (5-1) | +++ | − | − |
| Citraconic anhydride | +++[e] | ++[d] | ++[d] |
| Cyanate (5-2, A-1) | +++ | +++[d] | ++[d] |
| Cyanogen bromide (10-3) | − | + | − |
| 1,2-Cyclohexanedione (10-1, A-5) | ± | − | − |
| Diazoacetates (7-1) | − | ++ | − |
| Diazonium salts (9-3, A-7) | +++ | + | +++ |
| Dinitrofluorobenzene (6-5) | +++ | +++ | ++ |
| 5,5′-Dithiobis(2-nitrobenzoic acid) (8-3, A-2) | − | +++[e] | − |
| Ethoxyformic anhydride (5-1, A-7) | +++ | − | − |
| *N*-Ethylmaleimide (6-2, A-2) | ±± | +++ | − |
| Ethyl thiotrifluoroacetate (5-1) | +++[d] | − | − |
| Formaldehyde (6-7, A-1) | +++ | +++ | +++ |
| Haloacetates (6-1, A-2) | + | +++ | − |
| Hydrogen peroxide (8-7, A-9) | − | +++ | − |
| α-Hydroxycarbonyls/$NaBH_4$ | +++[f] | − | − |
| 2-Hydroxy-5-nitrobenzyl bromide (6-6, A-8) | − | ++ | − |

**Reactivities**[a, b]

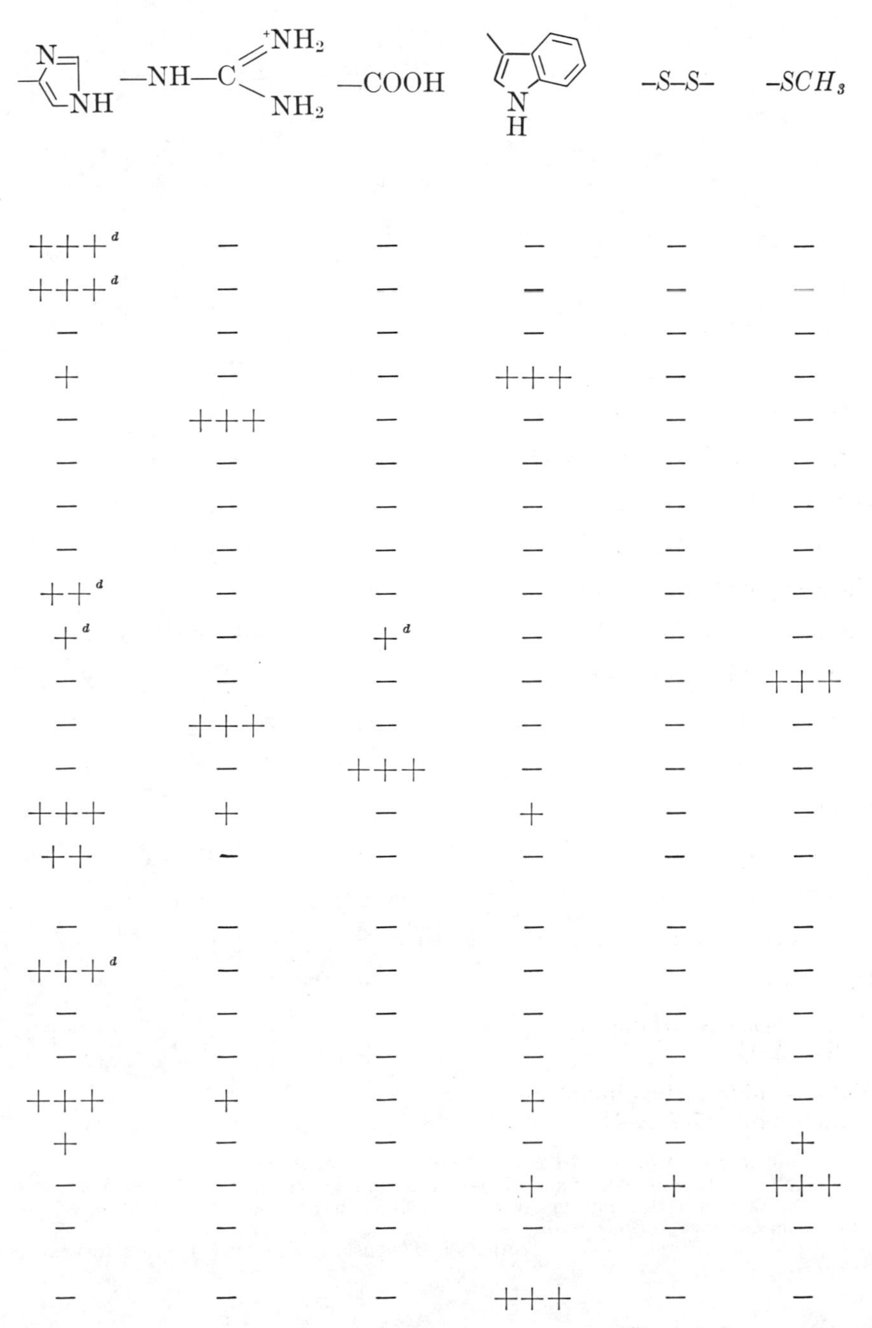

|  | $-NH-C(=\overset{+}{N}H_2)NH_2$ | $-COOH$ |  | *–S–S–* | *–SCH$_3$* |
|---|---|---|---|---|---|
| +++[d] | − | − | − | − | − |
| +++[d] | − | − | − | − | − |
| − | − | − | − | − | − |
| + | − | − | +++ | − | − |
| − | +++ | − | − | − | − |
| − | − | − | − | − | − |
| − | − | − | − | − | − |
| − | − | − | − | − | − |
| ++[d] | − | − | − | − | − |
| +[d] | − | +[d] | − | − | − |
| − | − | − | − | − | +++ |
| − | +++ | − | − | − | − |
| − | − | +++ | − | − | − |
| +++ | + | − | + | − | − |
| ++ | − | − | − | − | − |
| − | − | − | − | − | − |
| +++[d] | − | − | − | − | − |
| − | − | − | − | − | − |
| − | − | − | − | − | − |
| +++ | + | − | + | − | − |
| + | − | − | − | − | + |
| − | − | − | + | + | +++ |
| − | − | − | − | − | − |
| − | − | − | +++ | − | − |

*Continued on next page.*

**Table I.**

| *Reagent*[c] | $-NH_2$ | $-SH$ | —⟨phenyl⟩—OH |
|---|---|---|---|
| *p*-Hydroxyphenylglyoxal | + | − | − |
| Iodine (9-1, A-6) | − | +++ | +++ |
| *o*-Iodosobenzoate (8-4) | − | +++ | − |
| Maleic anhydride (5-1, A-1) | +++[e] | ++[d] | ++[d] |
| *p*-Mercuribenzoate (10-2, A-2) | − | +++ | − |
| Methyl acetimidate (5-3, A-1) | +++ | − | − |
| *O*-Methylisourea (5-3, A-1) | +++ | − | − |
| Performic acid (8-6) | − | +++ | − |
| Phenylglyoxal (10-1, A-5) | ++ | − | − |
| Photooxidation (8-8, A-7) | − | +++ | ±± |
| Sodium borohydride (8-1) | − | − | − |
| Sodium periodate | − | ±±± | ±±± |
| Succinic anhydride (5-1, A-1) | +++ | +++[d] | ++[d] |
| Sulfenyl halides (10-4) | − | +++ | − |
| Sulfite (8-2, A-3) | − | ±±±[e] | − |
| Sulfonyl halides (5-4) | +++ | +++ | +++ |
| Tetranitromethane (9-2, A-6) | − | +++ | +++ |
| Thiols (8-1, A-3) | − | − | − |
| Trinitrobenzenesulfonic acid (6-5, A-1) | +++ | ++[d] | − |
| Water-soluble carbodiimide and nucleophile (7-2, A-4) | ± | ± | ± |

[a] Adapted from Means and Feeney (*1*) with additional data.
[b] The symbols −, +, ++, and +++ indicate relative reactivities; ±, ±±, and ±±± likewise indicate relative reactivities which may or may not be attained depending upon the conditions used.
[c] Numbers in parentheses are sections in Means and Feeney (*1*) where reagent is discussed in detail.

**Continued**

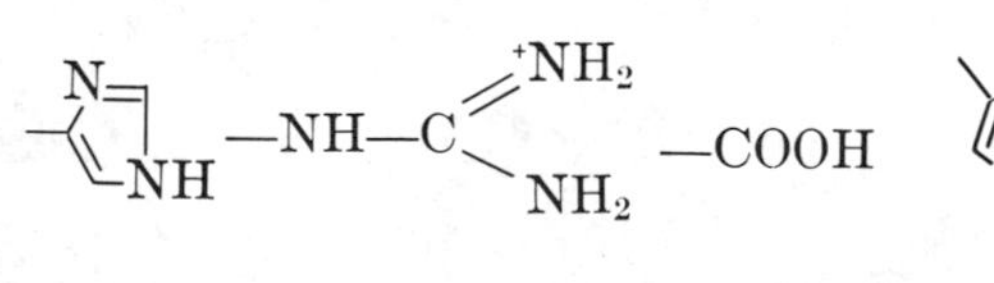

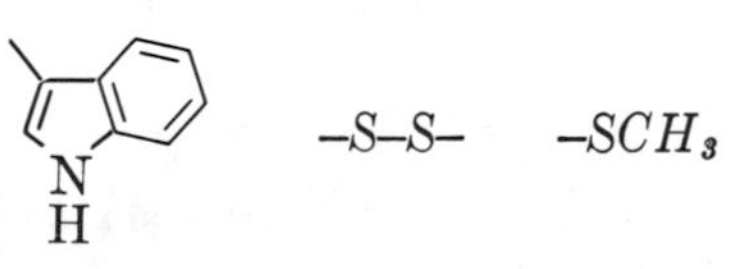

| | $-NH-C(=^{+}NH_2)NH_2$ | —COOH | | –S–S– | $-SCH_3$ |
|---|---|---|---|---|---|
| − | +++ | − | − | − | − |
| +++ | − | − | − | − | − |
| − | − | − | − | − | − |
| ++[d] | − | − | − | − | − |
| − | − | − | − | − | − |
| − | − | − | − | − | − |
| − | − | − | − | − | − |
| − | − | − | ++ | +++ | +++ |
| − | +++ | − | − | − | − |
| +++ | − | − | +++ | ± | +++ |
| − | − | − | − | +++ | − |
| ±±± | − | − | ±±± | − | ±±± |
| ++[d] | − | − | − | − | − |
| − | − | − | +++ | − | − |
| − | − | − | − | +++[e] | − |
| +++ | − | − | − | − | − |
| − | − | − | + | − | + |
| − | − | − | − | +++ | − |
| − | − | − | − | − | − |
| − | − | +++ | − | − | − |

[d] Spontaneously reversible under the reaction conditions or upon dilution, regenerating original group.
[e] Easily reversible, regenerating original group.
[f] Reversible upon mild treatment with sodium periodate.

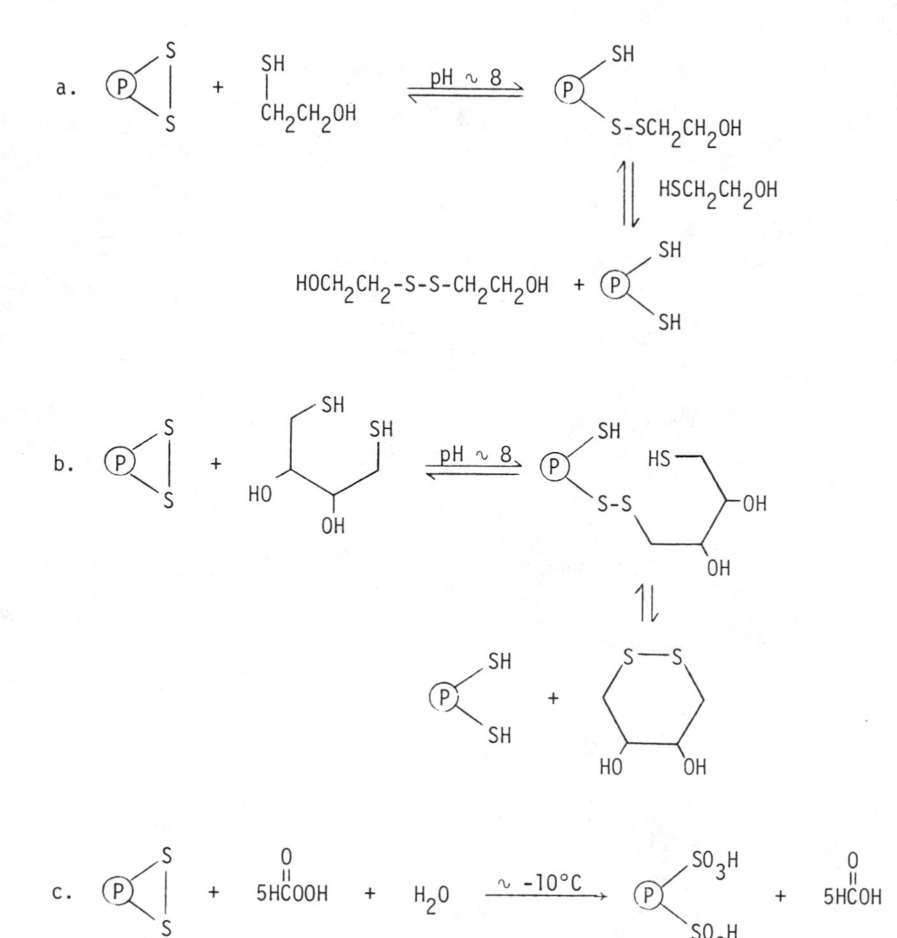


*Figure 2. Modifications of disulfide groups: (a) 2-mercaptoethanol; (b) DTT or DTE (Cleland's reagent); (c) performic acid.*

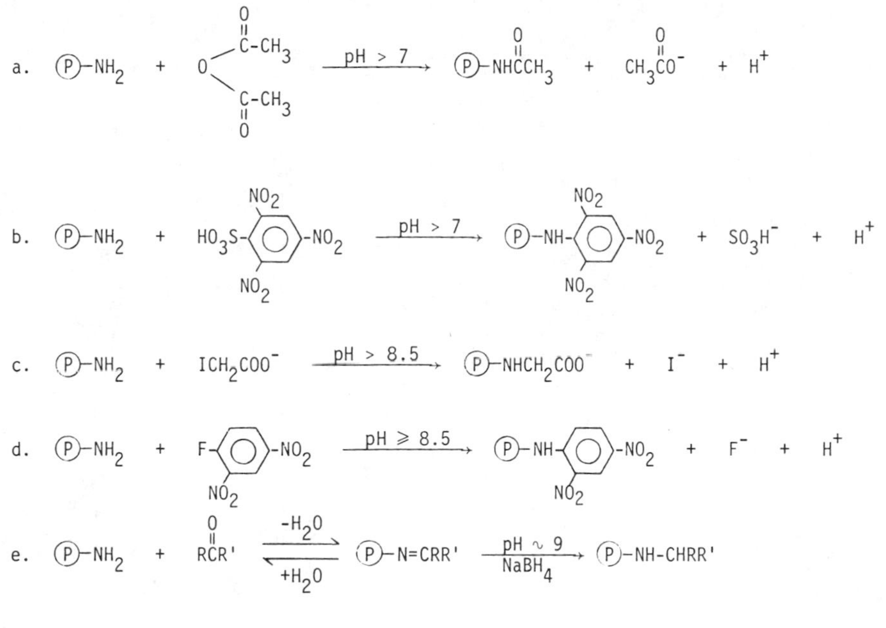


*Figure 3. Modifications of amino groups: (a) acetic anhydride; (b) TNBS; (c) iodoacetic acid; (d) DNFB or Sanger's reagent; (e) reductive alkylation.*

Methods also have been developed to increase the number of hydrolyzable bonds by converting cysteinyl residues to trypsin-susceptible S-(2-aminoethyl)cysteinyl residues using 2-bromoethylamine (*23*) or ethylenimine (*24*).

**Disulfide Groups.** Like the sulfhydryl group, the disulfide has unique and distinctive properties that have aided the development of specific, usually reductive, methods for its modification (*see* Figure 2). These usually are combined with one of the sulfhydryl methods either to block reoxidation back to the disulfide or to estimate the number of disulfides that have been cleaved. More often than not disulfides are reduced to free sulfhydryls by 2-mercaptoethanol (*see* Figure 2a), which has the advantages of high specificity for disulfides and long shelf life. In order to get complete reduction of disulfides the reaction should take place in the presence of a denaturing agent, and because the equilibrium constants of these reactions are near unity, a large excess of 2-mercaptoethanol must be used.

The problem of using a large excess of reducing agents appears to be alleviated by dithiothreitol (and its epimer, dithioerythritol), also known as Cleland's reagent (*25*) which is shown in Figure 2b. The

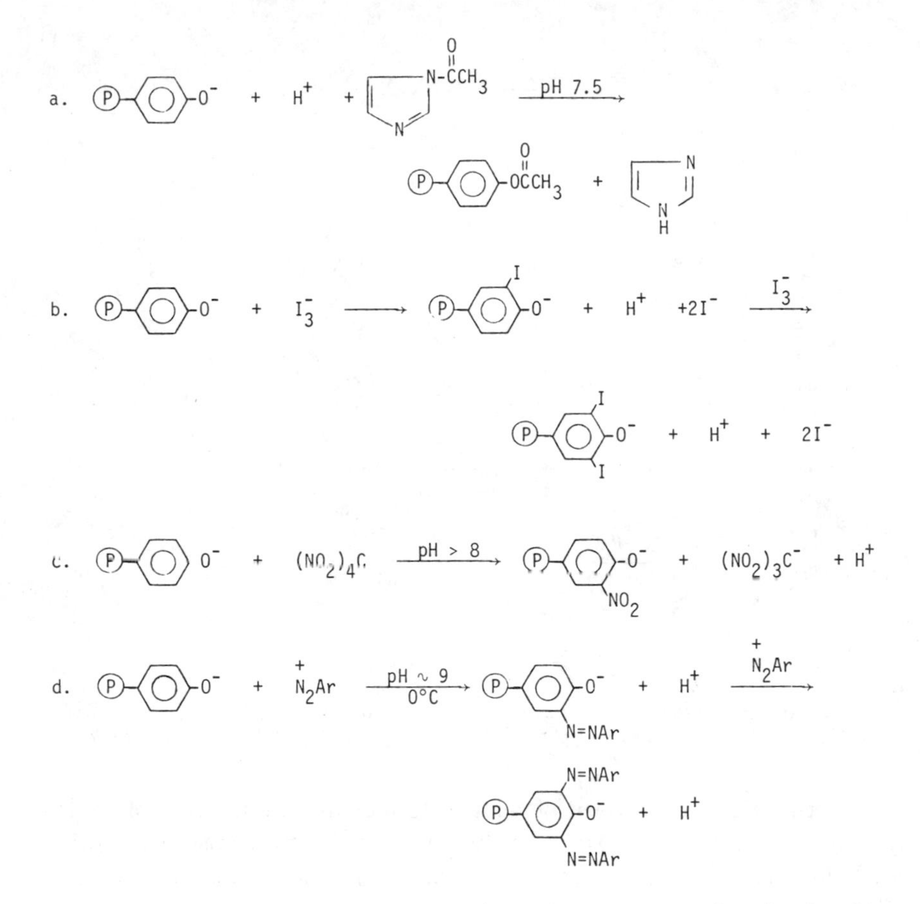


*Figure 4. Modifications of phenolic groups: (a)* N-*acetylimidazole; (b) iodination; (c) TNM; (d) diazotization.*

equilibrium is shifted towards reducing the protein disulfide because the second reduction step is intramolecular and the reagent forms a sterically favorable cyclic disulfide. The hydroxy groups make Cleland's reagent water soluble with little odor from the thiol groups.

**Amino Groups.** The nucleophilic character of unprotonated amino groups has been the key to developing a multitude of methods for their chemical modification (*see* Figure 3). Many of these have been acylations giving amide derivatives with properties depending on the reagent that is used. Recently alkylations have become prominent; these include reactions with haloacetates, aryl halides, and aromatic sulfonic acids, and reductive alkylations achieved by exposing the protein to an aldehyde or ketone in the presence of a reducing agent of the hydride donor type (such as sodium borohydride, cyanoborohydride, or amine boranes).

The effect of using different sized substituents on the carbonyl compound in reductive alkylations has received some attention recently (*26*). The amino groups of ovomucoid, lysozyme, and ovotransferrin were alkylated extensively (40–100%) with various carbonyl reagents in the presence of sodium borohydride. Monosubstitution was observed with acetone, cyclopentanone, cyclohexanone, and benzaldehyde, while 20–50% disubstitution was observed with 1-butanal and nearly 100% disubstitution was observed with formaldehyde. The methylated and isopropylated derivatives of all three proteins were soluble and retained almost full biochemical activities. Recently amine boranes have been shown to be possible alternative reducing agents for reductive alkylation (*27*). The structures of two successfully used amine boranes, dimethylamine borane and trimethylamine borane, are shown in Figure 11.

**Phenolic and Aliphatic Hydroxyl Groups.** The phenolic groups of tyrosyl residues can be modified either at the hydroxyl group or by substitution at the aromatic ring (*see* Figure 4). Modifications at the hydroxyl are generally reversible, some so easily that they have only a transitory existence or are reversed during removal of the reagent and purification of the product. Simple esterification is one of the commonly used methods, with reversal achieved in mildly alkaline solution or with hydroxylamine. Ring substitution, on the other hand, results in more stable products. One of the older and best established methods is

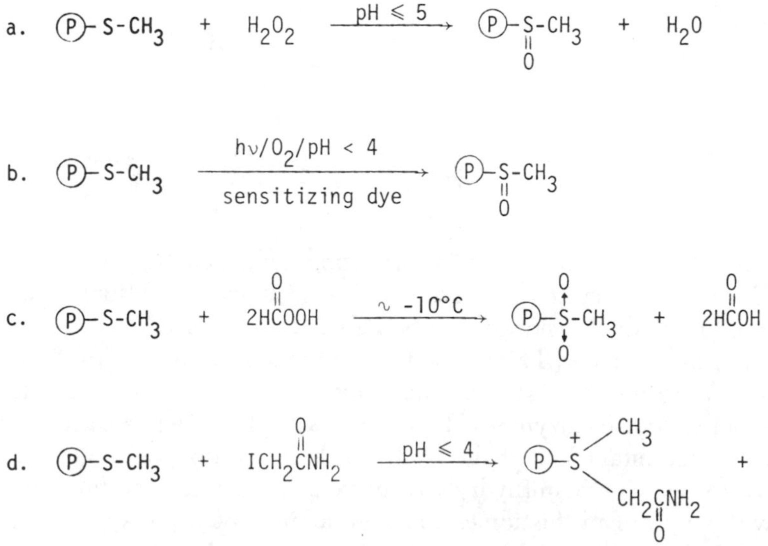

a. $\text{(P)-S-CH}_3 + H_2O_2 \xrightarrow{pH \leqslant 5} \text{(P)-S(=O)-CH}_3 + H_2O$

b. $\text{(P)-S-CH}_3 \xrightarrow[\text{sensitizing dye}]{h\nu/O_2/pH < 4} \text{(P)-S(=O)-CH}_3$

c. $\text{(P)-S-CH}_3 + 2HC(=O)OOH \xrightarrow{\sim -10°C} \text{(P)-S(}\rightarrow\text{O)}_2\text{-CH}_3 + 2HC(=O)OH$

d. $\text{(P)-S-CH}_3 + ICH_2C(=O)NH_2 \xrightarrow{pH \leqslant 4} \text{(P)-S}^{+}(CH_3)(CH_2C(=O)NH_2) + I^-$

*Figure 5. Modifications of thioether groups: (a) hydrogen peroxide; (b) photooxidation; (c) performic acid; (d) iodoacetamide.*

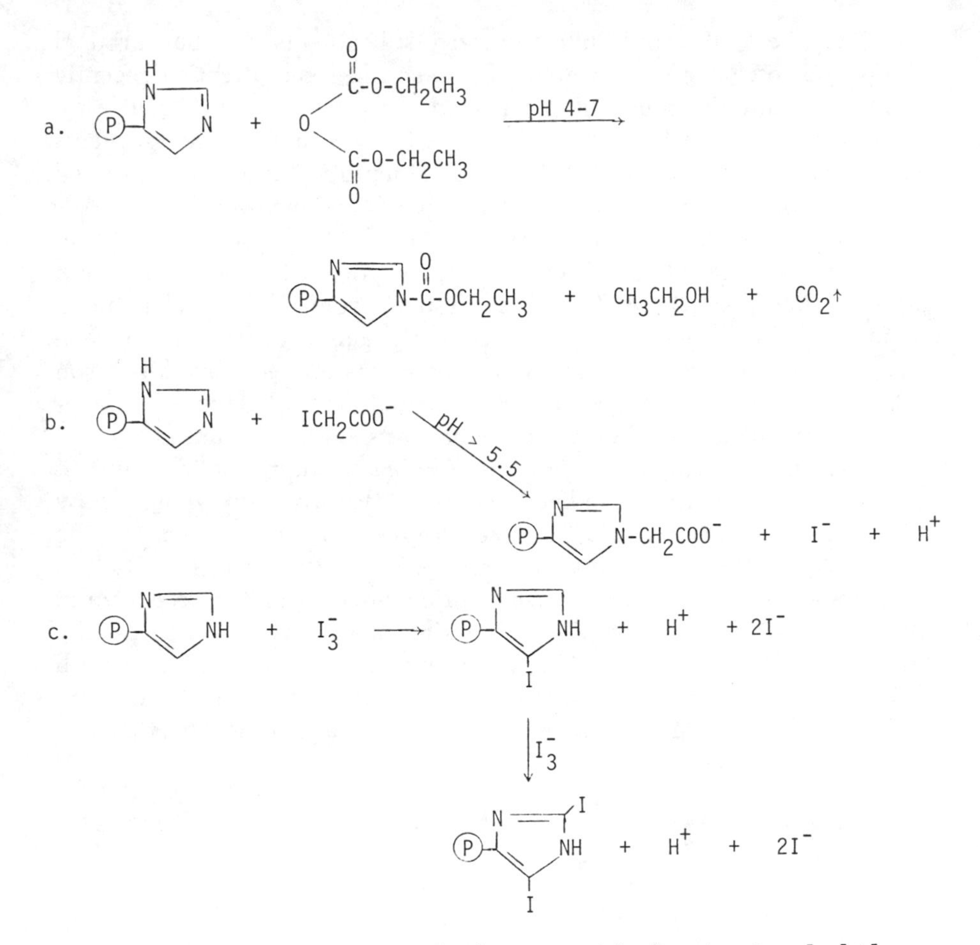


*Figure 6. Modifications of imidazole groups: (a) ethoxyformic anhydride (diethylpyrocarbonate); (b) iodoacetic acid; (c) iodination.*

iodination. Under mild conditions, reacting iodine with proteins also affects histidyl and cysteinyl residues. Under harsh conditions, tryptophanyl and methionyl residues may be affected also (*28*).

Tetranitromethane (TNM) reacts with tyrosyl residues in proteins to give 3-nitrotyrosine derivatives. Nitrating tyrosyl residues in rat liver microsomal stearyl coenzyme A desaturase with TNM leads to inactivation with concomitant loss of the iron prosthetic group (*29*). TNM also is known to react with sulfhydryls in proteins and, under certain conditions, with tryptophan, histidine, and methionine (*30*).

The aliphatic hydroxyl groups of threonine and serine side chains generally are modified by reagents that modify phenolic hydroxyls, but only under more severe conditions. However, once formed, the products are more stable than those with the phenolic hydroxyl.

**Thioether Groups.** Mild oxidizing agents modify the thioether group of methionine residues to the sulfoxide; more vigorous oxidizing agents form the sulfone (*see* Figure 5a–5c). The sulfoxide is easily overlooked, primarily because it is converted to methionine during the acid hydrolysis that usually precedes amino acid analysis.

At neutral and slightly alkaline pH, *N*-chlorosuccinimide (NCS) and *N*-chloro-*p*-toluenesulfonamide (chloramine-T) oxidize methionine residues in peptides and proteins to methionine sulfoxides (*31*). Chloramine-T is more selective than NCS; it does not react with tryptophan whereas NCS does. However both of these reagents react with cysteine. Treating the $Ca^{2+}$-dependent protein modulator with NCS in the presence of $Ca^{2+}$ resulted in selective oxidation of methionines 71, 72, 76, and possibly 109 in the modulator sequence with concomitant loss in interaction with cyclic nucleotide phosphodiesterase (*32*). Methionine residues have been implicated in the activation of cyclic nucleotide phosphodiesterase by the $Ca^{2+}$-dependent protein modulator.

A second type of methionine modification is alkylation by haloalkylamides (*see* Figure 5d). Since the sulfur atom remains unprotonated in rather strongly acidic solutions (such as pH 2), methionine usually can be alkylated selectively at pH values below 4.

**Imidazole Groups.** The imidazole group of histidine side chains is modified by alkylations of the nitrogens or by nucleophilic attack of the

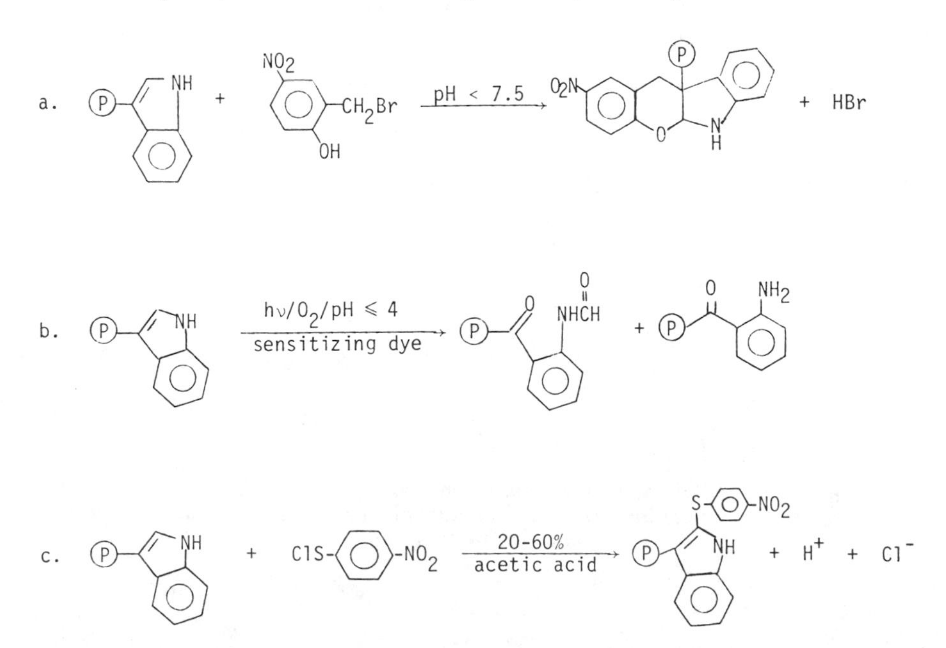


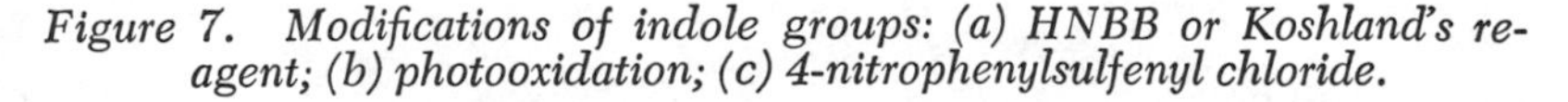

*Figure 7. Modifications of indole groups: (a) HNBB or Koshland's reagent; (b) photooxidation; (c) 4-nitrophenylsulfenyl chloride.*

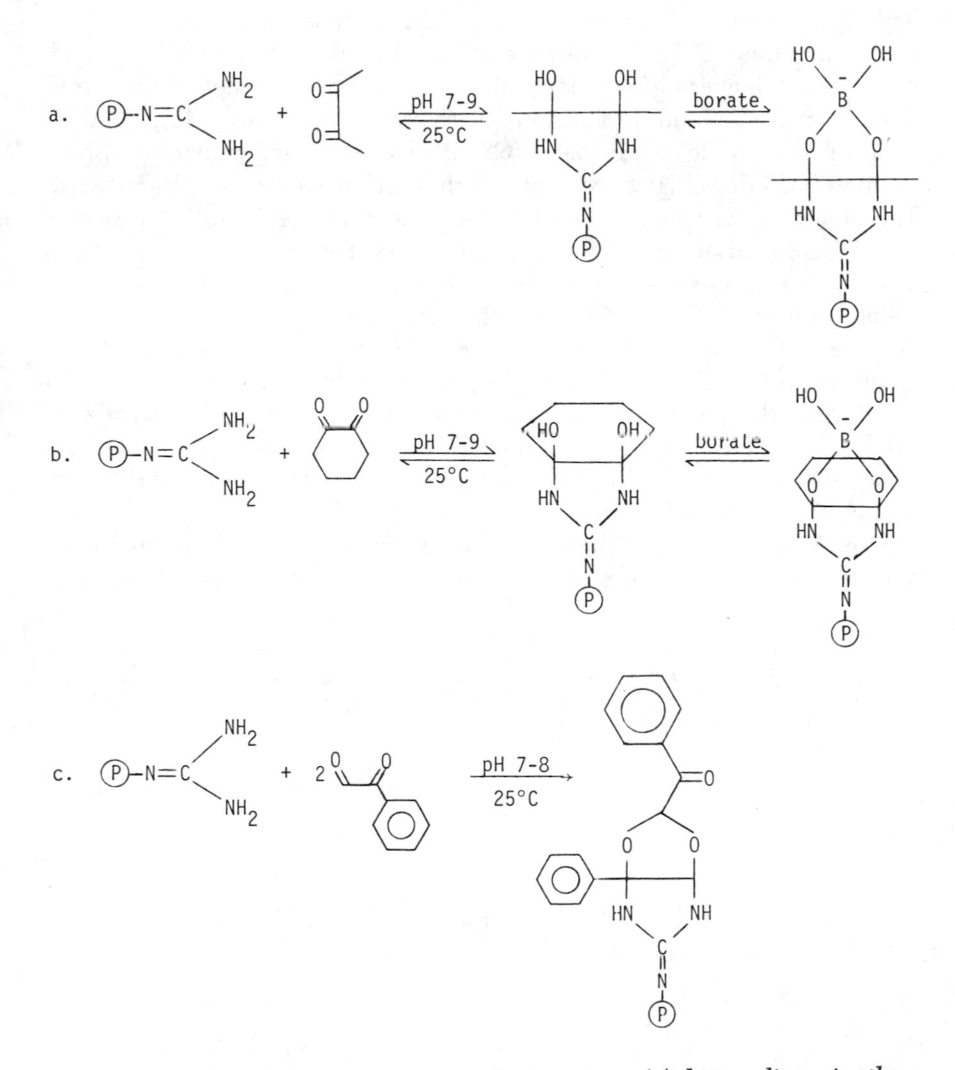


*Figure 8. Modifications of guanidino groups: (a) butanedione in the presence of borate ions; (b) 1,2-cyclohexanedione in the presence of borate ions; (c) phenylglyoxal.*

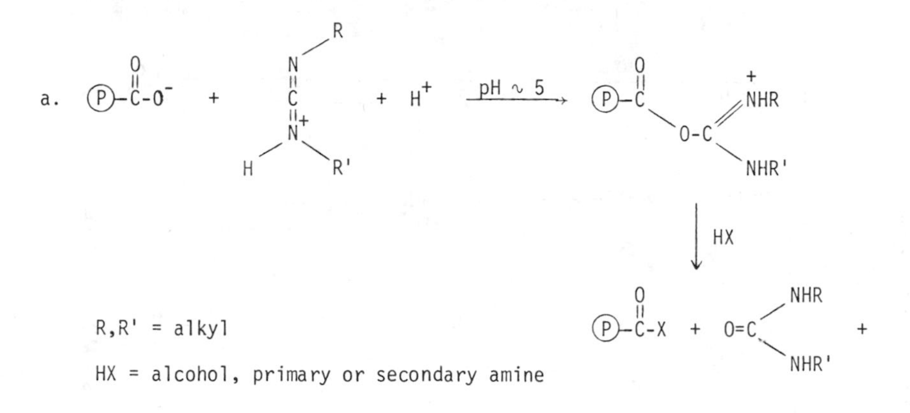


b. $(P)-C(=O)-O^- + (CH_3)_3O^+\ BF_4^- \xrightarrow{\text{pH 4-5}} (P)-C(=O)-OCH_3 + CH_3OCH_3 + BF_4^-$

c. $(P)-C(=O)-OH + CH_3OH \xrightarrow[\text{0-25°C}]{\text{0.02-0.1 } \underline{M} \text{ HCl}} (P)-C(=O)-OCH_3 + H_2O$

*Figure 9. Modifications of carboxyl groups: (a) esterification or amidation via water-soluble carbodiimides; (b) trimethyloxonium fluoroborate; (c) esterification with methanol–HCl.*

carbons (*see* Figure 6). Since histidines are involved in the active centers of many enzymes, there is a large volume of literature on their derivatizations.

Treating ovotransferrin and human serum transferrin with 170–400 molar excess of ethoxyformic anhydride resulted in complete ethoxyformylation of histidines with complete loss in iron-binding activity (*33*). The binding of each iron (two iron-binding sites per protein molecule) protected two histidines from ethoxyformylation, and in both cases the proteins remained completely active. These results plus kinetic analyses of the inactivations indicated two essential histidines in each binding site. Ethoxyformic anhydride also may react with amino groups.

**Indole Groups.** The indole group of tryptophan is susceptible to substitutions and to oxidative cleavage by several different reagents (*see* Figure 7). Less reactive than such nucleophiles as thiol and amino groups, and frequently located within the interior of protein molecules, tryptophans are unaffected by many commonly used reagents.

Recently the heterobifunctional reagent 2-nitro-4-azidophenylsulfenyl chloride, an analog to 4-nitrophenylsulfenyl chloride shown in Figure 7c,

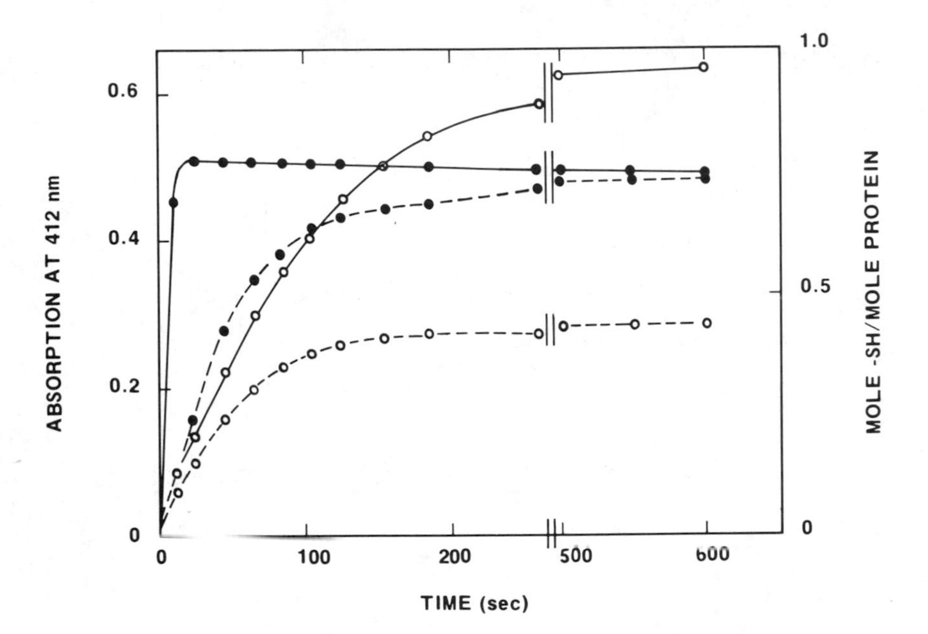


*Figure 10. Reaction of protein thiol groups as measured by DTNB: effects of cystamine intermediary. Penalbumin (○) and BSA (●) were each ($5.0 \times 10^{-5}$M) incubated with $2.0 \times 10^{-3}$M cystamine (——) and without cystamine (– – –) in 0.1M sodium phosphate, pH 8.0, in the presence of $4.0 \times 10^{-4}$M DTNB. Results indicate that penalbumin contains 2 mol of sulfhydryls per mole of protein (19). BSA contains 1 mol of sulfhydryl per mole of protein (20), but preparations may have lesser amounts due to oxidations.*

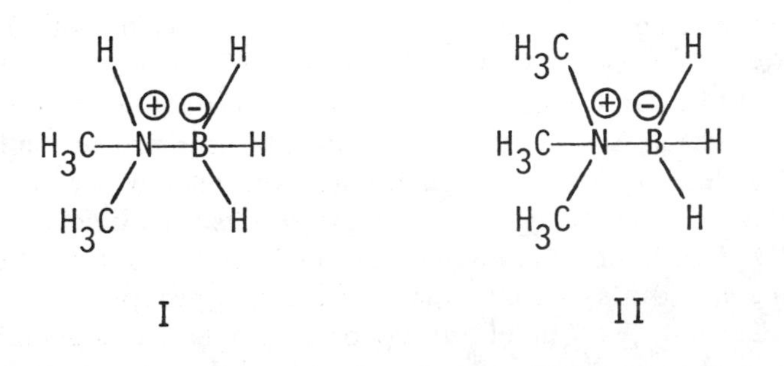


International Journal of Peptide and Protein Research

*Figure 11. Alternative reducing agents for reductive alkylation: (I) dimethylamine borane; (II) trimethylamine borane (27)*

was used to specifically attach a photoactivatable nitrophenyl azide group to the single tryptophan of glucagon (*34*). Glucagon contains no cysteine and only the tryptophan residue was modified. Sulfenyl chlorides also are known to react with sulfhydryls (*35, 36*).

**Guanidino Groups.** Arginine residues are difficult to modify with most reagents because of their extreme basicity, requiring a pH so high that the structure of the protein is in jeopardy. A number of dicarbonyl compounds (*37, 38, 39*) work well under neutral or mildly alkaline conditions (*see* Figure 8). Butanedione and 1,2-cyclohexanedione react with guanidino groups to produce a vicinal diol structure that is stabilized by combining with borate (*see* Figures 8a and 8b). Pande et al. (*40*) recently have introduced camphorquinone-10-sulfonic acid hydrate and camphorquinone-10-sulfonylnorleucine as specific, reversible reagents for modifying arginine residues. The adducts of the camphorquinone derivatives with the guanidyl group are stable to $0.5M$ hydroxylamine at pH 7 but are cleaved by *o*-phenylenediamine. These adducts are also unstable to hydrolysis in $6N$ HCl at 110°C.

It has been suggested that, as a general rule, enzymes acting on anionic substrates or cofactors probably will contain positively charged arginine residues as components of their ligand binding sites (*41*). Evidence supporting this hypothesis based on modifications using phenylglyoxal, butanedione, and 1,2-cyclohexanedione rapidly is accumulating in the literature. Phenylglyoxal, butanedione, and 1,2-cyclohexanedione were used in this laboratory to modify the arginines in ovotransferrin and human serum transferrin (*42*). Analysis of the reaction data suggested that there is one essential arginine involved in each iron binding site.

Recently attention has been given to developing chromophoric arginine reagents that work under mild conditions. The compound 4-hydroxy-3-nitrophenylglyoxal has been used at slightly alkaline pH to show essential arginines in creatine kinase (*43*) and yeast copper, zinc superoxide dismutase (*44*). The modified arginine absorbs near 405 nm and apparently is composed of two products. The compound *p*-hydroxyphenylglyoxal modifies arginine at pH 7–9 and 25°C to give a single product that can be quantitated at 340 nm using the molar absorption coefficient of $1.83 \times 10^4\ M^{-1}\ cm^{-1}$ at pH 9.0 (*45*). The compound *p*-nitrophenylglyoxal is particularly useful for analyzing the content of arginine because of the optical absorption of the product (*46*). Unlike 4-hydroxy-3-nitrophenylglyoxal and *p*-hydroxyphenylglyoxal, *p*-nitrophenylglyoxal undergoes a large change in absorption maximum when reacted with arginine so that unreacted reagent need not be removed to measure the product spectrophotometrically.

**Carboxyl Groups.** Carboxyl groups have a limited chemistry in aqueous media, and this is reflected in the rather small range of chemical methods for their modification in proteins. Existing methods give esters or amides (*see* Figure 9). Under some circumstances they may form amide linkages with the ε-amino groups of lysine to form cross-links.

Water-soluble carbodiimides (*see* Figure 9a) have become a standard means for modifying protein carboxyl groups because of the relatively mild conditions used. The carboxyl terminus of antifreeze glycopeptide 8 was converted to the *N*-(2-hydroxyethylamide) derivative by attaching ethanolamine via 1-ethyl-3-(3-dimethylaminopropyl)carbodiimide (*47*). The new amide derivative retained 90% of the potentiation activity, suggesting that the negative charge at the single carboxyl group is not essential.

Carboxyls are converted to methyl esters by reacting with trimethyloxonium fluoroborate (Figure 9b). Pepsin was modified with [$^{14}$C]trimethyloxonium fluoroborate at pH 5.0 with loss in catalytic activity (*48*). The relationship between the number of methyl groups that are incorporated and the remaining catalytic activity suggested that at least two carboxyl groups are essential for activity.

**The Effect of Conditions on Chemical Modifications.** The molecular properties of a protein respond continuously to changes in the disposition of its constituent amino acids. Changes in the ionization of functional groups lead in turn to adjustments in the complex network of attractive and repulsive forces that maintain the molecule's integrity, and hence to effects on the protein's biological activity. Since many chemical modification reactions are governed in a major way by the protein's overall structure, as well as by the disposition of the individual group or class of groups to be modified, it follows that such modifications are extremely sensitive to the conditions under which they are performed. Careful selection of the conditions is an essential step in the design of any chemical modification experiment.

An important feature of the reactions that are described above for the individual amino acid side groups is that they mostly depend on reagents much milder than the majority used in general organic chemistry. Mild conditions are usually necessary to prevent denaturation. Exceptions to this arise in cases where the protein is being modified for analysis, in which case it may be discarded finally or when the product is intended for a purpose quite different from that for which the protein was produced originally by a living organism. For example, proteins are used as raw materials for producing some commercial adhesives.

Among the more important factors affecting reactions with proteins, pH is the most important since it controls the distribution of potentially reactive side chains between reactive and unreactive ionization states (*see* Table II). Iodoacetic acid is a commonly used reagent in protein modifications and serves as an example. At low pH values (such as 2–5)

**Table II. The pK Values of Titratable Groups in Ribonuclease[a]**

| *Group* | *Residue* | *$pK_a$ (expected)* | *From Titration Curve* *Number of Groups* | *$pK_a$ (found)* |
|---|---|---|---|---|
| $\alpha$-$CO_2H$ | carboxy-terminal | 2.1–2.4 | | |
| $\beta$-$CO_2H$ | aspartate | 3.7–4.0 | 11 | 4.7 |
| $\gamma$-$CO_2H$ | glutamate | 4.2–4.5 | | |
| -Imidazolium | histidine | 6.7–7.1 | 4 | 6.5 |
| $\alpha$-$NH_3^+$ | amino-terminal | 7.6–8.0 | 1 | 7.8 |
| -SH | cysteine | 8.8–9.1 | 0 | — |
| $\epsilon$-$NH_3^+$ | lysine | 9.3–9.5 | 10 | 10.2 |
| -Phenolic | tyrosine | 9.7–10.1 | 3 + 3[b] | 9.95, > 12 |
| $—NH—C(=^+NH_2)NH_2$ | arginine | > 12 | 4 | > 12 |

[a] Data from Tanford and Hauenstein (*49*).
[b] Six phenolic side chains titrate in two groups, three side chains in each group having the indicated p$K$ values.

Chemical Modification of Proteins

it can be made specific for methionine residues since the other groups that also react with iodoacetic acid (sulfhydryls, imidazoles, and amino groups) are largely protonated and hence made unreactive. Some reagents will react with a group to give different products at different pH values. Between pH 8 and 9, 1,2-cyclohexanedione reacts with L-arginine to give $N^7$,$N^8$-(1,2-dihydroxycyclohex-1,2-ylene)-L-arginine (*38*), but at pH 11 the principal product is $N^5$-(1-hydroxy-5-oxo-2,4-diazabicyclo-[4.3.0]non-2-ylidene)-L-ornithine (*50*).

Temperature is another parameter that must be dealt with carefully since it, too, may affect the microenvironment of reactive groups. Some competitive side reactions may be minimized or prevented by thoughtful choices of temperature. Performic acid oxidations of proteins generally are done at low temperatures ($\sim -10°C$) to limit reaction to cysteines, cystines, and methionine. Higher temperatures may lead to reactions with tryptophan, tyrosine, serine, and threonine.

Some reagents are insoluble or only sparingly soluble in aqueous solution and require some organic solvent to assist dissolution. Unfortunately, for most proteins, few organic solvents can be used because they tend to be denaturing, frequently causing precipitation. Although most protein structures are designed mainly to allow solubility in aqueous media, some proteins exist in a membrane-bound form and can be solubilized only with the aid of detergents.

### *Special Types of Chemical Modifications*

**Relative Reactivity of Side-Chain Functional Groups of a Single Type.** In the earlier investigations of protein chemistry, differences in the reactivities of groups of a single type frequently were interpreted purely in terms of the degree of exposure to reagents approaching from outside of the protein. Thus one often heard poorly reactive groups described as masked. However, as our knowledge of protein structure has increased, it has become clear that the chemical reactivity of different groups is influenced additionally in a major way by other factors, particularly the polarity and charge of neighboring side chains. This awareness has been accompanied by the realization that understanding the reasons for different relative reactivities can be a considerable advance in understanding the structural features of the protein molecules that directly govern its biological activity. The changing patterns of reactivity obtained with different reagents can be especially instructive (*see* Table III).

As mentioned earlier, guanidino groups are found in the active centers of many enzymes and binding proteins acting upon anionic substrates or ligands. Positively charged arginine residues generally are thought to reside on the surface of proteins, exposed to solvent and reagents. The crystal structures of numerous proteins indicate that, as a rule, this is true. A large number of enzymes acting upon anionic substrates or cofactors with active-center arginines are modified selectively by dicarbonyl reagents at these residues. In most cases activity is lost concomitant with the modification of a single arginine residue. Since exposure is not a feature that distinguishes the functional arginine from all or most of the other arginines, the special microenvironment of the crevices accommodating the functional arginines appears to be responsible for the increased reactivity of these residues (*41, 54*). The reactivity of guanidino groups toward dicarbonyls is determined primarily by the $pK_e$ value of the guanidinium groups (*54*). Patthy and Thész (*54*) proposed that the $pK_a$ value of arginine residues of anion-binding sites is lower than that of other arginine residues, due to the strong positive electric potential of the anion-binding sites, and that the lower $pK_a$ of arginine leads to its enhanced reactivity towards dicarbonyls.

Some modifying reagents present the difficulty that modification of a first group can lead to alterations in the relative reactivity of other groups to the same or other reagents. For example, Cueni and Riordan (*55*) showed that the modification of Tyr-248 of carboxypeptidase A with

**Table III. The Effects of Different Reagents on the Relative Reactivity of Amino and Sulfhydryl Groups in Proteins**

| *Protein* | *Modification* | | | | *Ref.* |
|---|---|---|---|---|---|
| | *Reagent* | *Group Modified* | *Number Reacted*[a] | *Total Present* | |
| Lima bean inhibitor | TNBS[b] | amino | 3.1 | 5 | *51* |
| | *O*-methylisourea | amino | 3.8 | 5 | *52* |
| | ethyl acetimidate | amino | 5.0 | 5 | *52* |
| Chicken ovomucoid | formaldehyde + $NaBH_4$ | amino | 6.0 | 6 | *26* |
| | butyraldehyde + $NaBH_4$ | amino | 5.5 | 6 | *26* |
| | acetone + $NaBH_4$ | amino | 3.7 | 6 | *26* |
| | benzaldehyde + $NaBH_4$ | amino | 4.6 | 6 | *26* |
| | cyclopentanone + $NaBH_4$ | amino | 2.4 | 6 | *26* |
| | cyclohexanone + $NaBH_4$ | amino | 4.3 | 6 | *26* |
| Chicken ovalbumin | DTNB | sulfhydryl | 0.1 | 4 | *53* |
| | 2-PDS | sulfhydryl | 0.3 | 4 | *53* |
| | 4-PDS | sulfhydryl | 1.4 | 4 | *53* |
| | *N*-ethyl-maleimide | sulfhydryl | 0.2 | 4 | *53* |
| | CMB | sulfhydryl | 1.2 | 4 | *53* |
| | iodine | sulfhydryl | 3.7 | 4 | *53* |

[a] Number of groups modified upon treatment of the native protein.
[b] Abbreviations used: TNBS = 2,4,6-trinitrobenzenesulfonic acid; DTNB = 5,5′-dithiobis-(2-nitrobenzoic acid); 2-PDS = 2,2′-dipyridyl disulfide; 4-PDS = 4,4′-dipyridyl disulfide; CMB = *p*-chloromercuribenzoic acid.

diazotized 5-amino-1*H*-tetrazole makes the previously unreactive Tyr-198 susceptible to nitration with tetranitromethane, whereas treating the native protein with tetranitromethane results only in modification at Tyr-248 (*56*).

Recently methods of modification have been sought that have only small effects on the overall character of most proteins so that modification of a single group or class of groups can be studied without complications of this sort. Among those developed is the reductive methylation procedure for amino groups (*see* Figure 3e) (*27, 57, 58*). Kraal and Hartley (*59*) have described the use of this technique to map the amino groups of a protein (ribosomal elongation factor EF-Tu) that takes part in a number of multimolecular complexes. In their method, double labeling with reagents containing radioactive isotopes was used to distinguish readily modified amino groups from those which are less reactive. The

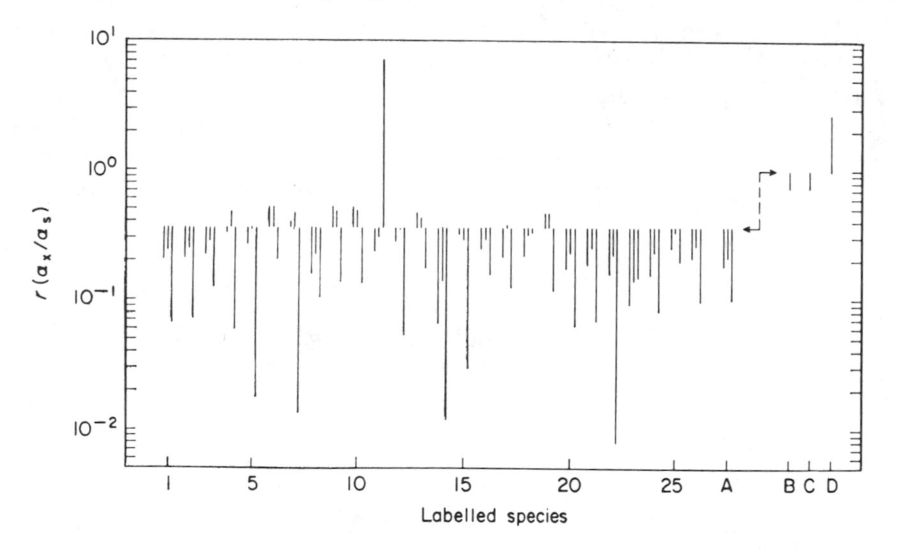


Journal of Molecular Biology

*Figure 12. Competitive labeling of amino groups from various EF–Tu complexes with regard to phenylalanine as the internal standard. Along the logarithmic vertical axis values for* r($a_x/a_s$) *are plotted as calculated from the* $^3H/^{14}C$ *ratios of labeled species and the corresponding ones of phenylalanine. On the horizontal axis Species 1 to 26 represent 26 radioactive tryptic spots, Species A represents undigested EF–Tu, Species B is Phe–tRNA, C is Phe–tRNA in the ternary EF–Tu complex, and D is the reference compound Phe–Gly. Results for Peptides 1 to 26 and undigested EF–Tu (A) are plotted in groups of 3 lines, representing the relative labeling of corresponding species derived from trace-labeled EF–Tu·GDP, EF–Tu·GTP, and EF–Tu·GTP·Phe–tRNA, respectively. The length of the lines is a measure of the deviation from the calculated extent of labeling of the reference ε- $H_2$ groups of benzoyl-Gly–Lys, the position of which is indicated by the left arrow. The length of the lines for B, C, and D represents the deviation from the extent of labeling of the standard α-$NH_2$ group of phenylalanine (right arrow)* (59).

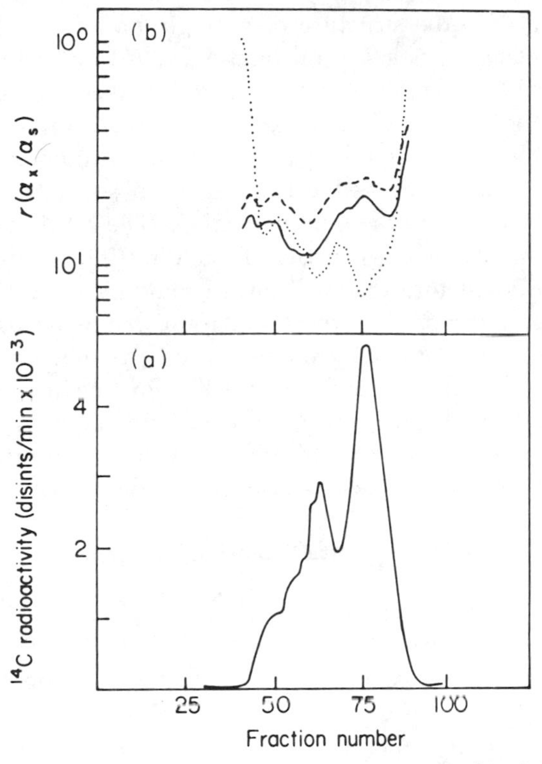


Journal of Molecular Biology

*Figure 13. Fractionation of the tryptic digests of the three competitively labeled EF–Tu preparations on a column of Sephadex G50. For details see Ref.* 59. *(a) Elution pattern of the $^{14}$C-labeled peptides of reference EF–Tu added to each of the $^{3}$H trace-labeled EF–Tu preparations. (b) A plot of the relative labeling (compare with Figure 12) as calculated from the $^{3}$H/$^{14}$C ratios of the corresponding column fractions in (a). EF–Tu·GDP (——); EF–Tu·GTP (– – –); EF–Tu·GTP·Phe–tRNA (· · ·).*

less reactive amino groups were protected perhaps by the elongation factor's association with another molecule. Reactive (presumably exposed) amino groups first were modified by treating the protein with relatively low levels of formaldehyde and a reducing agent that contained tritium which became incorporated into the methyl groups attached to the amino groups that were modified. Subsequently the protein was modified with higher levels of nonradioactive reducing agent and [$^{14}$C]-formaldehyde. The relative reactivities of the lysines then can be calculated directly from the ratio of $^{3}$H to $^{14}$C in each lysine-containing peptide of a digest (*see* Figures 12 and 13).

**Affinity Labeling.** As described in the previous section, differences in the reactivity to chemical modification of groups of a single type can be interpreted in terms of differences in their locations and environments within the protein molecule. A similar approach, but one that probes more specifically into the structure of a single protein, is that in which the modifying chemical is a natural ligand (substrate or inhibitor) or a reagent designed to resemble a natural ligand of that protein (*see* Table IV). Because the reagent binds to a specific region of the protein, the chances of a group in the binding region being positioned to encounter and react with it are enhanced greatly over those of identical groups that are located away from the binding region. The art of designing such a reagent—an affinity label—consists of building into a single molecule the characteristic structure that will ensure binding to the active center along with some appropriately reactive group that will ensure covalent coupling of the reagent to the protein. When the modifying reagent is a natural ligand, binding is indeed specific for the active center. In substrate labeling the bonds between the protein and its ligand, usually labile and/or transitory, are stabilized then by appropriate chemical treatment (such as reducing Schiff base intermediates with sodium boro-

**Table IV. Site-Selective Reagents**

| | *Example* | | |
|---|---|---|---|
| *Type* | *Reagent* | *Protein Labeled* | *Ref.* |
| Substrate | acetoacetate[a] | acetoacetate decarboxylase | *1* |
| | dihydroxyacetone phosphate[a] | aldolase | *1* |
| Pseudosubstrate | methotrexate | dihydrofolate reductase | *60* |
| | phenylmethanesulfonyl fluoride | chymotrypsin | *61* |
| Affinity reagent: classical | *N*-bromoacetyl-D-glucosamine | brain hexokinase | *62* |
| | tosyllysinechloromethyl ketone | trypsin | *63* |
| photoaffinity | 8-azidoguanosine 3′,5′-cyclic monophosphate | cyclic GMP-dependent protein kinase | *64* |
| | *o*-azidophenethyl pyrophosphate | phenyltransferase | *65* |
| $K_{cat}$ ("suicide") | α-hydroxybutynoate | D-lactate dehydrogenase | *66* |
| | $\Delta^{(3,4)}$-decynoyl-*N*-acetyl cysteamine | β-hydroxydecanoyl thioester dehydrase | *67* |

[a] In the presence of sodium borohydride.

hydride). Pseudosubstrates operate similarly to substrates except that the product is a poor leaving group, thus forming a stable linkage without additional chemical treatment. For the purpose of a basic study, the modified protein can be analyzed later by digestion and peptide mapping to determine which residue has reacted with the affinity label. Another application of the same strategy is the design of drugs to inactivate a specific enzyme by covalent attachment to the active site.

Affinity labeling demands a flexible approach, since no single strategy can be relied upon in every case. Those that are used fall into three main groups. In the first, which can be described as classical affinity labeling, the specificity of modification is due entirely to preferential binding of a reactive molecule to the protein's active center. Such a reagent is the well-known affinity label for chymotrypsin, tosylphenylalaninechloromethylketone (TPCK). TPCK combines affinity for the active site of chymotrypsin, due to the blocked phenylalanine portion of the molecule, with a highly reactive chloromethylketone structure that permits rapid alkylation of the nucleophilic active center histidine. In this and similar cases, the reagent's structure dovetails neatly with the arrangement of active-site residues to produce an efficient process of labeling.

In photoaffinity labeling, a second major strategy, ultrareactive nitrenes and carbenes are added to the range of reactive groups through which modifying reagents can be coupled to proteins. These are not present in the reagent as it is added to the protein but they are generated by exposing the protein–reagent mixture to light of the correct wavelength. Molecules of reagent that are bound to the protein can label the binding region even at hydrocarbon side chains, but the short lifetime of the reactive species holds nonspecific labeling to a low level. (For a review, *see* Chowdhry and Westheimer (*68*)).

The third main type of affinity labeling is that which is accomplished with a "suicide" or $k_{cat}$ reagent (*67, 69, 70*). The enzyme's own catalytic capacity is used to generate a reactive product, frequently an allene, that rapidly reacts with nearby active-site residues (*see* Figure 14). Accordingly, these are reagents of unsurpassed specificity, although it seems likely that they can be designed only for a relatively small number of enzymes. Nevertheless, the many superior characteristics of such reagents are exemplified by their uses in inactivating enzymes concerned with steroid metabolism (*9*). For a general discussion of these enzyme-activated irreversible inhibitors, *see* Seiler et al. (*7*) and Benisek et al. (*9*).

Inactivations also may be obtained by a reagent that has a strong affinity for an active area of a protein although the reagent is not an affinity label in the same sense as discussed in the previous paragraphs. An example of this is the inactivations of native serum transferrin or

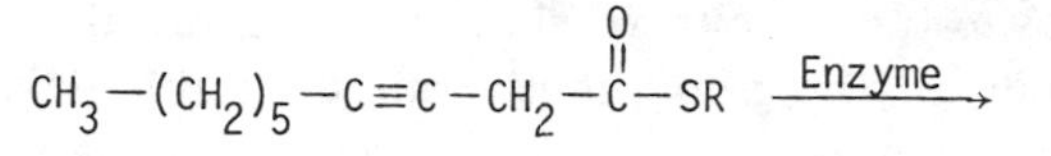


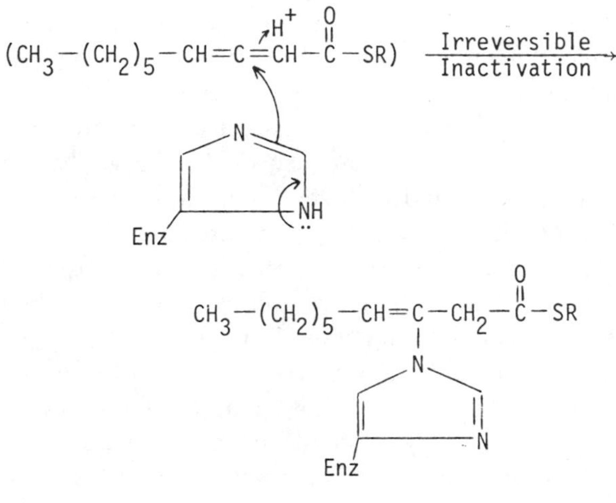


Science

*Figure 14. Irreversible inactivation of* E. coli *β-hydroxydecanoyl thioester dehydrase by a suicide reagent, $\Delta^{(3,4)}$-decanoyl* N-*acetyl cysteamine (67)*

chicken ovotransferrin by periodate. The transferrins require an anion (in blood it's carbonate or bicarbonate) for forming their iron complexes. Periodate will not substitute for bicarbonate as part of the iron complex, but it apparently binds somewhere in the vicinity of the bicarbonate binding area. Our laboratory (*71*) recently has extended an earlier observation of Azari and Phillips (*72*), that periodate inactivates chicken ovotransferrin with losses of tyrosines but does not inactivate its iron complex. However, in the presence of denaturing concentrations of urea, the inactivations and loss of tyrosine in both ovotransferrin and human serum transferrin were retarded severely (*see* Figure 15), showing the requirement of native structure for the inactivations. The activity was assayed after removing urea where, under the conditions used, unmodified transferrin renatures to give fully active proteins (*74*). It generally is agreed that two or three tyrosines are essential per binding site in transferrin (*75*). Although the inactivation of native transferrin by periodate requires the intact binding site, this is not an example of a true affinity label since periodate will not substitute for bicarbonate in the iron complex. A similar example was reported by Lee and Benisek (*76*) where ferrate, a highly reactive analog of phosphate, inactivated rabbit muscle phosphorylase b by abolishing the enzyme's ability to bind 5′-AMP.

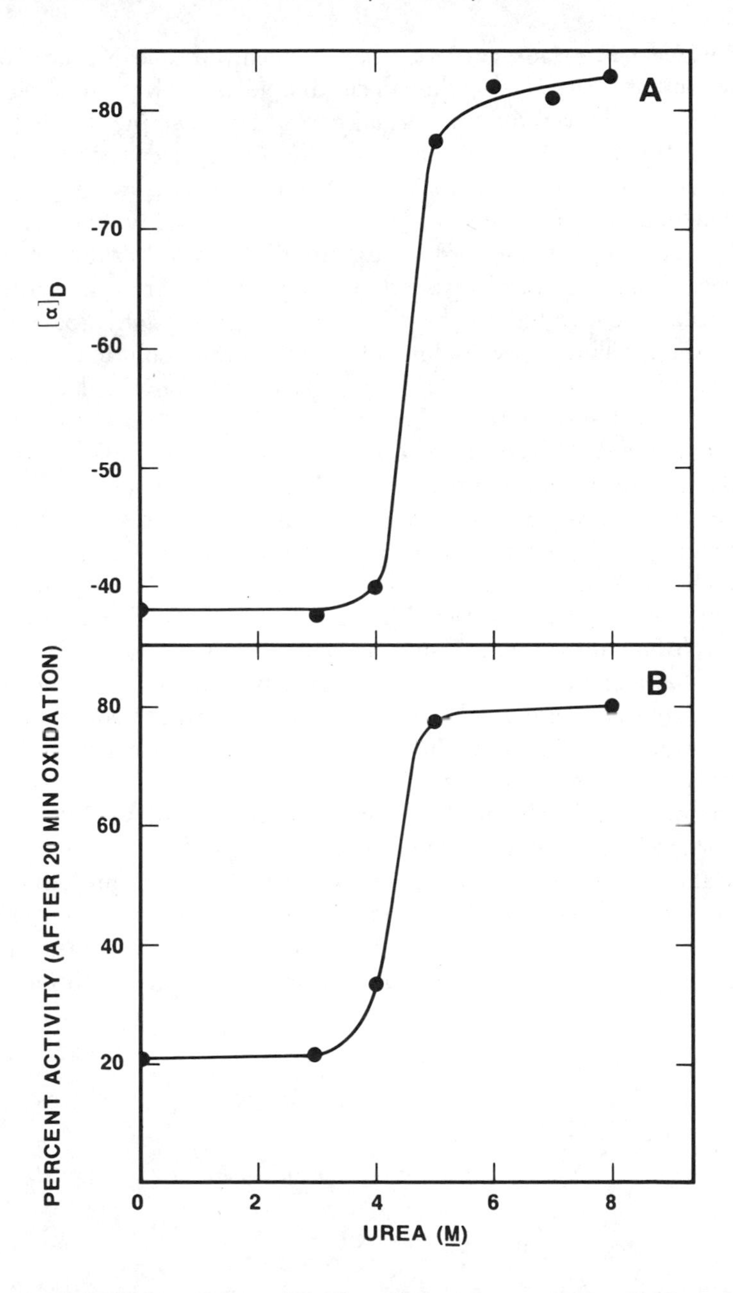


*Figure 15. Effects of urea. (A) The effect of urea on the optical rotation of ovotransferrin. Conditions were 10.5 g/100 mL at 30°C in 0.1M phosphate buffer, pH 7.9 (73). (B) The effect of urea on the reaction of periodate with ovotransferrin (activity remaining after 20 min of oxidation) (71).*

**Replacement Modifications.** Replacement of one amino acid by another amino acid is a highly desired objective of protein chemists, but has been achieved only in special cases. For example, the active-site serine of the bacterial proteinase subtilisin has been converted to a cysteine by modifying the serine with an active-site-directed reagent, phenylmethanesulfonyl fluoride, and treating the derivative with thioacetate ion (*77*). In several cases, an active site residue has been replaced by an unnatural residue. The active-site serine of serine proteases can be converted to the unsaturated analog, dehydroalanine, by esterification followed by β-elimination in alkaline solution (*78*). The active-center aspartic acid of lysozyme has been converted to a homoserine by reducing the ester of the carboxyl group and an affinity reagent, 2′,3′-epoxypropyl β-glycoside of bis(*N*-acetyl-D-glucosamine) (*79*).

Other conversions to unnatural residues occur when most proteins are exposed to high pH (*80, 81, 82*). The high pH causes a β-elimination of a cystine (*see* Figure 16) or *O*-substituted serine or threonine, with the formation of a dehydroalanine or a dehydro-α-aminobutyrate. Such products are subject to nucleophilic attack by the ε-amino group of a lysine to form a cross-linkage, such as lysinoalanine, or attack by cysteine to form lanthionine. Walsh et al. (*81*) have taken advantage of the formation of these cross-links to produce avian ovomucoids that have nonreducible cross-links and have lost the antiprotease activity of one of their two inhibitory sites (*see* Figure 17).

The majority of chemical modifications consists of the addition of a group or groups to a protein. However, some other interesting and useful modifications are those in which a component of the protein's native structure is removed and then replaced by an analogous but slightly different component. For example, the study of metalloenzymes has been advanced greatly by the technique of replacing a nonchromophoric

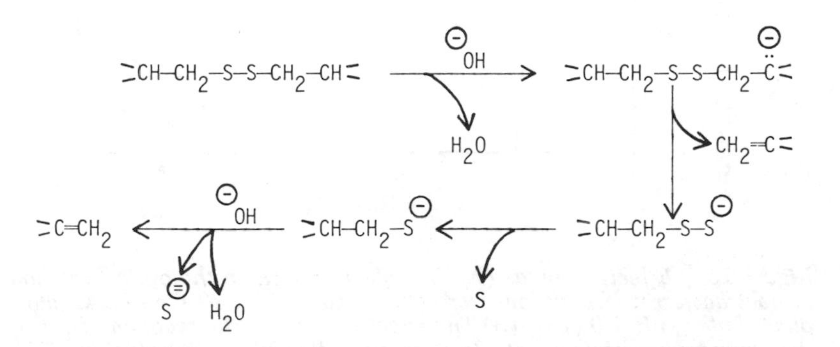


*Figure 16. β-Elimination scheme for disulfides on alkali treatment* (13)

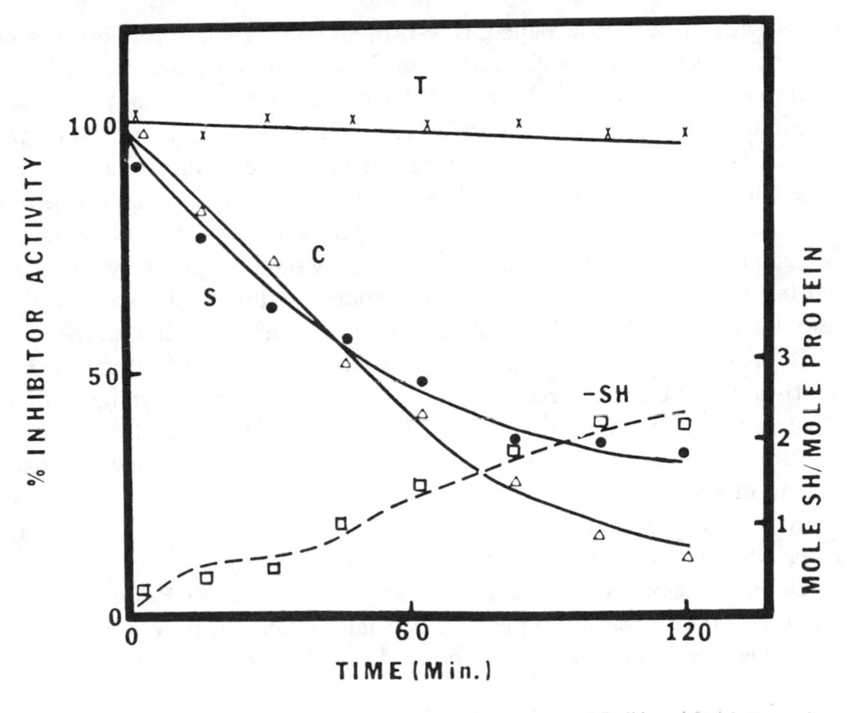


International Journal of Peptide and Protein Research

*Figure 17. The effect of disulfide bond modification of turkey ovomucoid by alkali on inhibitory activity against trypsin (T), α-chymotrypsin, (C), and subtilisin (S). Turkey ovomucoid (0.10mM) was treated with alkali (100mM NaOH) at 23°C. Sulfhydryl content (moles per mole of protein) (-SH) is shown (81).*

metal (e.g., zinc) at the active site with another metal ion (e.g., cobalt) that preserves, with some changes, the enzyme's catalytic activity, while permitting a large amount of useful information to be gathered from its absorption, circular dichroic (CD), and magnetocircular dichroic spectra in the visible range. Using such chromophoric metal atoms as active-site probes has been discussed by Vallee and Riordan (*83*).

Replacement of amino acid residues in a polypeptide chain still is achieved most easily by isolating mutant proteins from bacteria, but Laskowski, Jr. and his colleagues at Purdue University have developed a method for enzymatically removing an essential arginine residue from the reactive site of soybean trypsin inhibitor (Kunitz) and replacing it with a lysine residue (*84*).

**Introduction of Reporter Groups.** A reporter group is one that is attached to (or sometimes noncovalently associated with) a larger molecule for conveying information about the larger molecule's changes in conformation, environment, or association with ligands. These include groups with optical absorbance, free-radical spin labels for electron spin resonance spectrometry, and fluorine- and $^{13}C$-containing groups for NMR spectrometry. Vallee and Riordan (*83*) summarized the results that are gained with a particularly successful example of a reporter group of the first kind—diazotized arsanilic acid—which selectively modifies the active-site residue Tyr-248 of carboxypeptidase A and provides a probe of the catalytic process itself. (The enzyme's activity is not seriously affected by the modification.) The most valuable property of such reporter groups is that they permit dynamic studies of proteins that complement the static portraits obtained from X-ray crystallography.

**Reversible Modifications.** Reversibility of a modification is frequently a desired characteristic in order to regain the protein in its original native form. A variety of methods (*1*) is available for this purpose. Sometimes a further intermediate modification is necessary as used recently in the reductive alkylation of amino groups with an α-hydroxy aldehyde or ketone and subsequent removal of the hydroxyalkyl group by periodate cleavage (*85*) (*see* Figure 18).

**Determinations of Numbers and Types of Essential Residues.** Determining the number and types of essential residues in proteins has been approached by a number of methods (*1*). The simplest, but perhaps

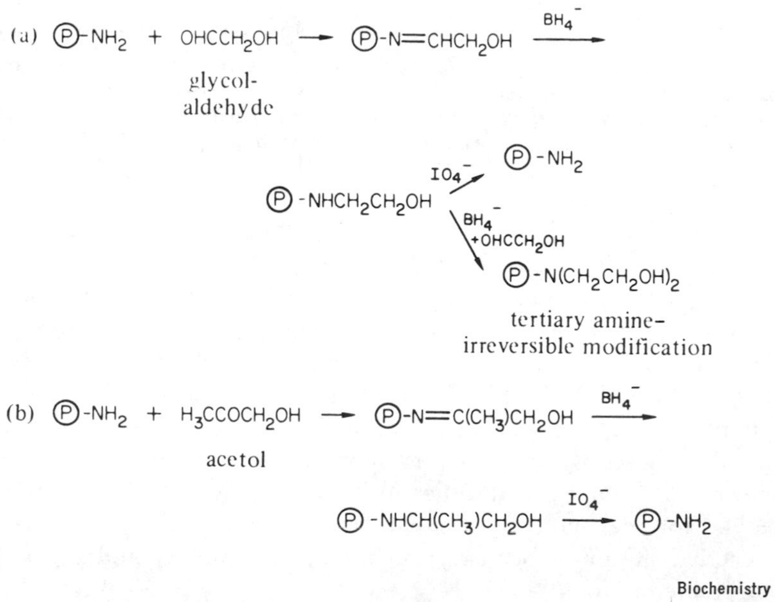

(a) $Ⓟ\text{-}NH_2 + OHCCH_2OH \longrightarrow Ⓟ\text{-}N{=}CHCH_2OH \xrightarrow{BH_4^-}$

glycol-
aldehyde

$Ⓟ\text{-}NHCH_2CH_2OH \xrightarrow{IO_4^-} Ⓟ\text{-}NH_2$

$Ⓟ\text{-}NHCH_2CH_2OH \xrightarrow{BH_4^- + OHCCH_2OH} Ⓟ\text{-}N(CH_2CH_2OH)_2$

tertiary amine-
irreversible modification

(b) $Ⓟ\text{-}NH_2 + H_3CCOCH_2OH \longrightarrow Ⓟ\text{-}N{=}C(CH_3)CH_2OH \xrightarrow{BH_4^-}$

acetol

$Ⓟ\text{-}NHCH(CH_3)CH_2OH \xrightarrow{IO_4^-} Ⓟ\text{-}NH_2$

Biochemistry

*Figure 18. Reversible reductive alkylation of amino groups. Amino groups are alkylated first by treatment with (a) glycolaldehyde or (b) acetol in the presence of sodium borohydride. Reversal of the modification is effected by treating the modified amino group with 10–20mM $NaIO_4$ for ~ 30 min (85).*

least applicable, is the use of active-site-directed reagents that are designed to interact specifically with a particular enzyme and covalently label an essential residue at the active site (i.e. affinity labeling). Of course, a nonessential residue may be modified, sterically preventing the substrate from reaching the active site, and hence making the nonessential residue appear to be essential.

A more widely used method for determining the number of essential residues in proteins was described by Ray and Koshland (*86*). Using a reagent that is specific for a given type of amino acid residue, the relationship between the pseudo-first-order rate constants for modification, and the concomitant protein inactivation gives information on the number of essential residues of that given type. The types of essential residues are determined by using different reagents that are specific for a different type. Modifying turkey ovomucoid with trinitrobenzenesulfonic acid caused losses in trypsin-inhibitory activity with losses in amino groups (*51*). Analysis of the data by the kinetic method of Ray and Koshland (*86*) showed a rate constant for inactivation approximately equal to the rate constant for modifying amino groups, implying that one amino group is essential (*see* Figure 19).

A statistical method for determining the number of essential residues in proteins is offered by Tsou (*87*). The number of essential groups is determined from quantitating the number of groups that are modified and the residual activity in samples of partially modified protein. If all of the residues of a given type, R, in a protein are equally reactive towards the modifying reagent, but only $i$ are essential, then the fraction of the $i$ essential residues of Type R remaining unmodified is equal to the fraction, $\chi$, of the total residues of Type R that remain unmodified. The fraction of the activity remaining, $a$, after partial modification therefore is given by $a^{1/i} = \chi$ with the assumption that modifying any one of the essential residues of Type R leads to complete loss of activity. The value of $i$, normally a small integer, is found from the plot of $a^{1/i}$ vs. $\chi$, where $i$ is selected as the integer that gives the best straight line. In a more complicated situation, where not all of the R residues react at the same rate but, instead, the essential and perhaps some of the nonessential ones react at a markedly different rate from that of the others, the general relation is given by $a^{1/i} = [n\chi - (n-p-s)]/p$, where $n$ is the total number of residues of Type R in the protein, $s$ residues of Type R react most rapidly, none of which are essential, followed by the slower modification of $p$ residues of Type R, $i$ of which are essential, and $(n-p-s)$ residues of Type R are unreactive. Again, the number of essential R residues, $i$, may be obtained from the best straight line of the plot $a^{1/i}$ vs. $\chi$. It is important to note that it is not necessary to have first-order kinetics since there is no explicit time factor that is involved in these two cases.

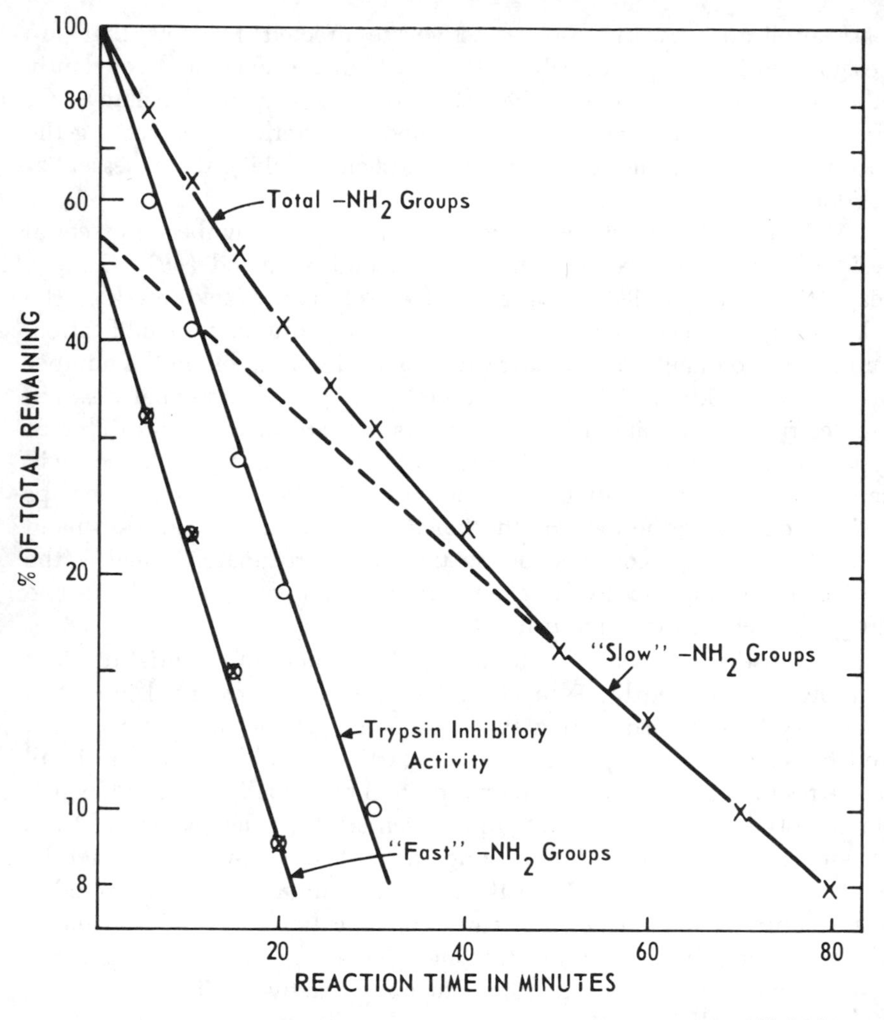


Biochemistry

*Figure 19. Semilogarithmic plot for the loss of amino groups and trypsin-inhibitory activity in turkey ovomucoid by modification with 2,4,6-trinitrobenzenesulfonic acid. The results suggest that one amino group is essential for activity (51).*

Failure to obtain a straight line for any reasonable values of $i$ may indicate that the essential and the nonessential R residues react with significantly but not greatly different rates. In this case, the mathematical relations are not only more complex than those in the previous cases that we considered but are also dependent upon the order of reaction between the reagent and R residues. According to Tsou (87) in this case, if the reaction is first order (or pseudo first order), then the relation can be written so that $i$ can be determined from the best straight line of the plot of log $[(n\chi/a^{1/i}) - p]$ vs. log $a$.

Analyzing the data of the modification of bovine pancreatic ribonuclease A with *p*-hydroxyphenylglyoxal (*45*) by the Tsou method (*87*) indicated one essential arginine for activity (*see* Figure 20). The method of Tsou (*87*) also has been used successfully by Paterson and Knowles (*48*), Jordan and Wu (*88*), and Rogers et al. (*42*), and has been reviewed by Horiike and McCormick (*89*).

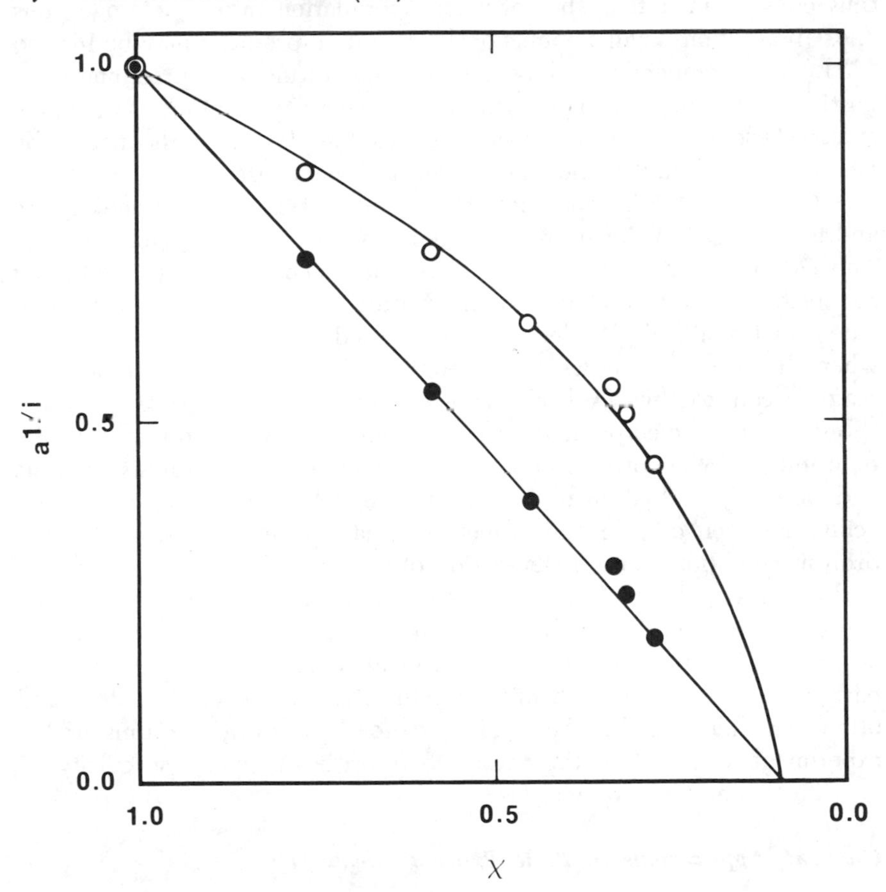


*Figure 20. Tsou plot. The graph shows the relation between the fraction of activity towards yeast ribonucleic acid remaining (a) and the fraction of arginine remaining (χ) for the reaction of* p*-hydroxyphenylglyoxal with bovine pancreatic ribonuclease A. The data are plotted assuming one essential arginine* (i = 1) (●) *and two essential arginines* (i = 2) (○)*. The plot shows one essential arginine for activity* (45).

### *Purification and Analysis*

**Purification of Chemically Modified Proteins.** The final stage of a chemical modification procedure is purification and analysis of the modified protein (*90*). In most cases the modified protein can be separated fairly easily from other components of the modification reaction mixture by means of a physical method (e.g., gel filtration, dialysis, or ion exchange resins), but it is important to note that modifications usually do not affect all of the molecules of a sample to exactly the same extent and thus convert an initially homogeneous population into a heterogeneous one. In addition, small amounts of deteriorative products may be formed (*13*). This frequently is illustrated by denaturing and precipitating a portion of the protein in the sample, perhaps representing those molecules whose structures have been distorted so far from the native by heavy modification that they unfold and become aggregated.

**Analysis.** At the analytical stage, interest centers on the nature and number of functional groups in the protein that have been modified. This can be assessed in several different ways. The amount of modifying agent that is incorporated into the protein can be measured directly when an isotopically labeled reagent is used, or by spectrophotometry when the reagent includes a chromophore. However, many modifying reagents can react with side chains of more than one type, so that the distribution of an incorporated reagent among different groups must be determined. When products are stable to acid or base hydrolysis, analysis is done easily by hydrolyzing the protein and then analyzing the amino acids; however, other chemical methods may be required when the hydrolysis conditions cause regeneration of the unmodified amino acids. It is always preferable to determine directly how much of a modified residue is present than to measure the disappearance of the residue in its unmodified form. Thoughtful planning for the analytical stage is an essential part of the design of a chemical modification study of a protein and often can influence the choice of a reagent. As ever, the aims of the experiment will dictate the order of priorities between specificity of modification and ease of analysis.

### *Current Applications in Basic Protein Chemistry*

Much of what has been discussed concerns the application of chemical modification to questions of basic protein chemistry. This is still the area of greatest activity. The reader should consult the many review articles and texts of the last decade for details (*1, 2, 3, 4, 91*). New methods published each year are gathered and discussed briefly in the Chemical Society's Specialist Periodical Reports on Amino Acids, Peptides, and Proteins.

Chemical modification is now a senior technique in protein science and has retained its importance despite the introduction of many powerful modern physical methods over the last 30 years. Some of the more useful areas of chemical modification have been reactive or active-site modifications, modifications to alter physical and biological properties, radioisotopic labeling, and modifications for analytical purposes.

Specific residues may be implicated in a protein's biological function by using group- or site-specific reagents. Affinity labels of various types may not only label the active site but also map its environment (the active center). Using radioactive tracers in biochemistry is now a truly classical technique. Applied to proteins, it usually has taken the form of iodination of tyrosine residues with radioactive iodine. Recently, proteins that have been reductively methylated using [$^{14}$C]formaldehyde have been made available commercially; this simple method for radio-labeling of proteins may become increasingly popular. Some reagents give a chromophoric product when they are reacted with proteins that allows a direct estimation to be made of the protein's content of the reacting residue. These include Ellman's reagent (DTNB) for thiol groups, TNBS for amino groups, and *p*-nitrophenylglyoxal for guanidino groups.

Today modification experiments continue to complement separate physical studies; e.g., affinity-labeling studies (*92, 93, 94*) of the enzyme in solution supported the suggestion of X-ray crystallographic studies that the carboxylate group of Glu-270 is located within the active site of carboxypeptidase A in a position where it could supply catalytic action as a general base. A further development is the direct study of modified proteins by physical methods. Perhaps the clearest example is the study of proteins that are modified by the attachment of spin labels by electron paramagnetic resonance spectroscopy, since the modification is actually the key to the study. The introduction of $^{13}$C by reductive methylation (*95*) for nuclear magnetic resonance (NMR) studies provides another way of obtaining a clear physical signal from selected groups in a protein.

### *Current Applications for Food and Nutritional Purposes*

Four major objectives in the chemical modification of food proteins are: (a) the blocking of deteriorative reactions; (b) improvement of physical properties; (c) improvement of properties related to acceptability, such as flavor and color; and (d) improvement of nutritional properties. Modifications also may aid the physical separation from crude animal, plant, or microbial material and may inactivate or remove relatively small amounts of undesirable substances. There is considerable interest among food technologists in chemical modification as a means of changing the physical characteristics of proteins to give, e.g., better solu-

**Table V. Average Weight Gain, Food Intake, and Nitrogen**
**Alkylated Caseins With or**

| *Casein Treatment* | *Weight Gain (grams)* | *Food Intake (grams)* | *Unadjusted PER* |
|---|---|---|---|
| Experiment 1 | | | |
| unmodified | 104.0 ± 6.3[†] | 306 ± 11[†] | 3.51 ± 0.19[†] |
| buffer control | 97.0 ± 6.4[†] | 294 ± 4[†] | 3.40 ± 0.22[†] |
| reagent control | 102.2 ± 7.4[†] | 298 ± 13[†] | 3.60 ± 0.19[†] |
| methylated | 11.6 ± 1.3[§] | 146 ± 8[§] | 0.85 ± 0.09[§] |
| isopropylated | −7.6 ± 0.4[★] | 108 ± 5[★] | −0.76 ± 0.05[★] |
| cyclopentylated | −8.6 ± 1.0[★] | 105 ± 5[★] | −0.87 ± 0.07[★] |
| Experiment 2 | | | |
| unmodified | 105.0 ± 5.4[φ] | 290 ± 12[φ,¶] | 3.67 ± 0.06[φ] |
| methylated | 13.2 ± 0.9[§] | 143 ± 4[§] | 0.97 ± 0.06[§] |
| methylated + 0.8% L-lysine · HCl | 81.6 ± 4.2[†] | 259 ± 12[†] | 3.08 ± 0.05[†] |
| 47% methylated | 111.0 ± 6.8[φ] | 313 ± 16[¶] | 3.81 ± 0.16[φ] |
| isopropylated | −8.8 ± 1.0[★] | 111 ± 4[★] | −0.89 ± 0.12[★] |
| isopropylated + 0.8% L-lysine · HCl | 107.2 ± 2.0[φ] | 276 ± 8[†,φ] | 3.70 ± 0.05[φ] |

[a] *From* Ref. *99.*
[b] Mean values of five rats ± SEM. Values within a column in an experiment not sharing a common superscript differ significantly ($P < 0.05$). Feeding periods were 19 and 21 d for Experiments 1 and 2, respectively.

bility or improved whipping performance. Several of these applications are discussed in other chapters in this volume (*11, 96, 97*) and elsewhere (*133*).

**Blocking of Deteriorative Reactions.** Proteins can undergo deterioration at a number of stages in their journey from their living source to the table. Harvesting, processing, cooking, and even subsequent storage can lead to changes in their structure. Some of these changes, of course, are considered desirable—the brown color of a bread crust, e.g. These reactions are so important that food researchers have done the bulk of the work on deteriorative processes. Recently, however, the importance of deteriorative reactions in general protein chemistry has become increasingly recognized (*see* the recent symposium volume by Whitaker and Fujimaki (*12*) on this subject). Since there is extensive literature on chemical deterioration of proteins and attempts to retard it by chemical means, only a short discussion will be added here to that which was published recently by our laboratory (*13*).

Perhaps the best-known chemical change commonly undergone by food proteins is the Maillard reaction, also known as nonenzymatic brown-

**Balance of Rats Fed Diets Supplying 10% Reductively Without Lysine Supplementation**[a,b]

| Weight Gain (grams) | Nitrogen Balance[c]: Nitrogen Intake (grams) | Fecal Nitrogen (grams) | Urinary Nitrogen (grams) | Retention (%) |
|---|---|---|---|---|
| 26.0 ± 1.9[φ] | 1.24 ± 0.06[†] | 0.048 ± 0.006 | 0.28 ± 0.05[§] | 73 ± 6[§] |
| 20.6 ± 1.2[†] | 1.08 ± 0.05[†] | 0.052 ± 0.007 | 0.32 ± 0.03[§] | 66 ± 3[§] |
| 23.4 ± 3.5[†,φ] | 1.20 ± 0.06[†] | 0.050 ± 0.002 | 0.29 ± 0.03[§] | 72 ± 5[§] |
| 4.2 ± 0.4[§] | 0.52 ± 0.02[§] | 0.053 ± 0.003 | 0.32 ± 0.04[§] | 29 ± 5[★] |
| −1.0 ± 0.7[★] | 0.30 ± 0.02[★] | 0.052 ± 0.004 | 0.19 ± 0.02[★] | 21 ± 2[★] |
| −1.4 ± 0.4[★] | 0.31 ± 0.01[★] | 0.055 ± 0.006 | 0.17 ± 0.01[★] | 27 ± 3[★] |
| 22.2 ± 1.7[†,φ] | 0.99 ± 0.05[†] | 0.050 ± 0.007[§,†] | 0.19 ± 0.01[§] | 76 ± 2[†] |
| 4.2 ± 0.4[§] | 0.42 ± 0.01[§] | 0.046 ± 0.004[§] | 0.18 ± 0.02[§] | 46 ± 5[§] |
| 19.4 ± 1.0[†] | 0.98 ± 0.05[†] | 0.055 ± 0.003[§,†] | 0.27 ± 0.04[†] | 67 ± 7[†] |
| 24.4 ± 1.2[φ] | 1.07 ± 0.05[†] | 0.059 ± 0.004[†] | 0.27 ± 0.03[†] | 69 ± 2[†] |
| −0.2 ± 0.2[★] | 0.26 ± 0.02[★] | 0.036 ± 0.004[★] | 0.14 ± 0.01[★] | 32 ± 5[★] |
| 24.6 ± 1.8[φ] | 1.00 ± 0.06[†] | 0.057 ± 0.006[†] | 0.26 ± 0.01[†] | 68 ± 2[†] |

[c] On last 4 d of each feeding trial.

Journal of Nutrition

ing (*6*). This is a carbonyl–amine reaction between protein amino groups and the carbonyl groups of sugars, usually glucose, fructose, or a pentose in foods, which is followed by secondary reactions. Proteins can become insolubilized and contribute to the presence of colored products and off-flavors. (The colors are usually brown to the naked eye, hence the expression "nonenzymatic browning"). Resulting nutritive losses and the possibility of toxicity of the new products are still under investigation (*96*).

There have been attempts to protect proteins against the Maillard reaction by chemical modification of the amino groups. Moderate levels of acetylation have been found to be beneficial (*98*). More recently, our laboratory has studied the effects of methylation, isopropylation, and cyclopentylation (*99*). Methylation is sufficient to prevent the interaction with sugars, and extensive methylation (covering as many as 50% of all the amino groups) of casein did not impair the growth of rats given the modified material as their dietary source of protein (*see* Table V). No growth was achieved when 90% of the amino groups had been methylated, although normal growth could be restored by supplementation with lysine. These results suggest that small amounts of protein added to a food product for its functional properties but subject to deterioration by way of the Maillard reaction might be protected efficiently and inexpensively by reductive methylation.

**Texturization.** In a recent review, Kinsella (*100*) described texturized food proteins as those in protein-rich products whose structure, shape, texture, flavor, and appearance are modified to resemble those of conventional food products. Most attention has been given recently to producing meat-like materials from plant proteins, and meat extenders and fillers are now common and accepted by many consumers. In fact, the idea of such foods is not a new one. John Harvey Kellogg, whose name appears on a number of familiar products, attempted to produce palatable meat-free foods as long ago as 1880.

In preparing textured proteins, crude protein mixtures usually are dispersed and exposed to concentrated alkali to raise the pH to 11 or 12. At this or some later stage, many other substances (flavoring agents, fiber strengtheners, compound lipids, and salts) are mixed with this alkaline solution. Subsequent neutralization and an extrusion or spinning process produces a final product with the desired physical properties. Individual chemical reactions that form part of such procedures have been studied in purified systems, but the large number of ingredients and the rather empirical nature of industrial versions make it difficult to account for all of the possible chemical events that may occur and give rise to compounds that are present in the ultimate product. Alkali treatment of proteins causes racemization (*101*) and chemical changes in serine, threonine, and cystine that can result in the formation of the unsaturated products of $\beta$-elimination and, by secondary reaction of these, new cross-links in the forms of lanthionine and lysinoalanine (*80, 82*).

With strict quality control and careful product analysis, the incidence of such deleterious reactions can be minimized. Reaction conditions and ingredient selection bear importantly on the character of the final product, and it must be emphasized that adding even a single new component to the mixture raises the possibility of new reactions whose effect on the food's nutritive value and safety must be assessed.

Despite the important contribution of the chemical steps to the overall process, it is probably fair to say that they are outweighed in importance by the physical opening and refolding of the protein. The denatured protein is renatured or otherwise refolded to give a bulk material with the desired properties. Many protein chemists see the development of such procedures as a rather unsophisticated process of trial and error, but it has proved to be challenging enough to attract many good minds among protein chemists. For example, Anson, formerly of Rockefeller Institute, made major contributions which Kinsella has described in the following terms, "The pioneering research of Anson and Pader (*102, 103, 104, 105*) in developing texturized chewy proteins stimulated innovation and changed the general attitudes toward the potential of plant protein foods" (*100*). As part of a long-term program, they studied the conditions of heating that are necessary to produce what they called structured gels.

In some processes sulfite is added and undoubtedly forms derivatives of sulfhydryls by sulfitolysis of disulfide bonds. Little is known about these products.

**Functional Properties.** Functional properties of food proteins (*106*) such as gelling ability, solubility, and the viscosity of their solutions obviously contribute to the nature of a food product in which they are present. Kinsella (*100*) has classified the functional properties of proteins in terms of their contribution to the complete food (*see* Table VI); his review and the papers from the recent symposium edited by Pour-El (*106*) contain a detailed discussion of the subject.

In the baking industry, proteins have long been modified chemically for improving the functional characteristics of food. Oxidants are used to modify and control the consistency and strength of dough. Sodium metabisulfite, e.g., has been used in cookie dough. The resulting enhancement of mechanical strength is probably the result of disulfide interchange, but the literature is not always in agreement as to mechanisms.

More controlled chemical modifications of food proteins can improve their performance as food ingredients. Most work has depended on the

**Table VI. Typical Functional Properties of Proteins Important in Food Applications[a]**

| *Property* | *Examples of Functional Properties[b]* |
|---|---|
| Organoleptic | color, flavor, odor, texture, mouthfeel |
| Hydration | solubility, dispersibility, wettability, water absorption, swelling, thickening, gelling, water-holding capacity, syneresis, viscosity, etc. |
| Surfactant | emulsification, foaming, aeration, whipping, protein/lipid film formation, lipid binding, flavor binding |
| Structural and rheological | elasticity, grittiness, cohesiveness, chewiness, viscosity, adhesion, network cross-binding, aggregation, stickiness, gelation, dough formation, texturizability, fiber formation, extrudability, etc. |

[a] Data from Kinsella (*100*).
[b] These functions vary with pH, temperature, protein concentration, protein species or source, prior treatment, ionic strength, and dielectric constant of the medium. They also are affected by other treatments, macromolecules in the medium, processing treatments, and modification, etc.

**CRC Critical Reviews in Food Science and Nutrition**

formation of acyl derivatives of ε-amino groups of lysine side chains, with acetylation and succinylation being the modifications examined most extensively (*107*). The emulsifying properties of milk proteins and others in margarine manufacturing were improved by acetylation (*108*). Similarly, *N*-succinylated egg yolk proteins had properties that allowed them to perform well in mayonnaise and salad dressings (*109*). In many studies with soy protein, the succinylated product was usually more soluble, whiter, and more dispersible than the unmodified protein (*110*). Succinylation greatly improved the dispersion characteristics and heat stability of fish myofibrillar protein in aqueous solution (*111*). Modifying egg white proteins with 3,3-dimethylglutaric anhydride, another dicarboxyanhydride, formed a product with increased heat stability and emulsifying capacity in acid solution (*112*). Finally, derivatization also has been used to assist the removal of undesirable components from a crude preparation; an example is presented in this volume (*11*).

**Increasing Nutritional Value.** Few direct attacks have been reported on increasing the nutritional value of proteins by those chemical modifications that incorporate substances in the protein. A successful exception has been the incorporation of amino acids into proteins. Since a separate presentation on this subject is given in this volume (*113*), only a brief discussion will be presented here.

Amino acids conveniently can be attached covalently to proteins by either of two routes (*see* Figure 21): (a) by forming an amide bond from the carboxyl group of the amino acid to be attached with the ε-amino group of lysine of the protein or (b) by attaching an amino group of the amino acid which is to be attached in an amide bond with the side-chain carboxyl group of aspartic acid or glutamic acid of the protein. Both of these types of attachments have been achieved. Because the linkage to the protein is unusual, they form derivatives called isopeptides. Using lysine and methionine, Puigserver et al. (*115, 116*) and others (*117, 118*) have found that these bonds are attacked by peptidases and that the incorporated amino acids, as well as the substituted amino acids that were originally in the protein backbone, are nutritionally available to the rat.

Some amino acids cannot be added to diets in the free form because of their bad taste, while the same amino acids covalently bound to the protein as isopeptides apparently have no taste and are therefore not objectionable. In addition, the covalently attached amino acids are not apt to be lost in some cookery practices where the soluble amino acids added to a protein preparation might be lost in drainage water on cooking. Although at this time the chemical reactions that are used would be inadequate for commercial usage, methods that are suitable to large-scale production should be possible.

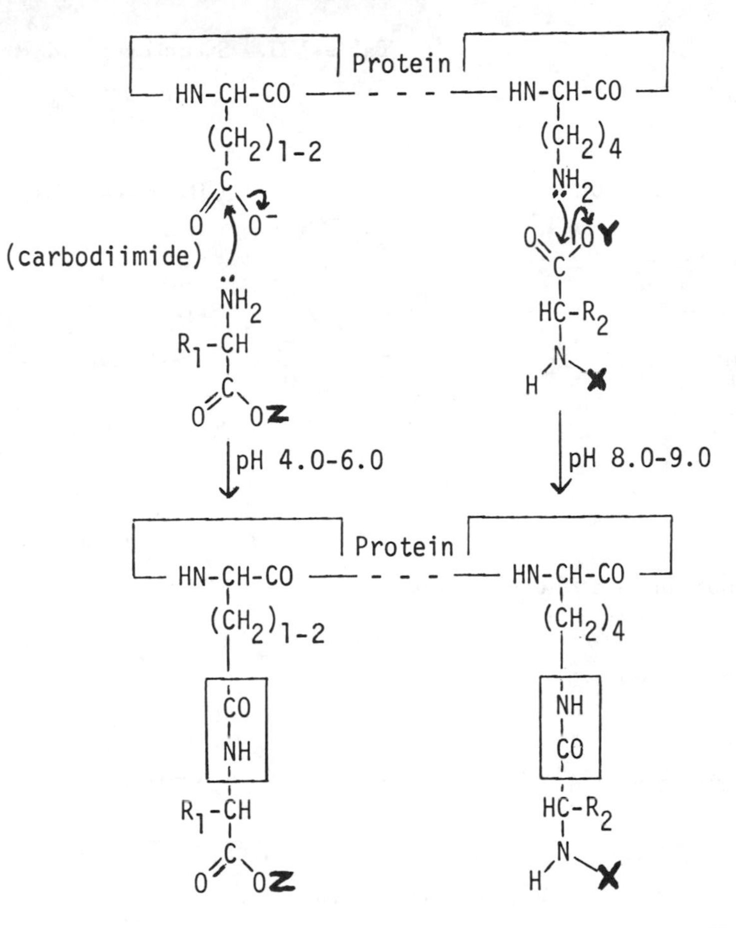


*Figure 21. General scheme for isopeptide bond formation between amino and carboxyl side groups of proteins and additional amino acids.* X *and* Z *are amino and carboxyl protecting groups, respectively;* Y *is a carboxyl-activating group* (114).

## *Current Applications in Pharmaceutical and Related Areas*

The interaction of proteins with chemical reagents is important to the pharmaceutical industry in a number of ways. Since an important strategy for drug action is the direct inactivation of a target molecule by covalent combination with the drug, there has long been interest in the means by which naturally occurring toxins achieve similar (although harmful) results in the body. An unfortunate aspect to this work has been the development of man-made toxic substances for chemical warfare. These chemicals are often alkylating agents (*see* Table VII). Devel-

**Table VII. Structures and Reaction**

| *Alkylating Agent* | *Structure* |
|---|---|
| Ethyl methane sulfonate | $CH_3CH_2{-}O{-}S(=O)_2{-}CH_3$ |
| Busulfan | $CH_3{-}S(=O)_2{-}O{-}(CH_2)_4{-}O{-}S(=O)_2{-}CH_3$ |
| Mechlorethamine (nitrogen mustard) | $CH_3N(CH_2CH_2Cl)_2$ |
| Phenoxybenzamine | 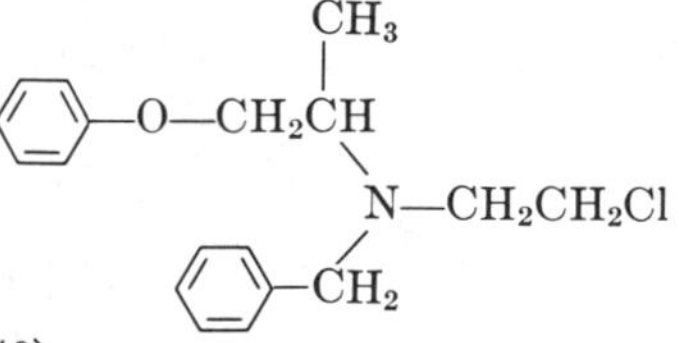 |

[a] Data from Goldstein, Aronow, and Kalman (*119*).

opment of a binary version of an existing weapon, the 155-mm artillery shell filled with the nerve gas Sarin (*O*-isopropylmethylphosphonofluoridate), is currently under political debate (*120*). Under the binary concept, two nonlethal (but far from harmless) precursors, isopropanol and methylphosphonyldifluoride, react after firing to form the lethal nerve gas Sarin.

An important consideration to the pharmacologist is the useful lifetime of a drug in the body. Protein hormones have been modified chemically to lower their rate of removal from the bloodstream, prolong the effect of the injected hormone, and thus reduce the required frequency of administration. The modification of animal insulin by coupling to protamine–zinc is probably the best-known example. Others are the studies by Marshall on the possibility of lengthening the lifetime in the bloodstream of injected proteins by supplying them as dextran conjugates

**Mechanisms of Some Alkylating Agents**[a]

| *Carbonium Ion Form* | *Product of Reaction with $R^-$* |
|---|---|
| $CH_3CH_2^+$ | $CH_3CH_2\text{–}R$ |

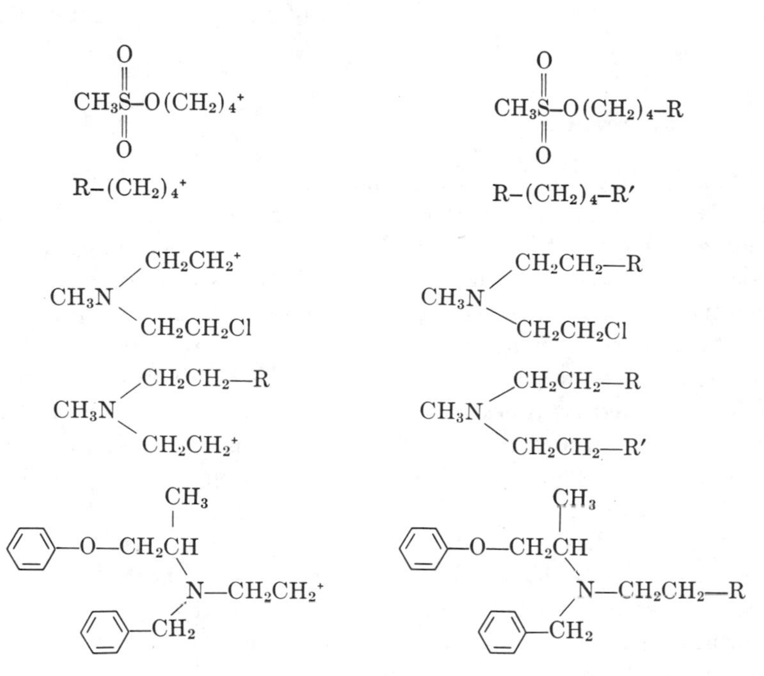


Principles of Drug Action: The Basis of Pharmacology

(*121*) and altering proteins' susceptibility to enzymic digestion by modifying side chains.

Substances also may be attached covalently to proteins for transporting the substances to targets in the body. Examples of this approach have been transporting radioactive iodine or toxic substances attached to antibodies in attempts to kill cancer cells and attaching radioactive metals to proteins and bleomycin for similar purposes (*122*). A recent novel approach has been that of Wu and Means (*123*) who have attached insulin to artificially formed liposomes by a reductive alkylation modification. The insulin–liposome aggregate reacts with insulin receptors.

As has been mentioned already, developing drugs that act as irreversible enzyme inhibitors represents an area of expanding interest. This subject is covered in other chapters of this book (*9, 124*) and need not be covered here. In passing, however, it is interesting to note that nature has shown the way in this area of scientific development as in so many others, since a number of naturally occurring low-molecular-weight toxins function by mechanisms of irreversible inactivations (*69*).

An area of chemical modification that is very important to the medical support industry is concerned with the modifications for altering biological properties. An application of glutaraldehyde's cross-linking effect on proteins is in the preparation of prostheses for internal organs, e.g. heart valves (*125*).

### *Developing Areas*

The chemical and enzymatic modification of proteins should continue to be an expanding and perhaps even burgeoning subject. New applications and cross-fertilization of the more established ones are evident today. Two examples of how efforts in the areas of food, nutritional, and pharmaceutical sciences have much in common will be cited here.

An older one, extensively used over 30 years ago, consisted of feeding iodinated casein and other proteins to animals for the thyroxin-type action of the iodinated tyrosines and their derivatives in the protein. Quantitative data concerning the amounts of this material fed to animals are unavailable or difficult to assess, but apparently the quantities were relatively large, at least as based on the reputed thyroxin-like activity. A newer link has been the report that peptides with opioid activity occurred in peptic hydrolysates of proteins (i.e. wheat glutin and alpha casein) (*126*). Zioudrou et al. (*126*) also isolated substances from gluten hydrolysates that stimulated adenylate cyclases and increased the contractions of the mouse vas deferens but did not bind the opiate receptors. It also was suggested that peptides derived from some food proteins may be physiologically important. These experiments were done, of course, where the proteins were treated enzymatically in vitro and were not the result of normal digestive processes in vivo. However their results also point to an ever-present requisite in the chemical modification of food proteins for testing for properties that are produced unexpectedly, either directly or indirectly, by the modification. As is evident, an indirectly produced substance might result from enzymatic hydrolysis of a chemically modified protein from which peptides could be produced that could have pharmacological action.

Perhaps the government grants realm is where many interrelationships can be seen most easily. The U.S. National Institutes of Health fund work that is basic to pharmacological sciences as well as to food and nutrition, which of course are all intimately related to health problems. Many of the basic studies on enzyme mechanisms include studies on affinity and suicide-type reagents, chemically modified proteins or peptides for pharmaceutical uses, and chemical studies on food proteins.

At this time, the universities in the United States also are building up their relationships with the pharmaceutical industry, a relationship which is both old and profitable in European countries. This is not only through private grants to educational institutions but also through direct funding by means of interlocking agreements.

Many examples of the natural interlock between these areas are visible in industry itself, reflecting the need for efficiently using biological assets and research personnel. For many years animal-products packing houses have been engaged deeply in producing and marketing biologicals. The German Bayer Company, an extremely large chemical concern, has been active for decades in the pharmaceutical area and more recently has acquired an interest in the field of nutrition and food technology. The Nestlé Corporation also has diversified from food products into nutritional and pharmaceutical products. In visits to the basic research laboratories of such concerns, one can see the impact of these diversifications on the research.

Obviously target-specific pharmaceuticals are usually the result of very sophisticated research, although accidental or serendipitous discoveries are made. There is a long list of requirements that the hypothetically perfect suicide reagent must meet if it is to be used as a pharmaceutical. In addition to those many requirements related to specificity, the reagent must be transported to a target site and then transferred into, and onto, the target itself. Developing this type of reagent therefore is tied intimately to the molecular biology of cell membranes and cellular constituents. What might be considered to be the opposite of the suicide-type drugs could be reagents developed to block, or prevent the action of, either naturally occurring toxins from foods or man-made toxic substances. Many of these reagents probably would be more noncovalent than covalent in nature in their mechanism of action because of the obvious deleterious effects of covalent interactions with many of the target areas. Yet this is an area requiring studies on the covalent interactions of the toxic substances as well as on the antidotes or prophylactic agents.

Still further types of noncovalently or covalently interacting drugs might be those involved in blocking neural responses. One such area of apparent current interest is the blocking of the action of proteolytic enzymes that remove peptides naturally inhibiting pain receptors (*127*). Of all of the drugs that have been consumed by the tons, aspirin most likely heads the list. Only recently has a major role been found for its function, the inhibition of prostaglandin synthesis by acetylating the $NH_2$-terminal serine of prostaglandin synthetase (*128, 129*). Yet some of its actions probably are still poorly understood or unknown. One of its suggested uses is for treating sickle-cell anemia. Recently a series of derivatives of aspirin has reacted more avidly with the sickle-cell hemoglobin than does aspirin and also is capable of passing through the red cell membrane, thereby suggesting its potential use for pharmaceuticals for treating sickle-cell anemia (*see* Table VIII) (*130*). [While this chapter was in galley proof, a further publication appeared describing a series of diaspirins that are bound in the $\beta$-cleft of hemoglobin and that show a wide range of $\beta'$-$\beta$ cross-linking effectiveness (*132*).]

Perhaps the toxoid research programs now being carried out in many research laboratories will provide a wide spectrum of toxoids for many different infectious diseases. Most of these toxoids should be prepared by specific-affinity-type reagents, including the suicide type, for the offending toxin from bacterial or viral agents.

The long-standing objectives of controlling deteriorative reactions in foods and pharmaceuticals and even in tissues should be a prime subject for continued study (*12*). Chemical derivatization offers a possible control of some of these deteriorative actions. A currently expanding area is the chemical stabilization of prosthetic implants such as heart valves and blood vessels (*125*). Of course, such prosthetic transplants must have physical stability, very low immunogenic capacity, and resistance to enzymatic and chemical (hydrolytic) attack. In spite of the widespread importance of the deteriorations of proteins, methods for determining deteriorations are presently inadequate (*13*). Methods must be developed for detecting and characterizing deteriorative products, rather than the usual approach of using methods that result merely in the report of losses.

### *Conclusions*

Chemical modification of proteins is an important everyday tool of basic protein chemistry. Its role in applied protein chemistry appears to be increasing and branching into different areas such as food, nutrition, and pharmacology.

If chemistry economically can change unpalatable or indigestible plant tissues into appealing and nutritious food items, such products probably will be made. The barriers are financial support for good nutritional and toxicological researches and the attitudes of the consumers. There is no need for a boon–bane dilemma (*131*). Although there is apparently little application of chemical modifications to food proteins anywhere in the world, this should change with future needs. Needs may arise particularly in less affluent regions of the world that are suffering from starvation. Animal feeds already receive satisfactory

**Table VIII. Aspirin Analog Modification of Hemoglobin[a]: Percentages of Modification of Intracellular and Extracellular Hemoglobins by Acylsalicylates**

| Compound | Modification of Hemoglobin: Intracellular[b,c] 1 Dose | Intracellular[b,c] 4 Doses | Extracellular (1 dose)[c,d] |
|---|---|---|---|
| Salicyl esters | | | |
| acetate ($C_2$) | < 5 | 16 | 7 |
| propionate ($C_3$) | < 5 | 20 | 5 |
| butyrate ($C_4$) | 0 | < 5 | < 5 |
| caproate ($C_6$) | 7 | 29 | 5 |
| caprylate ($C_8$) | 14 | 36 | 9 |
| caprate ($C_{10}$) | lysis | lysis | 13 |
| isovalerate (branched $C_5$) | 0 | < 5 | < 5 |
| 3,5-Dibromosalicyl esters | | | |
| acetate ($C_2$) | 77 | 100 | 77 |
| propionate ($C_3$) | 35 | 84 | 31 |
| butyrate ($C_4$) | 8 | 26 | 6 |
| caproate ($C_6$) | 9 | lysis | 16 |
| 4-Carboxyphenyl esters | | | |
| acetate ($C_2$) | < 5 | 22 | 10 |
| caproate ($C_6$) | 11 | 33 | 14 |
| *t*-butylacetate (branched $C_6$) | 5 | 22 | 5 |
| Bis(salicyl) diesters | | | |
| fumarate (unsaturated $C_4$) | 0 | 0 | 70 |
| succinate ($C_4$) | 0 | 0 | 14 |
| adipate ($C_6$) | < 5 | 15 | 16 |
| suberate ($C_8$) | 10 | 35 | 14 |
| sebacate ($C_{10}$) | 24 | 42 | 14 |
| dodecanedioate ($C_{12}$) | 35 | 65 | 17 |
| Bis(3,5-dibromosalicyl) diesters | | | |
| fumarate (unsaturated $C_4$) | 15 (1m*M*) | 93 (1m*M*) | 88 (1m*M*) |
| succinate ($C_4$) | 18 (1m*M*) | 64 (1m*M*) | 70 (1m*M*) |
| sebacate ($C_{10}$) | 29 (0.5m*M*) | 63 (0.5m*M*) | 30 (0.5m*M*) |

[a] Ref. *130*.
[b] Each dose involved exposure of 20% (v/v) erythrocyte suspensions in isotonic phosphate buffer, pH 7.2, to 5m*M* compound (unless otherwise noted) for 2 h at 37°C.
[c] Reproducibility of extent of modification was ± 5%.
[d] Dose involved treatment of 6 g/dL of hemoglobin solutions in 0.05*M* sodium phosphate, 0.01*M* sodium cyanide, pH 7.2, with 5m*M* compound (unless otherwise noted) for 2 h at 37°C.

Journal of Biological Chemistry

chemical treatments and further applications should be adopted for both human foods and animal feeds.

There are presently few available drugs that can be described as even nearly perfect active-site, target-specific chemical agents. However, current and future researches should provide agents more closely approaching such perfect ones.

### *Acknowledgments*

The authors appreciate the assistance of Chris Howland, Martha Jolley, and Clara Robison in editing and preparing the manuscript and they acknowledge the financial support from NIH Grant AM 26031.

### *Literature Cited*

1. Means, G. E.; Feeney, R. E. "Chemical Modification of Proteins"; Holden-Day: San Francisco, 1971.
2. Hirs, C. H. W.; Timasheff, S. N., Eds. *Methods Enzymol.* **1972,** *25.*
3. Ibid., **1977,** *47.*
4. Glazer, A. N.; DeLange, R. J.; Sigman, D. S. "Chemical Modification of Proteins: Selected Methods and Analytical Procedures"; North-Holland/American Elsevier: Amsterdam, 1975.
5. Trejo-Gonzalez, A.; Feria-Morales, A.; Wild-Altamirano, C. Chapter 9 in this book.
6. Feeney, R. E.; Blankenhorn, G.; Dixon, H. B. F. *Adv. Protein Chem.* **1975,** *29,* 135.
7. Seiler, N.; Jung, M. J.; Koch-Wester, J., Eds. "Enzyme-Activated Irreversible Inhibitors"; Elsevier/North Holland Biomedical: Amsterdam, 1978.
8. Metcalf, B. W. In "Chemical Deterioration of Proteins," *ACS Symp. Ser.* **1980,** *123,* 241.
9. Benisek, W. F.; Ogez, J. R.; Smith, S. B. Chapter 10 in this book.
10. Harkness, D. R. *Trends Biochem. Sci.* **1976,** *1,* 73.
11. Shetty, J. K.; Kinsella, J. E. Chapter 6 in this book.
12. Whitaker, J. R.; Fujimaki, M., Eds. "Chemical Deterioration of Proteins," *ACS Symp. Ser.* **1980,** *123.*
13. Feeney, R. E. In "Chemical Deterioration of Proteins," *ACS Symp. Ser.* **1980,** *123,* 1.
14. Ellman, G. L. *Arch. Biochem. Biophys.* **1959,** *82,* 70.
15. Nashef, A. S.; Osuga, D. T.; Feeney, R. E. *Anal. Biochem.* **1977,** *79,* 394.
16. Wilson, J. M.; Wu, D.; Motiu-DeGrood, R.; Hupe, D. J. *J. Am. Chem. Soc.* **1980,** *102,* 359.
17. Whitesides, G. M.; Lilburn, J. E.; Szajewski, R. P. *J. Org. Chem.* **1977,** *42,* 332.
18. Wilson, J. M.; Bayer, R. J.; Hupe, D. J. *J. Am. Chem. Soc.* **1977,** *99,* 7922.
19. Aminlari, M.; Osuga, D. T.; Ho, C.; Allison, R. G.; Feeney, R. E. *Fed. Proc. Fed. Am. Soc. Exp. Biol.* **1981,** *40,* 1837, Abstr. No. 1713.
20. Janatova, J.; Fuller, J. K.; Hunter, M. J. *J. Biol. Chem.* **1968,** *243,* 3612.
21. Allison, W. S. *Acc. Chem. Res.* **1976,** *9,* 293.
22. Lin, W. S.; Armstrong, D. A.; Gaucher, G. M. *Can. J. Biochem.* **1975,** *53,* 298.
23. Lindley, H. *Nature* **1956,** *178,* 647.

24. Raftery, M. A.; Cole, R. D. *Biochem. Biophys. Res. Commun.* **1963,** *10,* 467.
25. Cleland, W. W. *Biochemistry* **1964,** *3,* 480.
26. Fretheim, K.; Iwai, S.; Feeney, R. E. *Int. J. Pept. Protein Res.* **1979,** *14,* 451.
27. Geoghegan, K. F.; Cabacungan, J. C.; Dixon, H. B. F.; Feeney, R. E. *Int. J. Pept. Protein Res.* **1981,** *17,* 345.
28. Koshland, M. E.; Englberger, F. M.; Erwin, M. J.; Gaddone, S. M. *J. Biol. Chem.* **1963,** *238,* 1343.
29. Enoch, H. G.; Strittmatter, P. *Biochemistry* **1978,** *17,* 4927.
30. Sokolovsky, M.; Harell, D.; Riordan, J. F. *Biochemistry* **1969,** *8,* 4740.
31. Shechter, Y.; Burstein, Y.; Patchornik, A. *Biochemistry* **1975,** *14,* 4497.
32. Walsh, M.; Stevens, F. C. *Biochemistry* **1978,** *17,* 3924.
33. Rogers, T. B.; Gold, R. A.; Feeney, R. E. *Biochemistry* **1977,** *16,* 2299.
34. Demoliou, C. D.; Epand, R. M. *Biochemistry* **1980,** *19,* 4539.
35. Scoffone, E.; Fontana, A.; Rocchi, R. *Biochemistry* **1968,** *7,* 971.
36. Fontana, A.; Scoffone, E.; Benassi, C. A. *Biochemistry* **1968,** *7,* 980.
37. Riordan, J. F. *Biochemistry* **1973,** *12,* 3915.
38. Patthy, L.; Smith, E. L. *J. Biol. Chem.* **1975,** *250,* 557.
39. Takahashi, K. *J. Biol. Chem.* **1968,** *243,* 6171.
40. Pande, C. S.; Pelzig, M.; Glass, J. D. *Proc. Natl. Acad. Sci. U.S.A.* **1980,** *77,* 895.
41. Riordan, J. F.; McElvany, K. D.; Borders, C. L., Jr. *Science* **1977,** *195,* 884.
42. Rogers, T. B.; Børresen, T.; Feeney, R. E. *Biochemistry* **1978,** *17,* 1105.
43. Borders, C. L., Jr.; Pearson, L. J.; McLaughlin, A. E.; Gustafson, M. E.; Vasiloff, J.; An, F. Y.; Morgan, D. J. *Biochim. Biophys. Acta* **1979,** *568,* 491.
44. Borders, C. L., Jr.; Johansen, J. T. *Biochem. Biophys. Res. Commun.* **1980,** *96,* 1071.
45. Yamasaki, R. B.; Vega, A.; Feeney, R. E. *Anal. Biochem.* **1980,** *109,* 32.
46. Yamasaki, R. B.; Shimer, D. A.; Feeney, R. E. *Anal. Biochem.* **1981,** *111,* 220.
47. Geoghegan, K. F.; Osuga, D. T.; Ahmed, A. I.; Yeh, Y.; Feeney, R. E. *J. Biol. Chem.* **1980,** *255,* 663.
48. Paterson, A. K.; Knowles, J. R. *Eur. J. Biochem.* **1972,** *31,* 510.
49. Tanford, C.; Hauenstein, J. D. *J. Am. Chem. Soc.* **1956,** *78,* 5287.
50. Anzai, K. *Bull. Chem. Soc. Jpn.* **1969,** *42,* 3314.
51. Haynes, R.; Osuga, D. T.; Feeney, R. E. *Biochemistry* **1967,** *6,* 541.
52. Haynes, R.; Feeney, R. E. *Biochemistry* **1968,** *7,* 2879.
53. Aminlari, M., Ph.D. Thesis, Univ. of California, Davis, 1980.
54. Patthy, L.; Thész, J. *Eur. J. Biochem.* **1980,** *105,* 387.
55. Cueni, L.; Riordan, J. F. *Biochemistry* **1978,** *17,* 1834.
56. Muszynska, G.; Riordan, J. F. *Biochemistry* **1976,** *15,* 46.
57. Means, G. E.; Feeney, R. E. *Biochemistry* **1968,** *7,* 2192.
58. Means, G. E. *Methods Enzymol.* **1977,** *47,* 469.
59. Kraal, B.; Hartley, B. S. *J. Mol. Biol.* **1978,** *124,* 551.
60. Williams, J. W.; Morrison, J. F.; Duggleby, R. G. *Biochemistry* **1979,** *18,* 2567.
61. Ako, H.; Ryan, C. A.; Foster, R. J. *Biochem. Biophys. Res. Commun.* **1972,** *46,* 1639.
62. Swarup, G.; Kenkare, U. W. *Biochemistry* **1980,** *19,* 4058.
63. Shaw, E.; Mares-Guia, M.; Cohen, W. *Biochemistry* **1965,** *4,* 2219.
64. Geahlen, R. L.; Haley, B. E.; Krebs, E. G. *Proc. Natl. Acad. Sci. U.S.A.* **1979,** *76,* 2213.
65. Brems, D. N.; Rilling, H. C. *Biochemistry* **1979,** *18,* 860.
66. Olson, S. T.; Massey, V.; Ghisla, S.; Whitfield, C. D. *Biochemistry* **1979,** *18,* 4724.

67. Rando, R. R. *Science* **1974,** *185,* 320.
68. Chowdhry, V.; Westheimer, F. H. *Ann. Rev. Biochem.* **1979,** *48,* 293.
69. Rando, R. R. *Acc. Chem. Res.* **1975,** *8,* 281.
70. Rando, R. R. *Methods Enzymol.* **1977,** *46,* 28.
71. Geoghegan, K. F.; Dallas, J. L.; Feeney, R. E. *J. Biol. Chem.* **1980,** *255,* 11429.
72. Azari, P.; Phillips, J. L. *Arch. Biochem. Biophys.* **1970,** *138,* 32.
73. Glazer, A. N.; McKenzie, H. A. *Biochim. Biophys. Acta* **1963,** *71,* 109.
74. Yeh, Y.; Iwai, S.; Feeney, R. E. *Biochemistry* **1979,** *18,* 882.
75. Chasteen, N. D. *Coord. Chem. Rev.* **1977,** *22,* 1.
76. Lee, Y. M.; Benisek, W. F. *J. Biol. Chem.* **1976,** *251,* 1553.
77. Neet, K. E.; Koshland, D. E., Jr. *Proc. Natl. Acad. Sci. U.S.A.* **1966,** *56,* 1606.
78. Strumeyer, D. H.; White, W. N.; Koshland, D. E., Jr. *Proc. Natl. Acad. Sci. U.S.A.* **1963,** *50,* 931.
79. Eshdat, Y.; Dunn, A.; Sharon, N. *Proc. Natl. Acad. Sci. U.S.A.* **1974,** *71,* 1658.
80. Nashef, A. S.; Osuga, D. T.; Lee, H. S.; Ahmed, A. I.; Whitaker, J. R.; Feeney, R. E. *J. Agric. Food Chem.* **1977,** *25,* 245.
81. Walsh, R. G.; Nashef, A. S.; Feeney, R. E. *Int. J. Pept. Protein Res.* **1979,** *14,* 290.
82. Whitaker, J. R. In "Chemical Deterioration of Proteins," *ACS Symp. Ser.* **1980,** *123,* 145.
83. Vallee, B. L.; Riordan, J. F. In "Versatility of Proteins"; Li, C. H., Ed.; Academic: New York, 1978; p. 203.
84. Sealock, R. W.; Laskowski, M., Jr. *Biochemistry* **1969,** *8,* 3703.
85. Geoghegan, K. F.; Ybarra, D. M.; Feeney, R. E. *Biochemistry* **1979,** *18,* 5392.
86. Ray, W. J., Jr.; Koshland, D. E., Jr. *J. Biol. Chem.* **1962,** *237,* 2493.
87. Tsou, C.-L. *Sci. Sin.* **1962,** *11,* 1535.
88. Jordan, F.; Wu, A. *Arch. Biochem. Biophys.* **1978,** *190,* 699.
89. Horiike, K.; McCormick, D. B. *J. Theor. Biol.* **1979,** *79,* 403.
90. Feeney, R. E.; Osuga, D. T. In "Methods of Protein Separation"; Catsimpoolas, N., Ed.; Plenum: New York, 1975; Vol. 1, p. 127.
91. Jakoby, W. B.; Wilchek, M., Eds. *Methods Enzymol.* **1977,** *46.*
92. Riordan, J. F.; Hayashida, H. *Biochem. Biophys. Res. Commun.* **1970,** *41,* 122.
93. Hass, G. M.; Neurath, H. *Biochemistry* **1971,** *10,* 3535.
94. Pétra, P. H. *Biochemistry* **1971,** *10,* 3163.
95. Jentoft, J. E.; Jentoft, N.; Gerken, T. A.; Dearborn, D. G. *J. Biol. Chem.* **1979,** *254,* 4366.
96. Finot, P. A. Chapter 3 in this book.
97. Watanabe, M.; Arai, S. Chapter 7 in this book.
98. Carpenter, K. J. *Nutr. Abstr. Rev.* **1973,** *43,* 423.
99. Lee, H. S.; Sen, L. C.; Clifford, A. J.; Whitaker, J. R.; Feeney, R. E. *J. Nutr.* **1978,** *108,* 687.
100. Kinsella, J. E. *CRC Crit. Rev. Food Sci. Nutr.* **1978,** *10,* 147.
101. Masters, P. M.; Friedman, M. In "Chemical Deterioration of Proteins," *ACS Symp. Ser.* **1980,** *123,* 165.
102. Anson, M. L.; Pader, M. U.S. Patent 2 813 025, 1957.
103. Anson, M. L.; Pader, M. U.S. Patent 2 830 902, 1958.
104. Anson, M. L.; Pader, M. U.S. Patent 2 833 651, 1958.
105. Anson, M. L.; Pader, M. U.S. Patent 2 879 163, 1959.
106. Pour-El, A., Ed. "Functionality and Protein Structure," *ACS Symp. Ser.* **1979,** *92.*
107. Kinsella, J. E.; Shetty, K. J. In "Functionality and Protein Structure," *ACS Symp. Ser.* **1979,** *92,* 37.

108. Unilever N.V. Netherlands Patent Application 6 919 461, 1970; *Food Sci. Tech. Abstr.* **1971**, *3*, 10P1707.
109. Evans, M. T. A.; Irons, L. I. German Patent 1 951 247, 1970.
110. Melnychyn, P.; Stapley, R. B. South African Patent 6 807 706, 1969.
111. Groninger, H. S., Jr. *J. Agric. Food Chem.* **1973**, *21*, 978.
112. Gandhi, S. K.; Schultz, J. R.; Boughey, F. W.; Forsythe, R. H. *J. Food Sci.* **1968**, *33*, 163.
113. Puigserver, A. J.; Gaertner, H. F.; Sen, L. C.; Feeney, R. E.; Whitaker, J. R. Chapter 5 in this book.
114. Puigserver, A. J.; Sen, L. C.; Clifford, A. J.; Feeney, R. E.; Whitaker, J. R. In "Nutritional Improvement of Food and Feed Proteins"; Friedman, M., Ed.; Plenum: New York, 1978; p. 587.
115. Puigserver, A. J.; Sen, L. C.; Gonzales-Flores, E.; Feeney, R. E.; Whitaker, J. R. *J. Agric. Food Chem.* **1979**, *27*, 1098.
116. Puigserver, A. J.; Sen, L. C.; Clifford, A. J.; Feeney, R. E.; Whitaker, J. R. *J. Agric. Food Chem.* **1979**, *27*, 1286.
117. Li-Chan, E.; Helbig, N.; Holbek, E.; Chau, S.; Nakai, S. *J. Agric. Food Chem.* **1979**, *27*, 877.
118. Voutsinas, L. P.; Nakai, S. *J. Food Sci.* **1979**, *44*, 1205.
119. Goldstein, A.; Aronow, L.; Kalman, S. M. "Principles of Drug Action: The Basis of Pharmacology," 2nd ed.; John Wiley & Sons: New York, 1973.
120. Ember, L. R. *Chem. Eng. News* **1980**, *58*(50), 22.
121. Marshall, J. J. *Trends Biochem. Sci.* **1978**, *3*, 79.
122. Meares, C. F.; DeRiemer, L. H.; Leung, C. S.-H.; Yeh, S. M.; Miura, M.; Sherman, D. G.; Goodwin, D. A.; Diamanti, C. I. Chapter 13 in this book.
123. Wu, H.-L.; Means, G. E., unpublished data.
124. Powers, J. C. Chapter 12 in this book.
125. Woodruff, E. A. *J. Bioeng.* **1978**, *2*, 1.
126. Zioudrou, C.; Streaty, R. A.; Klee, W. A. *J. Biol. Chem.* **1979**, *254*, 2446.
127. Knight, M.; Klee, W. A. *J. Biol. Chem.* **1978**, *253*, 3843.
128. Roth, G. J.; Siok, C. J. *J. Biol. Chem.* **1978**, *253*, 3782.
129. Roth, G. J.; Stanford, N.; Jacobs, J. W.; Majerus, P. W. *Biochemistry* **1977**, *16*, 4244.
130. Zaugg, R. H.; Walder, J. A.; Walder, R. Y.; Steele, J. M.; Klotz, I. M. *J. Biol. Chem.* **1980**, *255*, 2816.
131. Feeney, R. E. In "Evaluation of Proteins for Humans"; Bodwell, C. E., Ed.; Avi: Westport, CT, 1977; p. 233.
132. Wood, L. E.; Haney, D. N.; Patel, J. R.; Clare, S. E.; Shi, G.-Y.; King, L. C.; Klotz, I. M. *J. Biol. Chem.* **1981**, *256*, 7046.
133. Cheftel, J. C. In "Nutritional and Safety Aspects of Food Processing"; Tannenbaum, S. R., Ed.; Marcel Dekker: New York, 1979; p. 153.

RECEIVED December 17, 1980.

# 2

# Fundamentals and Applications of Enzymatic Modifications of Proteins: An Overview

JOHN R. WHITAKER
University of California, Davis, CA

ANTOINE J. PUIGSERVER
Centre National de la Recherche Scientifique, Marseille, France

*Enzymes catalyze two types of modifications of proteins, that of hydrolytic and of nonhydrolytic modifications. These modifications, occurring in vivo during or after translation of the protein, are highly specific and may affect markedly the biological, chemical, and physical properties of the protein. Specific proteolysis can be a control process, allowing the biological activity of a number of proteins to be expressed. It also is an intracellular control process since proteins are continuously undergoing degradation at rates that are highly specific for each protein. Many highly specific proteases must occur in rather small amounts in vivo in order to account for the multiplicity of physiological roles in which they are involved. This chapter examines primarily some of the hydrolytic reactions that occur in vivo in the hope that some of the knowledge can be adapted to the hydrolytic modification of food proteins.*

Enzyme-catalyzed modifications of proteins deserve much more work and therefore must be considered as an important field for further scientific investigations. In spite of the prolific research carried out in this area during the past three decades, a number of enzymatic modifications not only are poorly investigated or understood but attempts to apply them to food protein systems are nonexistent. The fundamental aspects of enzymatic modification of proteins are of interest since potential applications for nutritional and functional improvements of food proteins appear to be numerous and promising. Enzymatic and chemical modifi-

0065-2393/82/0198-0057$07.75/0

cations of food and feed proteins (*1*), as well as genetic methods (*2, 3*) and fortification with either free (*4*) or covalently bound (*5, 6, 7, 8*) limiting essential amino acids, have been developed to improve biological quality and/or functional properties of proteinaceous foods. These methods should not be considered as competitive but rather complementary and should challenge biochemists and enzymologists involved in food-related research to discover the most efficient way to increase the acceptability of unconventional protein sources. Improvement of the functional and nutritional properties of food proteins still might be achieved should only one of these methods prove to be effective. For example, enzyme-catalyzed reactions, which take place under mild conditions, may be useful when chemical methods either are not feasible or may present potential health hazards due to possible toxicant formation.

### *Types of Enzymatic Modifications of Proteins*

For the purpose of this chapter, enzyme-catalyzed modifications of proteins will be divided into two groups: hydrolytic and nonhydrolytic reactions. Generally speaking, post-translational reactions occurring in vivo are catalyzed by highly specific enzymes under rather restricted conditions in contrast with in vitro modifications which are carried out under less specific conditions.

Modification in vivo by specific nonproteolytic enzymes adds groups to, or changes groups of, certain amino acid side chains in proteins and is important for biological function. Although only 20 amino acids are incorporated into proteins via translation, post-translational reactions lead to some 135 known modifications of the amino acid residues of proteins (*see* Table I). One of the most extensively modified proteins—post-translational—is collagen where more than 50 modifications have been described. The modifications listed in Table I fall into six types of reactions: phosphorylation, glycosylation, hydroxylation, acylation, methylation, and cross-linking (*see* Table II). In vivo modifications are carried out by specific enzymes and are often essential for expressing biological activity. In contrast to in vivo modification, only three types of nonhydrolytic modifications have been achieved in vitro with enzymes—phosphorylation, cross-linking, and tyrosine oxidation (*see* Table II). Some of the enzymatic modifications could improve the quality and the functional properties of proteins from plants, single cells, and other less conventional proteins. Such improvements might help to meet world protein needs.

These nonhydrolytic, post-translational enzymatic modifications of proteins, described in two recent publications (*9, 10*), will not be treated further in this chapter. Hydrolytic modification, in vivo and in vitro, is the single most frequently occurring enzymatic modification of proteins.

Of the many reactions taking place at a post-translational level, the major emphasis of this chapter will be a description of a few biological systems in which the restricted scission of one or a very few specific bonds in a precursor protein is involved in unmasking a biological function. Specific cleavage of the polypeptide chain of a presecretory protein or an integral membrane protein is required for segregating secretory proteins and for assembling cell membranes, respectively. Activation of biological precursors will be described for selected protein complexes and blood-clotting steps. In vitro hydrolytic modifications of proteins include reactions that already have been applied to food systems or that represent potential uses for nutritional improvements or changes in the functional properties of proteins (*11, 12*). Some examples of these reactions will be given later in this chapter.

The hydrolytic modifications of proteins catalyzed by enzymes include both generalized reactions, where a relatively large number of peptide bonds are split, and limited reactions where hydrolysis of one or only a few bonds are necessary in order to achieve the desired product (*see* Table III). Examples of both types of reactions will be presented below.

## *In Vitro Enzymatic Modification of Proteins Through Hydrolysis*

Some examples of in vitro hydrolytic modifications of proteins are shown in Table IV. The preparation of cheeses, chillproofing of beer, and the production of protein hydrolysates represent major uses of proteases. With the possible exception of cheese preparation, the application involves a substantial degree of hydrolysis. Therefore rather nonspecific proteases often are used.

**Coagulation of Casein.** The primary action of rennin on κ-casein to cause coagulation of the casein complex is a typical example of where a minor chemical change in one protein results in a major physical change in a food system. As shown in Figure 1, rennin hydrolyzes a specific Phe–Met bond in κ-casein to produce two fragments, the macropeptide and *p*-κ-casein (*13*). Neither fragment, alone or together, is able to stabilize the casein micelle. Therefore the caseins aggregate to form a clot. Although there are other peptide bonds that are hydrolyzed subsequently by rennin, the initial step is a highly specific one (*12*). The higher the ratio of the rate of hydrolysis of the specific Phe–Met peptide bond to that of general proteolysis, the firmer the clot will be.

**Chillproofing of Beer.** In chillproofing of beer, proteolytic enzymes, primarily papain, are used to hydrolyze the proteins in beer contributed by the grain, malt, and yeast. The process must be controlled so that the polypeptides produced are no longer able to form an insoluble precipitate

**Table I. In Vivo Nonhydrolytic Enzyme-Catalyzed Post-Translational Modifications of Amino Acid Residues of Proteins**[a]

| *Amino Acid (Group)* | *Typical Modifications* | *Total Modifications Known* |
|---|---|---|
| Arginine (guanidino) | methyl- (mono- and di-)<br>ADP-ribosyl-<br>citrulline<br>ornithine | 6 |
| Lysine ($\epsilon$-$NH_2$) | glucosyl-<br>phospho-<br>pyridoxyl-<br>biotinyl-<br>lipoyl-<br>acetyl-<br>methyl- (mono-, di-, tri-)<br>$\delta$-hydroxyl-<br>$\delta$-glycosyl-<br>crosslinks | 33 |
| Histidine (imidazole) | methyl- (1- and 3-)<br>phospho- (1- and 3-)<br>iodo-<br>flavin- | 6 |
| Proline | 4-hydroxy-<br>3-hydroxy-<br>3,4-dihydroxy-<br>4-glycosyloxy- | 5 |
| Phenylalanine (benzene ring) | $\beta$-hydroxy-<br>$\beta$-glycosyloxy- | 2 |
| Tyrosine (benzene ring/hydroxyl) | $\beta$-hydroxy-<br>$\beta$-glycosyloxy-<br>sulfono-<br>iodo- (mono-, di-)<br>bromo- (mono-, di-)<br>chloro- (mono-, di-)<br>bis-ether<br>adenylyl<br>uridylyl-<br>RNA | 18 |

**Table I. Continued**

| *Amino Acid (Group)* | *Typical Modifications* | *Total Modifications Known* |
|---|---|---|
| Serine (hydroxyl) | phospho-<br>glycosyl-<br>methyl-<br>phosphopantethcinc-<br>ADP-ribosyl | 8 |
| Threonine (hydroxyl) | phospho-<br>glycosyl-<br>methyl- | 6 |
| Cysteine (sulfhydryl) | cystine<br>glycosyl-<br>dehydroalanyl<br>heme<br>flavin<br>seleno- | 7 |
| Aspartic acid/glutamic acid (carboxyl) | $\gamma$-carboxyl-<br>$\beta$-phospho-<br>methyl | 3 |
| Asparagine/glutamine (amide) | glycosyl<br>R-NH-<br>pyrrolidone | 4 |
| Carboxyl terminal residue (carboxyl) | -amide<br>-amino acid | 11 |
| Amino terminal residue (amino group) | acetyl-<br>formyl-<br>glucosyl-<br>amino acyl-<br>pyruvyl-<br>$\alpha$-ketobutyryl-<br>methyl-<br>glycuronyl-<br>murein | 19 |
| Total modifications | | 135[b] |

[a] Adapted from Whitaker (*9*) and Uy and Wold (*10*).
[b] Including 3 each for aspartic acid and glutamic acid and 4 each for asparagine and glutamine.

**Table II. Enzyme-Catalyzed Nonhydrolytic Modifications of Proteins**

| *In Vivo* | *In Vitro* |
|---|---|
| Phosphorylation | phosphorylation |
| Glycosylation | |
| Hydroxylation | hydroxylation of tyrosine |
| Acylation | |
| Methylation | |
| Cross-linking | lipoxygenase-catalyzed cross-linkages |

**Table III. Hydrolytic Modifications of Proteins**

*Generalized*
- Extracellular digestion (GI tract)
- Intracellular digestion (lysosomes)
- Modification for food use

*Limited*
- Modification for food use
- Physiological modification

**Table IV. In Vitro Enzyme-Catalyzed Hydrolytic Modifications of Food Proteins**

- Preparation of cheeses and soy derivatives
- Solubilization of protein concentrates
- Production of protein hydrolysates
- Gluten modification in bread doughs
- Chillproofing of beer
- Plastein formation
- Tenderization of meats
- Quality determination of proteins

with tannins in the beer when it is chilled but they still are able to entrap the carbon dioxide required for head formation on the beer.

**Plastein Formation.** Plastein formation is another example of using proteases to modify high-protein food systems to drastically change the properties of that system (*11*). In the plastein reaction a protease such as papain is used to partially hydrolyze the proteins to about a 10,000–20,000-dalton size at a pH near neutrality. After concentrating the hydrolyzate to ~ 35% (based on protein) and a change in pH, the same protease or a different one is used to catalyze the resynthesis of a few peptide bonds. This may result in a decrease in the solubility of the protein.

Some potential uses of the plastein reaction are given in Table V. The plastein reaction has been proposed for removing bitter peptides formed through previous hydrolysis of proteins by facilitating the resyn-

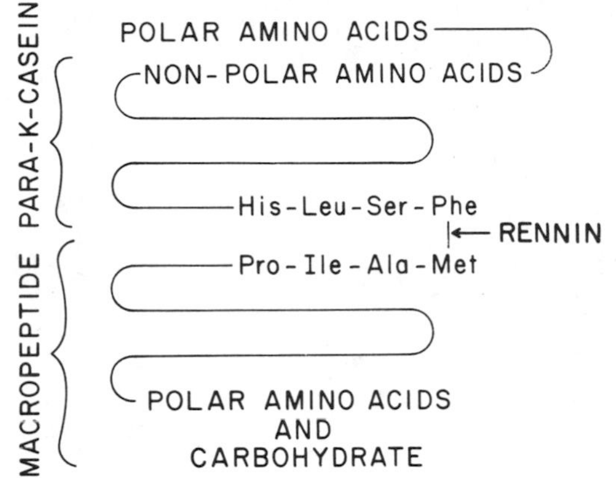


*Figure 1. Schematic of a proposed structure of bovine κ-casein A with attached carbohydrates* (12)

thesis and/or recombination of peptides to form products that are no longer bitter. More recently Arai (*14*) has used the plastein reaction to covalently incorporate glutamate into the protein hydrolysate, thereby alleviating the bitter taste.

More details of the plastein reaction and its application to remove pigments such as chlorophyll, or to remove off-flavor components such as the beany taste of soybeans, are shown in Figure 2. The protein of the food system is solubilized and denatured (in order to achieve proteolysis), a protease is added, and the hydrolytic reaction is allowed to proceed. On partial hydrolysis of the protein the pigments and flavor constituents are released from the protein; they are removed, the hydrolyzate is concentrated, and resynthesis and/or rearrangement of the amino acid sequence of the polypeptides is catalyzed by the same or a different protease. Resynthesis also can be carried out in the presence of added amino acid esters in order to improve the nutritional/functional properties of the protein.

The mechanism of the plastein reaction is shown by the equation on page 64.

**Table V. Some Potential Uses of the Plastein Reaction in High-Protein Foods**

Increase solubility
Change physical properties
Remove bitter peptides
Incorporate limiting essential amino acids
Remove color, flavor, and pigments
Remove unwanted amino acids
Prepare surfactants

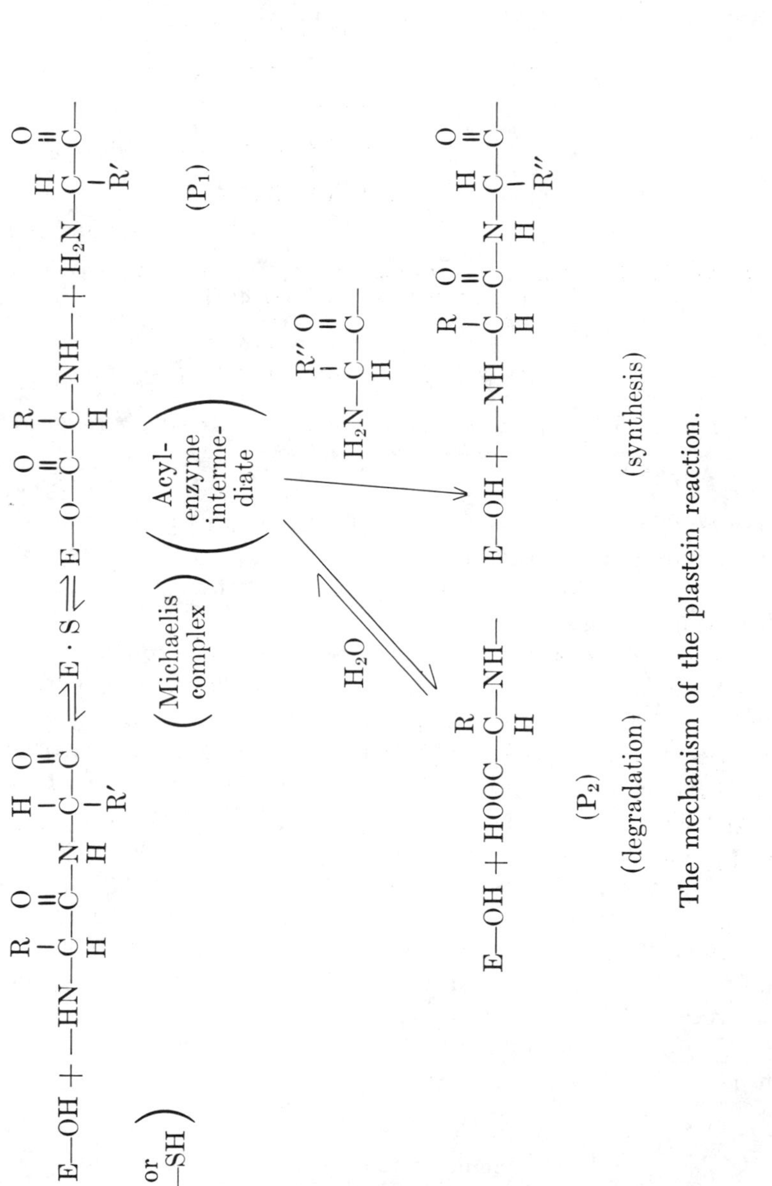


The mechanism of the plastein reaction.

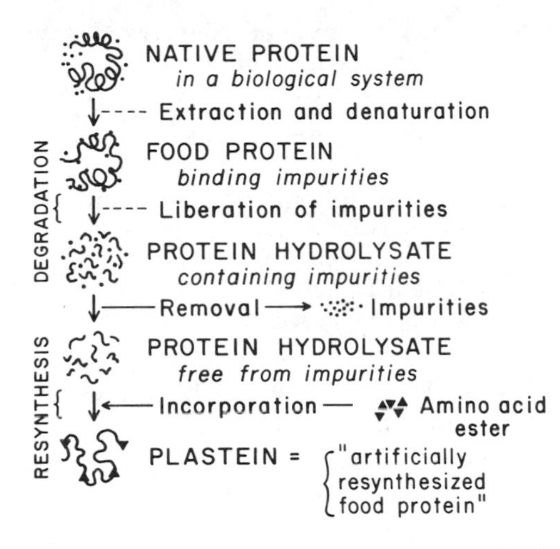


American Chemical Society

*Figure 2. A process of enzymatic protein degradation and resynthesis for producing a plastein with improved acceptability and an improved amino acid composition* (11)

Following formation of the Michaelis complex by association of enzyme and substrate, a covalent acyl-enzyme intermediate is formed with release of the other part of the polypeptide chain ($P_1$). The acyl-enzyme intermediate is hydrolyzed normally (degradation), releasing the other portion of the original polypeptide ($P_2$). However, two additional reactions may occur depending on the presence and concentration of other nucleophiles in addition to water. The synthesis of new peptide bonds may occur by the amino group of a peptide competing with water for the acyl group of the acyl-enzyme intermediate as shown (synthesis; a transpeptidation reaction) or by allowing the enzyme to react with a $P_2$-like compound to form an acyl-enzyme intermediate (reversal of hydrolysis) followed by transpeptidation as shown (synthesis). More recently Watanabe and Arai (*15*) have used the plastein reaction to synthesize surfactants from proteins and fatty acid esters for potential food use. While the plastein reaction is carried out readily, especially on a laboratory scale, and can change markedly the functional and nutritional properties of the original protein, it may not be economically attractive yet to the food-processing industry.

### *In Vivo Enzymatic Modification of Proteins Through Hydrolysis*

Enzymatic hydrolysis of proteins in vivo is a very important process and it is used biologically in many ways (*see* Table VI). Generally these reactions can be grouped as either important for protein turnover or for

**Table VI. In Vivo Enzyme-Catalyzed Hydrolytic Modifications of Proteins**

Gastric or intestinal digestion
Cellular protein turnover
Post-mortem tenderization of meat
Aggregation of polypeptide chains
Blood clotting
Activation of precursors
Complement action
Cleavage of presecretory proteins

expressing biological activity of a protein. Examples of each of these will be discussed.

**Protein Turnover.** GASTROINTESTINAL TRACT. In humans and in other animals with digestive systems, there are two systems for protein turnover—the gastrointestinal system and the cellular digestive system. In the gastrointestinal system, protein digestion begins in the stomach via the action of pepsin, an endoprotease that has maximum activity at peptide bonds involving the aromatic amino acids (*see* Table VII). Digestion is completed in the small intestine by several proteases of different specificities. Trypsin, chymotrypsin, and elastase are endoproteases acting to hydrolyze proteins and polypeptides to smaller peptides. Di- and tripeptides may enter the epithelial cells of the small intestine where they

**Table VII. Enzyme-Catalyzed Hydrolysis of Proteins in the Gastrointestinal Tract**

| *Enzyme* | *Location* | *Specificity* |
| --- | --- | --- |
| Pepsin | stomach[a] | aromatic amino acid residues |
| Trypsin | small intestine[b] | arginine and lysine |
| Chymotrypsin | small intestine[b] | phenylalanine, tyrosine, tryptophan |
| Elastase | small intestine[b] | alanine |
| Carboxypeptidase A | small intestine[b] | all amino acid residues except arginine, lysine and proline |
| Carboxypeptidase B | small intestine[b] | arginine and lysine |
| Aminopeptidases | small intestinal mucosa | broad |
| Tripeptidases | small intestinal mucosa | broad |
| Dipeptidases | small intestinal mucosa | broad |

[a] Secreted by the chief cells of stomach as the zymogen.
[b] Secreted by the pancreas as the zymogen.

become hydrolyzed by specific di- and tripeptidases. Carboxypeptidases A and B and/or aminopeptidases act on longer peptides to produce amino acids that can be transported across the small intestinal wall. Free amino acids resulting from either extracellular or intracellular digestion then cross the basal membrane of the epithelial cell and pass into the blood stream where they are transported to the liver and other organs for resynthesis into specific proteins.

INTRACELLULAR TURNOVER. There are many proteases involved in the turnover of proteins at the intracellular level, all of which have rather restricted specificity. These enzymes, while also present in the cytoplasm, are highly concentrated in the lysosomes. They generally are referred to as cathepsins and include both endo- and exosplitting hydrolases.

These proteases have been studied extensively because of their importance in the continuous turnover of proteins at the cellular level. The reader is referred to several recent excellent reviews (*16, 17, 18*). In this chapter, we shall concentrate on the specificity of protein turnover as an example of the potential that exists for selective hydrolytic modifications of food proteins.

One of the continuing puzzles of intracellular protein turnover is its specificity (*see* Table VIII)(*19*). If hydrolysis occurs in the lysosome as a result of phagocytosis one might expect that all of the proteins would be turned over at more or less the same rate, especially those of a single organ. When protein turnover as a class is examined, the mitochondrial protein has a half-life of 4–5 d. However, when the rates of turnover of individual proteins are determined the half-lives range from about 10 min for ornithine decarboxylase, 1.4 d for catalase, and 16 d for Isozyme 5 of lactate dehydrogenase to essentially no turnover of elastin and collagen. Therefore there must be some control mechanism(s) that permits proteases to distinguish among different proteins.

One possible explanation is that the rate of proteolysis of individual proteins is controlled by the equilibrium between the native and dena-

**Table VIII. Rates of Intracellular Turnover of Some Selected Proteins**[a]

| *Protein* | *Half-Life* |
|---|---|
| Ornithine decarboxylase | 10 min |
| δ-Amino levulinate synthetase | 60 min |
| Catalase | 1.4 d |
| $LDH_5$ | 16 d |
| Mitochondria protein as whole | 4–5 d |
| Rat protein, 70% | 4–5 d |
| Cultured cells | 1–2%/h |

[a] Adapted from Schimke and Bradley (*19*).

tured forms of a protein (*see* Table IX). It generally is accepted that peptide bonds are not susceptible to proteolysis when the polypeptide chain is folded in the native state. The equilibrium constant ($K_{eq}$) is expected to differ among different proteins because of amino acid sequence and size differences, etc. Other factors, such as those listed in Table IX, could influence the equilibrium between the native and denatured states of proteins so that their turnover could differ depending on the nutritional state of the organism, for example. The effects of all of these factors on the rate of proteolytic degradation of one or more proteins have been tested now and indeed they do affect, often markedly, the stability of a protein to proteolyses (*20*).

An exciting example of the possible relationship between intracellular proteolysis and the conversion of muscle to meats is that of the turnover of the individual proteins of the myofibrillar system of muscle. The key event could be a limited proteolysis of one or more proteins in the Z-line of the sacromere which might lead to dissociation of the sacromeres followed by disintegration of the thick and thin filaments into their constituent proteins (*see* Figure 3) (*21*). The individual proteins then could be hydrolyzed in the usual fashion by proteases of the lysosomes. In meat tenderization, limited proteolysis may lead only to the stage of dissociation of some of the sacromeres.

It is quite likely that one could take advantage of the different stabilities of food proteins as a function of temperature, pH, ionic strength, etc., to achieve selective hydrolysis.

**Post-Translational Cleavage of Proteins.** Highly specific cleavage of proteins following biosynthesis is involved in a wide variety of biological processes. Some well-known examples of specific scission of polypeptide chains include the activation of digestive zymogens (*22*), prohormones (*23, 24, 25*), and precursors of the blood coagulation cascade (*26*). Other examples are given in Table X. In a number of cases,

**Table IX. Factors Affecting Intracellular Turnover of Proteins**[a]

$$\text{Native protein} \overset{K_{eq}}{\rightleftharpoons} \text{denatured protein} \xrightarrow{\text{proteolysis}} \text{peptides} + \text{amino acids}$$

*Equilibrium* ($K_{eq}$) *controlled by:*

Interaction with substrates, coenzymes, and allosteric effectors
Covalent modification
Subunit interaction
Protein–protein interaction
Change in hydrophobicity
Incorporation of wrong amino acids
Nutritional state

[a] Adapted in part from Holzer and Heinrich (*20*).

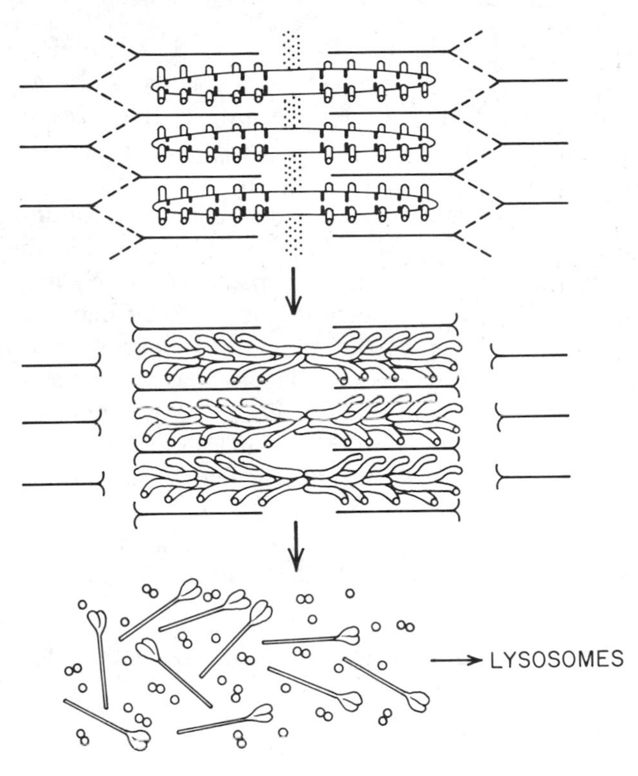


*Figure 3. Schematic illustrating the possible role of a calcium-activated protease in the degradation of the Z-line of sacromere of muscle and the dissociation of the proteins of the thick and thin filaments* (21)

highly specific proteases are involved in the key hydrolytic step (*see* Table XI) (*20*). Some examples follow.

Small Intestinal Sucrase–Isomaltase. An example of the role of polypeptide precursor cleavage has been described recently for small intestinal sucrase–isomaltese. This enzyme complex is a membrane-intrinsic, dimeric glycoprotein of the intestinal brushborder anchored by a hydrophobic segment located near the amino terminal end of the isomaltase subunit (*28, 29*). Studies of the biosynthesis of this binary complex of the rat's small intestinal microvillus membrane have shown that both subunits are synthesized as a single-polypeptide-chain precursor which then is split into the sucrose–maltose and isomaltose–maltose hydrolyzing subunits (*30*). These studies also suggested that pancreatic elastase may be responsible for the in vivo specific hydrolysis of the precursor as was shown for elastase in vitro. Therefore pancreatic enzymes may be involved in late post-translational processing of intestinal membrane enzymes (*30*).

ACTIVATION OF ZYMOGENS OF PROTEASES. The role of specific proteolysis in converting precursor proteins to proteases has been known since the discovery of intestinal enterokinase and its role in converting trypsinogen to trypsin (*see* Figure 4) (*31*). Intestinal enterokinase is a highly specific protease (*see* Table XI) with the seemingly single mission of splitting the lysine–isoleucine bond's six amino acid residues from the *N*-terminal end of trypsinogen (*31*). Hydrolysis of this single peptide bond permits the conformation of trypsinogen to change to that of trypsin with full activity. As shown in Figure 4, the activation of trypsin also controls the activation of the chymotrypsinogens, procarboxypeptidases A and B, proelastase, and prophospholipase to the active enzymes.

Some details of the activation of chymotrypsinogen A to $\alpha$-chymotrypsin by limited proteolysis are given in Figure 5 (*32*). The key step in the activation of chymotrypsinogen A is hydrolysis of the peptied bond involving $Arg^{15}$–$Ile^{16}$ by trypsin. The product of this step, $\pi$-chymotrypsin, is fully active (higher specific activity than $\alpha$-chymotrypsin) but it undergoes further proteolysis leading to the splitting of the $Leu^{13}$–$Ser^{14}$ (by

**Table X. Physiological Systems Controlled by Limited Post-Translational Proteolysis[a]**

| *Physiological System* | *Example* |
|---|---|
| Assembly | bacteriophage<br>virus<br>membrane<br>procollagen → collagen<br>fibrinogen → fibrin |
| Defense reactions | blood coagulation<br>fibrinolysis<br>complement reaction |
| Development | maturation of spermatozoa, release of ova, and fertilization (proacrosin → acrosin)<br>prochitin synthetase → chitin synthetase<br>proccharacteristic cocoonase → cocoonase |
| Digestion | zymogen → enzyme |
| Hormone production | proinsulin → insulin<br>angiotensinogen → angiotensin |
| Oncogenic transformations | division, growth, migration, and adhesion |
| Tissue injury | impairment of cell contact inhibition<br>prekallikrein → kallikrein<br>kininogen → kinin |
| Translocation | preprotein → protein |

[a] Adapted in part from Neurath (*27*).

**Table XI. Some Examples of Highly Specific Proteases Involved in Physiological Processes**[a]

Group-specific proteinase of mast cells which inactivates PALP-apoenzymes

Proteinases of *B. megaterium* spores which cleave spore protein

RecA gene product from *E. coli* which cleaves bacteriophage λ repressor

The acid proteinase, renin, which cleaves angiotenginogen to ansiotensin

The carboxydipeptidase that cleaves angiotensin

The neutral proteinase, enterokinase, which cleaves trypsinogen to trypsin

The albumin-degrading light subunit of rat kidney γ-glutamyl transpeptidase

The plasminogen-converting streptokinase

The proteinase which cleaves the T4 prehead precursor protein

Calcium-activated protease of myofibrils (?)[b]

Methionine aminopeptidase (?)[b]

[a] Adapted in part from Holzer and Heinrich (*20*).
[b] Preliminary data indicate that these are probably highly specific proteases.

π-chymotrypsin), $Tyr^{145}$–$Thr^{146}$, and $Asn^{147}$–$Ala^{148}$ bonds (by π- and δ-chymotrypsins(?)) to give α-chymotrypsin, the stable product of this limited proteolysis. In the presence of trypsin, π-, and δ-chymotrypsins, chymotrypsinogen A can be converted through the neochymotrypsinogens to α-chymotrypsin as shown in Figure 5. As a consequence of these reactions a protein, translated as a single polypeptide chain, is converted by post-translational modification into a protein having three polypeptide chains.

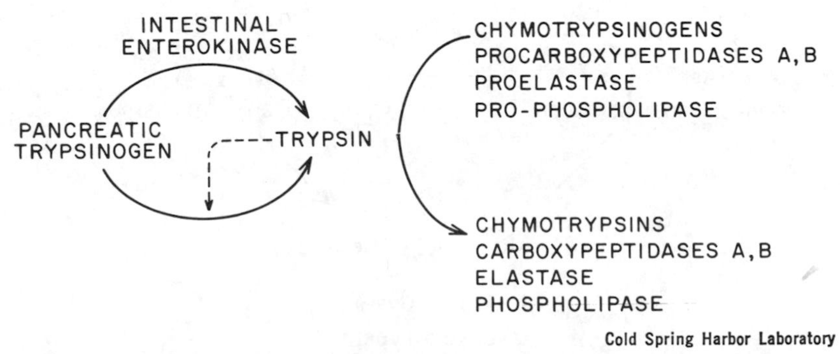


*Figure 4. The role of enterokinase in the activation of trypsinogen and subsequent activation of other protease zymogens by trypsin, a two-stage cascade (27)*

A much more complicated example of the role of limited proteolysis in a physiological process is that of the activation of the Hageman Factor (zymogen → protease conversion) by events triggered by tissue damage, such as cutting (*see* Figure 6) (*33*). Activation of the Hageman Factor triggers three systems: (1) kinin formation, which results in increased blood pressure; (2) the intrinsic blood-clotting system with its complex cascade of zymogen-to-protease conversions (*26*); and (3) the activation

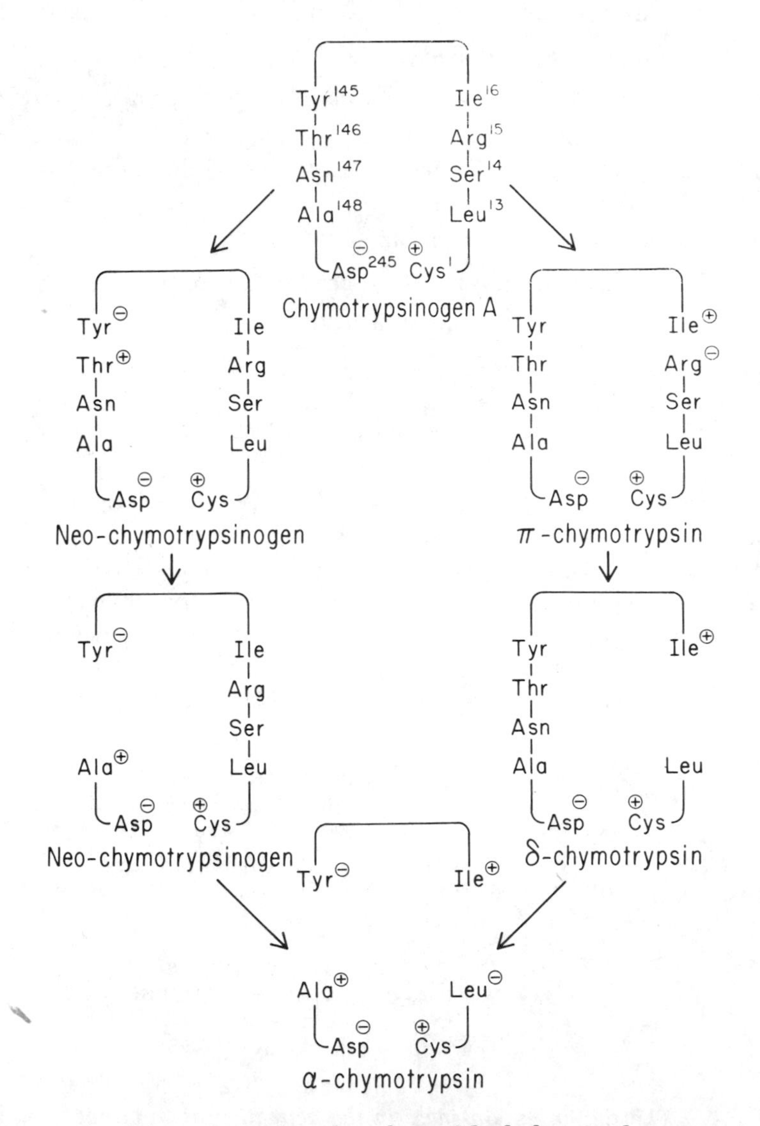


*Figure 5. Schematic of the role of limited hydrolysis in the conversion of bovine chymotrypsinogen A to π-, δ-, and α-chymotrypsins* (32)

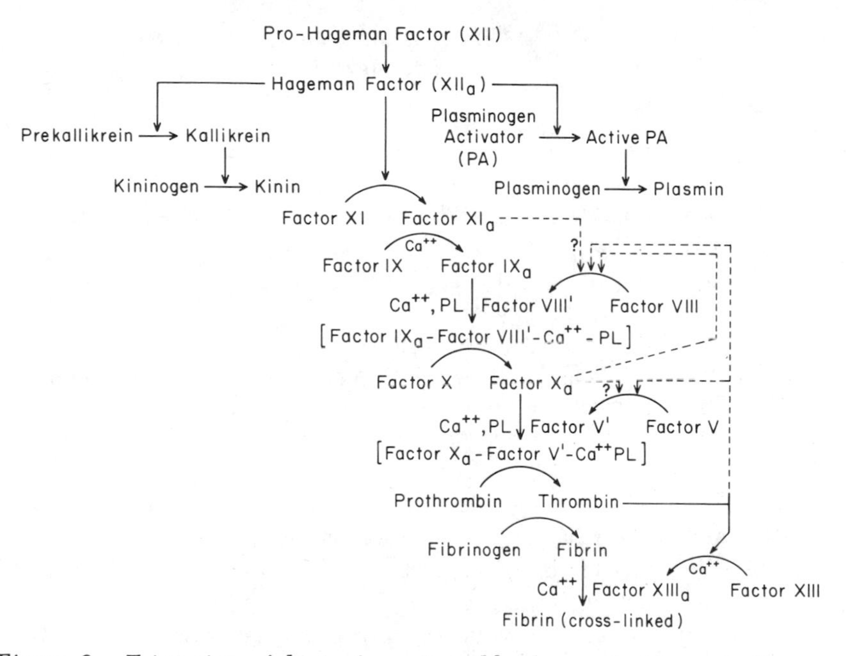


*Figure 6. Triggering of kinin formation, blood coagulation, and fibrinolysis through specific proteolytic activation of the Hageman Factor (Factor XII). In the cascades, the factor on the left side of the reaction (zymogen) is converted to an active enzyme by proteolysis. PL = phospholipids.*

of plasminogen to plasmin in preparation for dissolution (fibrinolysis) of blood clots that might break loose from the damaged area and be free in the blood. This is a truly remarkable example of the importance of limited proteolysis in the regulation of key physiological processes.

OTHER EXAMPLES. Limited specific proteolysis appears to be involved in forming the 7S nerve growth factor (NGF) complex. The NGF purified from the submaxillary gland of the adult male mouse is a dimer ($\beta_2$) contained in a larger multisubunit complex. As shown in Figure 7, two types of subunits, $\alpha$ and $\gamma$, are associated with the NGF, $\beta_2$, to form the 7S complex $\alpha_2\beta_2\gamma_2$ (*35, 36*). The $\alpha$ subunit is an acidic protein which, in association with $\beta_2$, inhibits the arginine esteropeptidase activity of the $\gamma$ subunits. There is considerable evidence (*37, 38, 39*) that the NGF complex is produced initially as a precursor (pro-NGF) since the $\beta$ subunits contain extension peptides at the carboxyl terminal end (pro-$\beta$ subunits) (*see* Figure 8). The $\gamma$ subunits hydrolyze the pro-$\beta$ subunits at a specific Arg–*X* peptide bond and remain associated to the resulting $\beta$ subunits to produce the 7S NGF complex (*40*).

The role of the intrinsic protease ($\gamma$) in NGF processing is reminiscent of the processing of the gag gene proteins. The precursor protein

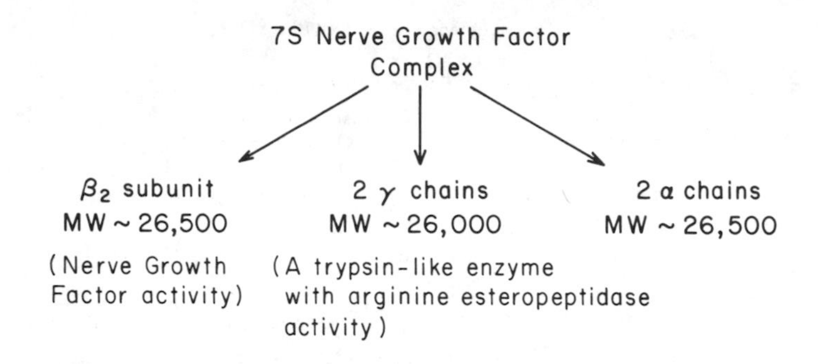


*Figure 7. Subunit structure of the 7S NGF complex (adapted from Ref. 34). The molecular weight of the complex is approximately 130,000 daltons and it is stable between pH 5.0 and 8.0.*

produced by avian sarcoma–leukosis virus has a molecular weight of 76,000 daltons (Pr 76) and is translated on a single-messenger RNA (*41, 42*). Cleavage of the Pr 76 protein occurs soon after its biosynthesis in a stepwise fashion. The first fragment having a molecular weight of 15,000 daltons is thought to be the protease that is responsible for the further proteolytic cleavage of the Pr 76. The final products are proteins with molecular weights of 12,000, 15,000, 19,000, and 27,000 daltons (*42*).

Several of the small physiologically active peptides now are known to be derived from a single protein precursor synthesized by the pituitary gland (*43*). As shown in Figure 9, ACTH (adrenocorticotropic hormone), β-LPH (β-lipotropic hormone), β-MSH (β-melanocyte-stimulating hormone), β-endorphin, and enkephalin result from the specific proteolytic cleavage of a precursor protein of approximately 8,000 daltons.

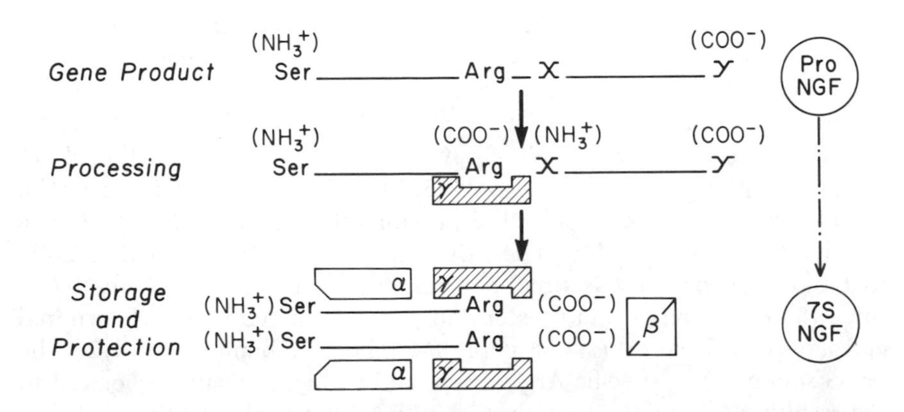


*Figure 8. Hypothesis for NGF biosynthesis as a precursor (pro-NGF) and its assembly into the 7S complex (adapted from Ref. 34). NGF complex consists of 2α, 2γ, and 2 pro-β chains. The pro-β chains are converted to β chains through specific proteolysis catalyzed by γ subunits.*

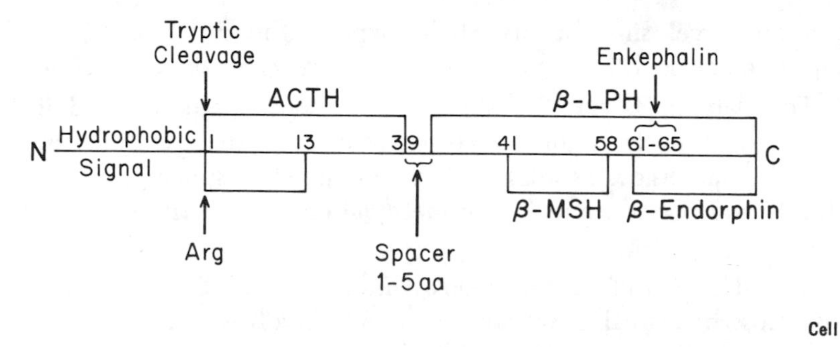


Cell

*Figure 9. The precursor protein giving rise to ACTH, β-LPH, β-MSH, β-endorphin, and enkephalin through specific proteolysis* (83)

Another general role of post-translational cleavage of polypeptide chains in viral systems is the specific processing of structural proteins leading to the assembly of the virus (*44, 45*). In this respect, the semliki–Forest virus may be considered as an interesting model to find out how proteins which have different cellular locations reach their final sites within the cell. A single messenger RNA directs the synthesis of the viral capsid protein and of two membrane proteins.

Similar proteins, such as the receptor immunoglobulin on the surface of B lymphocytes and the closely related serum immunoglobulins which differ only by the presence of a hydrophobic sequence near the carboxyl terminal end of the receptor immunoglobulin, are the result of RNA splicing at a post-transcriptional level and not from post-translational processing [*see* Figure 10 (*46*)]. However, such examples do not preclude the increasing importance of post-translational hydrolytic modification of proteins in several biological systems and especially at the plasma

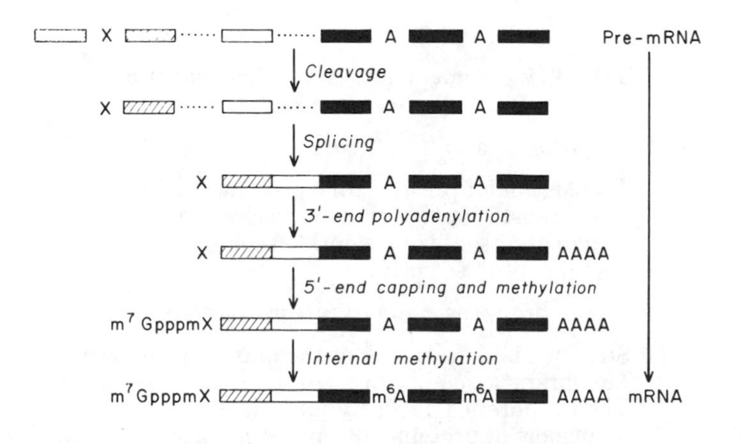


*Figure 10. Conversion of high-molecular-weight pre-mRNA to mRNA through cleavage, splicing, end-group capping, and methylation following transcription*

membrane level since it may help explain the difference in protein composition of the different membrane compartments of cells (*47, 48, 49*).

The data of Figure 10 illustrate that post-synthetic modification, including hydrolysis, is not limited to proteins. In this example, the large pre-mRNA undergoes post-transcriptional cleavage, splicing, modification of the 3′ and 5′ ends, and finally methylation to give the specific active mRNA.

**Translocation of Proteins Across Membranes.** The transfer of proteins across biological membranes generally involves a hydrolytic modification step of the precursor form of the mature protein. This processing has been shown clearly to occur during segregation of secretory proteins, transport of proteins into mitochondria, and entry of plant and microbial toxins into cells as shown in Table XII.

SEGREGATION OF SECRETORY PROTEINS. A large number of proteins, including precursors of digestive enzymes (such as prechymotrypsinogen A), milk proteins, ovalbumin, and several hormones, are secretory proteins that are synthesized in specialized cells. The pancreatic exocrine cell (*see* Figure 11) has been used as a model to describe the intracellular pathway of secretory proteins (*47*). Among the early events of biosynthesis of secretory proteins is the segregation of polypeptide chains into the cisternae of the rough endoplasmic reticlum. In vitro translation studies of mRNA coding for secretory proteins in the absence of microsomal membranes permitted Blobel et al. and others to elucidate the mechanism of transport of proteins across the plasma membrane (*50, 51, 52, 53*). A number of mammalian secretory proteins contain a short hydrophobic sequence (15–30 amino acid residues) at their amino terminus which is

**Table XII. Some Factors in Translocation of Proteins Across Membranes**

*Segregation of secretory proteins*

- Translation of presecretory proteins
- Post-translational precursor processing
- Removal of the signal peptide sequence
- Signal peptidase activity

*Cell membrane insertion of integral proteins*

- Structural organization of the plasma membrane
- Membrane assembly in vitro
- The membrane trigger hypothesis
- Transport of proteins into mitochondria

*Entry of toxins into plant and microbial cells*

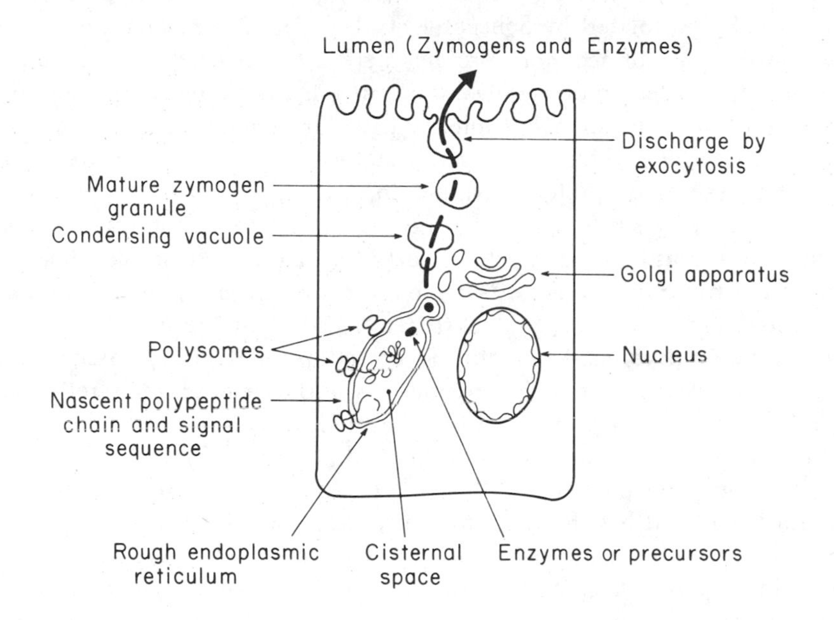


*Figure 11. Schematic of the intracellular transport pathway of enzymes and precursors in the pancreatic exocrine cell (based on the data presented in Ref.* 47)

assumed to allow vectorial discharge of the nascent polypeptide chain into a membrane-bounded compartment (*54, 55, 56, 57*). The hydrophobic extension at the *N*-terminus of a presecretory protein, designated as the signal peptide, binds specifically to the receptor on the membrane. The resulting membrane-bound ribosome still is able to synthesize the polypeptide chain with a co-translational transport of the nascent preprotein into the intracellular membrane-bounded compartment. The signal hypothesis stipulates that secretory proteins are transported into the endoplasmic cisternae only during translation (*58*).

The signal sequence was present only when mRNAs are translated in a cell-free system devoid of microsomal membranes. It is suggested in vivo that this sequence constitutes a metabolically short-lived peptide extension and becomes useless as soon as the secretory protein reaches the endoplasmic reticulum cisternae. The proteolytic enzyme(s) (signal peptidase) responsible for removing the amino terminal extension of presecretory proteins is confined to the rough endoplasmic reticulum (*59, 60, 61*). Translation of mRNAs for secretory proteins in wheat germ lysates or reticulocyte lysates in the presence of microsomal membranes resulted in the removal of the signal peptide (*62, 63, 64*). Sequences of the prepieces reported so far contain a high percentage of hydrophobic residues as originally postulated by Schechter et al. (*65*). The mechanism(s) for the translocation of secretory proteins across the plasma membrane has been highly conserved during evolution since hydrophobic signal peptides also were found in animals evolutionarily far removed from mammals, as in fish (*63, 66*), insects (*67*), and bacteria (*60, 68, 69*). However, at least one secretory protein, ovalbumin, is known to lack the amino terminal pre-piece (*70*), although the nascent protein contains the functional equivalent of a signal sequence (*71*). A detailed physicochemical analysis of the translocation of secreted and membrane-spanning proteins suggested that initiation of transfer of ovalbumin across the plasma membrane is different from other secreted proteins (*72*). Recently a model for insertion of signal peptide into the endoplasmic reticulum membrane has been proposed from the analysis of structural properties of the amino acid sequences from 22 signal peptides( *73*).

It should be pointed out that all of the early studies on in vitro translation of mRNA for secretory proteins in cell-free systems were carried out with canine pancreatic subfractions. The probable reason for using canine pancreatic subfractions is that canine pancreas has a very low ribonuclease level (*74*) allowing the isolation of undegraded polyribosomes (*75*). More recently isolation of undegraded mRNA from sources containing high levels of ribonuclease has become possible by using placental ribonuclease inhibitor (*76*) or the potent protein denaturant guanidine thiocyanate (*77*).

Collagen Biosynthesis. The pro-$\alpha$ chains (mol wt 150,000) of collagen are translated on mRNA followed by hydroxylation of lysine and proline by prolyl hydroxylase(s) and lysyl hydroxylase, respectively (*see* Figure 12). These hydroxylating enzymes appear to be associated almost exclusively with the rough microsomal fraction indicating that hydroxylation proceeds as the chain is translated. Carbohydrate also is attached at this point by the glycosyl transferases. Association of the pro-$\alpha$ chains to form the triple helix probably is facilitated by the *N*- and *C*-terminal peptides, especially by the latter which contain cysteine leading to formation of interchain disulfide bonds. The procollagen molecule then is secreted to the exterior where it undergoes specific proteolysis first by the *N*-protease and then by the *C*-protease with a reduction in molecular weight from 450,000 to 300,000 daltons. Before or after highly specific aggregation, lysyl oxidase actively deaminates certain Lys and/or Hylys residues to produce highly reactive aldehyde groups. The aldehyde groups then condense with reactive $\epsilon$-$NH_2$ groups, primarily from Hylys, to form a network of intermolecular aldimine and keto-imine bonds (*80*) thereby giving great stability and insolubility to the collagen fibrils.

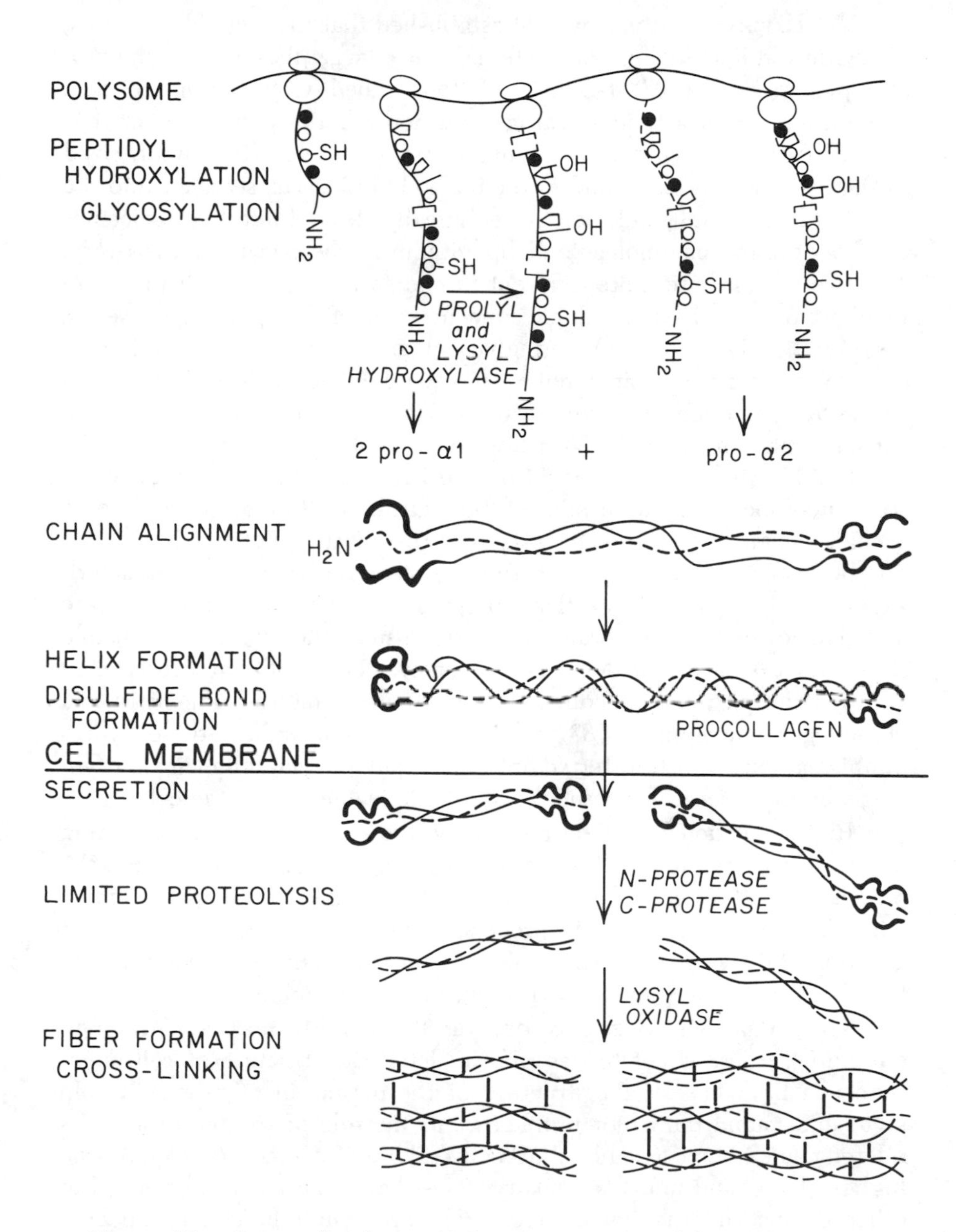


*Figure 12. Proposed scheme for in vivo post-translational enzymatic modifications involved in collagen formation (adapted from Refs. 78, 79, and 81)*

VITELLOGENIN. It is now well established that the egg yolk proteins, phosvitin and lipovitellin, are synthesized as a large phosphoglycolipoprotein precursor of 210,000–240,000 daltons called vitellogenin (*82, 83*). Vitellogenin has nearly identical composition regardless of whether it is isolated from egg-laying higher animals or from insects. It is synthesized in the liver in vertebrates and in the fat body in insects, secreted into the blood or hemolymph, and then proteolytically cleaved into two molecules of phosvitin and one molecule of lipovitellin in the oocyte or the ovary. The two proteins are quite different in composition (*84*) with phosvitin (mol wt 32,000) having a high concentration of phosphorylated serine residues (~ 120 residues) and glycosyl groups, only one methionine, and being lipid free. Lipovitellin (mol wt 120,000) on the other hand has several methionine residues, a small amount of serine, little or no carbohydrate, and contains 20% lipid by weight.

**Cell Membrane Insertion of Integral Proteins.** In the past few years structure–function relationships of biological membranes have become a major challenge to biochemists and biologists. Following a large number of interesting studies on the chemical and structural characterization of the lipid bilayer that surrounds a living cell (*85, 86*), the question now is how membranes are formed since they represent dynamic structures that are remodeled continuously. The fact that proteins, in addition to lipids and carbohydrates, are normal molecular constituents of biological membranes (*87*) raised the problem of how these hydrophobic proteins are transferred into the lipid bilayer following biosynthesis by the normal water-soluble synthetic machinery of the cell.

In this section we describe briefly the two models for inserting integral proteins into cell membranes with special emphasis on the protease(s)-catalyzed hydrolytic modification of these proteins associated with the membrane assembly process. These two models are (1) self assembly following translation of the proteins and (2) coupling of translation with insertion of the protein into the membrane.

STRUCTURAL ORGANIZATION OF THE PLASMA MEMBRANE. Although our purpose here is not to describe in detail the structure of cell membranes, a brief look at the structure of the plasma membrane will help us to understand the major problems and the role of specific proteolysis related to membrane assembly. All cells—those of bacteria (prokaryotes), higher plants, and animals (eukaryotes)—have plasma membranes, but other distinct internal membranes (*88*) are found in eukaryotic cells (nuclei, golgi bodies, mitochondria, endoplasmic reticula, and lysosomes).

Of the three molecular constituents of the plasma membrane shown in Figure 13, carbohydrates and proteins are mainly responsible for structural and functional asymmetry of the membrane (*89, 90, 91*). Carbohydrates, always found linked to lipids (glycolipids) or proteins

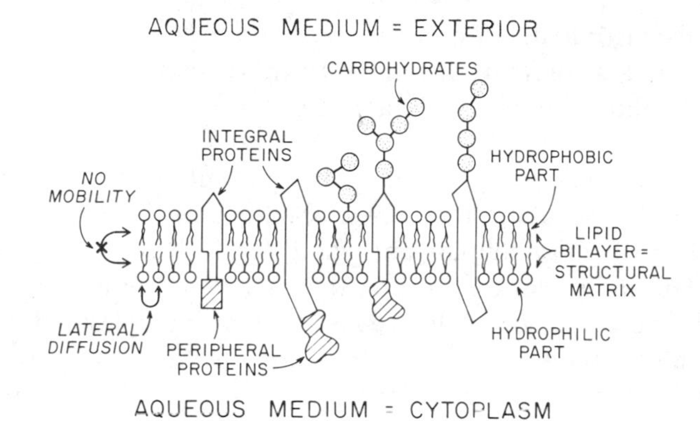


*Figure 13. Schematic of the plasma membrane with a structural and functional asymmetric orientation of the molecular constituents (adapted from Refs.* 89, 91, *and* 98*). Aqueous medium = exterior.*

(glycoproteins) of the bilayer, are less abundant than proteins or lipids and are always present on the outer surface of the membrane. Some integral proteins embedded in the lipid bilayer may only be exposed either on the outer surface or on the inner surface (cytoplasmic) of the cell. However, most proteins of the membrane are exposed on both sides of the membrane since they span the thickness of the bilayer (87). In contrast, peripheral membrane proteins always have been found on the cytoplasmic surface interacting with integral proteins. All of the protein molecules have a fixed orientation which may be different for each species; this gives the overall structure a definite asymmetry. Phospholipids and cholesterol, the latter being restricted to the plasma membrane of mammalian cells, contribute to the structural matrix of the membrane and to membrane asymmetry (92).

From the schematic model of the plasma membrane we have just depicted, it is clear that the two sides, the inner and the outer surfaces, should have different functions as a result of different structures. Moreover, the lipid bilayer may be considered as a hydrophobic barrier preventing diffusion of water-soluble molecules from both sides, thus maintaining a permanent distinction between the inside and outside of a cell. It also allows the membrane to form closed vessels, which appear to be an absolute requirement for maintaining the fixed asymmetric orientations of the cell membrane constituents.

Assembly of Proteins into Cell Membranes. The individual components of a biological membrane, after separation in the presence of a high concentration of detergent, still are able to associate into a membrane when the detergent is removed. However, such self-assembly of the normal constituents of a membrane fail to show asymmetry. This might be the result of a random insertion of integral proteins as the lipid bilayer is reconstituted. In order to maintain a specific orientation of a membrane integral protein, the amino terminal end of the protein molecule always must be inserted into the lipid bilayer from the same side (*93*).

Integral proteins are generally water-insoluble and because of this hydrophobic characteristic they are embedded in the hydrophobic part of the bilayer of phospholipids and cholesterol. However, the segments of the protein molecules that are on the inner and outer surfaces of the cell membrane are hydrophilic (*94, 95*) since cytoplasm and external fluid are mainly made of water-soluble components. The question raised is how the hydrophilic segments of an integral protein can penetrate the thickness of the hydrophobic core of the membrane.

Taking advantage of a virus membrane system, by far much simpler than eukaryotic or even prokaryotic cell membranes, Rothman and coworkers (*93, 95*) showed that a high degree of coupling existed between polypeptide synthesis and insertion into the membrane. Of the five proteins specified by the genome RNA of vesicular stomatitis virus, three are nucleoproteins, the fourth is the matrix protein, and the last protein (the G protein for glycoprotein) is a membrane-spanning protein. Most of the polypeptide chain of the G protein and all of the carbohydrate are on the outer surface of the membrane and only about 30 amino acids remain in the cytoplasm. The G protein is synthesized by ribosomes associated with the rough endoplasmic reticulum (*see* Figure 14) (*96, 97*) of infected cells and then transferred to the plasma membrane. Because of the steric hindrance of the ribosome, about 40 amino acids must be linked together before the signal sequence that represents the first 16 residues of the growing chain emerges thus allowing the polypeptide chain to span the lipid bilayer. As soon as the entire signal sequence,

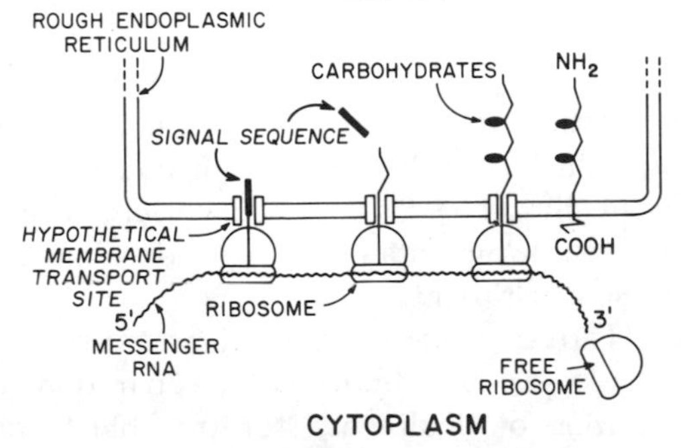


*Figure 14. Synthesis and assembly of an integral protein into the membrane of the endoplasmic reticulum (based on the data presented in Refs.* 93 *and* 95)

which contains a high amount of hydrophobic residues, reaches the lumen of the rough endoplasmic reticulum it is released by proteolytic enzymes associated with the reticulum; the biosynthesis of the remainder of the polypeptide chain and its glycosylation continues to completion (*see* Figure 14).

The role of catalysis in membrane assembly is emphasized again by the above model since the *N*-terminal sequence of the nascent polypeptide chain of a spanning protein is released by proteolysis as soon as it reaches the cytosol. The *N*-terminal polypeptide chain extension may help the chain penetrate the hydrophobic bilayer and solubilize the resulting hydrophobic *N*-terminal part of the chain in the aqueous medium of the cytoplasm. However, the role of the protease-catalyzed hydrolysis of the polypeptide chain in membrane assembly is minimized in the membrane trigger hypothesis (*99*). According to this model, the essential role of the leader sequence would be to modify, in association with the lipid bilayer, the folding pathway of the protein in such a way that the polypeptide chain could span the membrane.

A specific proteolytic conversion of precursors also is involved in the transport of proteins into mitochondria and chloroplasts (*100, 101*). Most of the proteins of these organelles are synthesized as larger precursors (*102, 103*) on cytoplasmic ribosomes and then transferred into the mitochondria or chloroplasts (*104*). While the precursors are being transported across membranes, they are processed to their mature size in the absence of protein synthesis (*105*) in an ATP-dependent process (*106*).

Microbial and plant toxins also are activated during translocation across the plasma membrane of sensitive cells (*107, 108*). Several proteinaceous toxins of plant or microbial origin such as abrin, ricin, cholera, and diphtheria toxins are either composed of two different polypeptide chains or two quite different segments of the same chain. In abrin or ricin, the B chain interacts with membrane receptors thereby facilitating the transfer of the A chain into the cytosol where it exerts its toxic effects (*109, 110, 111*). It generally is assumed that the transfer of such toxins across plasma membranes does not involve endocytosis but that it occurs as described above for secretory proteins, although translocation occurs after translation of the proteinaceous toxin. It generally is not established yet that a limited proteolytic step is needed for the translocation of toxins across membranes, although there is good evidence that this occurs in some cases.

### *Concluding Remarks*

Many more examples could be selected to illustrate the role of specific proteolysis in expressing the biological activity of proteins whether that activity is as an enzyme, a hormone, or as an assembly of protein molecules to form collagen, elastin, fibrin, or membranes. Proteolysis, once thought to be only a generalized degradative process for recovery of amino acids, is known now to be involved as a regulatory process in many key cellular events in animals, plants, and microorganisms. It is hoped that some of the highly specific proteases, along with knowledge of factors controlling the specific and limited hydrolysis of proteins, can be adapted to the modification of proteins for food use.

### *Acknowledgment*

This work was supported in part by NIH Grant 26031.

### *Literature Cited*

1. Feeney, R. E.; Whitaker, J. R., Eds. "Food Proteins: Improvement Through Chemical and Enzymatic Modification," *Adv. Chem. Ser.* **1977,** *160.*
2. Misra, P. S.; Mertz, E. T.; Glover, D. V. *Cereal Chem.* **1976,** *53,* 699.
3. Johnson, V. A.; Mattern, P. J. In "Nutritional Improvement of Food and Feed Proteins," *Adv. Exp. Med. Biol.* **1978,** *105,* 301.
4. Altschul, A. M. *Nature (London)* **1974,** *248,* 643.
5. Bjarnason-Baumann, B.; Pfaender, P.; Siebert, G. *Nutr. Metab.* **1977,** *21* (1), 170.
6. Puigserver, A. J.; Sen, L. C.; Clifford, A. J.; Feeney, R. E.; Whitaker, J. R. *J. Agric. Food Chem.* **1979,** *27,* 1286.
7. Yamashita, M.; Arai, S.; Imaizumi, Y.; Amano, Y.; Fujimaki, M. *J. Agric. Food Chem.* **1979,** *27,* 52.
8. Li-Chan, E.; Helbig, N.; Holbek, E.; Chau, S.; Nakai, S. *J. Agric. Food Chem.* **1979,** *27,* 877.
9. Whitaker, J. R. In "Food Proteins: Improvement Through Chemical and Enzymatic Modification," *Adv. Chem. Ser.* **1977,** *160,* 95.
10. Uy, R.; Wold, F. In "Chemical Deterioration of Proteins," *ACS Symp. Ser.* **1980,** *123,* 49.
11. Fujimaki, M.; Arai, S.; Yamashita, M. In "Food Proteins: Improvement Through Chemical and Enzymatic Modification," *Adv. Chem. Ser.* **1977,** *160,* 156.
12. Richardson, T. In "Food Proteins: Improvement Through Chemical and Enzymatic Modification," *Adv. Chem. Ser.* **1977,** *160,* 185.
13. Swaisgood, H. E. *J. Dairy Sci.* **1975,** *58,* 583.
14. Arai, S. Abstracts, AGFD No. 22, American Chemical Society. August 25, 1980.
15. Watanabe, M.; Arai, S., Chap. 7 in this book.
16. Barrett, A. J. In "Proteases and Biological Control"; Reich, E.; Rifkin, D. B.; Shaw, E., Eds.; Cold Spring Harbor Laboratory: New York, 1975; Vol. 2, p. 467.
17. Goldberg, A. L.; Dice, J. F. *Ann. Rev. Biochem.* **1974,** *43,* 835.
18. Goldberg, A. L.; St. John, A. C. *Ann. Rev. Biochem.* **1976,** *45,* 747.
19. Schimke, R. T.; Bradley, M. O. In "Proteases and Biological Control"; Reich, E.; Rifkin, D. B.; Shaw, E., Eds.; Cold Spring Harbor Laboratory: New York, 1975; Vol. 2, p. 515.
20. Holzer, H.; Heinrich, P. C. *Ann. Rev. Biochem.* **1980,** *49,* 63.

21. Dayton, W. R.; Goll, D. E.; Stromer, M. H.; Reville, W. J.; Zeece, M. G.; Robson, R. M. In "Proteases and Biological Control"; Reich, E.; Rifkin, D. B.; Shaw, E., Eds.; Cold Spring Harbor Laboratory: New York, 1975; Vol. 2, p. 551.
22. Neurath, H.; Walsh, K. A.; White, W. P. *Science* **1967**, *158*, 1638.
23. Goltzman, D.; Peytremann, E.; Callahan, E.; Tregear, G. W.; Potts, J. T., Jr. *Endocrinology* **1975**, *96*, 71.
24. Steiner, D. F.; Kemmler, W.; Tager, H. S.; Peterson, J. D. *Fed. Proc.* **1974**, *33*, 2185.
25. MacGregor, R. R.; Chu, L. L. H.; Cohn, D. V. *J. Biol. Chem.* **1976**, *251*, 6711.
26. Davie, E. W.; Fujikawa, K.; Kurachi, K.; Kisiel, W. *Adv. Enzymol. Relat. Areas Mol. Biol.* **1979**, *48*, 277.
27. Neurath, H. In "Proteases and Biological Control"; Reich, E.; Rifkin, D. B.; Shaw, E., Eds.; Cold Spring Harbor Laboratory: New York, 1975; Vol. 2, p. 51.
28. Frank, G.; Brunner, J.; Hauser, H.; Wacker, H.; Semenza, G.; Zuber, H. *FEBS Lett.* **1978**, *96*, 183.
29. Brunner, J.; Hauser, H.; Braun, H.; Wilson, K. J.; Wacker, H.; O'Neill, B.; Semenza, G. *J. Biol. Chem.* **1979**, *254*, 1821.
30. Hauri, H-P.; Quaroni, A.; Isselbacher, K. *Proc. Natl. Acad. Sci. USA* **1979**, *76*, 5183.
31. Maroux, S.; Baratti, J.; Desnuelle, P. *J. Biol. Chem.* **1971**, *246*, 5031.
32. Desnuelle, P. *The Enzymes* **1960**, *4*, 93.
33. Ulevitch, R. J.; Cochrane, C. G.; Revak, S. D.; Morrison, D. C.; Johnston, A. R. In "Proteases and Biological Control"; Reich, E.; Rifkin, D. B.; Shaw, E., Eds.; Cold Spring Harbor Laboratory: New York, 1975; Vol. 2, p. 85.
34. Server, A. C.; Shooter, E. M. *Adv. Protein Chem.* **1977**, *31*, 339.
35. Varon, S.; Nomura, J.; Shooter, E. M. *Proc. Natl. Acad. Sci. USA* **1967**, *57*, 1782.
36. Varon, S.; Nomura, J.; Shooter, E. M. *Biochemistry* **1968**, *7*, 1296.
37. Angeletti, R. H.; Bradshaw, R. A. *Proc. Natl. Acad. Sci. USA* **1971**, *68*, 2417.
38. Berger, E. A.; Shooter, E. M. *J. Biol. Chem.* **1978**, *253*, 804.
39. Greene, L. A.; Shooter, E. M. *Ann. Rev. Neurosci.* **1980**, *3*, 353.
40. Moore, J. B., Jr.; Mobley, W. C.; Shooter, E. M. *Biochemistry* **1974**, *13*, 833.
41. Weiss, S. R.; Varmus, H. E.; Bishop, J. M. *Cell* **1977**, *12*, 983.
42. Eisenman, R. N.; Vogt, V. M. *Biochim. Biophys. Acta* **1978**, *473*, 187.
43. Roberts, J. L.; Herbert, E. *Proc. Natl. Acad. Sci.* **1977**, *74*, 5300.
44. Hershko, A.; Fry, M. *Ann. Rev. Biochem.* **1975**, *44*, 775.
45. Garoff, H.; Simons, K.; Dobberstein, B. *J. Mol. Biol.* **1978**, *124*, 587.
46. Singer, P. A.; Singer, H. H.; Williamson, A. R. *Nature* **1980**, *285*, 294.
47. Palade, G. *Science* **1975**, *189*, 347.
48. Weiser, M. M.; Neumeier, M. M.; Quaroni, A.; Kirsch, K. *J. Cell. Biol.* **1978**, *77*, 722.
49. Oda, K.; Ikehara, Y.; Kato, K. *Biochim. Biophys. Acta* **1979**, *552*, 225.
50. Blobel, G.; Sabatini, D. In "Biomembranes"; Masson, L. A., Ed.; Plenum: New York, 1971; Vol. 2, p. 193.
51. Milstein, C.; Brownless, G. G.; Harrison, T. M.; Mathews, M. B. *Nature, New Biol.* **1972**, *239*, 117.
52. Blobel, G.; Dobberstein, B. *J. Cell. Biol.* **1975**, *67*, 835.
53. Campbell, P. N.; Blobel, G. *FEBS Lett.* **1976**, *72*, 215.
54. Devillers-Thiery, A.; Kindt, T.; Scheele, G.; Blobel, G. *Proc. Natl. Acad. Sci. USA* **1975**, *72*, 5016.
55. Scheele, G.; Dobberstein, B.; Blobel, G. *Eur. J. Biochem.* **1978**, *82*, 593.

56. Chan, S. J.; Keim, P.; Steiner, D. F. *Proc. Natl. Acad. Sci. USA* **1976,** *73,* 1964.
57. Strauss, A. W.; Bennett, C. D.; Donohue, A. M.; Rodkey, J. A.; Alberts, A. W. *J. Biol. Chem.* **1977,** *252,* 6846.
58. Blobel, G.; Dobberstein, B. *J. Cell. Biol.* **1975,** *67,* 852.
59. Jackson, R. C.; Blobel, G. *Proc. Natl. Acad. Sci. USA* **1977,** *74,* 5598.
60. Chang, C. N.; Blobel, G.; Model, P. *Proc. Natl. Acad. Sci. USA* **1978,** *75,* 351.
61. Strauss, A. W.; Zimmerman, M.; Boime, I.; Ashe, B.; Mumford, R. A.; Alberts, A. W. *Proc. Natl. Acad. Sci. USA* **1979,** *76,* 4225.
62. Lingappa, V. R.; Devillers-Thiery, A.; Blobel, G. *Proc. Natl. Acad. Sci. USA* **1977,** *74,* 2432.
63. Shields, D.; Blobel, G. *Proc. Natl. Acad. Sci. USA* **1977,** *74,* 2059.
64. Birken, S.; Smith, D. L.; Canfield, R. E.; Boime, I. *Biochem. Biophys. Res. Commun.* **1977,** *74,* 106.
65. Schechter, I.; McKean, D. J.; Guyer, R.; Terry, W. *Science* **1975,** *188,* 160.
66. Rapoport, T. A.; Höhne, W. E.; Klatt, D.; Prehn, S.; Hahn, V. *FEBS Lett.* **1976,** *69,* 32.
67. Suchanek, G.; Kreil, G.; Hermodson, A. *Proc. Natl. Acad. Sci. USA* **1977,** *75,* 701.
68. Inouye, S.; Wang, S. S.; Sekizawa, J.; Halegoua, S.; Inouye, M. *Proc. Natl. Acad. Sci. USA* **1977,** *74,* 1004.
69. Inouye, H.; Beckwith, J. *Proc. Natl. Acad. Sci. USA* **1977,** *74,* 1440.
70. Palmiter, R. D.; Gagnon, J.; Walsh, K. A. *Proc. Natl. Acad. Sci. USA* **1978,** *75,* 94.
71. Lingappa, V. R.; Shields, D.; Woo, S. L. C.; Blobel, G. *J. Cell. Biol.* **1978,** *79,* 567.
72. von Heijne, G. *Eur. J. Biochem.* **1980,** *103,* 431.
73. Garnier, J.; Gaye, P.; Mercier, J.-C.; Robson, B. *Biochimie* **1980,** *62,* 231.
74. Barnard, E. A. *Nature* **1969,** *221,* 340.
75. Dickman, S. R.; Bruenger, E. *Biochemistry* **1965,** *4,* 2335.
76. Scheele, G.; Blackburn, P. *Proc. Natl. Acad. Sci. USA* **1979,** *76,* 4898.
77. Chirgwin, J. M.; Przybyla, A. E.; MacDonald, R. J.; Rutter, W. J. *Biochemistry* **1979,** *18,* 5294.
78. Bornstein, P. *Ann. Rev. Biochem.* **1974,** *43,* 567.
79. Cardinale, G. J.; Udenfriend, S. *Adv. Enzymol.* **1974,** *41,* 245.
80. Bailey, A. J.; Robins, S. P.; Balian, G. *Nature* **1974,** *251,* 105.
81. Grant, M. E.; Heathcote, J. G.; Cheah, K. S. E. In "Processing and Turnover of Proteins and Organelles in the Cell"; Rapoport, S.; Schewe, T., Eds.; FEBS 12th Meet.: Dresden, 1978; Vol. 53, p. 29.
82. Clemens, M. J. *Prog. Biophys. Mol. Biol.* **1974,** *28,* 69.
83. Tata, J. R. *Cell* **1976,** *9,* 1.
84. Tata, J. R. In "Processing and Turnover of Proteins and Organelles in the Cell"; Rapoport, S.; Schewe, T., Eds.; FEBS 12th Meet.: Dresden, 1978; Vol. 53, p. 11.
85. Bangham, A. D. *Ann. Rev. Biochem.* **1972,** *41,* 753.
86. Cronan, J. E., Jr. *Ann. Rev. Biochem.* **1978,** *47,* 163.
87. Singer, S. J. *Ann. Rev. Biochem.* **1974,** *43,* 805.
88. DePierre, J. W.; Ernster, L. *Ann. Rev. Biochem.* **1977,** *46,* 201.
89. Bretscher, M. S. *Science* **1973,** *181,* 622.
90. Steck, T. L. *J. Cell. Biol.* **1974,** *62,* 1.
91. Rothman, J. E.; Lenard, J. *Science* **1977,** *195,* 743.
92. Op den Kamp, J. A. F. *Ann. Rev. Biochem.* **1979,** *48,* 47.
93. Rothman, J. E.; Lodish, H. F. *Nature* **1977,** *269,* 775.
94. Lenard, J.; Compans, R. W. *Biochim. Biophys. Acta* **1974,** *344,* 51.

95. Katz, F. N.; Rothman, J. E.; Lingappa, V. R.; Blobel, G. *Proc. Natl. Acad. Sci. USA* **1977**, *74*, 3278.
96. Morrison, T.; Lodish, H. F. *J. Biol. Chem.* **1975**, *250*, 6955.
97. Grubman, M. J.; Moyer, S. A.; Banerjee, A. K.; Ehrenfeld, E. *Biochem. Biophys. Res. Commun.* **1975**, *62*, 531.
98. Singer, S. J.; Nicolson, G. L. *Science* **1972**, *175*, 720.
99. Wickner, W. *Ann. Rev. Biochem.* **1979**, *48*, 23.
100. Maccecchini, M. L.; Rudin, Y.; Blobel, G.; Schatz, G. *Proc. Natl. Acad. Sci. USA* **1979**, *76*, 343.
101. Chua, N. H.; Schmidt, G. W. *J. Cell. Biol.* **1979**, *81*, 461.
102. Highfield, P. E.; Ellis, R. J. *Nature* **1978**, *271*, 420.
103. Cashmore, A. R.; Broadhurst, M. K.; Gray, R. E. *Proc. Natl. Acad. Sci. USA* **1978**, *75*, 655.
104. Schatz, G.; Mason, T. L. *Ann. Rev. Biochem.* **1974**, *43*, 51.
105. Chua, N. H.; Schmidt, G. W. *Proc. Natl. Acad. Sci. USA* **1978**, *75*, 6110.
106. Nelson, N.; Schatz, G. *Proc. Natl. Acad. Sci. USA* **1979**, *76*, 4365.
107. Tagg, J. R.; Dajani, A. S.; Wannamaker, L. W. *Bacteriol. Rev.* **1976**, *40*, 722.
108. Pappenheimer, A. M., Jr. *Ann. Rev. Biochem.* **1977**, *46*, 69.
109. Olsnes, S.; Fernandez-Puentes, C.; Carrasco, L.; Vazquez, D. *Eur. J. Biochem.* **1975**, *60*, 281.
110. Sahyoun, N.; Cuatrecasas, P. *Proc. Natl. Acad. Sci. USA* **1975**, *72*, 3438.
111. Boquet, P.; Silverman, M. S.; Pappenheimer, A. M., Jr.; Vernon, W. B. *Proc. Natl. Acad. Sci. USA* **1976**, *73*, 4449.

RECEIVED November 18, 1980.

# FOOD AND NUTRITIONAL ASPECTS

3

# Nutritional and Metabolic Aspects of Protein Modification During Food Processing

P. A. FINOT

Research Department, Nestlé Products Technical Assistance Co., Ltd., Ch-1814 La Tour de Peilz, Switzerland

*During the processing of food proteins, uncontrolled chemical reactions affect the side chain of some amino acids, which leads to changes in their biological availability and induces specific physiological properties. Some chemical reactions lead to irreversible derivatives inducing nutritional changes because the living organisms cannot regenerate the amino acids involved, e.g. ε-deoxyfructosyllysine, ε(β-aspartyl)-lysine, lysinoalanine, and lysine–polyphenolic complexes. Some other reactions lead to derivatives that the living organisms can transform biochemically in order to regenerate the initial amino acids, e.g. Schiff's bases of lysine, methionine sulfoxide, ε(γ-glutamyl)lysine, and the oxidation products of tryptophan. The metabolic transit of the synthetic and protein-bound chemically modified amino acids is described in order to explain the nutritional and physiological changes.*

Many reviews (*1, 2, 3, 4*) and symposia (*5, 6*) already have been devoted to the chemical reactions affecting the amino acids and the proteins during the processing of food proteins. These reviews mention the different chemical reactions occurring with proteins, their mechanisms, the physical parameters involved, the structure of the derivatives formed, the in vitro and in vivo analytical methods used to evaluate the importance of the damage, and the nutritional effects. We refer to these reviews for details.

The nutritionists are interested not only by the decrease in the nutritional value induced by these chemical modifications but also more often by the physiological effects produced by the new molecules that are formed. These effects depend on the behavior of these molecules in the organism.

0065-2393/82/0198-0091$08.50/0

Along with these conventional studies, it was necessary to develop a new field of investigation aimed at knowing the behavior of these new molecules in the organism—release by the intestinal enzymes, absorption, urinary and fecal excretions, retention in the organs and tissues, biochemical modification, and structure of the urinary catabolites. The incorporation of radioactivity in the organs and tissues is also a means to measure and control the biological utilization of the modified amino acids. This new approach which we call metabolic transit is very helpful for understanding the nutritional and physiological effects observed by the conventional nutritional approaches.

According to the complexity of the problems and the degree of knowledge we want to obtain, different substrates can be studied: synthetic free molecules, or the protein-bound molecules if they are easily detectable and not transformed in the organism; synthetic radioactive molecules for studying the biochemical modifications; and protein-bound radioactive molecules for extrapolating to the normal feeding conditions. Experiments can be performed on whole living animals, organ homogenates, or perfused organs. The combination of all of these approaches leads to a more precise knowledge of the real metabolism of the molecules being studied.

This chapter deals with the Maillard reaction, the oxidation of sulfur-containing amino acids, isopeptide bonds, alkaline treatments, the interaction of proteins with polyphenols, and the oxidation and heat treatment of tryptophan.

## *The Maillard Reaction*

**Chemical Reactions: Structure of the Compounds Formed.** The Maillard reaction, which involves an amino group and a reducing sugar, occurs in foods between the free amino acids, the free amino groups of proteins ($\alpha$-end amino groups and the $\epsilon$-amino group of lysine) and numerous reducing sugars (glucose, maltose, fructose, lactose, pentoses, etc.). This reaction occurs in two steps (7).

(a) The early Maillard reaction which produces colorless products:

$$\begin{matrix} R\text{–}NH_2 \\ + \\ \text{reducing sugar} \end{matrix} \rightleftarrows \begin{matrix} \text{Schiff's base} \\ \upharpoonleft\downharpoonright \\ \text{aldosylamine} \end{matrix} \rightarrow \text{Amadori compound}$$

(b) The advanced Maillard reaction leading to brown pigments:

$$\text{Amadori compound} \xrightarrow[\text{Strecker degradation}]{\text{dehydration scission}} \underbrace{\left\{ \begin{matrix} \text{aldehydes} \\ \text{ketones} \\ \text{reductones} \end{matrix} \right\}}_{\text{premelanoidins}} \xrightarrow[\text{reaction}]{\text{polymerization}} \underbrace{\text{Brown pigments}}_{\text{melanoidins}}$$

The intermediary chemical forms of the early Maillard reaction (Schiff's base and aldosylamine) cannot be isolated easily because they are chemically unstable (*8*), but their presence is detected by reduction with $NaBH_4$ (*9*). The Amadori compounds are stable, their synthesis is easy (*10, 11, 12, 13, 32*), and their presence can be detected by their reducing power towards ferricyanide (*14*) and other reagents such as triphenyltetrazolium chloride (*15*) or methylene blue (*14*) and also by the degradation derivatives that they give under acid hydrolysis. For example, the presence of ε-deoxyfructosyllysine or ε-deoxylactulosyllysine in a protein resulting, respectively, from the reaction between the ε-amino group of lysine and glucose or lactose can be detected from the furosine (*16, 17, 18*) and pyridosine (*19*) they form during acid hydrolysis.

The numerous decomposition products of the Amadori compounds are responsible for the taste and aromas developed by the Maillard reaction (*20, 48*). Because of their high chemical reactivity (aldehydes, ketones, diketones, and reductones, etc.) these degradation products react together and polymerize to give soluble and insoluble brown polymers that may or may not contain nitrogen. According to their molecular weight and their color, these groups of substances are called premelanoidins or melanoidins (*20*).

**Nutritional and Physiological Effects.** The ingestion of Maillardized foods induces nutritional and physiological effects depending on the extent and on the intensity of the reaction.

NUTRITIONAL EFFECTS DUE TO THE BLOCKAGE OF LYSINE. The most important Maillard reaction in food proteins occurs with the ε-amino group of lysine. Since lysine is an essential amino acid, nutritional consequences can be expected. These depend on the chemical structure of the lysine derivatives formed.

- Unmodified lysine is, in principle, biologically available and will be determined as such by animal assays (*21*) and enzymatic methods (*22, 23, 24*), or determined as reactive lysine by the Carpenter method (*23, 24, 25, 26*) and by guanidination (*24, 27*).
- Lysine present as a Schiff's base or aldosylamine is biologically available as demonstrated with synthetic free Schiff's bases (ε-benzylidenelysine and ε-salicylidenelysine) (*9, 28*) (*see* Table I). This can be explained by the reversibility of the reaction with the subsequent regeneration of lysine, probably in the acid medium of the stomach.
- The Amadori derivative of lysine and glucose (ε-deoxyfructosyllysine) is not biologically available to rats (*9, 28*). The same has been observed also with the other Amadori compounds of lysine (α- and α-ε-di-derivatives) and those Amadori compounds derived from other amino acids such as methionine (*29*), tryptophan (*12, 30, 32*), and leucine (*12*).

**Table I. Biological Availability in Rats of the Amino Acids Chemically Modified During Food Processing**

| | *Percentage of Biological Availability of the Derivatives* | | |
|---|---|---|---|
| *Maillard reaction products* | *Free Synthetic* | *Protein Bound* | *Regeneration of the Amino Acid in* |
| Schiff's bases | | | |
| ε-Benzylidene- and ε-salicylidenelysine (*9, 28*) | 100 | 100 | stomach |
| Amadori compounds | | | |
| ε-Fructosyl-, α-fructosyl-, and α-ε-difructosyllysine (*9, 28*) | 0 | 0 | large intestine by microflora |
| Other fructosyl amino acids: methionine (*29*), leucine (*30*) tryptophan (*12*) | 0 | | |
| *Cross-linkings* | | | |
| Isopeptides (*118, 81, 83*) | | | |
| ε(γ-glutamyl)lysine | 100 | ? | kidneys |
| ε-(β-aspartyl)lysine | 0 | 0 | |
| Alkaline treatments | | | |
| lysinoalanine (lysine) (*100*) | 0 (rats) partly (chicks) | | kidneys ? |
| lanthionine (cysteine) (*100*) | 0 (rats) partly (chicks) | | |
| *Oxidation Products of* | | | |
| Methionine (*59–62, 70–72*) | | | |
| methionine sulfoxide | 90–100 | 90–100 | liver, kidneys, intestine ? |
| methionine sulfone | 0 | 0 | |
| Cystine cysteine (*59–62*) | | | |
| cystine disulfoxide | ? | ? | |
| cystine disulfone | 0 | 0 | |
| cysteine sulfenic acid | ? | ? | |
| cysteine sulfinic acid | 0 | 0 | |
| cysteine sulfonic acid | 0 | 0 | |
| Tryptophan | | | |
| oxidation by $H_2O_2$ (*132, 137*) | partly | partly | |
| *Others* | | | |
| Lysine polyphenols (*119, 123*) | 0 | 0 | |
| Heat treatment of tryptophan (*132, 137*) | 0 | 0 | |

This leads to a decrease in the biological value of the protein. Milk is a good model to study the nutritional effects of the Maillard reaction because of its high levels of lysine and lactose. In processed milks, unavailable lysine is present only as ε-deoxylactulosyllysine (*9, 24*); its level which depends on the type of treatment can be measured by the furosine method (*24, 31*).

- Lysine in Advanced Maillard Products. In food proteins, the advanced Maillard reaction results from the chemical degradation of the Amadori compounds of lysine; the lysine residues resulting from this degradation are present as chemically unidentified derivatives. Kinetic studies on the progress of the advanced Maillard reaction in a dried milk showed an increase in the Amadori compounds to a maximum that is reached when the browning appears. As long as browning develops, the level of the Amadori compound decreases at the same time as the level of unidentified lysine resulting from the decomposition of the Amadori compound increases (*see* Figure 1).

NUTRITIONAL EFFECTS DUE TO THE PRESENCE OF THE MAILLARD PRODUCTS. Many physiological or antinutritional effects have been attributed to the Maillard products. Specific effects have been attributed to the Amadori products; deoxyfructosylphenylalanine (a model substance not likely to be present in large quantities in foods) appears to depress the rate of protein synthesis in chicks (*32*) and to partially inhibit in vitro and in vivo the absorption of tryptophan in rats (*33*). The compound ε-deoxyfructosyllysine inhibits the intestinal absorption of threonine, proline, and glycine and induces cytomegaly of the tubular cells of the rat kidneys (*34*) as does lysinoalanine. In parenteral nutrition the infusion of the various Amadori compounds formed during sterilization of the amino acid mixture with glucose is associated with milk dehydration in infants and excessive excretion of zinc and other trace metals in both infants and adults (*35, 36, 37*).

Other effects have been attributed to the advanced Maillard products. Adrian (*20*) described a favorable effect of low doses of the premelanoidins formed between glucose and glycine on appetite; at higher doses, the same material decreases the nitrogen retention. The effects of these products diminish when the heating time increases; the insoluble melanoidins obtained after 10 h heating had no effect.

The effects of premelanoidins on nitrogen retention is explained by a reduction in protein digestibility due to (a) the inactivation of the proteolytic enzymes, (b) the formation of indigestible peptides (*20, 38*), (c) the inhibitory effects on amino acid absorption (*33*) and (d) by a decrease in the efficiency of the protein synthesis. The Maillard products also have an inhibitory effect on the intestine saccharidases (*39*).

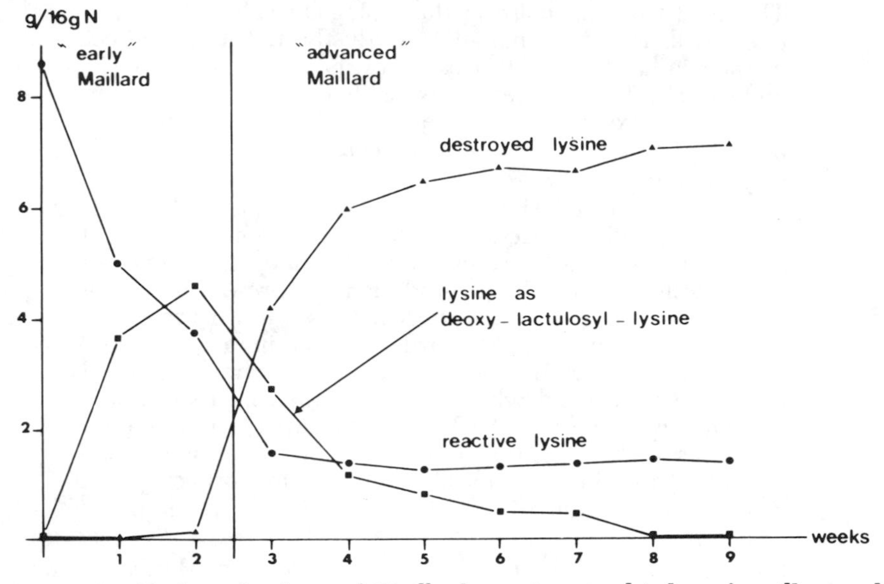


*Figure 1. Early and advanced Maillard reactions in dried cow's milk stored at 70°C—evolution of reactive lysine, lysine as deoxylactulosyllysine, and destroyed lysine (by difference)* (4)

In addition, Adrian also observed a hypertrophy of the caecum, liver, and kidneys, and a decrease in fertility (*20, 133*) attributed to a certain toxicity or to nutritional deficiencies. Lee et al. (*33*) observed a hypertrophy of the liver and kidney and an increase of the serum transaminases (GOT and GPT) with Maillard fractions extracted from browned apricots, and Kimiagar et al. (*47*) observed larger changes in biological parameters after a long-term study with browned egg albumin.

**Metabolic Transit.** FREE AMADORI COMPOUNDS. It is well known that the synthetic Amadori compounds of the free amino acids are absorbed by the intestine and excreted unchanged in the urine (*9, 28, 30*). The transport is not active as observed with deoxyfructosyltryptophan (*30*) and ε-deoxyfructosyllysine (*40*), and the level of absorption depends on the nature of the amino acid and on the conditions of ingestion. Nutritional assays and metabolic transit studies performed with radioactive Amadori compounds of tryptophan (*12, 30*), leucine (*12*), and lysine (*9, 28, 41*) given orally or intravenously on normal or antibiotics-treated animals have shown that the intestinal microflora can regenerate part of the amino acid. This can be absorbed subsequently at a very low level by the caecum or the large intestine and incorporated into the tissue proteins or utilized by the intestinal microflora. Barbiroli (*13*) showed also that some intestinal enzymes were able to liberate some amino acids from their Amadori compounds but to a very small extent. This observation could explain why the Amadori products often have a slight positive effect on growth (*9, 28*) or on protein synthesis (*12*).

PROTEIN-BOUND ε-DEOXYFRUCTOSYLLYSINE. The urinary and fecal excretions of protein-bound ε-deoxyfructosyllysine and ε-deoxylactulosyllysine were compared with the urinary and fecal excretions of free ε-deoxyfructosyllysine (*41, 42*) in rats. For this, an industrial spray-dried preparation made of casein and glucose and a roller-dried supplementary protein food made of wheat flour and milk were used (*see* Table II).

Protein-bound ε-deoxyfructosyllysine and ε-deoxylactulosyllysine were excreted in the urines at lower levels (9–13%) than free ε-deoxyfructosyllysine (about 65%); they were excreted as free and not as peptide-bound molecules. Their fecal excretion was very low, demonstrating the degrading effect of the intestinal microflora.

CASEIN-BOUND $^3$H LYSINE IN UNTREATED, EARLY, AND ADVANCED MAILLARD CASEIN (*41*). Goat sodium caseinate biologically labelled with L-4,5-$^3$H-lysine was treated with glucose under conditions inducing early and advanced Maillard reactions. The two preparations contained about the same reactive lysine (*see* Table III); the early Maillard $^3$H-casein contained only ε-deoxyfructosyllysine as Maillard products and the advanced Maillard $^3$H-casein contained some ε-deoxyfructosyllysine and unidentified derivatives of lysine and glucose as brown pigments.

Untreated, early Maillard and advanced Maillard ($^3$H-lysine)-casein were given to groups of three rats each in a single meal and the radioactive urinary and fecal excretions were measured for 74 h. After this period, the animals were killed and the radioactivity of different organs was determined.

For the untreated ($^3$H-lysine) casein, the radioactive urinary and fecal excretions were low, about 10% and 4%, respectively, after 74 h. The unexcreted radiactivity was retained in the organism and incorporated into the proteins. The radioactivity measured in the liver, muscle, and kidneys (expressed as percentage of the ingested dose per gram of tissue) can be considered as a function of the quantity of lysine incorporated into the protein of these tissues and proportional to the biological availability of $^3$H-lysine in the casein samples. The value of radioactivity measured in the liver, muscle, and kidneys can be considered as being given by a protein whose lysine is 100% available (*see* Table III, Column 5).

For the early Maillard ($^3$H-lysine) casein, the urinary excretion was higher than for the untreated casein. This value represents the quantity of protein-bound ε-deoxyfructosyllysine absorbed by the intestine. It is the same value as that determined in the previous experiment

Table II. Urinary and Fecal Excretion of Free and Protein-

| *Test Material* | *ε-Deoxyketosyllysine Content (Grams of Lysine Equivalent/16 g N)* |
|---|---|
| Free ε-deoxyfructosyllysine | |
| Spray-dried preparation made of casein + glucose | 3.75 (as deoxyfructosyllysine) |
| Supplementary protein food made of wheat flour + milk | 2.63 (as deoxylactulosyllysine) |

[a] Refs. *28* and *41*.

Table III. Urinary Excretion, Fecal Excretion, and Retention

| *Sample of $^3H$-lysine Casein* | *Percentage of Ingested Radioactivity* | | |
|---|---|---|---|
| | *Feces* | *Urines* | *Total Excretion* |
| Untreated | mean— 3.8 | 10.3 | 14.1 |
| | S.E.— 0.6 | 0.7 | 1.0 |
| Early Maillard | mean—13.5 | 21.7 | 35.0 |
| | S.E.— 0.5 | 1.5 | 2.0 |
| Advanced Maillard | mean—47.0 | 15.3 | 62.0 |
| | S.E.— 6.0 | 0.6 | 6.0 |

[a] Time was 74 h after a meal containing $^3$H-lysine casein, untreated, and heated under conditions for the early and advanced Maillard reaction (*41*).

by the chemical analysis (*see* Table II). The radioactivity measured in the liver and the muscle corresponds to about 60% of the value obtained with the untreated casein, suggesting that in our early Maillard ($^3$H-lysine) casein, the lysine availability is 60% of that of the untreated casein, a value very close to that obtained for reactive lysine by chemical analysis (Carpenter's method) (*see* Table III, Column 4).

For the advanced Maillard ($^3$H-lysine) casein, the urinary excretion was low and the fecal excretion high, suggesting that lysine engaged in advanced Maillard products is absorbed at a very low extent by the intestine and that the intestinal microflora does not metabolize these products. The radioactivity measured in the liver and in the muscle indicates that the biological availability of lysine in the advanced Maillard ($^3$H-lysine)-casein is lower than that determined by the chemical analysis (reactive lysine). About 50% of reactive lysine should be not biologically available (*see* Table III, Columns 4 and 5). The level

**Bound ε-Deoxyketosyllysines (% of Ingested Quantity) [a]**

| ε-Deoxyketosyllysines Found in: | | |
|---|---|---|
| *Urines* | *Feces* | *Total Recovery* |
| mean—64 | extreme values | 77 |
| S.E.— 7 | 3–26 | 9 |
| mean— 9 | 3 | 11 |
| S.E.— 3 | 1 | 3 |
| mean—13 | 2 | 16 |
| S.E.— 2 | 1 | 1 |

**of Radioactivity [a] in Rat Liver, Muscle, and Kidneys**

| *Percentage of Reactive Lysine* | *Percentage of Radioactivity Retained/Grams of* | | |
|---|---|---|---|
| | *Liver* | *Muscle* | *Kidneys* |
| 100 | 1.02 | 0.66 | 0.99 |
| | 0.06 | 0.03 | 0.03 |
| | (100%) [b] | (100%) [b] | (100%) [b] |
| 62.7 | 0.61 | 0.39 | 0.74 |
| | 0.05 | 0.12 | 0.05 |
| | (59.8%) [b] | (59.1%) [b] | (74.0%) [b] |
| 57.3 | 0.25 | 0.20 | 0.49 |
| | 0.03 | 0.01 | 0.05 |
| | (24.5%) [b] | (30.3%) [b] | (49.5%) [b] |

[b] Percentage of the value obtained by the untreated ($^3$H-lysine) casein.

of the fecal excretion indicates that reactive lysine that is not available probably is not absorbed because it is present in indigestible peptides.

The radioactivity measured in the kidneys of the rats fed on the early and advanced Maillard casein was higher than in the liver and in the muscle due to a certain retention of the Maillard products, which can be observed in the whole body autoradiography (*see* Figure 2).

PROTEIN-BOUND MELANOIDINS (*28, 41*). Sodium caseinate and U-$^{14}$C-glucose were heated in a water solution under conditions that induce an advanced Maillard reaction. The protein was precipitated at its isoelectric point and washed with water to remove traces of glucose. The residue was hydrolyzed by pronase and chromatographed on Biogel P4 according to the method proposed by Clark and Tannenbaum (*43, 44*). One brown fraction of high molecular weight and another slightly colored fraction of low molecular weight were isolated and given to rats by stomach tubing for a 24-h metabolic transit study. These two fractions

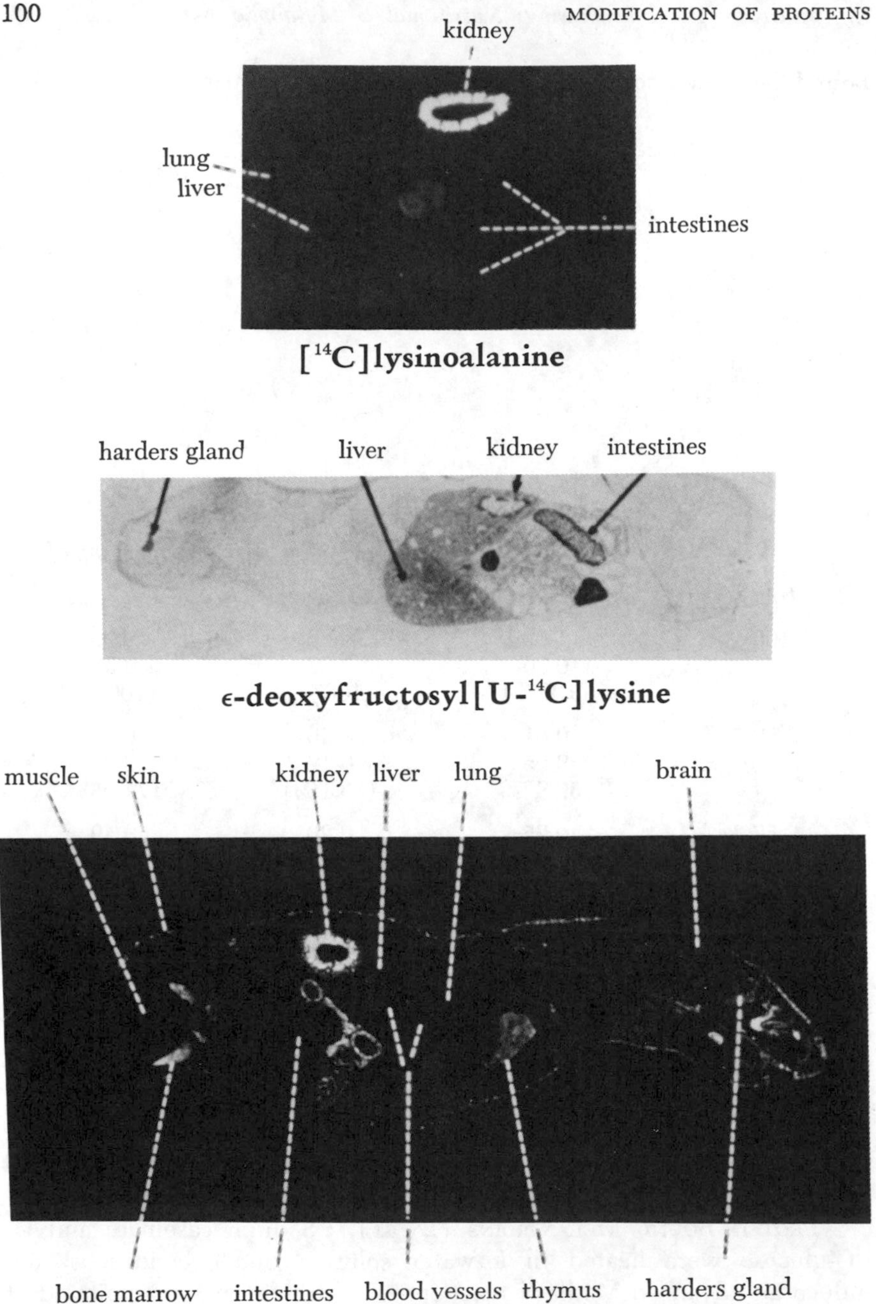


*Figure 2. Whole-body autoradiographies of rats, 24 h after oral ingestion of derivatives of U-$^{14}$C-L-lysine: lysinoalanine, ε-deoxyfructosyllysine, ε-formyllysine, ε-(γ-glutamyl)lysine, and α-formyllysine.*

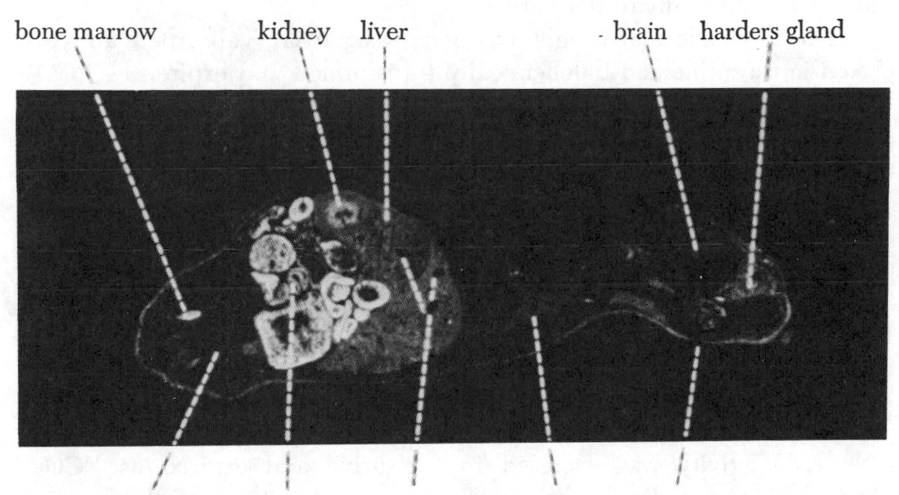


**α-formyl) [U-$^{14}$C] L-lysine**

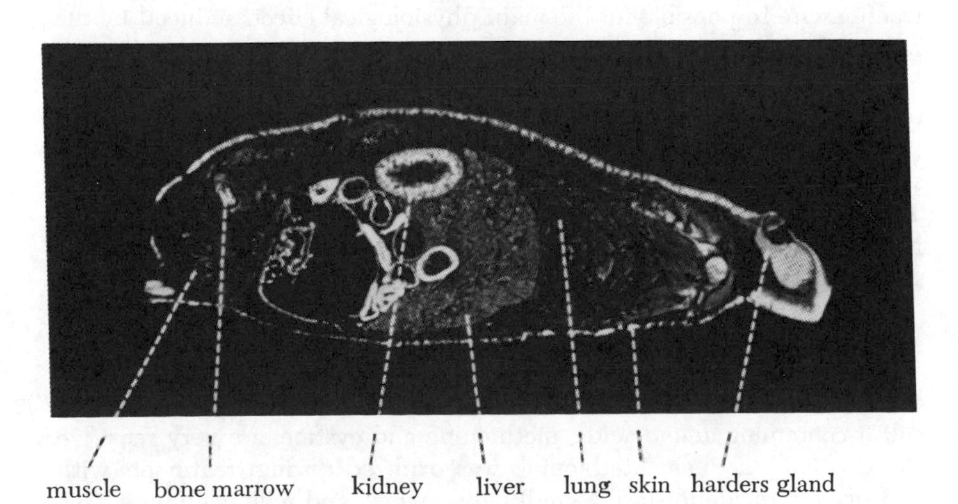


**ε(γ-L-glytamyl) [U-$^{14}$C] L-lysine**

*Figure 2. Continued.*

did not produce any furosine upon acid hydrolysis as a proof of the absence of ε-deoxyfructosyllysine; they therefore can be considered as pure advanced Maillard derivatives.

The low-molecular-weight fraction was partially absorbed and excreted in the urine and biochemically transformed and expired as $^{14}CO_2$ (*see* Table IV). The high-molecular-weight fraction was slightly absorbed and excreted in the urine and transformed into $^{14}CO_2$, with most of this fraction being recovered in the feces (*see* Table IV).

FREE MELANOIDINS (*28, 41*). Free melanoidins were prepared by a reaction between glycine and U-$^{14}$C-glucose in the presence of water according to Adrian's method (*45*). The brown preparation was chromatographed on Sephadex G-15 according to Yamaguchi and Fujimaki (*46*). The radioactive brown peak, treated again under the same chromatographic conditions to remove traces of glucose, was given to rats by stomach tubing for a 24-h metabolic transit experiment. As for the protein-bound melanoidins of the high-molecular-weight fraction, very little radioactivity was excreted in the urine and expired as $^{14}CO_2$. Most of the ingested radioactivity was recovered in the feces because of its indigestibility (*see* Table IV).

These results indicate that the high-molecular-weight melanoidins are metabolically inert because they are practically not absorbed and not biochemically transformed. In contrast, the low-molecular-weight melanoidins and the premelanoidins are partially absorbed and metabolically transformed either by the microflora or the tissues. These last fractions are responsible for the main physiological effects induced by the Maillard products.

WHOLE-BODY AUTORADIOGRAPHY. Whole-body autoradiography performed 8 h after intravenous injection of ε-deoxyfructosyl-U-$^{14}$C-lysine shows that the radioactivity is localized in the bladder and in the kidneys since the excretion is not complete, and a little in the pancreas (*41*). Twenty-four h after oral ingestion, the radioactivity is localized mainly in the large intestine and a little in the cortex of the kidneys, giving the same pattern as $^{14}$C-lysinoalanine (*see* Figure 2).

### *Oxidation of the Sulfur-Containing Amino Acids*

**Chemical Reactions: Structure of the Compounds Formed.** The sulfur-containing amino acids, methionine and cystine, are very sensitive to oxidation reactions. Methionine was oxidized during treatments with hydrogen peroxide to sterilize milk (*49*), whey, and milk containers (*49, 50*), with sulfites in the presence of oxygen and manganese ions (*51*), during the lipid oxidation (*52, 70*), and in the presence of quinones or polyphenols plus oxygen at pH 7 with polyphenol oxidase and at alkaline pH (*53*).

Two oxidation products can be formed—methionine sulfoxide ($NH_2$–CH–(COOH)–$CH_2$–$CH_2$–SO–$CH_3$), which can be determined after alkaline hydrolysis (*54*) but not after acid hydrolysis, which regenerates methionine, and methionine sulfone ($NH_2$–CH–(COOH)–$CH_2$–$CH_2$–$SO_2$–$CH_3$), which can be determined after acid hydrolysis since it is acid stable.

Cystine and cysteine also can be oxidized under the same conditions but the analytical methods to detect these oxidation products in proteins have not been developed yet. The different oxidation products of cystine, which have been synthesized (*55–58*), are: cystine disulfoxide ($NH_2$–CH–(COOH)–$CH_2$–SO–SO–$CH_2$–(COOH)–CH–$NH_2$), cystine disulfone ($NH_2$–CH–(COOH)–$CH_2$–$CO_2$–$SO_2$–$CH_2$–(COOH)–CH–$NH_2$), cysteine sulfenic acid ($NH_2$–CH–(COOH)–$CH_2$–SOH), cysteine sulfinic acid ($NH_2$–CH–(COOH)–$CH_2$–$SO_2H$), and cysteine sulfonic acid or cysteic acid ($NH_2$–CH–(COOH)–$CH_2$–$SO_3H$). It is doubtful that all of these derivatives described in 1935 by Bennett are present in proteins.

**Nutritional Effects of Oxidized Sulfur Amino Acids.** In 1937, Bennett (*59, 60, 61, 62*) already showed that the different oxidation products of methionine and cystine did not have the same biological effects to promote the growth of rats fed on diets deficient in sulfur-containing amino acids. Methionine sulfone and cysteic acid did not promote growth while the lower oxidation products had a positive effect and could replace methionine and cystine to a certain extent (*see* Table I).

It was confirmed later that free cysteic acid and free methionine sulfone were not biologically available (*63, 64, 138*) and that free methionine sulfoxide was partly available. Miller and Samuel (*64*) observed that the food efficiency of a mixture of free amino acids was lower when the methionine source was replaced by methionine sulfoxide. The food efficiency was restored when 50% of the methionine sulfoxide was replaced by free methionine. Gjoen and Njaa (*66*) confirmed that free methionine sulfoxide was nearly as available as methionine when the amino acid mixture contained cystine. This suggests that methionine sulfoxide is reduced before it is used for protein synthesis. In order to elucidate this point, we have compared the metabolic transit of free methionine sulfoxide with that of free methionine.

**Metabolic Transit of U-$^{14}$C-L-Methionine Sulfoxide.** U-$^{14}$C-L-methionine sulfoxide was prepared according to the method of Lepp and Dunn (*67*) and its metabolic transit on rats was compared with that of free U-$^{14}$C-L-methionine of the same specific activity (*28*).

As expected from the nutritional effects observed by Bennett (*59, 60, 61, 62*), Miller et al. (*65*), and Gjoen and Njaa (*66*), no significant differences were observed at the level of the excretion and incorporation in the tissues of the radioactivity (*see* Table V).

**Table IV. Total Excretion and Retention of the Radioactivity High-Molecular-Weight Fractions, and**

Protein bound melanoidins (casein + U-$^{14}C$-glucose)
low-molecular-weight fraction

high-molecular-weight fraction

Free melanoidins (glycine + U-$^{14}C$-glucose)
high-molecular-weight fraction

[a] Refs. *28, 41,* and *42.*

**Table V. Metabolism of $^{14}C$-L-Methionine and $^{14}C$-L-Methionine Sulfoxide in Rats[a]**

| | *$^{14}C$-L-Methionine* | | *$^{14}C$-L-Methionine Sulfoxide* | |
|---|---|---|---|---|
| | *Mean* | *S.E.* | *Mean* | *S.E.* |
| (a) 24 h Living animals | (% of recovered radioactivity) | | | |
| $^{14}CO_2$ | 29.8 | 0.3 | 24.7 | 3.1 |
| Urines | 4.8 | 0.4 | 7.1 | 0.8 |
| Feces | 5.2 | 0.2 | 2.0 | 0.4 |
| Carcass | 60.3 | 0.3 | 66.5 | 3.5 |
| | (% of recovered radioactivity per g of tissue) | | | |
| Liver | 2.6 | 0.3 | 2.7 | 0.5 |
| Kidneys | 1.35 | 0.1 | 2.25 | 0.4 |
| (b) 2 h Liver perfusion | (% of initial specific activity) | | | |
| Perfusate | 36.5 | 4.6 | 86.6 | 9.5 |
| | (% of initial radioactivity) | | | |
| $CO_2$ | 17.3 | 6.0 | 1.5 | 0.9 |
| Liver homogenate—protein fraction | 35.9 | | 3.4 | 1.3 |

[a] Distribution of the radioactivity (a) after a 24-h experiment on living animals (oral ingestion) and (b) after a 2-h liver perfusion (*28*).

**24 h After Ingestion of Protein-Bound Melanoidins of Low- and After Ingestion of Free Melanoidins**[a]

| | *Percentage of Recovered Radioactivity in* | | |
|---|---|---|---|
| *$^{14}CO_2$* | *Urines* | *Feces + Intestinal Content* | *Carcass* |
| mean—7.7 | 26.9 | 62.3 | 2.6 |
| S.E.—1.4 | 4.3 | 5.1 | 0.6 |
| mean—1.8 | 4.0 | 93.5 | 1.1 |
| S.E.—1.1 | 2.3 | 3.4 | 0.9 |
| mean—1.5 | 1.0 | 92.7 | 1.6 |
| S.E.—0.3 | 0.1 | 4.6 | |

However, small differences appear, showing that perhaps these two molecules do not have the same metabolism. The compound $^{14}$C-methionine seems to be less absorbed by the intestine, to be transformed into $^{14}CO_2$ in a larger proportion, and to be retained in the tissues at a lower level than $^{14}$C-methionine sulfoxide. These differences are small and could be attributed either to the normal variations between animals or to effective differences in their respective metabolism. The peak of $^{14}CO_2$ appears 60–75 min after ingestion of $^{14}$C-methionine, but 90–100 min after ingestion of $^{14}$C-methionine sulfoxide (*see* Figure 3). This delay, which is observed in the transformation into $^{14}CO_2$, can be due to a lower absorption rate or to the normal delay for the reduction of methionine sulfoxide into methionine. This delay could be beneficial for a better utilization of the molecule.

The pattern of the urinary excretion was the same for the two molecules and the radioactive catabolites separated by thin-layer chromatography (TLC) and paper electrophoresis were comparable, indicating no abnormal retention and no detoxifying process.

The delay of about 30 min in the appearance of the peak of $^{14}CO_2$ and the high level of utilization of methionine sulfoxide as source of methionine indicate that this molecule readily and easily is reduced into methionine. The reduction process has been studied on perfused rat liver (*68*).

The prefused rat liver is able to reduce $^{14}$C-methionine sulfoxide, and to use it as $^{14}$C-methionine, but with a low yield. In a 2-h perfusion, the levels of the incorporation into the protein and of the oxidation into $^{14}CO_2$ were much lower for methionine sulfoxide than for methionine (*see* Table V). The radioactivity remaining in the perfusion

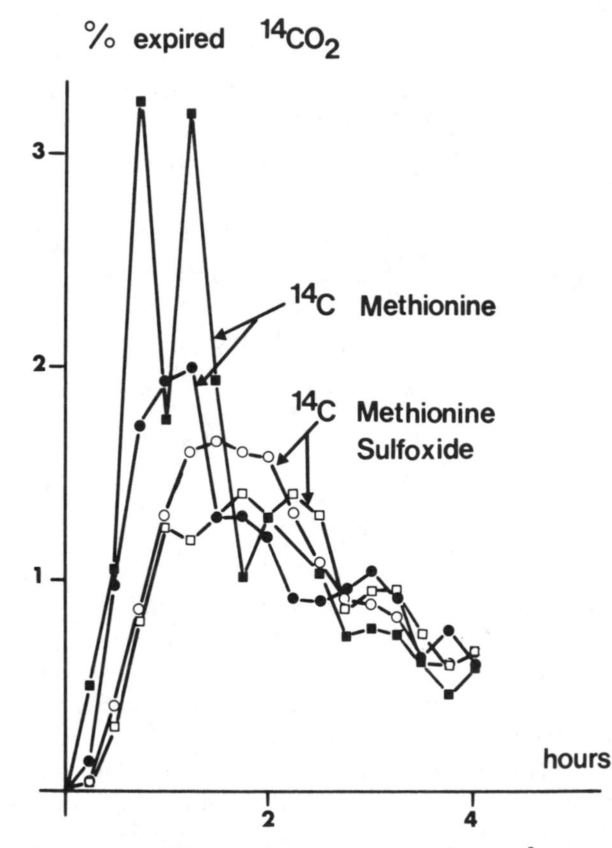


*Figure 3. Kinetics of $^{14}CO_2$ expiration in rats after oral ingestion of U-$^{14}$C-L-methionine and U-$^{14}$C-L-methionine sulfoxide (each curve corresponds to one experiment)*

fluid was much higher for methionine sulfoxide indicating that the molecule does not easily enter the liver. No free methionine sulfoxide was found in the liver cells indicating that the reduction rate is rapid. Other organs therefore must be involved in the reduction of methionine sulfoxide to explain its complete utilization as a source of methionine in the whole animals.

The localization at the cellular and subcellular levels of the liver and kidney enzymes responsible for this reduction as well as the mechanism of action have been studied by Aymard et al. (69); the liver enzymic activity is localized mainly in the cytosol while the kidney enzymatic activity appears to be associated mainly with the cell membranes and nuclei. In both cases, the enzyme is thermolabile and its activity is enhanced by adding NADH.

Other tissues probably are involved in the reduction process of methionine sulfoxide. The metabolic studies performed on the living animals suggest that methionine sulfoxide is transported actively through the intestinal wall and reduced during its passage or reduced in the lumen and actively transported as free methionine.

**Nutrition Effect of Protein-Bound Methionine Sulfoxide.** There is some discrepancy in the results concerning the biological availability of protein-bound methionine sulfoxide. Ellinger and Palmer (*71*) found that oxidized casein had a lower NPU (Net Protein Utilization) than normal casein. Slump and Schreuder (*72*) concluded that there was a positive biological availability of peptide-bound methionine sulfoxide.

The most convincing results are those obtained by Gjoen and Njaa (*66*) and Cuq et al. (*70*) with growth tests in rats fed on oxidized protein. The weight gain of rats fed on oxidized fish meal was 85% of the value obtained with nonoxidized protein (*66*). The PER (Protein Efficiency Ratio) and the NPU values of oxidized casein (in which methionine was oxidized completely into methionine sulfoxide) were 10–15% lower than the values given by nonoxidized casein and in some experiments the values were not statistically different (*70*). These authors concluded that protein-bound methionine sulfoxide was slightly less biologically available than protein-bound methionine. This suggests that protein-bound methionine sulfoxide behaves in the digestion process as protein-bound methionine despite the difference in chemical structure.

**Physiological Effects.** In spite of very good biological usage of free and protein-bound methionine sulfoxide and slight differences in their metabolic transit, this amino acid induced biological changes; e.g., important differences were observed in the levels of free methionine in the blood plasma (*65, 66*) and in the muscles (*70*) in rats fed on diets containing methionine sulfoxide. Moreover, the presence of methionine sulfoxide in the tissues can modify the activity of some enzymes such as glutamine synthetase (*73*) and $\gamma$-glutamylcysteine synthetase (*74*) which are inactivated in vitro by this molecule.

### *Isopeptides*

**Chemical Reactions: Analytical Methods.** During severe heat treatments of food proteins, chemical reactions occur between the $\epsilon$-amino group of lysine and the carboxamide group of the glutaminyl and asparaginyl residues (*75*).

$$\text{lysyl residues} \left\{ \begin{array}{l} + \text{ glutaminyl residue} \rightarrow \epsilon\text{-}(\gamma\text{-glutamyl})\text{lysyl} \\ + \text{ asparaginyl residue} \rightarrow \epsilon\text{-}(\beta\text{-aspartyl})\text{lysyl} \end{array} \right.$$

These reactions create cross-links between the protein chains, which are responsible for the loss in the nutritive value of the severely heated proteins. This loss can be attributed either to the reduction of the total nitrogen digestibility or to the reduction of the availability of the lysine residues engaged in such isopeptides.

Such isopeptides also have been found naturally present in keratin (*77, 78, 85*) and in polymerized fibrin (*79*). Their quantitative determination requires an enzymic hydrolysis using pepsin, pronase, aminopeptidase, and prolidase as described by Cole et al. (*135*) followed by a chromatographic separation using an amino acid analyzer under very specific conditions (*80*).

The hypothesis of the loss of lysine availability engaged in such peptides has been tested in animal assays with the free synthetic molecules.

**Biological Availability of the Free Isopeptides.** The biological availability of free $\epsilon$-($\gamma$-glutamyl)lysine was tested first by Mauron (*118*), who found that this isopeptide's growth-promoting effect was as good as that of free lysine. This result was confirmed later by Waibel and Carpenter (*81*) and Finot et al. (*83*) (*see* Table I). In contrast, free $\epsilon$-($\beta$-aspartyl)lysine was not found to be biologically available (*83*) (*see* Table I).

We therefore can conclude easily that protein-bound $\epsilon$-($\beta$-aspartyl)-lysine is not biologically available, but the question that arises is whether protein-bound $\epsilon$-($\gamma$-glutamyl)lysine is biologically available. If it is, what is the biological mechanism involved in the release of lysine from this isopeptide, free and protein-bound?

**Metabolic Transit of Free Radioactive $\epsilon$-($\gamma$-Glutamyl)lysine.** The first metabolic study of this isopeptide was made by Waibel and Carpenter (*81*) who showed that this molecule was present in the blood plasma of chicks and rats receiving it in their diet. Using $\epsilon$-($\gamma$-glutamyl)-[4,5-$^3$H]-lysine, Raczynski et al. (*82*) confirmed that the isopeptide passed unchanged across the intestinal wall into the serosal fluid in everted sacs and found that the kidneys were very active in hydrolyzing this isopeptide. These authors also found small hydrolytic activities in the intestinal mucosa and the liver.

We have incubated these two isopeptides with homogenates of intestinal mucosa, liver, and kidneys and showed tht $\epsilon$-($\beta$-aspartyl)lysine was not at all hydrolyzed while $\epsilon$-($\gamma$-glutamyl)lysine was hydrolyzed only by the kidneys' homogenate. On perfused rat liver, the radioactive free $\epsilon$-($\gamma$-glutamyl)-$^{14}$C-lysine was not transformed biochemically at all.

The kidney enzyme involved in the release of lysine from $\epsilon$-($\gamma$-glutamyl)lysine is not the $\epsilon$-acyllysinase found by Leclerc and Benoiton (*84*) to liberate lysine from $\epsilon$-formyllysine and $\epsilon$-acetyllysine. This was confirmed by our autoradiographic pictures which show an accumulation of radioactivity in the outer part of the kidneys with $\epsilon$-($\gamma$-glutamyl)-$^{14}$C-lysine and in the cortex with $\epsilon$-formyl-$^{14}$C-lysine (*see* Figure 2); that is to say in the cells where lysine probably is regenerated from its derivatives.

We have compared the metabolic transit in rats of $\epsilon$-($\gamma$-glutamyl)-U-$^{14}$C-lysine with that of U-$^{14}$C-lysine, after ingestion by stomach tubing and intravenous injection (*83*).

In contrast with the feeding experiments in which free $\epsilon$-($\gamma$-glutamyl)lysine added to a lysine-deficient diet was used completely as a source of lysine, we have found that the isopeptide given by stomach tubing was excreted partially in the feces (30–40%) and that under these experimental conditions it was not used completely. This difference in the intestinal absorption of the isopeptide is probably due to a too rapid intestinal transit time of the isopeptide when given in a water solution by stomach tubing in fasting animals as compared with the normal feeding experiments and also because the isopeptide is not transported actively through the intestinal wall. (*See* Table VI).

The kinetics of the expiration of $^{14}CO_2$ resulting from the oxidation of $^{14}$C-lysine and $\epsilon$-($\gamma$-glytamyl)-$^{14}$C-lysine after oral and intravenous administration provide information that can explain the biological mechanism involved in the release of lysine from this isopeptide (*see* Figure 4).

The peaks of $^{14}CO_2$ appeared 30 min and 180 min, respectively, after stomach tubing of $^{14}$C-lysine and $\epsilon$-($\gamma$-glutamyl)-$^{14}$C-lysine. This delay was not observed after intravenous injection of these two compounds.

**Table VI. Metabolic Transit of Free $^{14}$C-Lysine and Free $\epsilon$:($\gamma$-Glutamyl)-$^{14}$C-Lysine in Rats[a]**

| *Percentage of Administered Radioactivity Found in* | *$^{14}$C-Lysine* | | *$\epsilon$-($\gamma$-Glutamyl)-$^{14}$C-lysine* | |
|---|---|---|---|---|
| | *Stomach Tubing (24-h Experiment)* | *Intravenous (8-h Experiment)* | *Stomach Tubing (24-h Experiment)* | *Intravenous (8-h Experiment)* |
| Feces | 7.5 | | 35.6 | |
| Urine | 5.5 | 8.3 | 2.8 | 7.3 |
| $^{14}CO_2$ | 18.5 | 28.0 | 28.0 | 30.1 |
| Appearance of the peak of $^{14}CO_2$: minutes after administration | 35 | 35 | 190 | 60 |

[a] Mean of two experiments (*28, 83*).

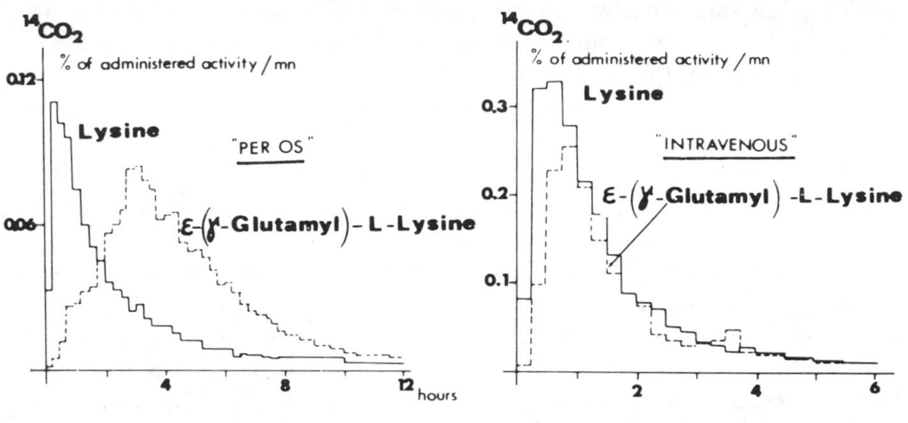


Advances in Experimental Medicine and Biology

*Figure 4. Kinetics of $^{14}CO_2$ expiration in rats after oral and intravenous administration of U-$^{14}$S-L-lysine and ε(γ-glutamyl)-U-$^{14}$C-L-lysine* (83)

The delay is probably due to the late absorption of the isopeptide (in the distal part of the small intestine) while free lysine is absorbed in the duodenum. After its late absorption, the isopeptide is transported readily to the kidneys which hydrolyze it very rapidly (as indicated by the kinetics of the expiration of $^{14}CO_2$ after intravenous injection) (*see* Figure 4). Since no free isopeptide was found in the urines, this hydrolysis is complete.

**Behavior of the Protein-Bound Isopeptides.** The problem of protein-bound isopeptide absorption was investigated by Hurrell et al. (*76*) who found that these two isopeptides were not present in greater concentration than the other nitrogen compounds in the ileal and fecal contents of rats that were fed heat-treated protein containing aspartyl- and glutamyllysine. They proposed that the loss in the nutritive value of severely heated proteins was mainly due to a decrease in the nitrogen digestibility. The isopeptides appeared to be either absorbed by the intestine or metabolized by the intestinal microflora.

### *Alkaline Treatments*

**Reactions: Analytical Methods.** Alkaline treatments are used in the food and feeds technology for solubilization and purification of proteins, to destroy toxic contaminants, and to obtain functional properties. These alkaline treatments induce many chemical modifications on the side chains of the amino acid residues, which have been described by many workers.

| | | |
|---|---|---|
| Cysteine and cystine | → | dehydroalanine (*86, 87, 92*) |
| Serine and serine phosphate | → | dehydroalanine (*89, 104*) |
| Dehydroalanine + lysine | → | lysinoalanine (*89, 90*) |
| Dehydroalanine + cysteine | → | lanthionine (*88, 90*) |
| Dehydroalanine + $H_2O$ | → | pyruvic acid (*86, 91*) |
| Arginine | → | ornithine + urea (*90*) |
| Dehydroalanine + ornithine | → | ornithinoalanine (*90*) |
| Dehydroalanine + $NH_3$ | → | β-aminoalanine (*93*) |
| Isoleucine | → | alloisoleucine (*94*) |
| L-lysine | → | D,L-lysine (*94*) |

The most important modifications are the formation of lysinoalanine and lanthionine. Lysinoalanine can be detected or measured by TLC (*95*) or ion-exchange chromatography (*96, 97*). Lanthionine, which is eluted at the same time as glycine by ion-exchange chromatography (*1, 2*) can be detected by TLC (*98*). Lysinoalanine was formed also upon heat treatments and has been detected in many heated-food proteins (*95*). The problems linked to the presence of this amino acid consequently have been generalized to all the heated-food proteins.

The nutritional and physiological effects of the alkali-treated food proteins have been studied extensively. Efforts have been concentrated mainly on the effects of lysinoalanine.

**Nutritional and Physiological Effects of Alkali-Treated Proteins.** The first effect of the alkaline treatment of food proteins is a reduction in the nutritive value of the protein due to the decrease in (a) the availability of the essential amino acids chemically modified (cystine, lysine, isoleucine) and in (b) the digestibility of the protein because of the presence of cross-links (lysinoalanine, lanthionine, and ornithinoalanine) and of unnatural amino acids (ornithine, alloisoleucine, β-aminoalanine, and D-amino acids). The racemization reaction occurring during alkaline treatments has an effect on the nitrogen digestibility and the use of the amino acids involved.

These effects were described by de Groot and Slump (*99*) followed by Provansal et al. (*94*). Later, it was confirmed that the lysine moiety of lysinoalanine was completely unavailable to the rat and only partly available to the chick, while a certain part of the cysteine moiety of lanthionine (32% to 52% according to the racemic mixture) was available to chicks (*100*) (*see* Table I).

The most outstanding effect of the alkali-treated proteins that has been reported is due to the presence of lysinoalanine, which induces renal alterations in rats, designated nephrocytomegaly, which consists of enlarged nuclei and increased amounts of cytoplasm in epithelial cells of the straight portion of the proximal tubules (*101, 102, 103, 105, 106*). This effect has never been found in other species such as mice, hamsters, quails, rabbits, dogs, and monkeys (*103*).

After some controversies concerning the effects of the diet composition and the strain of the animals on the development of nephrocytomegaly, it was admitted that free as well as protein-bound lysinoalanine were really responsible for this effect. It also was shown that this nephrocytomegaly was reversible (*109*) and that the dose response was not the same for the different stereoisomers of lysinoalanine (*107, 108, 110*); the LD isomer is more active than the LL, DL, and DD isomers.

Moreover, the nephrocytomegalic alterations are not limited to the presence of lysinoalanine only since ε-deoxyfructosyllysine (*34*), 2,3-diaminopropionic acid (*136*), and ornithinoalanine (*107*) also induce the same renal alterations in the rat.

In order to elucidate the mechanism of the induction of cytomegaly, we investigated the metabolic transit of lysinoalanine.

**Metabolic Transit of Lysinoalanine.** URINARY AND FECAL EXCRETION OF PROTEIN-BOUND LYSINOALANINE (*113*). Three different alkali-treated proteins (lactalbumin, fish protein isolate, and soya protein isolate) containing, respectively, 1.79, 0.38, and 0.14 g of lysinoalanine/16 g nitrogen were given to rats and the urines and feces were collected. Lysinoalanine was measured before and after acid hydrolysis. The fecal excretion varied from 33 to 51% of the total ingested lysinoalanine and the urinary excretion varied from 10 to 25%. The higher level of lysinoalanine found after acid hydrolysis indicates that a certain quantity is excreted in the urines as combined lysinoalanine (*see* Table VII). The total recovery was inferior to the ingested quantity (50 to 71%) indicating that the molecule is transformed or retained in the body of the rat.

METABOLIC TRANSIT OF $^{14}$C-LYSINOALANINE IN RATS. The radioactive $^{14}$C-lysinoalanine was synthesized from uniformly labelled $^{14}$C-lysine. The compound α-*N*-formyl-$^{14}$C-lysine, prepared according to

**Table VII. Urinary and Fecal Excretions of Protein-Bound Lysinoalanine in Rats[a] (percentage of ingested quantity) (*113*)**

| *Alkali-Treated Proteins* | | *Urines* | | *Feces After Hydrolysis* | *Total Recovery After Hydrolysis* |
|---|---|---|---|---|---|
| | | *Before Hydrolysis* | *After Hydrolysis* | | |
| Lactalbumin | mean | 6.4 | 10.5 | 51 | 61.5 |
| | S.E. | 0.5 | 0.2 | 5 | 5.4 |
| Fish protein isolate | mean | 7.1 | 16.4 | 33 | 49.4 |
| | S.E. | 0.9 | 0.4 | 6 | 6.6 |
| Soya protein isolate | mean | 18.1 | 24.5 | 47 | 71.5 |
| | S.E. | 3.6 | 0.8 | 5 | 3.8 |

[a] Percentage of ingested quantity (*113*).

**Table VIII. Metabolic Transit of $^{14}C$-Lysinoalanine and Retention in the Liver and the Kidneys[a]**

*24-h Experiments on Rats*
*Percentage of Ingested Radioactivity Recovered in*

| $CO_2$ | Feces | Urines | Total Excretion |
|---|---|---|---|
| 10.9–24.3 | 10.4–33.8 | 26.1–64.1 | 69.6–91.8 |

*24-h Experiments on Other Species*
*Percentage of Ingested Radioactivity Retained per Grams of Tissue*

| | Rat | Mouse | Hamster | Quail |
|---|---|---|---|---|
| Liver | 0.1–0.23 | 0.2–0.4 | 0.2–0.5 | 1.2–2.1 |
| Kidneys | 1.1–7.3 | 1.1–1.3 | 0.3–0.3 | 0.5–0.6 |
| Kidneys/Liver | 9–33 | 3.3–5.5 | 0.6–1.5 | 0.3–0.4 |

[a] Ref. *113*.

Hoffmann et al. (*114*), was reacted with acetaminoacrylic ethyl ester in alkaline medium (*115*) followed by acid hydrolysis and purification by ion-exchange chromatography (*113*).

Ten to 15 mg/kg by body weight of $^{14}C$-lysinoalanine were given orally to rats and the radioactive excretions were measured for 24 h. The values obtained were very widely dispersed, indicating that the animals do not respond uniformly (*see* Table VIII). The fecal excretion varied between 10 and 34% indicating that this amino acid is not transported actively. The total radioactive urinary excretion varied between 26 and 64% and the expired $^{14}CO_2$ varied between 11 and 24% indicating that the molecule is oxidized biochemically. This transformation occurs in the body of the animals and not in the intestine by the microflora since $^{14}CO_2$ is formed also after intravenous injection (*113*). The urinary catabolites were separated by ion-exchange chromatography with continuous recording of the radioactivity by an anthracene scintillation cell (Chroma/cell-Nuclear Chicago). More than 10 peaks were detected. Lysinoalanine was the major urinary constituent (60–65%); the other peaks were eluted since there was more acid in character than lysinoalanine (*see* Figure 5).

Twenty-four h after injecting radioactive lysinoalanine, the kidneys remain very radioactive with a specific radioactivity 10 to 30 times higher than that of the liver (*see* Table VIII). This radioactivity is localized in the medular part of the kidney (*see* Figure 2) and in the tubular cells in which the nephrocytomegaly develops. This radioactivity

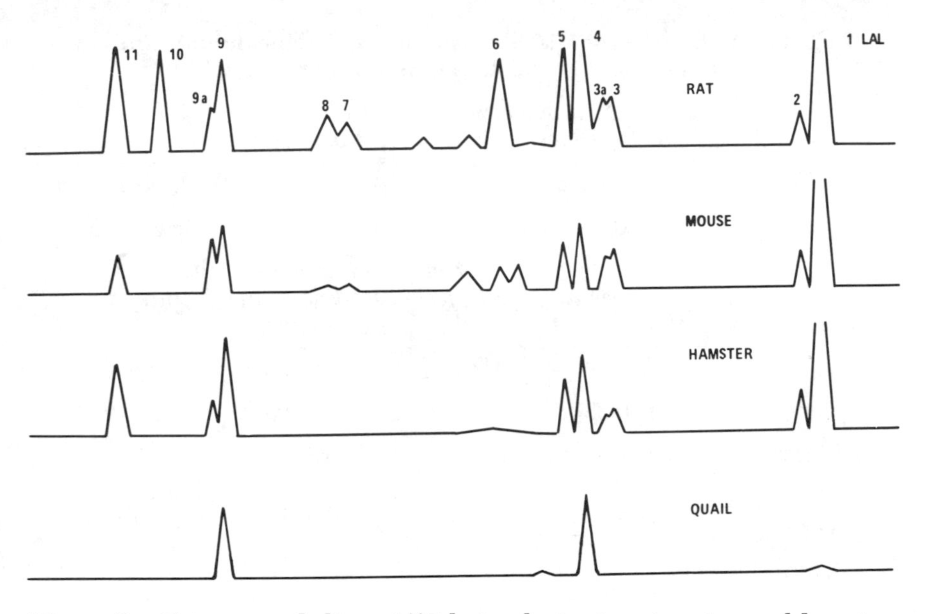


*Figure 5. Urinary catabolites of* $^{14}C$*-lysinoalanine in rats, mice, and hamsters. Fecal catabolites in quails. Separation on a Beckman amino acid analyzer. Short column for the separation of lysinoalanine (LAL) followed by the normal column for the acidic and neutral amino acids (to read from the right to the left).*

is protein-bound (*28*) and corresponds to the incorporation of the radioactive lysine moiety generated from radioactive lysinoalanine by the kidney enzymes (*116*). It is well known that the rat kidneys contain an ε-alkyllysinase active on derivatives such as ε-methyllysine (*117*). This enzyme may be involved in the regeneration of lysine from lysinoalanine; later on the generated lysine would be incorporated into the proteins of the cells containing this enzyme.

METABOLIC TRANSIT OF $^{14}C$-LYSINOALANINE IN OTHER SPECIES. In mice and in hamsters, the metabolic transit of lysinoalanine is not very different from that of rats except in the composition of some urinary catabolites (*113*) and the level of the radioactivity remaining in the kidneys (*see* Table VIII). In quails the feces contained small amounts of lysinoalanine (6–15%) and only two important catabolites. The urines of mice and hamsters contained two catabolites less than those of rats (*see* Figure 5), and although mice and hamsters do not develop nephrocytomegaly, they presented the same kind of retention of radioactivity in the kidney cells as rats. This retention in the medular part of the kidneys is well observed on the whole-body autoradiographies of Figure 2; however, in mice and hamsters, this retention is quantitatively less important than in rats (*see* Table VIII).

**Mechanism of Induction of Nephrocytomegaly.** The mechanism of induction of nephrocytomegaly is not understood yet; several hypotheses can be discussed.

(a) Lysinoalanine would be a substrate like others such as ε-deoxyfructosyllysine (*34*), ornithinoalanine (*107*), or 2,3-diaminopropionic acid (*136*) capable of inducing nephrocytomegaly. The cationic character of these molecules parallels that of the polyamines like spermidine and spermine and would bind nucleic acids and interfere in the protein synthesis. The question is, why are rats more sensitive than the other species?

(b) Is there any relationship between nephrocytomegaly and the retention of lysine in the kidney cells after regeneration from its derivatives? Two derivatives, lysinoalanine and ε-frucosyllysine, both nephrocytomegaly inducers, are retained in the kidney cells. For lysinoalanine, we know that this is due to the regeneration of lysine which is incorporated later on into the proteins. The high level of lysine in the cells where it is regenerated could be responsible for the renal defect. For ε-fructosyllysine, we do not know if the radioactivity retained in the kidneys is due to the incorporation of lysine into the proteins through a mechanism identical to that observed with lysinoalanine.

Other derivatives of lysine, ε-(γ-glytamyl)lysine and ε-formyllysine, are hydrolyzed in the kidneys, and the lysine moiety is regenerated and incorporated into the kidney proteins in the cells where hydrolysis occurs. No retention of radioactivity occurs with derivatives such as α-formyl-$^{14}$C-lysine which is not hydrolyzed in the kidneys (*83*) (*see* Figure 2). At present, we do not know yet whether these derivatives of lysine induce nephrocytomegaly.

(c) In contrast with ε-deoxyfructosyllysine, which is excreted unchanged in the urines, the urinary catabolites of lysinoalanine are numerous; more than 10 different radioactive peaks were separated by ion-exchange chromatography. Some of them appear to correspond to the acetylated derivatives of lysinoalanine (*113*) and another one has been identified as a bicyclic derivative (1,7-diazabicyclo(4.3.0)nonane-6,8-dicarboxylic acid) formed by the action of L-amino acid oxidase in the presence of catalase (*112*). Such a derivative formed only from the L,L or L,D stereodiomers would explain why the induction of nephrocytomegaly is dependent on the stereoisomerism of lysinoalanine.

If we compare the elution pattern of the urinary catabolites of the different species, we observe that the rat urines contain two catabolites not present in the mouse and hamster urines (*see* Figure 5). The structure of these catabolites is not known yet.

Complementary studies are needed in order to investigate these different hypotheses.

***Protein–Polyphenol Interaction***

**Reactions.** Polyphenols, such as caffeic acid or chlorogenic acid, are present in various plant foods (fruits, leaves, oilseeds, etc.) and can be oxidized during processing into quinones. This oxidation, involving oxygen, occurs at neutral pH through the action of pohlyphenol oxidases or directly at alkaline pH. Quinones spontaneously polymerize into brown pigments and when proteins are present, they covalently bind amino acid residues such as lysine, cysteine, and possibly tryptophan. During this process, sulfur amino acids and tryptophan also can be oxidized.

We have studied the reaction between lysine and polyphenols in a model system containing casein and caffeic acid, incubated at room temperature at pH 7 and 10 with and without tyrosinase. In the absence of tyrosinase and oxygen, reactive lysine does not change; in the absence of tyrosinase and in the presence of oxygen, reactive lysine decreases when the pH is higher than 9. In the presence of tyrosinase and oxygen, the level of reactive lysine decreases when the pH is 7, which is the optimum pH of the enzymic activity. At pH 7, with tyrosinase, the reaction rate is low, while at pH 10, the reaction rate is more rapid (*119*).

**Metabolic Transit of Lysine Bound to Caffeic Acid.** In order to follow the metabolism of lysine bound to caffeic acid, goat casein, biologically labelled with tritiated lysine, was treated with caffeic acid at pH 7 with tyrosinase and at pH 10 without tyrosinase and was given to rats. The urinary and fecal excretions and the incorporation of lysine in the tissues were measured (*120*) (*see* Figure 6).

The fecal excretion is about twice as high with casein treated at pH 7 plus tyrosinase than with untreated casein. At pH 10, the fecal excretion is about 5 times higher. The urinary excretion was about the same magnitude and the radioactive urinary catabolites separated by ion-exchange chromatography were the same for the three test materials. This indicates that the reaction products are excreted mainly in the feces and not absorbed.

In the case of casein treated at pH 10 with caffeic acid, the level of the radioactivity incorporated in the liver, kidney, and muscle proteins were lower than that measured for untreated casein and casein treated at pH 7, indicating a higher loss in available lysine as expected from the high fecal excretion measured.

The decrease in available lysine due to reaction with polyphenols can occur during the preparation of protein isolates or concentrates from polyphenol-rich materials such as alfalfa leaves or sunflower seeds.

Preliminary experimentation showed that lysine is not the only amino acid involved in the reaction process with polyphenols. Methio-

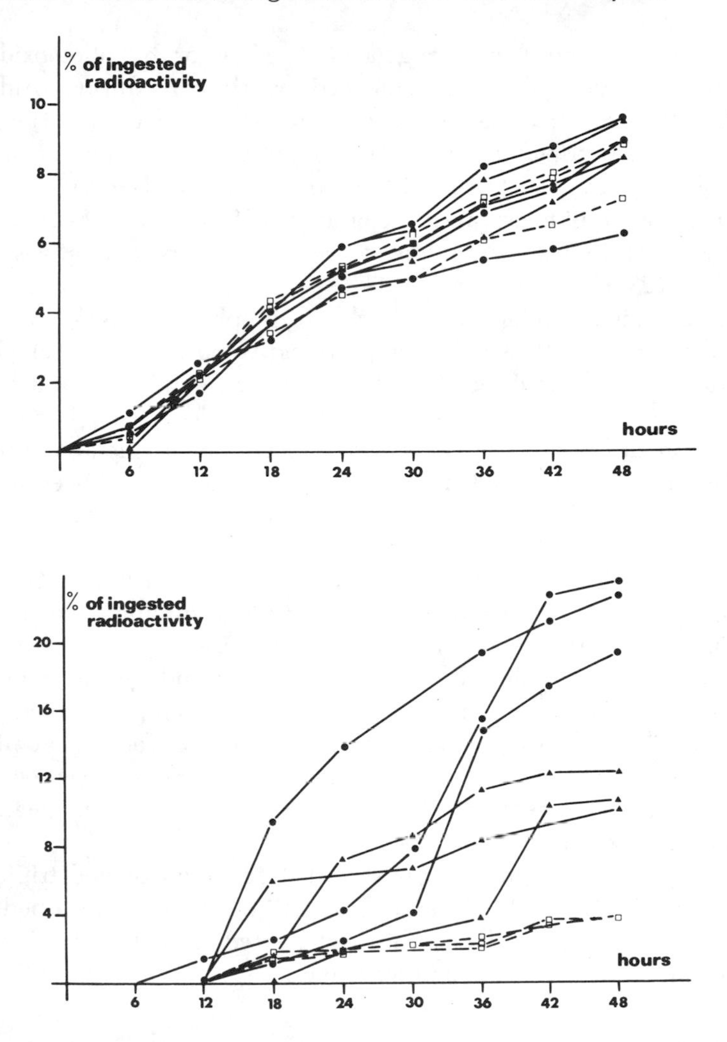


*Figure 6. Urinary* (top) *and fecal* (bottom) *excretion of the radioactivity after oral administration in rats of* ($^3$*H-lysine*) *casein untreated* (□) *and treated with caffeic acid at pH 7 plus tyrosinase* (▲) *and pH 10* (●). *Each curve corresponds to one rat; each product was tested on three rats.*

nine, cystine, and tryptophan seem to be implicated also in chemical reactions that modify the nutritional properties (*119*).

## *Effects of Heat Treatments and Oxidation on Tryptophan*

**Reactions: Nutritional Consequences.** Tryptophan is very sensitive to heat treatments and to oxidation by hydrogen peroxide (*123, 129,*

*130, 131, 137*), atmospheric oxygen (*121, 123*), or by photooxidation (*127, 128*). Its destruction is influenced by the pH during oxidation (*121, 123, 137*) and by the presence of reducing sugars (*122*). After oxidation of the tripeptide glycyltryptophylglycine by hydrogen peroxide followed by hydrolysis with methanesulfonic acid, six to nine derivatives of tryptophan were separated (*123, 131, 137*) while after heat treatment in water solution, only three to six derivatives were separated (*123, 137*).

The chemical modifications of the tryptophan residues lead to a decrease in the nutritive value of proteins as observed in autoclaved soja meals (*124*), heated meats (*125*), heated casein (*126*), and heated skim milk (*122*); this last reference is probably the most reliable work published in this field. The nutritional effects and the metabolic transit of heat-treated and oxidized tripeptide (gly–try–gly) have been investigated (*123, 132, 137*) recently; only the metabolic transit study is related here.

**Metabolic Transit of Peptide Bound Heat-Treated and Oxidized Tryptophan.** The radioactive glycyl-L-($^3$H)-tryptophylglycine was synthesized by the method proposed by Zimmerman et al. (*134*) (*N*-hydroxysuccinimide + dicyclohexylcarbodiimide) and treated under the following conditions: (a) oxidation by 0.2*M* hydrogen peroxide at pH 7 for 30 min at 50°C and (b) heat treatment (130°C for 3 h at pH 7) (*137*). The untreated, oxidized and heat-treated radioactive peptides were given to rats accustomed to eating two meals a day in metabolic cages to collect the urines and feces.

The results (*see* Figure 7) indicate that (a) the oxidized tripeptide is excreted mainly in the feces and a certain part is absorbed and excreted in the urines and (b) the heat-treated tripeptide is absorbed almost completely and excreted in the urines.

### *Conclusions*

During the processing of foods, several chemical reactions affect the structure of some amino acids: reactions with reducing sugars and polyphenols, interaction of the side chains of the amino acids, oxidation, etc. In some cases, these chemical modifications do not seem to change the nutritional availability of the amino acids involved, while in other cases, they do. In all cases, the presence of the newly formed molecules induces biochemical and physiological changes.

The metabolic transit studies of the molecules formed provide information complementary to those obtained by the conventional nutritional approaches. This information can be used to control the analytical methods, to explain the nutritional responses, and also to foresee the possible physiological effects. The extrapolation to humans of the results

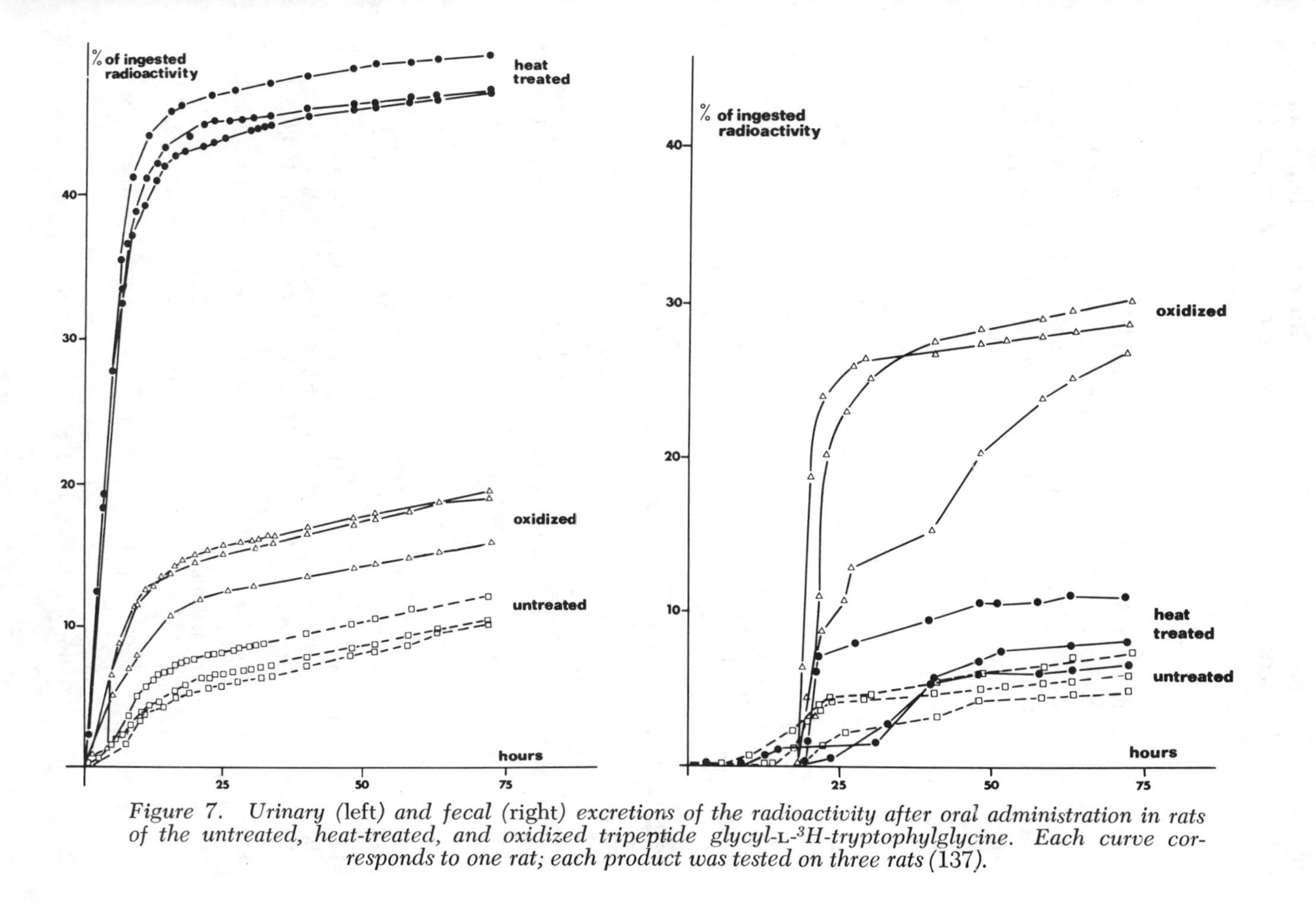


*Figure 7. Urinary (left) and fecal (right) excretions of the radioactivity after oral administration in rats of the untreated, heat-treated, and oxidized tripeptide glycyl-*L*-$^3$H-tryptophylglycine. Each curve corresponds to one rat; each product was tested on three rats (137).*

obtained on animals with molecules labelled with radioactive isotopes will have to be made in the future, using molecules labelled with stable isotopes.

### *Acknowledgments*

We are grateful to D. Salter of the National Institute for Research in Dairying, Shinfield, Reading, England for the labelling of $^3$H-lysine casein and R. Jost for its purification. Many thanks to I. Bracco for the whole-body autoradiographies and R. F. Hurrell for his advice and the correction of the manuscript.

### *Literature Cited*

1. Mauron, J. In "Protein and Amino Acid Functions"; E. J. Bigwood, Ed.; Pergamon: Oxford, 1972; Vol. 2, p. 417.
2. Cheftel, J. C. In "Food Proteins"; Whitaker, J. R.; Tannenbaum, S. R., Eds.; Avi: Westport, CT, 1977; p. 401.
3. Mauron, J. *Food Sci. Technol., Proc. Int. Congr., 4th,* **1974,** *1,* 564–577.
4. Hurrell, R. F. In "Food and Health: Science and Technology"; Birch, G. G.; Parker, K. J., Eds.; *Appl. Sci:* Essex, 1980; pp. 369–388.
5. Friedman, M., Ed. "Protein Crosslinking: Nutritional and Medical Consequences," *Adv. Exp. Med. Biol.* **1977,** *86A–86B.*
6. Eriksson, C., Ed. "Maillard Reactions in Food: Chemical, Physiological, and Technological Aspects," Pergamon: Oxford, in press.
7. Hodge, J. E. *J. Agric. Food Chem.* **1953,** *1,* 928–943.
8. Weitzel, G.; Geyer, H. V.; Fretzdorff, A. M. *Chem. Ber.* **1957,** *90,* 1153–1161.
9. Finot, P. A.; Bujard, E.; Mottu, F.; Mauron, J. In "Protein Crosslinking: Nutritional and Medical Consequences," *Adv. Exp. Med. Biol.* **1977,** *86B,* 343–365.
10. Gottschalk, A. *Biochem. J.* **1952,** *52,* 455.
11. Finot, P. A.; Mauron, J. *Helv. Chim. Acta* **1969,** *52,* 1988.
12. Sgarbieri, V. C.; Amaya, J.; Tanaka, M.; Chichester, C. O. *J. Nutr.* **1973,** *103,* 657–663.
13. Barbiroli, G.; Mazzaracchio, P.; Ciusa, W. In "Protein Crosslinking: Biochemical and Molecular Aspects," *Adv. Exp. Med. Biol.* **1977,** *86B,* 449–470.
14. Borsook, H.; Abrams, A.; Lowy, P. H. *J. Biol. Chem.* **1955,** *215,* 111.
15. Anet, E. F. L. J.; Reynolds, T. M. *Aust. J. Chem.* **1957,** *10,* 182.
16. Finot, P. A.; Bricout, J.; Viani, R.; Mauron, J., *Experientia* **1968,** *24,* 1097–1099.
17. Brueggemann, J.; Erbersdobler, H. *Z. Lebensm.-Unters.* **1968,** *137,* 137–143.
18. Erbersdobler, H. *Milchwissenschaft* **1970,** *25,* 280–284.
19. Finot, P. A.; Viani, R.; Bricout, J.; Mauron, J. *Experientia* **1969,** *25,* 134–135.
20. Adrian, J. In "World Review of Nutrition and Dietetics"; Karger: Basel, 1974; Vol. 19, 71–122.
21. Mottu, F.; Mauron, J. *J. Sci. Food Agric.* **1969,** *18,* 57–62.
22. Mauron, J.; Mottu, F.; Bujard, E.; Egli, R. H. *Arch. Biochem. Biophys.* **1955,** *59,* 433–451.

23. Bujard, E.; Handwerck, V.; Mauron, J. *J. Sci. Food Agric.* **1967,** *18,* 52–57.
24. Bujard, E.; Finot, P. A. *Ann. Nutr. Alim.* **1978,** *32,* 291–305.
25. Carpenter, K. J. *Biochem. J.* **1960,** *77,* 604–610.
26. Carpenter, K. J. *Nutr. Abstr.* **1973,** *43,* 423–451.
27. Bujard, E.; Mauron, J. *Proc. Int. Congr. Nutr., 6th* **1964,** 489.
28. Finot, P. A.; Magnenat, E.; Mottu, F.; Bujard, E. *Ann. Nutr. Alim.* **1978,** *32,* 325–338.
29. Horn, M. J.; Lichtenstein, H.; Womack, M. *J. Agric. Food Chem.* **1968,** *16,* 741–745.
30. Tanaka, M.; Lee, T. C.; Chichester, C. O. *J. Nutr.* **1975,** *105,* 989–994.
31. Finot, P. A.; Deutsch, R.; Bujard, E. In "Maillard Reactions in Food: Chemical, Physiological, and Technical Aspects," Pergamon: Oxford, in press.
32. Johnson, G. H.; Baker, D. H.; Perkins, E. G. *J. Nutr.* **1977,** *107,* 1659–1664.
33. Lee, C. M.; Lee, T. C.; Chichester, C. O. *Comp. Biochem. Physiol. A* **1977,** *56*(4A), 473–476.
34. Erbersdobler, H. F.; von Wangenheim, B.; Haenichen, T. *Proc. Int. Congr. Nutr., 11th,* Rio de Janeiro, Summary No. 531, p. 330.
35. Stegink, L. D.; Shepherd, J. A.; Fry, L. K.; Filer, J. *Pediatr. Res.* **1974,** *8,* 396.
36. Stegink, L. D.; Freeman, J. B.; Meyer, P. D.; Filer, L. J.; Fry, L. K.; Denbesten, L. *Fed. Proc.* **1975,** *34,* 931.
37. Freeman, J. B.; Stegink, L. D.; Meyer, P. D.; Fry, L. K.; Denbesten, L. *J. Surg. Res.* **1975,** *18,* 463–469.
38. Tanaka, M.; Amaya, J.; Lee, T. C.; Chichester, C. O. *Proc. Int. Congr. Food Sci. Technol. 4th* **1974,** *1,* 632–640.
39. Lee, C. M.; Chichester, C. O.; Lee, T. C. *J. Agric. Food Chem.* **1977,** *25,* 775–778.
40. Antonioli, J.; Finot, P. A., unpublished data.
41. Finot, P. A.; Magnenat, E. In "Maillard Reactions in Food: Chemical, Physiological, and Technical Aspects," Pergamon: Oxford, in press.
42. Finot, P. A. In "Protein in Human Nutrition"; Porter, J. W. G.; Rolls, B. A., Eds.; Academic: London, 1973; p. 501–514.
43. Clark, A.; Tannenbaum, S. R. *J. Agric. Food Chem.* **1970,** *18,* 891–894.
44. Ibid., **1973,** *21,* 40–43.
45. Adrian, J.; Frangne, R.; Petit, L.; Godon, B.; Barbier, J. *Ann. Nutr. Alim.* **1966,** *20,* 257–277.
46. Yamaguchi, M.; Fujimaki, M. *J. Food Sci. Technol.* **1970,** *17,* 147–192.
47. Kimiagar, M.; Lee, T. C.; Chichester, C. O. *J. Agric. Food Chem.* **1980,** *28,* 150–155.
48. Hurrell, R. F. In "Food Flavours"; Morton, I. D.; MacLeod, A. J., Eds.; Elsevier: Amsterdam, 1980; p. 399.
49. Fox, P. F.; Kosikowski, F. V. *J. Dairy Sci.* **1967,** *50,* 1.
50. Naguib, Kh.; Hussein, L. *Milchwissenschaft* **1972,** *27,* 758.
51. Yang, S.-F. *Biochemistry* **1970,** *9,* 5008.
52. Tannenbaum, S. R.; Barth, H.; Le Roux, J. P. *J. Agric. Food Chem.* **1969,** *17,* 1353.
53. Cuq, J. C.; Hurrell, R.; Finot, P. A., unpublished data.
54. Neumann, N. P. In "Methods in Enzymology"; Colowick, S. P.; Kaplan, N. O., Eds.; Academic: New York, 1967; Vol. 11, p. 487.
55. Toennies, G.; Lavine, T. F. *J. Biol. Chem.* **1934,** *105,* 107–113.
56. Ibid., 115–121.
57. Lavine, T. F. *J. Biol. Chem.* **1936,** *113,* 583–597.
58. Toennies, G.; Lavine, T. F. *J. Biol. Chem.* **1936,** *113,* 576–582.

59. Bennett, M. A. *Biochem. J.* **1937,** *31,* 962.
60. Ibid., **1939,** *33,* 88T.
61. Bennett, M. A. *J. Biol. Chem.* **1941,** *141,* 573–578.
62. Bennett, M. A. *Biochem. J.* **1939,** *34,* 1794.
63. Njaa, L. R. *Br. J. Nutr.* **1962,** *16,* 571–577.
64. Miller, D. S.; Samuel, P. *Proc. Nutr. Soc.* **1968,** *27,* 21A.
65. Miller, S. A.; Tannenbaum, S. R.; Seitz, A. W. *J. Nutr.* **1970,** *100,* 909–916.
66. Gjoen, A. U.; Njaa, L. R. *Br. J. Nutr.* **1977,** *37,* 93–105.
67. Lepp, A.; Dunn, M. S. *Biochem. Prep.* **1965,** *4,* 8083.
68. Bartosek, I.; Guaitani, A.; Garattini, S. In "Isolated Liver Perfusion and Its Applications"; Bartosek, I.; Guaitani, A.; Miller, L. L., Eds.; Raven: New York, 1973; pp. 63.
69. Aymard, C.; Seyer, L.; Cheftel, J. C. *Agric. Biol. Chem.* **1979,** *43,* 1869–1872.
70. Cuq, J. L.; Besancon, P.; Chartier, L.; Cheftel, J. C. *Food Chem.* **1978,** *3,* 85–102.
71. Ellinger, G.; Palmer, R. *Proc. Nutr. Soc.* **1969,** *28,* 42A.
72. Slump, P.; Schreuder, M.-A. W. *J. Sci. Food Agric.* **1973,** *24,* 657.
73. Ronzio, R. A.; Rowe, W. B.; Meister, A. *Biochemistry* **1969,** *8,* 1066–1075.
74. Richman, P. G.; Orlowski, M.; Meister, A. *J. Biol. Chem.* **1973,** *248,* 6684–6690.
75. Bjarnason, J.; Carpenter, K. J. *Br. J. Nutr.* **1970,** *24,* 313.
76. Hurrell, R. F.; Carpenter, K. J.; Sinclair, W. J.; Otterburn, M. S.; Asquith, R. S. *Br. J. Nutr.* **1976,** *35,* 383.
77. Asquith, R. S.; Otterburn, M. S.; Buchanan, J. H.; Cole, M.; Fletcher, J. C.; Gardner, K. L. *Biochim. Biophys. Acta* **1970,** *221,* 342.
78. Asquith, R. S.; Otterburn, M. S.; Gardner, K. L. *Experientia* **1971,** *27,* 1388.
79. Pisano, J.; Finlayson, J. S.; Peyton, M. P. *Science* **1968,** *160,* 892.
80. Otterburn, M.; Healy, M.; Sinclair, W. In "Protein Crosslinking, Nutritional and Medical Consequences," *Adv. Exp. Med. Biol.* **1977,** *86B,* 239.
81. Waibel, P. E.; Carpenter, K. J. *Br. J. Nutr.* **1972,** *27,* 509.
82. Raczynsky, G.; Snochowski, M.; Buraczewski, S. *Br. J. Nutr.* **1975,** *34,* 291.
83. Finot, P. A.; Mottu, F.; Bujard, E.; Mauron, J. In "Nutritional Improvement of Food and Feeds Proteins"; Friedman, M., Ed.; Plenum: London, 1978; p. 549.
84. Leclerc, J.; Benoiton, L. *Can. J. Biochem.* **1968,** *46,* 471.
85. Sugawara, K. *Agric. Biol. Chem.* **1979,** *43,* 2543–2548.
86. De Marco, C.; Coletta, M.; Cavallini, D. *Arch. Biochem. Biophys.* **1963,** *100,* 51.
87. Gawron, O.; Odstrchel, G. *J. Am. Chem. Soc.* **1967,** *89,* 3263.
88. Horn, M. J.; Breese-Jones, D.; Ringel, S. J. *J. Biol. Chem.* **1941,** *138,* 141.
89. Bohak, Z. *J. Biol. Chem.* **1964,** *239,* 2878.
90. Ziegler, K.; Melchert, I.; Luerken, C. *Nature* **1967,** *214,* 404.
91. Dann, J. R.; Olivier, G. L.; Gates, J. W. *J. Am. Chem. Soc.* **1957,** *79,* 1644.
92. Swan, J. M. *Nature* **1957,** *179,* 965.
93. Asquith, R. S.; Booth, A. K.; Skinner, J. D. *Biochem. Biophys. Acta* **1969,** *181,* 164.
94. Provansal, M. P.; Cuq, J. L.; Cheftel, J. C. *J. Agric. Food Chem.* **1975,** *23,* 938.

95. Sternberg, M.; Kim, C. Y.; Schwende, F. J. *Science* **1975**, *190*, 992.
96. Slump, P. *J. Chromatogr.* **1977**, *135*, 502.
97. Raymond, M. L. *J. Food Sci.* **1980**, *45*, 56.
98. Sternberg, M.; Kim, C. Y.; Plunkett, R. A. *J. Food Sci.* **1975**, *40*, 1168.
99. De Groot, A. P.; Slump, P. *J. Nutr.* **1969**, *98*, 45.
100. Robbins, K. R.; Baker, D. H.; Finley, J. W. *J. Nutr.* **1980**, *110*, 907–915.
101. Woodard, J. C.; Short, D. D. *J. Nutr.* **1973**, *103*, 569–575.
102. Van Beek, L.; Feron, V. J.; De Groot, A. P. *J. Nutr.* **1974**, *104*, 1630–1636.
103. De Groot, A. P.; Slump, P.; Feron, V. J.; Van Beek, L. *J. Nutr.* **1976**, *106*, 1527–1538.
104. Freimuth, U.; Schlegel, B.; Gahner, E.; Nötzold, H. *Nahrung* **1974**, *181*, k5–k8.
105. Struthers, B. J.; Dahlgreen, R. R.; Hopkins, D. T. *J. Nutr.* **1977**, *107*, 1190–1199.
106. Gould, D. H.; Mac Gregor, J. T. In "Protein Crosslinking: Nutritional and Medical Consequences," *Adv. Exp. Biol. Med.* **1977**, *86B*, 29.
107. Feron, V. J.; Van Beek, L.; Slump, P.; Beems, R. B. *Proc. FEBS Meet.* **1977**, *44*, 139–147.
108. Slump, P.; Kraaikamp, W. C.; Schreuder, H. A. W. *Ann. Nutr. Alim.* **1978**, *32*, 271–279.
109. Struthers, B. J.; Hopkins, D. T.; Dahlgren, R. R. *J. Food Sci.* **1978**, *43*, 616–618.
110. Tas, A. C.; Kleipool, R. J. C. *Lebensm.-Wiss. Technol.* **1976**, *9*, 360–362.
111. Engelsma, J. W.; Meulen, J. D.; Slump, P.; Haagsma, N. *Lebensm.-Wiss. Technol.* **1979**, *12*, 202–203.
112. Leegwater, D. C.; Tas, A. C. *Lebensm.-Wiss. Technol.* **1980**, *14*, 87–91.
113. Finot, P. A.; Bujard, E.; Arnaud, M. In "Protein Crosslinking: Nutritional and Medical Consequences"; Friedman, M., Ed.; Plenum: London, 1977; p. 51–71.
114. Hofmann, K.; Stutz, E.; Spühler, G.; Yajima, H.; Schwartz, E. T. *J. Am. Chem. Soc.* **1960**, *82*, 3727.
115. Okuda, T.; Zahn, H. *Chem. Ber.* **1965**, 1164.
116. Struthers, B. J.; Dahlgren, R. R.; Hopkins, D. T.; Raymond, M. L. In "Soy Protein and Human Nutrition"; Wilcke, H. L.; Hopkins, D. T.; Waggle, D. H., Eds.; Academic: New York, 1979; p. 235.
117. Kim, S.; Benoiton, L.; Paik, W. K. *J. Biol. Chem.* **1964**, *239*, 3790.
118. Mauron, J. *J. Int. Vitaminol.* **1970**, *40*, 209–227.
119. Cuq, J. L.; Hurrell, R. F.; Finot, P. A., unpublished data.
120. Finot, P. A.; Hurrell, R. F.; Cuq, J. L., unpublished data.
121. Stewart, M.; Nicholls, C. H. *Aust. J. Chem.* **1974**, *27*, 205–208.
122. Dworschak, E.; Orsi, F. *Acta Alim.* **1977**, *6*, 59–71.
123. Cuq, J. L.; Aymard, C.; Dame-Cahagne, M.; Cheftel, J. C.; Hurrell, R. F.; Finot, P. A., presented at the 40th Annual IFT Meeting, New Orleans, June, 1980 (available on magnetic tape from IFT).
124. Sarwar, G.; Bowland, P. *Can. J. Anim. Sci.* **1976**, *56*, 433–437.
125. Osner, R. C.; Johnson, R. M. *J. Food Technol.* **1958**, *3*, 81–86.
126. Miller, F. L.; Carpenter, K. J.; Milner, C. K. *Br. J. Nutr.* **1965**, *19*, 547–565.
127. Jori, G.; Galliazzo, G.; Marzotto, A.; Scoffone, E. *Biochem. Biophys. Acta* **1968**, *154*, 1.
128. Holt, L. A.; Milligan, B.; Rivett, D. E.; Stewart, F. H. C. *Biochem. Biophys. Acta* **1977**, *499*, 131–138.
129. Hachimori, Y.; Horiniski, H.; Horihara, K.; Shibata, K. *Biochem. Biophys. Acta* **1964**, *93*, 346–360.
130. Steinhart, H. *Z. Tierphysiol. Tierernähr. Futtermittelkd.* **1978**, *41*, 48–56.

131. Cuq, J. L.; Aymard, C.; Besançon, P.; Cheftel, C. *Ann. Nutr. Alim.* **1978**, *32*, 346–372.
132. Hurrell, R. F.; Dame, M.; Finot, P. A., unpublished data.
133. Adrian, J.; Susbielle, H. *Ann. Nutr. Alim.* **1975**, *29*, 151–158.
134. Zimmerman, J. E.; Anderson, G. W.; Callahan, F. M. *J. Am. Chem. Soc.* **1964**, *86*, 1839.
135. Cole, M.; Fletcher, J. C.; Garkner, K. L.; Corfield, M. C. *Appl. Polym. Symp.* **1971**, *18*, 147.
136. Reyniers, J. P.; Pleasants, J. R., presented at the *40th Annual IFT Meeting, New Orleans, June, 1980;* Abstract No. 413.
137. Dame, M., Ph.D. Thesis, Université des Sciences et Techniques du Languedoc, Montpellier, France, 1980.
138. Anderson, G. H.; Ashley, D. V. M.; Jones, J. D. *J. Nutr.* **1976**, *106*, 1108–1114.
139. Friedman, M., Ed. "Nutritional Improvement of Food and Feed Proteins," *Adv. Exp. Med. Biol.* **1978**, *105*.

RECEIVED November 18, 1980.

# 4

# [14]C-Methyl-Labeled Milk Proteins for Studies of Protein Interactions and Proteolysis in Milk

B. O. ROWLEY, W. J. DONNELLY[1], D. B. LUND, and T. RICHARDSON

Department of Food Science, University of Wisconsin—Madison, Madison, WI 53706

*$^{14}$C-Methyl-β-lactoglobulin (M-β-L), $^{14}$C-methyl-κ-casein (M-κ-C), and $^{14}$C-methyl-β-casein (M-β-C) were prepared from purified proteins by reductive methylation of lysine residues using $^{14}$C-HCHO. Partition of added M-β-L or M-κ-C between the serum (supernatant) and micellar (particulate) phases of skim milk was measured after ultracentrifugal fractionation at 110,000 × gravity. Heat treatment caused association of most M-β-L with the particulate fraction; this association was inhibited partially when* N*-ethylmaleimide (NEM) was added before heating. Urea at 1 mg/mL, when added before heating, also reduced the fraction of M-β-L or M-κ-C recovered in casein micelles. Partial methylation did not reduce the susceptibility of β-casein to proteolysis by plasmin. Using $^{14}$C-methyl caseins as tracers provided valuable information on postsecretory degradation of caseins in normal and mastitic milks.*

Many methods used to study protein interactions and proteolysis in complex mixtures are often unsuitable for application to food systems, where severe heat treatments or enzymic modification may change the properties of proteins and their reactivity in an analytical procedure. Chromatographic, electrophoretic, and immunochemical methods are excellent tools for fractionating protein mixtures and measuring the concentrations of the individual components, but heat treatment or enzymic

[1] Current address: Principal Research Officer, The Agriculture Institute, Moorepark Research Centre, Fermoy, Co. Cork, Republic of Ireland.

0065-2393/82/0198-0125$05.75/0

hydrolysis may cause denaturation and aggregation or proteolysis, respectively, of proteins, thereby changing their size, charge, and rate of migration on electrophoresis, and modifying or even eliminating recognition by antibodies specific for the native proteins.

A simple procedure has been described for preparing $^{14}$C-methyl-labeled proteins (*1, 2*). Such derivatives and their fragments still may be detected and measured after the proteins have been denatured or hydrolyzed. In this chapter, we describe experiments that illustrate the use of $^{14}$C-methyl-labeled milk proteins to study protein interactions in heated skim milk and to follow enzymic hydrolysis of milk proteins in various milks.

### *Preparation of $^{14}$C-Methyl-Labeled Milk Proteins*

**Reductive Methylation.** In the procedure described by Means and Feeney (*3*), aqueous protein solutions were treated with small amounts of aliphatic aldehydes or ketones and sodium borohydride. At pH 9.0 and 0°C, this resulted in formation of a Schiff base between the protein amino groups and the carbonyl groups, followed by reduction to give the *N*-alkyl derivative. With formaldehyde as the reagent, *N,N*-dimethyl-lysine residues in the protein were the major products of methylation. Methylation caused little change in the physical properties of the proteins, although enzymes may be inactivated if essential lysine residues are modified. More recently sodium cyanoborohydride has been used as the reducing agent (*4, 5*) instead of sodium borohydride, thereby allowing reductive alklation of proteins at pH values ca. 7 without side reactions.

Using $^{14}$C-HCHO as the methylating agent gives the $^{14}$C-methyl-proteins. This procedure, when applied to milk proteins, compares favorably with the alternative approach of Mills (*6*). Following injection of cows with a mixture of $^{14}$C-amino acids, milk was obtained which contained $^{14}$C-labeled milk proteins (*6*). From this milk, randomly labeled milk proteins could be isolated. However, unless an application for $^{14}$C-labeled milk proteins specifically requires uniform labeling of all amino acid residues rather than labeling of methyl groups bound to the $\epsilon$-amino group of lysine residues, radiolabeling by reductive methylation using $^{14}$C-HCHO is preferable as a simpler and cheaper method for preparing $^{14}$C-labeled milk proteins with high specific activity.

**$^{14}$C-Methyl-Labeled $\kappa$-Casein.** The effect of reductive methylation on the properties of $\alpha_{s1}$-, $\beta$-, and $\kappa$-caseins was investigated in this laboratory (*7*). The purified proteins were treated to obtain either 20% or 60% modification of lysine residues. Electrophoretic mobilities at pH 8.5 or 6.5 of the modified proteins were indistinguishable from those of the native caseins.

The effect of protein methylation (20% modification) on the stability of casein micelles was evaluated. The stability of the $\alpha_s/\kappa$ micelle system towards $Ca^{2+}$ precipitation was decreased when either or both proteins were methylated (*see* Figure 1). Methylation also resulted in a slight reduction in the stability of the $\beta$–$\kappa$ casein micelle system (*see* Figure 2).

The temperature-sensitive precipitation of unmodified and methylated $\beta$-caseins in the presence of calcium was measured also (*see* Figure 3). Methylation caused an increase of up to 3°C in the precipitation temperature of calcium $\beta$-caseinate. Results from rennet clotting of an $\alpha_{s1}$-$\kappa$ casein micelle system indicated that replacing native $\alpha_{s1}$-casein with the reductively methylated protein had little influence on clotting time, while replacing $\kappa$-casein with its reductively methylated derivative re-

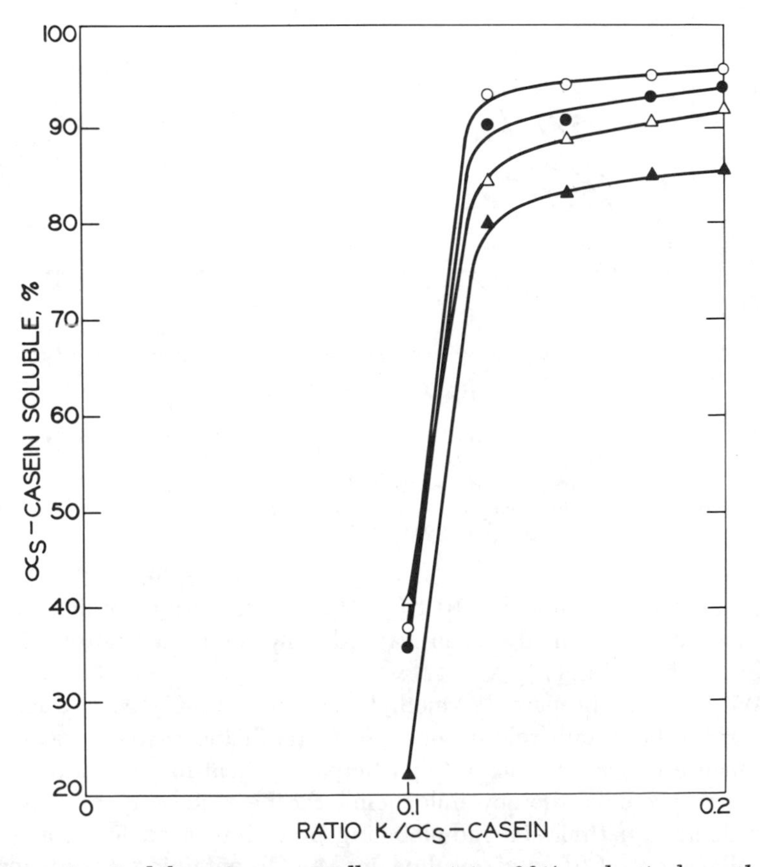


*Figure 1. Stability of $\alpha_{s1}$-$\kappa$ micelles containing 20% reductively methylated (RM) components towards precipitation by $Ca^{2+}$: ○, $\alpha_{s1}$-$\kappa$ micelles; △, RM$\alpha_{s1}$-$\kappa$ micelles; ●, $\alpha_{s1}$-RM$\kappa$ micelles; ▲, RM$\alpha_{s1}$-RM$\kappa$ micelles. The $\alpha_{s1}$-casein concentration was 7.4 mg/mL; pH 6.6; $Ca^{2+}$ concentration, 0.021M; 37°C (7).*

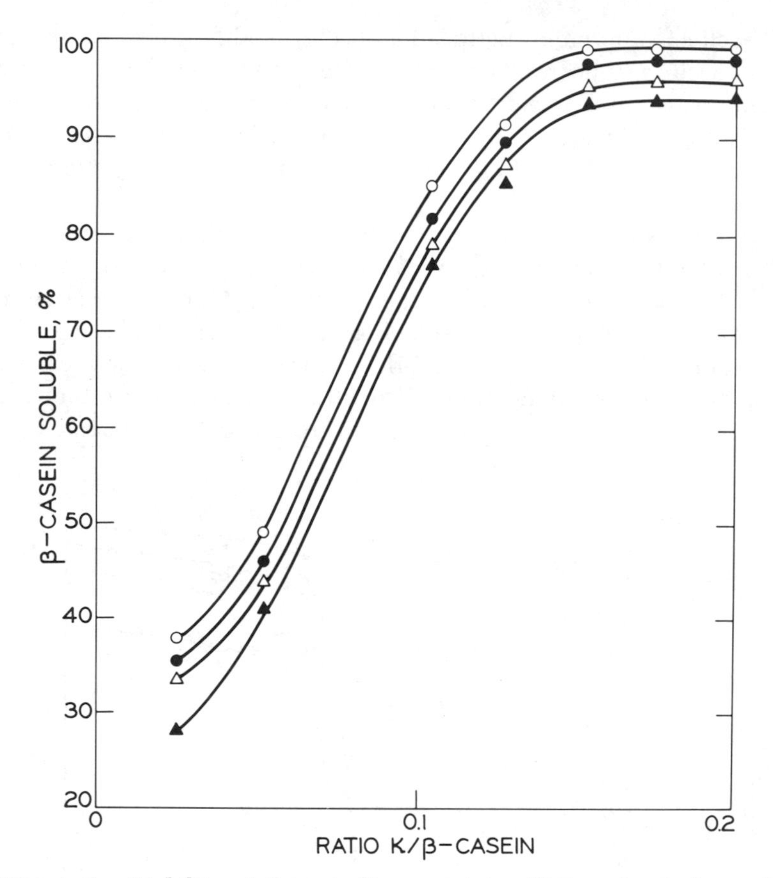


*Figure 2. Stability of β-κ micelles containing 20% reductively methylated (RM) components towards precipitation by $Ca^{2+}$: ○, β-κ micelles; △, RMβ-κ micelles; ●, β-RMκ micelles; ▲, RMβ-RMκ micelles. The β-casein concentration was 4.29 mg/mL; pH 6.6; $Ca^{2+}$ concentration, 0.022*M*; 30°C (7).*

duced the clotting time by 10 to 15%. These results indicated that extensive reductive methylation caused only slight alterations in the properties of caseins (*7*).

We chose to prepare $^{14}C$-methyl-κ-casein (M-κ-C) as a tracer because of the important role of κ-casein in stabilizing casein micelles (*8*) and because κ-casein is known to participate in heat-induced interactions with whey proteins, thereby influencing the heat stability of milk (*9*). The reductive methylation radiolabeling procedure used low concentrations of reagents (*10*) and resulted in M-κ-C containing approximately 1 μmol of $^{14}C$-methyl groups for every micromole of protein monomer (about 3 μCi/mg). When tracer M-κ-C was added to skim milk, and trichloroacetic acid was added to a concentration of 2%, about 1% of the radioactivity remained soluble. After clotting of the milk with excess rennet, 12 to 14% of the radioactivity remained soluble in 2% trichloroacetic acid. This is consistent with the known role of κ-casein, in which rennet action releases a fragment of κ-casein which remains soluble in relatively dilute trichloroacetic acid solutions (*11*). Thus, tracer M-κ-C behaves similarly to native κ-casein in milk when treated with rennet.

**$^{14}$C-Methyl-Labeled β-Lactoglobulin.** Because β-lactoglobulin is the whey protein present in highest concentration in milk and is the major source of milk free-sulfhydryl groups, it was chosen for study. Reductive methylation of β-lactoglobulin (containing both A and B variants) over a range of conditions causing modification of 5 to 80% of the amino groups resulted in unchanged free-sulfhydryl content and

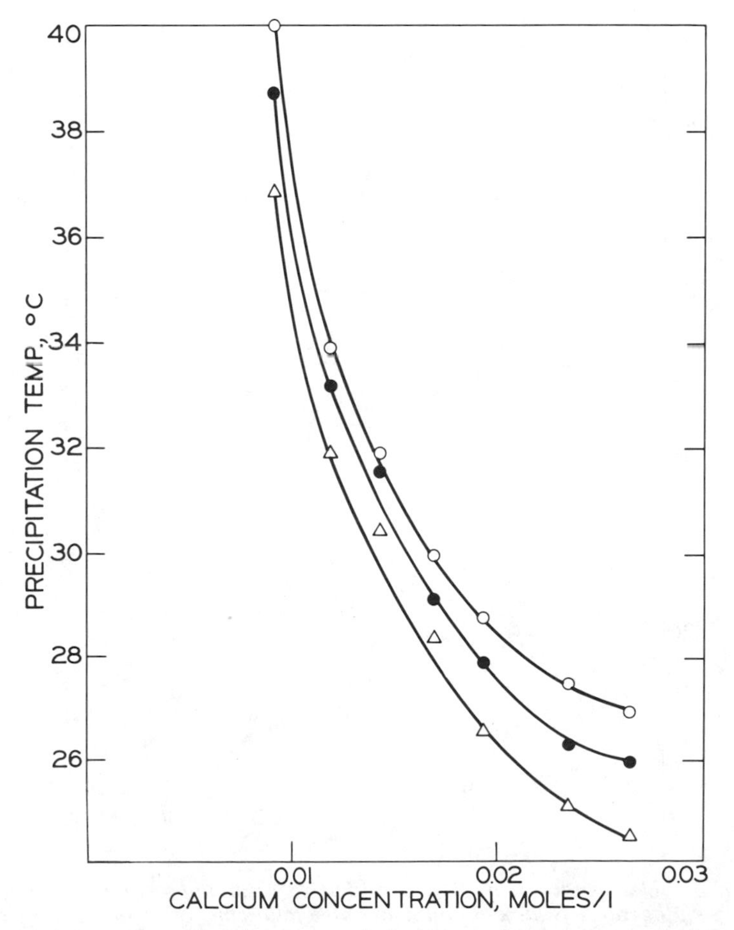


*Figure 3. Effect of reductive methylation on the temperature of precipitation of Ca-β-caseinate: △, β-casein; ●, 20% reductively methylated β-casein; ○, 60% reductively methylated β-casein; β-casein concentration, 8.69 mg/mL; pH 6.6 (7).*

electrophoretic mobility. There was a slight change in migration on isoelectric focusing (*10*). $^{14}$C-Methyl-β-lactoglobulin (M-β-L) was prepared, at low reagent concentration, to contain approximately 1 μmol of $^{14}$C-methyl groups for every micromole of protein monomer (about 3 μCi/mg). Tracer M-β-L was added to skim milk and heated for 30 min over a range of temperatures. The fraction of the radioactivity remaining in the undenatured fraction (i.e. in the filtrate at pH 4.6) was determined then (*10*).

Heat denaturation characteristics of M-β-L were compared with those of endogenous β-lactoglobulin (*see* Figure 4). Values for the latter

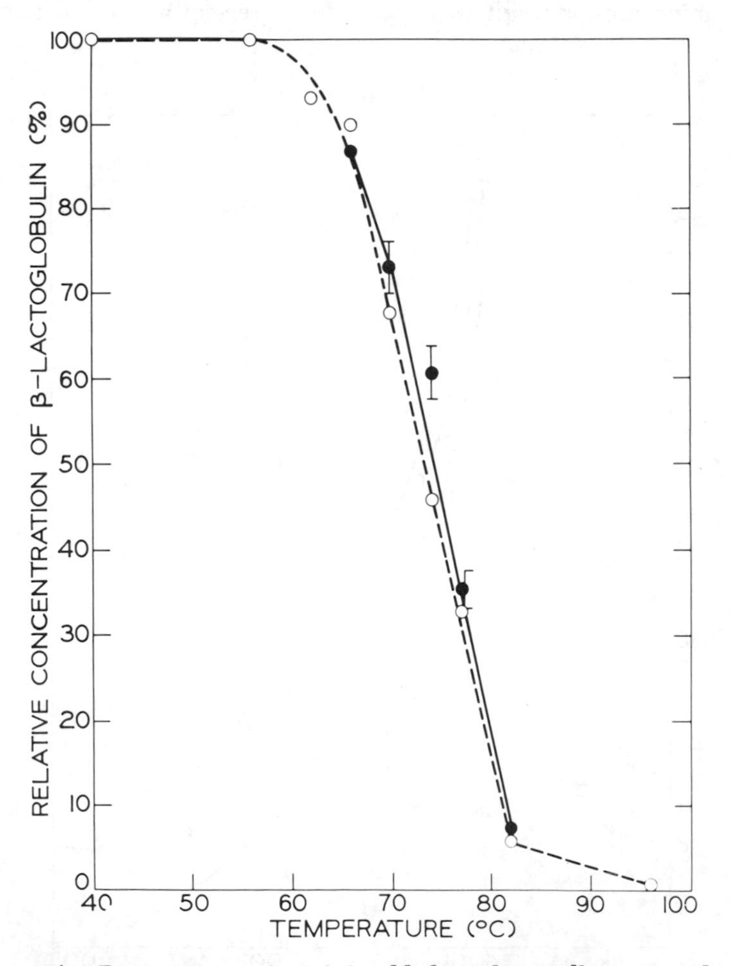


*Figure 4. Denaturation of M-β-L added to skim milk compared with that of endogenous β-lactoglobulin in milk following 30 min heating at various temperatures: ○, denaturation of endogenous β-lactoglobulin (12); ●, denaturation of M-β-L added to skim milk. A range for triplicates of more than 2% is indicated by a bar (10).*

are replotted from the data of Larson and Rolleri (*12*). The similarity of the two curves in Figure 4 indicates that methylation has little effect on heat denaturation of β-lactoglobulin.

**$^{14}$C-Methyl-β-Casein.** $^{14}$C-Labeled proteins prepared by reductive methylation have potential as substrates in the study of proteolytic enzymes. A serious limitation is that complete methylation of lysine residues results in inhibition of proteolysis by enzymes with trypsin-like specificity (*13*). It was interesting to determine whether this problem could be overcome by incomplete methylation which left unaltered most of the lysine residues in more or less random distribution throughout the protein. β-Casein was selected as a suitable protein for this study since it is cleaved by trypsin-like enzymes to well characterized fragments, the γ-caseins, in addition to less well characterized fragments, the proteose peptones. We anticipated that this type of study could provide a basis for a general investigation of milk protein transformation by the native milk proteinase which has a specificity similar to trypsin (*14*).

β-Casein was prepared from whole casein isolated by isoelectric precipitation from the mixed milk of Friesian cows. This casein was fractionated on DEAE–cellulose using the method of Davies and Law (*15*). The β-casein obtained from the initial column chromatography was rechromatographed under identical conditions to yield a purified β-casein consisting predominantly of genetic Variant A with a small amount of Variant B. The γ-casein standards for electrophoresis were prepared from chromatography of individual cow milk types β casein A and AB using the method of Groves and Gordon (*16*).

The purified β-casein was methylated reductively to yield a product in which 10 to 12% of the lysine residues were methylated, with a specific activity of 4.42 μCi/mg. This was diluted with varying amounts of cold β-casein before use.

### *Heat Treatment of Milk Containing $^{14}$C-Methyl-Labeled κ-Casein or β-Lactoglobulin as Tracers*

**Protein Composition of Milk.** Skim milk is a colloidal suspension of extreme complexity. The particulate phase, the casein micelles, consists primarily of a mixture of $\alpha_{s1}$, $\alpha_{s2}$, β, and κ-caseins combined with calcium ions and an amorphous calcium–phosphate–citrate complex. The soluble phase contains lactose, a fraction of the caseins and calcium, and, in raw milk, the whey proteins, which are predominantly β-lactoglobulin and α-lactalbumin. When milk is centrifuged at high speed (in our experiments, 30 min at 110,000 × gravity), the casein micelles sediment. This permits one to separate the two physical phases of skim milk and to measure changes in composition of the phases resulting from various treatments. In cold, raw milk, nearly one-third of all protein is found in the soluble phase. This work presents experiments that illustrate how protein interactions that result in shifts of specific proteins between the particulate and soluble phases in skim milk may be examined using $^{14}C$-labeled milk proteins as tracers.

**Effect of Forewarming and Concentration on $^{14}C$-Methyl-β-Lactoglobulin Distribution in Milk.** The stability of milk that is to be concentrated before sterilization is improved by a preliminary heat treatment (forewarming), but concentrated milk may be destabilized by the same heat treatment. An interaction of serum proteins with κ-casein may be involved; more detail is available in a recent review on the heat stability of milk (*17*).

To study the effect of forewarming, concentration, and redilution on the distribution between milk phases of β-lactoglobulin, tracer M-β-L was added to raw milk. Before any heat treatment, at least 95% of the C-14 label was recovered in the supernatant fraction after ultracentrifugation. Table I summarizes the findings after this milk was submitted to a forewarming treatment (for these experiments, 20 min at 95°C), with or without concentration at 40°C to one-half the original volume. After a 24-h storage period at 5°C, portions of concentrate also were diluted with cold water and analyzed by centrifugation at 5°C. After forewarming, about two-thirds of the tracer became associated with the particulate fraction. Virtually all of the fraction remaining in the supernatant was denatured in that it coprecipitated with the casein at pH 4.6. In forewarmed milk concentrated to one-half volume, a larger fraction of M-β-L and of protein as measured by the Lowry protein assay (*18*) was found in the supernatant. The original distribution was restored by dilution. Concentration may result in dissociation of a part of the β-lactoglobulin and total protein (i.e. caseins) from the micelles, with reformation of particles upon dilution. Such reversible changes in protein distri-

**Table I. Ultracentrifugal Fractionation of M-β-L in Skim Milk (The Effect of Forewarming, Concentration, and Redilution)**

| | | *Recovery of C-14* | | *Protein (Lowry)* | |
|---|---|---|---|---|---|
| *Treatment* | *Volume* | *Supernatant Fluid* | *Pellet* | *Supernatant Fluid* | *Pellet* |
| | | *Percentages* | | *Milligrams/Milliliter* | |
| Forewarmed milk | 1X | 30.6 ± 0.6[a] | 67.2 ± 1.2 | 7.0 | 26 |
| Concentrated milk | 0.5X | 43.6 | 51.3 | 19.0 | 44 |
| Rediluted milk | 1X | 31.8 ± 0.3 | 67.8 ± 1.1 | 7.0 | 27 |
| Rediluted milk | 2X | 31.5 | 67.2 | — | — |

[a] When mean ± S.D. is given, $n = 4$. Otherwise, data are the mean for $n = 2$.

**Table II. Ultracentrifugal Fractionation of M-β-L in Skim Milk (Effect of NEM Added Before, During, and After Sterilization)**

| Treatment | Recovery of C-14: Supernatant Fluid | Recovery of C-14: Pellet |
|---|---|---|
| | *Percentages* | |
| No NEM added | 54.4 ± 2.8[a] | 43.0 ± 2.9 |
| 3m*M* NEM added | | |
| before heating | 70.2 ± 2.4[b] | 27.0 ± 1.8[b] |
| during heating | 51.6 ± 0.7 | 45.1 ± 1.8 |
| after heating | 55.5 ± 2.9 | 42.1 ± 1.8 |

[a] Mean ± S.D., $n = 3$.
[b] Differs from control value (*t*-test), $P < 0.01$ (*33*).

bution may be related to changes in pH and concentration of ionic calcium. Alternatively, concentration may have increased the viscosity of the skim milk thereby retarding sedimentation of particulate proteins.

**Effect of *N*-Ethylmaleimide on $^{14}$C-Methyl-β-Lactoglobulin Distribution in Heated Milk.** The role of sulfhydryl groups in the formation of a β-lactoglobulin–casein complex in milk and on heat stability has been explored in various ways. The older work was reviewed thoroughly in 1969 (*9*), and a more recent review also has referred to this aspect of milk stability (*17*). Many of these studies have used sulfhydryl-blocking reagents, especially *N* ethylmalcimide (NEM), as probes for sulfhydryl group mediation in these interactions. As noted by others (*19*), NEM exhibits several disadvantages as a sulfhydryl reagent: under certain conditions it may be nonspecific for cysteine, chemically unstable, and too bulky to react with hindered groups. Nevertheless, it has been used widely. The evidence suggests that complex formation between isolated β-lactoglobulin and κ-casein may be prevented by NEM. Thiol-blocking reagents retard the heat denaturation of β-lactoglobulin in milk. Whey proteins apparently confer stability during forewarming towards more intense heat treatment of the concentrate, and the effect is inhibited by NEM.

The experiments to be described tested the effect of *N*-ethylmaleimide on the distribution between soluble and micellar phases of milk of tracer M-β-L. NEM was added to unconcentrated skim milk containing tracer M-β-L either 20 min before sterilization for 14 min at 121°C; during sterilization, after 5 min heating; or after sterilization and 5 min cooling (*see* Table II). After storing the samples for 24 h at room temperature, they were analyzed after centrifugation at 25°C. Only addition of NEM before heating changed the distribution by significantly reducing the fraction of M-β-L which became associated with the particulate phase. When milk is heated, the concentration of reactive sulfhydryl groups increases greatly, then declines during storage, with the major decrease being observed during the first day at room temperature (*20*). If this decline were due to formation of disulfide linkages between proteins, one would expect to see a change in the degree of association of β-lactoglobulin with casein micelles when NEM is added immediately after heating, but no difference was observed.

Similar experiments were used to examine the forewarming process. Milk containing tracer M-β-L was forewarmed, then NEM was added to part, and the unconcentrated skim milks were sterilized and analyzed after 1-d storage at room temperature. The data in Table III show that when skim milk was first forewarmed, the presence of NEM during the second, more severe heat treatment did not change the distribution of M-β-L. These observations may be of relevance to experiments on evaporated milk reported by Trautman and Swanson (*21*). When *p*-chloromercuribenzoate, which also blocks sulfhydryl groups, was added to milk after forewarming, it had no effect on the stability of the evaporated milk product. But, if the reagent were added before forewarming, the samples were unstable to sterilizing temperatures. We added NEM to our system before forewarming and again before sterilization (*see* Table IV). Even when NEM was present during forewarming, further addition before sterilization resulted in an increase in the fraction of M-β-L remaining in the soluble phase. NEM is chemically unstable, and it is likely that little or none of it survives the forewarming step. After NEM has alkylated a sulfhydryl group, the resulting derivative is reported to be stable to vigorous treatments such as acid hydrolysis. Alkylated cysteine is recovered as S-(2-succinyl)cysteine (*22*). But it also has been reported that alkylation by NEM may be gradually reversible under mild conditions (*23*). Based on these limited data, we do not know whether this

**Table III. Ultracentrifugal Fractionation of M-β-L in Skim Milk (Effect of NEM Added Before Sterilization to Previously Forewarmed Milk)**

| *Treatment* | *Recovery of C-14* — *Supernatant Fluid* | *Pellet* |
|---|---|---|
| | *Percentages* | |
| Forewarmed milk without NEM | 38.1 ± 1.0[a] | 58.0 ± 0.8 |
| Sterilized milk | | |
| without NEM | 44.1 ± 1.4 | 51.0 ± 0.2 |
| with 1m*M* NEM | 44.5 ± 1.6 | 49.7 ± 1.2 |

[a] Mean ± S.D., $n = 4$.

**Table IV. Ultracentrifugal Fractionation of M-β-L in Skim Milk (Effect of NEM Added Before Forewarming and Again Before Sterilization)**

| Treatment | *Recovery of C-14* | | *Protein (Lowry)* | |
|---|---|---|---|---|
| | *Supernatant Fluid* | *Pellet* | *Supernatant Fluid* | *Pellet* |
| | *Percentages* | | *Milligrams/Milliliter* | |
| Milk forewarmed with 1m*M* NEM | 39.3 ± 0.8[a] | 56.6 ± 0.7 | 7.2 ± 0.2 | 28.6 ± 0.4 |
| Sterilized milk | | | | |
| without additional NEM[b] | 37.6 ± 5.3 | 58.0 ± 5.4 | 9.5 ± 0.6 | 26.5 ± 0.6 |
| with 1m*M* NEM[b] | 47.4 ± 5.4 | 48.3 ± 5.6 | 10.7 ± 0.5 | 25.3 ± 0.7 |

[a] Mean ± S.D., $n = 4$.
[b] Mean values for these two groups differed ($t$-test), $P < 0.05$ (*33*).

sensitivity to addition of more NEM before the second heat treatment may be due to regeneration of free sulfhydryls by reversal of the initial alkylation, or to alkylation of sulfhydryls generated during forewarming after the first aliquot of reagent has been destroyed by heat.

Current thinking on the mechanism of the heat-induced interaction of β-lactoglobulin with micellar κ-casein favors a thiol–disulfide exchange reaction since β-lactoglobulin and κ-casein contain cysteine residues. As mentioned previously, this interaction is blocked by thiol reagents such as NEM and *p*-chloromercuribenzoate (*9*). However, it is unclear whether the thiol-blocking reagents prevent a thiol-mediated aggregation of β-lactoglobulin prior to a thiol-independent interaction with κ-casein or whether these reagents inhibit a thiol–disulfide exchange reaction between the two proteins or both. Much evidence suggests that a heat-induced thiol-dependent aggregation of β-lactoglobulin precedes an interaction of the aggregate with κ-casein (*9*).

In Figure 5, the data in Tables II, III, and IV are summarized for comparative purposes in the light of the foregoing mechanism. These data indicate that when raw skim milk containing radiomethyl-β-lactoglobulin is forewarmed in the presence or absence of NEM approximately the same quantity of radioactivity is sedimented. This would be contrary to accepted thinking that the thiol-blocking agent should inhibit an interaction. However, if the heat-induced aggregation of β-lactoglobin is not mediated by thiol groups, then the β-lactoglobulin aggregates may merely cosediment with micelles upon ultracentrifugation. Perhaps the judicious and selective use of solubilizing, reducing, and alkylating reagents while using radiolabeled proteins could distinguish between simple denaturation of β-lactoglobulin and its interaction with κ-casein. An

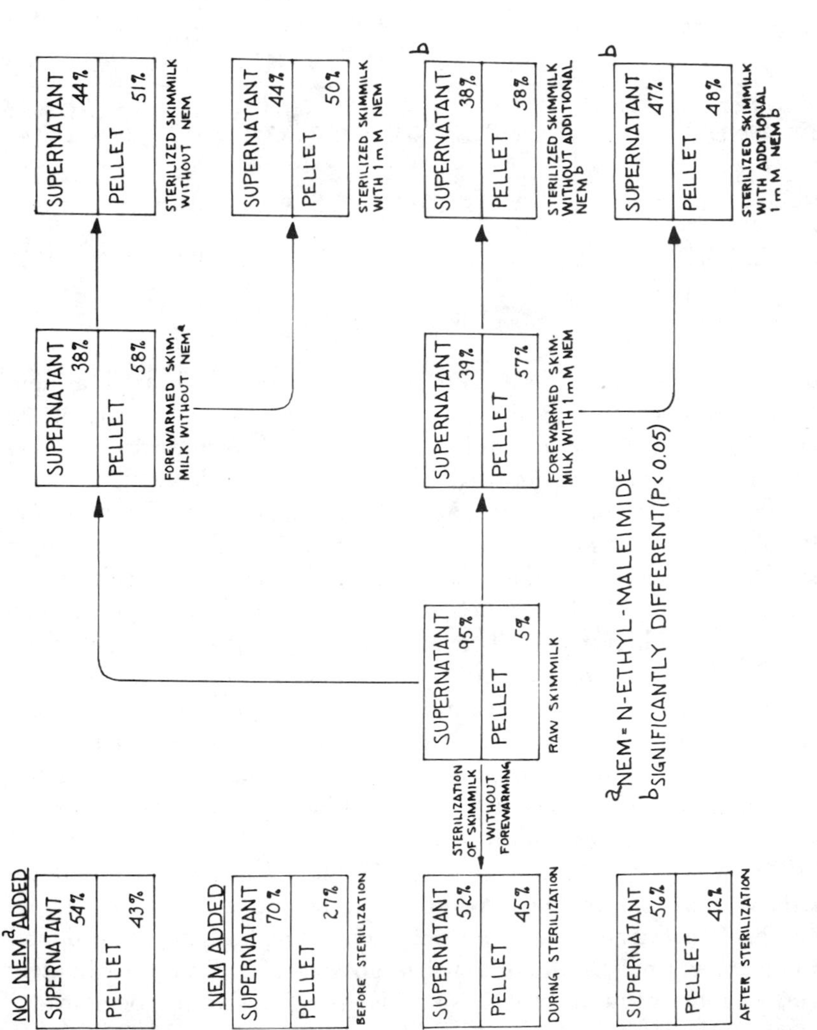

*Figure 5. Effect of various heat treatments on the distribution of radioactivity in milk with added M-β-L. This is a summary of Tables II, III, and IV.*

extreme interpretation of the forewarming data, on the other hand, might lead one to conclude that in skim milk an interaction between β-lactoglobulin and κ-casein would not necessarily involve thiol groups.

The sterilization of unconcentrated skim milk without prior forewarming yields more radioactivity in the supernatant fluid (*see* Figure 5). This may indicate less interaction between β-lactoglobulin and κ-casein as a result of direct sterilization thus pointing out the need to forewarm milk to maximize the interaction. On the other hand, direct sterilization may result in thermally induced fragmentation of proteins not protected by a previous interaction thereby yielding more radioactivity in the supernatant fluid. Vigorous thermal treatments such as those used in the present study lead to protein destruction (*24*). It is evident, however, that NEM added before direct sterilization results in 70% of the radioactivity in the supernatant fraction indicating that thiol-mediated reacions are important in this case.

Another apparent anomaly in the data stems from inspecting the radioactive distributions resulting from the forewarming treatments given in Tables I and III. Although the precision within each experiment is acceptable, the difference in radioactive distribution between experiments is rather large (30.6–38.1%). The two experiments were carried out approximately 8 weeks apart. This points out the variation possible due to season of the year, handling of the milk prior to acquisition, etc. in trying to quantify milk protein interactions.

It is obvious from the difficulties in interpreting these preliminary data in the light of current thinking that much work remains to be done to sort out the various thermally induced interactions during the heat treatments of milk. However, using radiolabeled milk proteins can play an important role in studying and defining these interactions and holds out the promise for quantifying them.

It should be emphasized that the foregoing experiments were not designed to study the mechanisms of heat-induced protein interactions in milk. They merely serve to illustrate the utility of radiolabeled milk proteins in studying this very complex phenomenon. The radiolabeling of other milk proteins along with double-labeling experiments should prove very instructive in future research on the heat stability of milk.

**Effect of Urea on the Distribution of $^{14}$C-Methyl-β-Lactoglobulin or $^{14}$C-Methyl-κ-Casein in Heated Milk.** Muir and Sweetsur (*25*) exploited the interesting observations by Robertson and Dixon (*26*) that the heat stability of milk proteins is related to the urea level of the milk, and that adding relatively small amounts of urea increased the heat stability of milk (*27*). To examine the urea effect further, we added urea to milk containing either M-β-L or M-κ-C and measured the ultracentrifugal distribution of C-14 after heat treatment. The experiments using

M-κ-C could be complicated by the fact that, in milk, κ-casein is primarily a micellar protein, while the tracer is soluble. However, if equilibration between κ-casein and M-κ-C occurs prior to the analysis, then the distribution of M-κ-C will be indicative of the whole. On adding tracer M-κ-C to cold, raw milk, a significant fraction (25 to 30%) became associated with the particulate fraction. After gentle warming (to 37°C) and cooling (back to 5°C), about 60% of the C-14 was particulate, while forewarming rendered more than 80% of the C-14 particulate. There appears to be an exchange between soluble and micellar κ-casein in milk, but this cannot be interpreted to mean that tracer M-κ-C is in equilibrium with that κ-casein in milk which may be unavailable for exchange.

First, raw milk was forewarmed with tracer M-β-L or M-κ-C in parallel experiments. Then 1 mg/mL of urea was added and the milks were sterilized. The data in Table V indicate that urea caused a small increase in pH, perhaps due to partial hydrolysis of the urea, with ammonia formation. It also caused an increase of about 7% in the fraction of each tracer recovered in the supernatant fluid and shifted about 1.5 mg/mL of protein from the pellet to the supernatant fractions. In subsequent experiments, urea was present during both the forewarming and sterilization (*see* Table VI). Urea at 1 mg/mL caused a slightly larger (10%) shift of M-β-L to the supernatant fluid, but the changes in distribution

**Table V. Ultracentrifugal Fractionation of M-β-L or M-κ-C in Skim Milk (Effect of Urea Added Before Sterilization of Forewarmed Milk)**

| | | *Recovery of C-14* | | *Protein (Lowry)* | |
|---|---|---|---|---|---|
| *Treatment* | *pH* | *Supernatant Fluid* | *Pellet* | *Supernatant Fluid* | *Pellet* |
| | | *Percentages* | | *Milligrams/Liter* | |
| M-β-L | | | | | |
| no urea added | 6.46 | 45.9 ± 1.2[a] | 50.2 ± 0.8 | 10.0 ± 0.3 | 25.3 ± 0.4 |
| 1 mg/mL urea added | 6.53 | 53.0 ± 2.4[b] | 42.4 ± 1.3[c] | 11.4 ± 0.5[b] | 23.7 ± 0.4[c] |
| M-κ-C | | | | | |
| no urea added | 6.47 | 29.5 ± 1.0 | 65.8 ± 1.1 | 10.1 ± 0.2 | 25.3 ± 0.0 |
| 1 mg/mL urea added | 6.53 | 36.2 ± 1.6[c] | 58.8 ± 1.5[c] | 11.7 ± 0.6[c] | 23.1 ± 0.5[c] |

[a] Mean ± S.D., $n = 3$.
[b] Differs from the control (no urea) value ($t$-test), $0.01 < P < 0.05$ (*33*).
[c] Differs from the control (no urea) value ($t$-test), $P < 0.01$ (*33*).

**Table VI. Ultracentrifugal Fractionation of M-β-L or M-κ-C in Skim Milk (Effect of Urea Added Before Forewarming and Sterilization)**

| Treatment | Recovery of C-14: Supernatant Fluid | Recovery of C-14: Pellet | Protein (Lowry): Supernatant Fluid |
|---|---|---|---|
| | *Percentages* | | *Milligrams/Milliliter* |
| M-β-L | | | |
| no urea added | 40.3 ± 1.7[a] | 55.8 ± 1.1 | 9.7 ± 0.4 |
| 0.25 mg/mL urea added | 42.0 ± 2.2 | 53.8 ± 0.9[b] | 10.0 ± 0.5 |
| 1.0 mg/mL urea added | 50.5 ± 2.4[c] | 44.9 ± 1.6[c] | 11.7 ± 0.5[c] |
| M-κ-L | | | |
| no urea added | 28.1 ± 0.8 | 67.7 ± 2.1 | 9.6 ± 0.2 |
| 0.25 mg/mL urea added | 29.3 ± 0.7 | 67.2 ± 1.0 | 9.9 ± 0.2 |
| 1.0 mg/mL urea added | 35.0 ± 0.5[c] | 61.1 ± 0.5[c] | 11.4 ± 0.1[c] |

[a] Mean ± S.D., $n = 4$.
[b] Differs from the control (no urea) value ($t$-test), $0.01 < P < 0.05$.
[c] Differs from the control (no urea) value ($t$-test), $P < 0.01$.

of M-κ-C and of protein (Lowry) were similar to those of Table V. A lower concentration (0.25 mg/mL) of urea caused a similar but statistically insignificant shift in the same direction.

### $^{14}$C-Methylated β-Casein as a Substrate for Plasmin and Its Application to the Study of Milk Protein Transformations

Details of the experimental procedures relating to the use of $^{14}$C-methylated β-casein (M-β-C) as a substrate for plasmin can be found in the article by Donnelly et al. (*28*).

**Plasmin Hydrolysis of β-Casein.** Studies on the susceptibility of partially methylated β-casein to cleavage by trypsin-like enzymes were carried out using the enzyme porcine plasmin. In preliminary investigations, we confirmed that the γ-caseins produced by plasmin hydrolysis of native β-casein were identical with those occurring naturally (*28*). These are designated $\gamma_1$-, $\gamma_2$-, and $\gamma_3$-casein according to the nomenclature recommendations of Whitney et al. (*29*) and correspond to residues 29–109, 106–209, and 198–209 of β-casein. Presumably the proteose peptone products of plasmin hydrolysis are identical with their natural counterparts, but the latter were not available for comparison.

**Plasmin Hydrolysis of $^{14}$C-Methyl-β-Casein.** On the basis of a study by Ottesen and Svensson (*2*), which described the relative amounts of mono- and dimethyllysinyl derivatives produced by reductive methylation, we calculated that M-β-C of specific activity 4.42 μCi/mg had 10–12% of its lysine residues in modified form. The γ-caseins produced by plasmin hydrolysis of the modified protein were identical with those from the native protein. Thus, γ-caseins from both sources had identical electrophoretic patterns. Furthermore, when partially hydrolyzed M-β-C was cochromatographed with whole casein on DEAE–cellulose, the elution behavior of the principal fragments was entirely characteristic of native γ-caseins (*see* Figure 6). The profile in Figure 6 also sheds light on the nature of minor protein peaks that appear in the earliest eluted fractions of whole caseins. It is clear that these may be identified, in part, as part of the γ-casein complex. It appears from Figure 6 that partial methylation has little effect on the chromatographic properties of a protein since the radioactive peaks and the corresponding protein peaks are virtually coincidental. This is consistent with previous reports that reductive methylation of proteins has little effect on their physicochemical

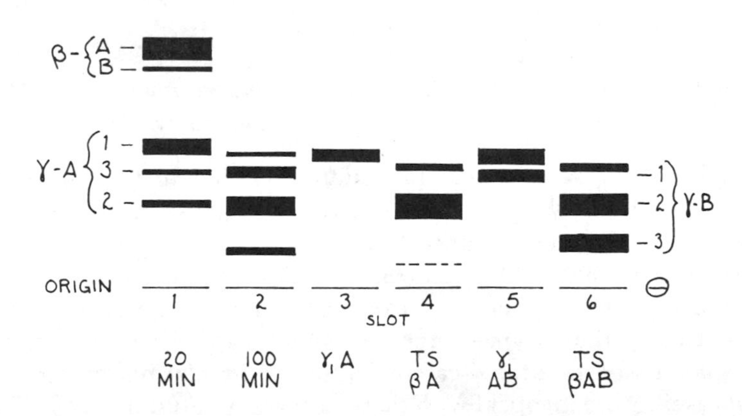


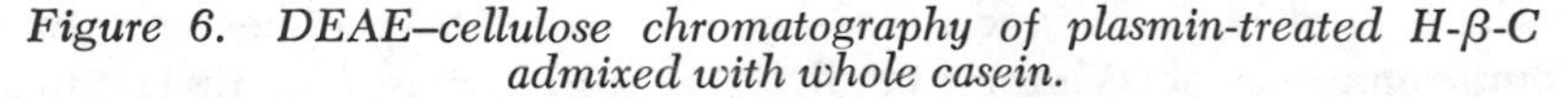


*Figure 6. DEAE–cellulose chromatography of plasmin-treated H-β-C admixed with whole casein.*

*β-Casein (13 mg) containing 0.036 μCi M-β-C was subjected to plasmin hydrolysis for 45 min and the reaction mixture was dissolved in 7 mL column buffer (5mM Tris—3mM NaCl—6M urea, pH 8.55) together with 100 mg whole casein that had been alkylated with iodoacetamide. The sample solution was applied to a column (1.6 × 50 cm) of DEAE–cellulose equilibrated with column buffer. Elution was with a NaCl gradient (3.0–155mM) in column buffer (gradient volume, 1.0 L); 5.0 mL fractions were collected. Under the conditions used as $\alpha_s$-caseins remained adsorbed to the column; κ-caseins were eluted in Fractions 35–56; $A_{280}$ measurement (——); radioactivity (– – –) (28).*

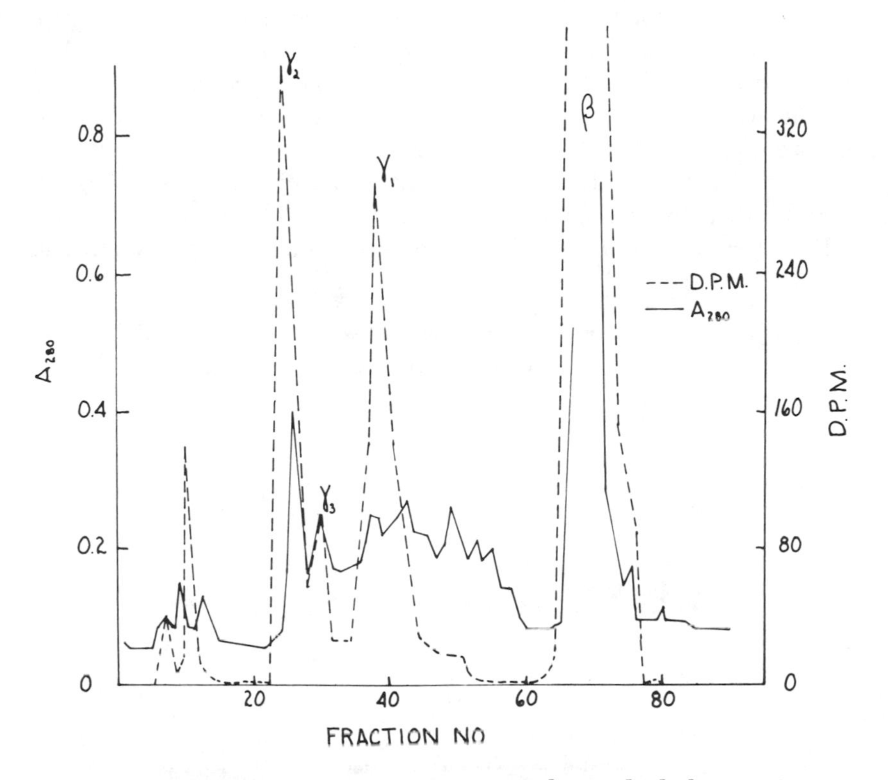


*Figure 7. PAGE of reaction mixture after plasmin hydrolysis of β-casein for 0, 20, 30, 40, and 70 min (Slots 1–5) and Fractions 1 and 2 from hydroxyapatite chromatography of hydrolysis products after plasmin treatment of β-casein for 70 min (Slots 6 and 7). Fraction 2 contains a further phosphopeptide that is not visible in Slot 7 but appears on disc gels in the expected position for proteose peptone component 8F (cf. Ref. 32) (28).*

properties (*7, 10, 22*). No radioactive peaks corresponding to proteose peptones can be seen in Figure 6 and it is probable that they are not resolved from β-casein.

**Rate of Plasmin Hydrolysis.** It may be concluded from the above electrophoretic and chromatographic data that partial methylation of a protein does not interfere with the site specificity of plasmin hydrolysis or result in the production of artifactual hydrolytic products. It remained to show what effect methylation had on the rate of hydrolysis by plasmin. M-β-C was diluted approximately 300-fold with unlabeled protein and subjected to plasmin hydrolysis. The reaction mixture was sampled at various intervals up to 70 min, by which time most of the β-casein was transformed (*see* Figure 7, Slots 1–5). The extent of transfer of radioactivity to the reaction products was examined to determine the suitability of the radiolabeled protein as a substrate for plasmin. A partial separation of $\beta$-casein from products was achieved by stepwise chromatography on a column of hydroxyapatite (*see* Figure 8). The fraction unadsorbed by hydroxyapatite (Fraction 1) contained $\gamma_2$- and $\gamma_3$-caseins whereas the remaining products, together with residual $\beta$-casein, were eluted with 300m*M* phosphate buffer (Fraction 2, Figure 8, Slots 6 and 7). Protein in Fraction 1 was quantified by $A_{280}$ measurement and radioactivity in Fractions 1 and 2 by liquid scintillation counting. Data (*see* Figure 9) indicate Fraction 1 radioactivity and protein

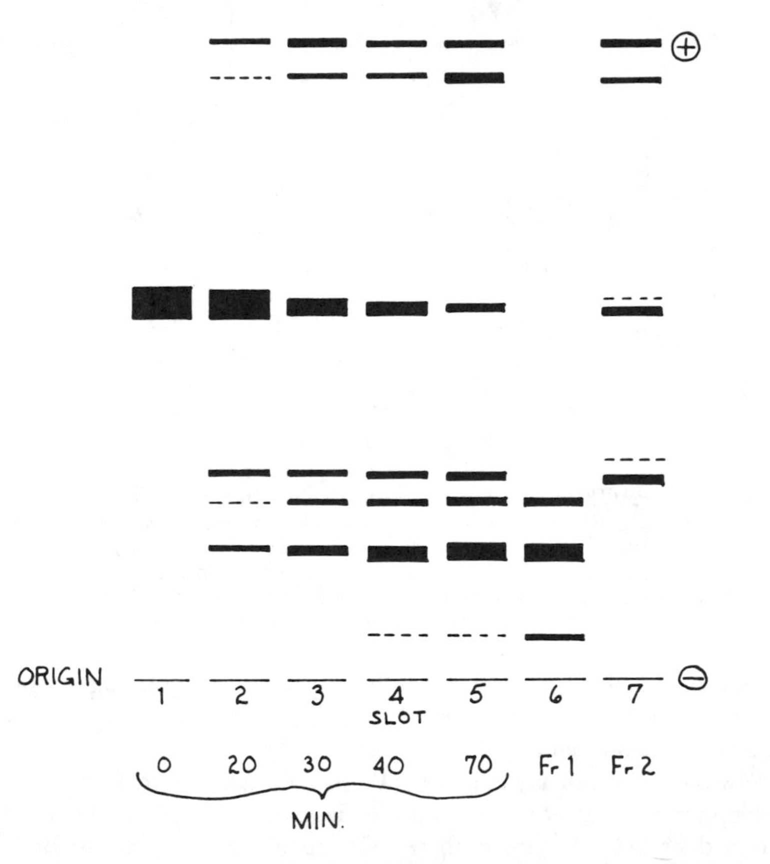


*Figure 8. Hydroxyapatite chromatography of plasmin-treated $\beta$-casein. The $\beta$-casein (110 mg) containing 1.62 $\mu$Ci M-$\beta$-C was incubated with plasmin and one-fifth of the reaction mixture was withdrawn at t = 0, 20, 30, 40 and 70 min. Samples were dissolved in 2.0-mL column buffer (10mM $Na_2HPO_4$–$NaH_2PO_4$, pH 6.8, containing 6M urea) and a known amount (90–100% of each) was applied to columns of hydroxyapatite (1.6 × 10 cm) equilibrated with the same buffer. Columns were eluted at a flow rate of 20 mL/h with column buffer (40 mL) followed by 300 mM $Na_2HPO_4$–$NaH_2PO_4$, pH 6.8, containing 6M urea (110 mL); 5.0 mL fractions were collected. The profile shown is for a reaction time of 40 min (28).*

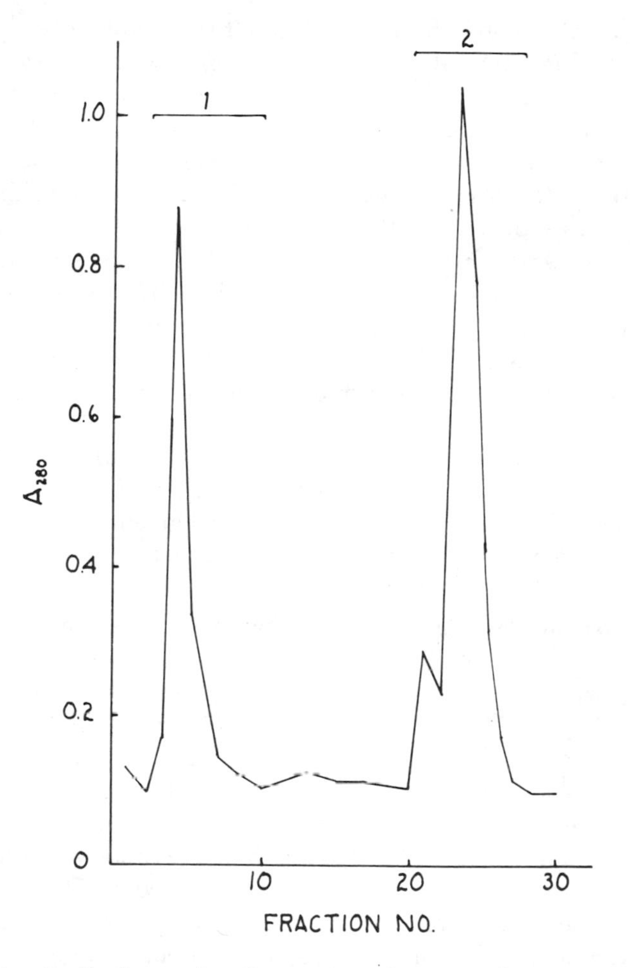


*Figure 9. Radiochemical and protein analysis of column fractions after hydroxyapatite column chromatography of plasmin-treated β-casein for various reaction times: ●, total activity or specific activity, Fraction 1; ■, total activity, Fraction 2; □, specific activity, residual β-casein; ○, total protein, Fraction 1. Values for total protein and total activity have been adjusted to correspond to the same amount of total protein (21.92 mg) applied to each column (28).*

increased linearly throughout the reaction period with a corresponding decrease in the radioactivity of Fraction 2. The specific activities of $\gamma_2 + \gamma_3$-caseins and of residual β-caseins were approximately constant throughout the period of hydrolysis (*see* Figure 9). Within the limits of experimental error this shows that no preferential hydrolysis of native β-casein had occurred, i.e. that partial methylation had not reduced the substrate susceptibility towards plasmin. This result extends the versatility of $^{14}$C-methyl proteins as substrates for hydrolytic enzymes and overcomes the limitation of resistance to trypsin-like enzymes previously reported for fully methylated proteins (*13*). The usefulness of the partially methylated C-14 derivatives is enhanced by applying a modified reductive methylation procedure that ensures high product-specific activities compared with those obtained previously (*13*) although the level of substitution was reduced greatly. The average specific activity of $\gamma_2 + \gamma_3$-caseins was 39,200 dpm/mg compared with 35,200 dpm/mg for the parent $\beta$-casein (*see* Figure 9). This is higher than expected from the lysine content of the $\gamma_2 + \gamma_3$-caseins (approximately $3.2 \times 10^{-4}$ mol/g compared with $4.6 \times 10^{-4}$ mol/g for $\beta$-casein) and indicates that reductive methylation of $\beta$-casein occurred in a manner that was selective for lysine residues in the $\gamma$-casein portion of the molecule. This behavior has been observed previously for other proteins (*30*). Radioactivity of protein fragments per se therefore is not sufficient to quantify proteolytic activity unless the specific activity of the fragments is known. This limitation does not apply in studies using $^{14}$C-methyl protein substrates where relative activities only are required.

**Proteolysis in Mastitic Milk.** An example of the use of $^{14}$C-methyl protein substrates for qualitative analysis of proteolysis in complex systems is the study of casein degradation in stored milk from a subclinically infected quarter of a Friesian cow (*31*). $^{14}$C-Methyl caseins ($\alpha_{s1}$, $\alpha_{s2}$, and $\beta$) were incubated separately in portions of the same milk at 37°C for 8 h. The casein then was isolated and fractionated on DEAE–cellulose. Figure 10 shows the protein and radiochemical profiles of casein from milk in which M-$\beta$-C was incubated. Extensive fragmentation of tracer occurred giving fragments that contained 30.7% of the total radioactivity recovered from the column. The elution behavior of the principal fragments is consistent with their identification as $\gamma$-caseins (cf Figure 6), and, in agreement with this, inclusion of soybean trypsin inhibitor suppressed greater than 90% of the fragmentation. A similar type of breakdown of M-$\beta$-C occurred in a healthy control milk from the same animal but in that case fragments accounted for only 2.8% of the recovered activity. Fragmentation patterns for $\alpha_{s1}$- and $\alpha_{s2}$-caseins were obtained in a similar manner. In the case of $\alpha_{s1}$-casein transformation, one of the principal fragments eluted as a well-defined peak centered on Fraction 43 (indicated by an arrow in Figure 10). This is the location of uncharacterized minor components of casein first described by Davies and Law (*15*). The tracer studies suggest that these components may be identified at least partly as proteolytic fragments of $\alpha_{s1}$-casein. Overall the results permitted the conclusion that mastitic infection is associated with a marked increase in milk of a plasmin-like enzyme activity and that this is responsible for most of the changes in casein composition observed in mastitic milk.

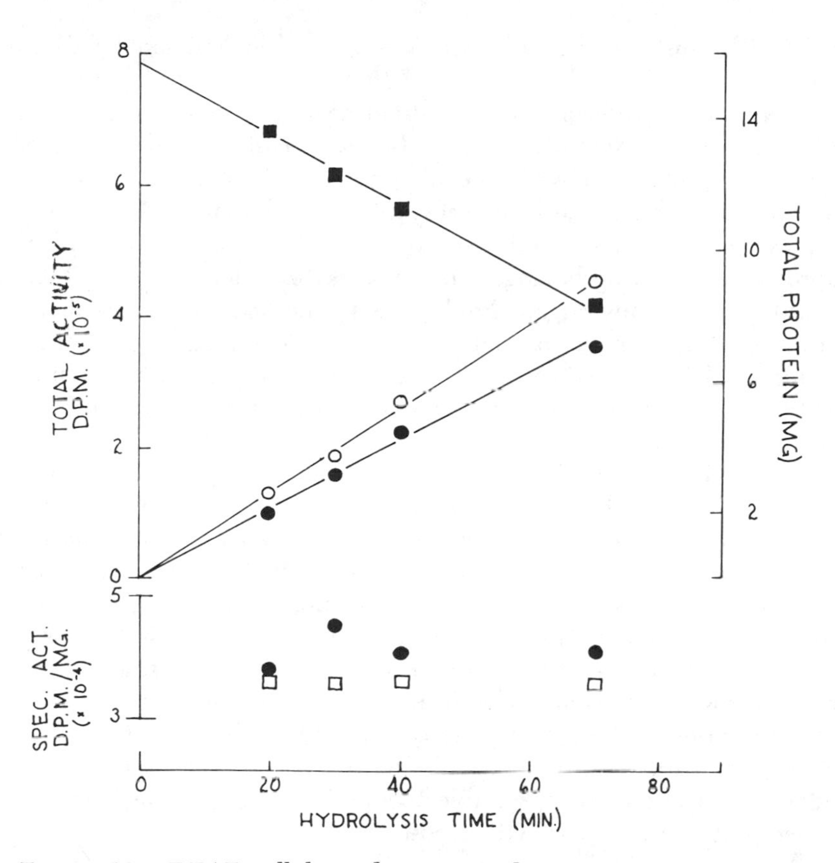


*Figure 10. DEAE–cellulose chromatography of casein from mastitic milk containing M-β-C.*

*About 2 × 10⁶ dpm of the C-14 labeled protein was incubated in the milk for 8 h before casein precipitation. Chromatography was conducted as described in the legend to Figure 6 but with the following alterations: column dimensions, 1.6 × 25 cm; sample loading, 70 mg; elution was with column buffer (10 fractions), followed by column buffer adjusted to 30mM NaCl (8 fractions), followed by a linear NaCl gradient (60–250mM) in column buffer (gradient volume, 500 mL). Total activity, Fraction 2 (■); total protein, Fraction 1 (○); total activity or specific activity, Fraction 1 (●); specific activity residual β-casein (□).*

## Conclusions

The specific activities of the different radiomethylated milk proteins varied between 3 and 4.4 μCi/mg. This degree of labeling, resulting from the use of $^{14}$C-HCHO of high specific activity, allows the proteins to be used as true tracers. For example, the addition of 1 mg of labeled protein with a specific activity of 4 μCi/mg to a liter of milk should provide excellent counting statistics on less than 0.5 mL of the milk. Also in the experiments on the plasminolysis of β-casein, the radioactive β-casein was diluted 100–300-fold with cold β-casein before experimentation, which illustrates the sensitivity of using radiomethylated milk proteins as tracers.

These studies demonstrate that treatments that have been reported to modify the heat stability of milk also may change the distribution of tracer milk proteins between the physical phases of milk. Analysis of these interactions may provide useful information about the mechanisms of these effects. In unpublished experiments, we have extended this approach for studying homogenized milk-based systems, also containing a lipid phase, to investigate lipid–protein interactions. Dual-label experiments, for example in milk containing M-$\beta$-L and $^{3}$H-methyl-$\kappa$-casein, could be applied to the isolation and characterization of protein complexes in milk. Labeled caseins could prove valuable as probes for elucidating micelle structure.

The results on the hydrolysis of partially methylated $\beta$-casein by plasmin indicate that proteins radiomethylated to a low level can serve as substrates for trypsin-like enzymes and probably for proteinases in general. Because it is likely that methylation will interfere with enzymatic attack at lysine residues, the complete hydrolysis of $\beta$-casein probably would not be possible. Studies on mastitic milk demonstrate the usefulness of $^{14}$C-methyl proteins for qualitative examination of protein hydrolysis in complex multiprotein systems where resolution and characterization of individual protein fragments is difficult. The requirements in such studies are the availability of pure samples of the proteins under investigation and a suitable technique for separating the radiolabeled protein from hydrolytic products.

### *Acknowledgments*

This work was supported in part by Ross Laboratories (Columbus, OH), the College of Agricultural and Life Sciences, University of Wisconsin—Madison, the Cooperative State Research Service, USDA, and by the Irish Agricultural Institute. T. Richardson was a Fulbright–Hays Scholar at the Agriculural Institute in Fermoy, County Cork, Ireland, where the hydrolysis of methylated $\beta$-casein by plasmin was studied. The authors are grateful for helpful discussions with L. K. Creamer.

### *Literature Cited*

1. Rice, R. H.; Means, G. E. *J. Biol. Chem.* **1971,** *246,* 831.
2. Ottesen, M.; Svensson, B. *C. R. Trav. Lab. Carlsberg* **1971,** *32,* 445.
3. Means, G. E.; Feeney, R. E. *Biochemistry* **1968,** *7,* 2192.
4. Dottavio-Martin, D.; Ravel, J. M. *Anal. Biochem.* **1978,** *87,* 562.
5. Jentoft, N.; Dearborn, D. G. *J. Biol. Chem.* **1979,** *254,* 4359.
6. Mills, O. E. *N. Z. J. Dairy Sci. Technol.* **1976,** *11,* 164.
7. Olson, N. F.; Richardson, T.; Zadow, J. G. *J. Dairy Res.* **1978,** *45,* 69.

8. Waugh, D. F.; von Hippel, P. H. *J. Am. Chem. Soc.* **1956,** *78,* 4576.
9. Sawyer, W. H. *J. Dairy Sci.* **1969,** *52,* 1347.
10. Rowley, B. O.; Lund, D. B.; Richardson, T. *J. Dairy Sci.* **1979,** *62,* 533.
11. MacKinlay, A. G.; Wake, R. G. In "Milk Proteins—Chemistry and Molecular Biology"; H. A. McKenzie, Ed.; Academic: New York, 1971; Vol. 2, p. 175.
12. Larson, B. L.; Rolleri, G. D. *J. Dairy Sci.* **1955,** *38,* 351.
13. Drucker, H. *Anal. Biochem.* **1972,** *46,* 598.
14. Humbert, G.; Alais, C. *J. Dairy Res.* **1979,** *46,* 559.
15. Davies, D. T.; Law, A. J. R. *J. Dairy Res.* **1977,** *44,* 213.
16. Groves, M. L.; Gordon, W. G. *Biochim. Biophys. Acta* **1969,** *194,* 421.
17. Fox, P. F.; Morrissey, P. A. *J. Dairy Res.* **1977,** *44,* 627.
18. Lowry, O. H.; Rosebrough, N. J.; Farr, A. L.; Randall, R. J. *J. Biol. Chem.* **1951,** *193,* 265.
19. Wilson, G. A.; Wheelock, J. V.; Kirk, A. *J. Dairy Res.* **1974,** *41,* 37.
20. Patrick, P. S.; Swaisgood, H. E. *J. Dairy Sci.* **1976,** *59,* 594.
21. Trautman, J. C.; Swanson, A. M. *J. Dairy Sci.* **1959,** *42,* 895.
22. Means, G. E.; Feeney, R. E. "Chemical Modification of Proteins"; Holden-Day: San Francisco, 1971.
23. Beutler, E.; Srivastava, S.; West, C. *Biochem. Biophys. Res. Commun.* **1970,** *38,* 341.
24. Creamer, L. K.; Matheson, A. R. *N. Z. J. Dairy Sci. Technol.* **1980,** *15,* 80.
25. Muir, D. D.; Sweetsur, A. W. M. *J. Dairy Res.* **1976,** *43,* 495.
26. Robertson, N. H.; Dixon, A. *Agroanimalia* **1969,** *1,* 141.
27. Muir, D. D.; Sweetsur, A. W. M. *J. Dairy Res.* **1977,** *44,* 249.
28. Donnelly, W. J.; Barry, J. G.; Richardson, T. *Biochim. Biophys. Acta* **1981,** *626,* 117.
29. Whitney, R. McL.; Brunner, J. R.; Ebner, K. E.; Farrell, H. M., Jr.; Josephson, R. V.; Morr, C. V.; Swaisgood, H. E. *J. Dairy Sci.* **1976,** *59,* 795.
30. Feeney, R. E.; Blankenhorn, G.; Dixon, H. B. F. In "Advances Protein Chemistry"; Anfinsen, C. B.; Edsall, J. T.; Richards, F. M., Eds.; Academic: New York, 1975; Vol. 29, pp. 135–203.
31. Barry, J. G.; Donnelly, W. J. *J. Dairy Res.*, in press.
33. Bhattacharyya, G. K.; Johnson, R. A. "Statistical Concepts and Methods"; John Wiley & Sons: New York, 1977.

RECEIVED October 6, 1980.

5

# Covalent Attachment of Essential Amino Acids to Proteins by Chemical Methods: Nutritional and Functional Significance

ANTOINE J. PUIGSERVER and HUBERT F. GAERTNER
Centre National de la Recherche Scientifique, Marseille, France

LOURMINIA C. SEN, ROBERT E. FEENEY, and JOHN R. WHITAKER
University of California, Davis, CA

*Limiting essential amino acids covalently attached to proteins by using activated amino acid derivatives can improve the nutritional quality and change the functional properties of proteins. The best chemical methods for incorporating amino acids into water-soluble proteins involve using carbodiimides,* N-*hydroxysuccinimide esters of acylated amino acids, or* N-*carboxy-α-amino acid anhydrides. The last two methods can give up to 75% incorporation of the amount of amino acid derivative used. With the anhydride method, as many as 50 residues of methionine have been linked to the 12 lysine residues of casein. The newly formed peptide and isopeptide bonds are hydrolyzed readily by intestinal aminopeptidase, making the added amino acids and the lysine from the protein available nutritionally.*

It is well recognized that the nutritional value of dietary proteins depends primarily on the content of their constituent amino acids, especially of their essential amino acids. Because of deficiencies of lysine and methionine, and to a lesser extent of a few other amino acids, proteins from plants and other alternative sources have low biological quality. Moreover, incomplete digestion of the protein also may result in a lack of complete availability (*1*) of the essential amino acids and may further reduce its value. In many raw plant foodstuffs such as soybeans, common beans, or unprocessed protein foods, undenatured

0065-2393/82/0198-0149$05.00/0

proteins are resistant to enzymatic hydrolysis (*2*). The presence of biologically active substances, i.e. protease inhibitors and lectins, also may have an adverse effect either on the digestibility or the availability of amino acids (*3, 4*). Chemical deteriorative changes of proteins, which occur during processing or storage, frequently lead to a decreased content of essential amino acids or to lower rates of hydrolysis by proteolytic enzymes because of the formation of cross-linked products (*5, 6, 7, 8*).

All of the above-mentioned factors affecting the biological quality of proteins should be taken into consideration when dealing with the nutritional improvement of food and feed proteins. Furthermore, the covalent attachment of essential amino acids to proteins by chemical methods must avoid damage to the biological quality of proteins if it is to have potential applications.

A number of enzymatic (*9, 10, 11*), microbial (*12, 13*), physical (*14, 15*), and chemical methods (*16, 17, 18, 19*) have been applied already to the modification of food proteins and other less conventional sources of proteins or have been suggested for improving their biological value and functional properties. To achieve the same goals, other studies have been directed primarily towards the covalent attachment of limiting essential amino acids through enzymatic or chemical reactions (*20, 21, 22, 23, 24*). It is beyond the scope of this chapter to review all of these methods but a few brief comments on some aspects of fortifying foods with either free or covalently bound amino acids is needed before discussing our own work on this topic.

### *General Aspects of Fortification of Proteins with Amino Acids*

Supplementing foods and feeds with free essential amino acids is well documented (*25, 26, 27, 28, 29*) but, as shown in Table I, it has some disadvantages. The decreased biological utilization of free amino acids when fed at high levels should be emphasized among the possible disadvantages of such a method. The intestinal absorption of free amino acids is thought to involve a mechanism that is different from that of small polypeptide absorption and subsequent hydrolysis in the villi (*30, 31*). Moreover a close relationship between protein hydrolysis catalyzed by intestinal enzymes and active transport of the resulting small peptides or free amino acids has been suggested (*30, 31*). The functional properties of protein-enriched foods will not be changed when free amino acids are added while altered physical characteristics may be expected from the covalent attachment of amino acids. Proteins from plants and other less conventional sources must possess appropriate functional properties in order to increase consumer acceptability and consequently direct consumption of plant products (*32*). The need for proteins with multiple

**Table I. Improvement of the Nutritional Value of Food and Feed Proteins**

| | |
|---|---|
| *Supplementation with Free Amino Acids* | |
| Advantage: | An easy and relatively efficient way of improving the nutritional value. |
| Disadvantages: | Possible losses of the added amino acid during processing and food preparation. |
| | Side reactions occurring during processing or storage. |
| | Modification of the sensory properties. |
| | Detrimental effects of lysine modification (Maillard reactions) are not precluded. |
| | Decreased biological utilization of free amino acids fed at high levels. |
| *Supplementation Through Covalent Attachment of Amino Acids* | |
| Aims: | To prevent deteriorative reactions of lysyl residues. |
| | To improve nutritional quality and functional properties. |
| | To avoid some, if not all, of the above-mentioned disadvantages. |
| Methods: | Enzyme-catalyzed incorporation of amino acids into peptides (two successive processes) or proteins (a one-step process). |
| | Chemical modification of peptides or proteins leading to single or multiple addition at the same amino group. |

functional properties resulted in the development of various texturization methods and other processes which may lead to chemical deteriorations of proteins (*33*). Fortifications of food proteins with free amino acids usually does not prevent deteriorative changes. By contrast, supplementation through covalent attachment of amino acids should be useful with respect to protecting essential amino acid residues and improving both nutritional and functional properties.

Enzyme-catalyzed attachment of amino acids to proteins represents an attractive and interesting way for improving the nutritional value of food proteins. The enzymes that participate in the gastrointestinal digestion of food proteins catalyze exclusively hydrolytic reactions under physiological conditions. However the synthetic activity of proteolytic enzymes was reported first by Danilewski in 1886, and more recently a number of studies have been devoted to plastein formation from concentrated protein hydrolysates (*34, 35, 36, 37, 38*). Among possible uses of the plastein reaction, the improvement of nutritional and functional properties of food proteins should be mentioned as well as removal of bitter peptides or other unwanted bound compounds.

Recently a simplified process was developed for incorporating L-methionine directly into soy proteins during the papain-catalyzed hydrolysis (*21*). The covalent attachment of the amino acid requires a very high concentration of protein and occurs through the formation of an acyl-enzyme intermediate and its subsequent aminolysis by the methionine ester added in the medium. From a practical point of view, the main advantage of enzymatic incorporation of amino acids into food proteins, in comparison with chemical methods, probably lies in the fact that racemic amino acid esters such as D,L-methionine ethyl ester can be used since just the L-form of the racemate is used by the stereospecific proteases. On the other hand, papain-catalyzed polymerization of L-methionine, which may occur at low protein concentration (*39*), will result in a loss of methionine because of the formation of insoluble polyamino acid chains greater than 7 units long.

Most of the experiments on incorporating amino acid esters into proteins during the plastein reaction have been carried out with papain, indicating that it is one of the best enzymes for this purpose. Other enzymes such as chymotrypsin (*40*) or carboxypeptidase Y from *Saccharomyces cerevisiae* (*41*) are potent catalysts for peptide synthesis in homogeneous systems using *N*-acylamino acid esters of peptides as substrates and amino acid derivatives or peptides as nucleophile components. Adding organic co-solvents favored peptide bond synthesis (*42, 43*).

The chemical methods already available for the covalent attachment of amino acids to proteins include those used to modify carboxyl and amino groups of proteins.

### *Covalent Attachment of Amino Acids to Protein Carboxyl Groups*

Carbodiimides, particularly the water-soluble derivatives, were used first as coupling reagents for peptide synthesis (*44, 45*) and then for modifications of groups (*46, 47, 48, 49*). As shown in Figure 1, the reaction of a carboxyl group of a protein with a water-soluble carbodiimide involves forming an *O*-acylisourea intermediate that then may rearrange to give the corresponding *N*-acyl urea, be hydrolyzed to regenerate the carboxyl group, or react with a nucleophile, i.e. amino acid or polypeptide, leading to the formation of an amide or isopeptide bond.

Recently, limiting essential amino acids have been attached covalently to soy protein isolates (*24*) and wheat gluten (*52*) by the water-soluble

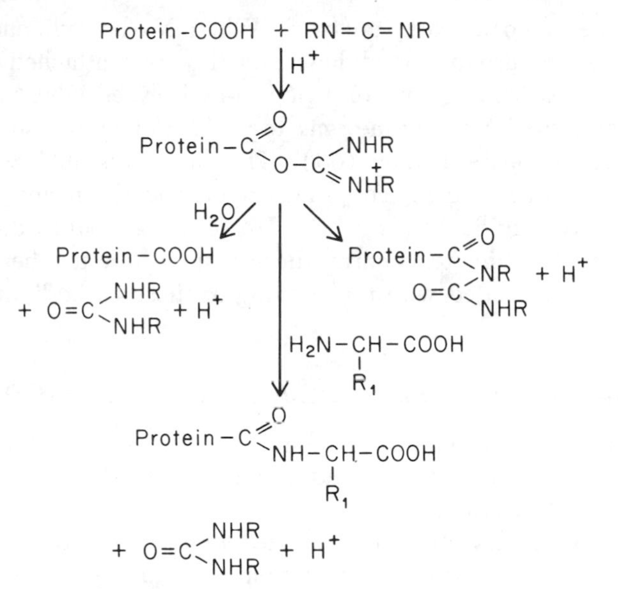


*Figure 1. Covalent binding of amino acids to proteins using a water-soluble carbodiimide as a carboxyl group activating reagent* (50, 51)

carbodiimide condensation reaction. As indicated in Table II, tho methionine or tryptophan content of soy protein isolate was increased by 6.3- and 11.3-fold, respectively, as compared with the control protein. Prior partial hydrolysis of the protein before derivatization further increased the levels of covalently bound amino acids. These newly attached amino acids, via isopeptide bonds, were hydrolyzed easily in

**Table II. Content of Some Amino Acids (Gram/16 Grams Nitrogen) of Soy Protein Modified by Covalent Binding of Methionine and Tryptophan Using the Water-Soluble 1-Ethyl-3-(3-dimethylaminopropyl)carbodiimide (*24*)**

| *Protein* | *Aspartic Acid* | *Glutamic Acid* | *Lysine* | *Methionine* | *Tryptophan* |
|---|---|---|---|---|---|
| Control soy protein | 11.41 | 19.40 | 5.83 | 0.94 | 0.95 |
| Soy protein + methionine | 10.84 | 18.34 | 5.71 | 5.92 | 0.90 |
| Soy protein hydrolysate + methionine | 10.57 | 18.21 | 4.45 | 7.22 | 0.94 |
| Soy protein + tryptophan | 10.32 | 17.50 | 5.25 | 0.83 | 10.74 |
| Soy protein hydrolysate + tryptophan | 8.82 | 15.37 | 5.17 | 0.94 | 17.05 |

vitro by a pepsin–pancreatin mixture. When *N*-ε-benzylidenelysine, in which the ε-amino group is modified reversibly, was attached covalently to the α- and γ-carboxyl groups of a pepsin-hydrolyzed wheat gluten, the chemical reactivity of the lysine was high (91%) as measured by the dinitrobenzenesulfonate method (*53*). The value was 48% when underivatized lysine was used for coupling, suggesting the formation of γ–α, γ–ε, and α–ε isopeptide bonds (*52*). However, animal-feeding studies must be done in order to confirm these results and to check possible health hazards associated with the feeding of these carbodiimide-treated samples.

### *Covalent Attachment of Amino Acids to Amino Groups of Proteins*

Proteins may be modified covalently via the ε-amino group of lysyl residues using two types of activated amino acids: *N*-hydroxysuccinimide esters and *N*-carboxyanhydrides.

**The Active Esters Method.** Selected amino-protecting groups and conditions required for their deprotection in organic solvents are shown in Table III. Active *N*-hydroxysuccinimide esters of *t*-butyloxycarbonyl-L-

**Table III. Protection and Deprotection of Amino Acids in Peptide Synthesis**

| *Amino Protecting Group* | *Deprotection Conditions* |
| --- | --- |
| 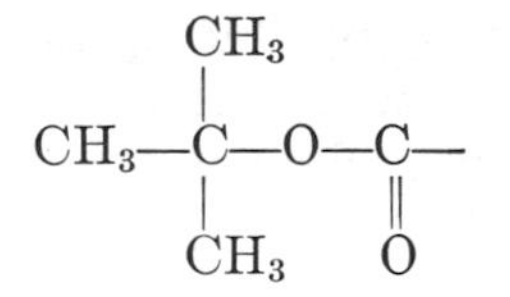 *t*-Butyloxycarbonyl (*t*-Boc) | pure trifluoroacetic acid or formic acid (drastically acidic)[a] |
| 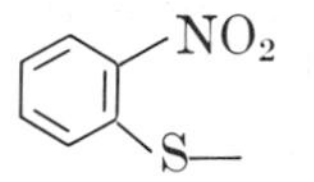 *o*-Nitrophenylsulfenyl (NPS) | dilute HCl solution (mildly acidic)[b] |
| 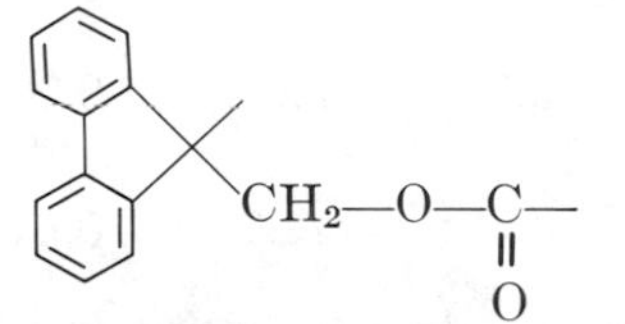 9-Fluorenylmethoxycarbonyl | 50% piperidine in $CH_2Cl_2$ or DMF[d] (mildly basic)[c] |

[a] Ref. *54*.
[b] Ref. *55*.
[c] Ref. *56*.
[d] DMF = dimethylformamide.

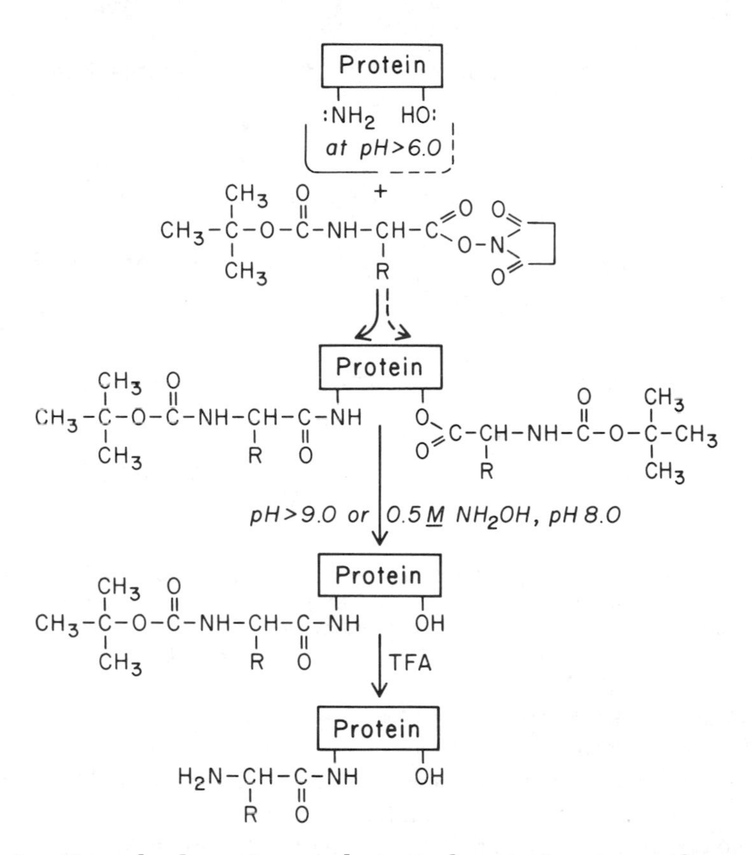


*Figure 2. General scheme for covalent attachment of amino acids to proteins by the active-esters method. Reaction conditions are described in Ref. 22; TFA = trifluoroacetic acid.*

amino acids have been used widely in structure–function relationship studies of enzymes and biologically active proteins (*57, 58, 59, 60, 61*). Using the other two protecting groups shown in Table III has not been mentioned yet in chemical modification studies of proteins in aqueous solution. The conditions required for deblocking of *t*-butyloxycarbonyl derivatives of proteins are now well known (*22*) whereas no data are presently available for removing the other two groups when they are bound to proteins. The stability of peptide bonds to trifluoroacetic acid treatment is rather high; however, using more labile amino-protecting groups would make the modification of food proteins easier.

The general scheme for coupling *N*-hydroxysuccinimide esters of *t*-butyloxycarbonyl-protected amino acids to lysyl residues of proteins is shown in Figure 2 and the structure of the resulting isopeptide bond is depicted in Figure 3. The other isopeptide bonds involving $\beta$- or $\gamma$-carboxyl groups of aspartic and glutamic acid, respectively, and $\alpha$ or $\epsilon$ amino groups of lysine are formed in carbodiimide condensation reactions (*52*).

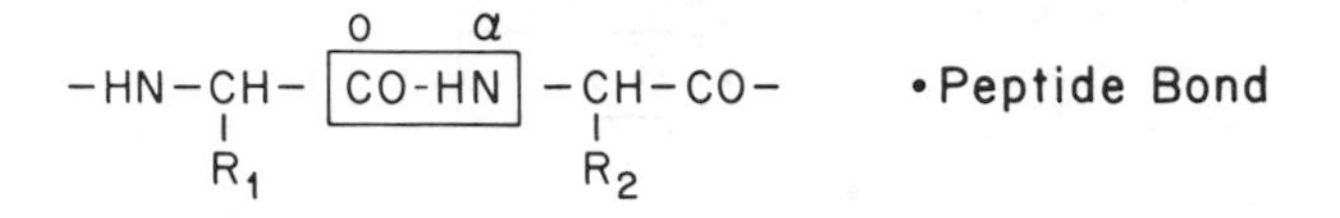


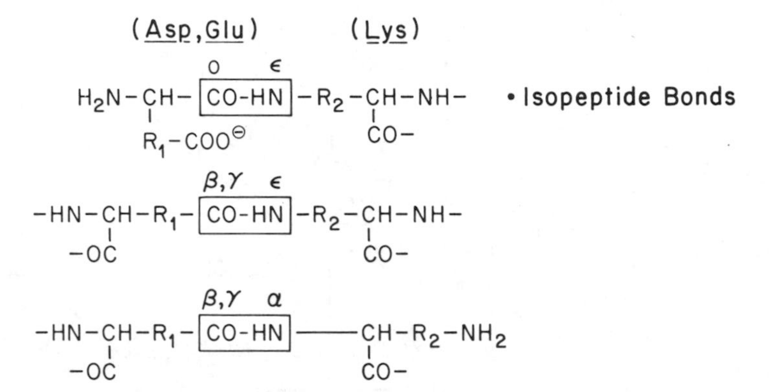


*Figure 3. Structure of peptide and isopeptide bonds resulting from covalent attachment of amino acids to proteins by chemical methods. In isopeptide bond formation $R_1 = -CH_2-$ or $-CH_2CH_2-$ of aspartic or glutamic acid and $R_2 = -(CH_2)_n-$ of lysine.*

Some of these bonds may not be hydrolyzed enzymatically and therefore may not be available biologically.

**Hydrolysis of the Isopeptide Bond.** The bioavailability of methionine covalently linked to casein has been studied by the response (protein efficiency ratio) of rats that have been fed the modified protein (*23, 63*). The covalently bound methionine appears to be as available as the free amino acid. This finding indicates that there is an efficient enzymatic

**Table IV. Hydrolysis of the Isodipeptide [$^{14}$C-methyl]-ε-*N*-L-Methionyl-L-lysine by the Proteolytic Enzymes of the Alimentary Tract**

| *Enzymes* | *Extent of Hydrolysis (%)*[a] | *E/S (w/w)* |
|---|---|---|
| Hog—pepsin | 2 | 1:1 |
| —pancreatic juice | 3 | 1:6 |
| —solubilized aminopeptidase | 95 | 1:500 |
| —membrane-bound aminopeptidase | 91 | 1:500 |
| Rat—membrane-bound aminopeptidase | 93 | 1:500 |

[a] Determined after a 3-h incubation period at 37°C. The concentration of the isopeptide was 5m*M* in 0.05*M* phosphate buffer, pH 7.0 (*62*).

**Table V. Hydrolysis of the Radiolabeled Isodipeptide ε-N-L-Methionyl-L-lysine by Rat Tissue Homogenates**

| *Tissue* | *AAN Units*[a] $\times 10^3$ | *Nanomoles Methionine Released*[b] $\times 10^6$ | *Methionine/AAN* $\times 10^2$ |
|---|---|---|---|
| Kidney | 60 | 1.3 | 0.2 |
| Liver | 17 | 1.9 | 1.1 |
| Intestine | | | |
| homogenate | 31 | 6.0 | 1.9 |
| supernatant | 3 | 0.6 | 2.0 |
| pellet | 44 | 5.2 | 1.2 |

[a] Hydrolysis of alanine *p*-nitroanilide according to Roncari and Zuber (*67*).
[b] Extent of hydrolysis determined after overnight incubation at 37°C of substrate with tissue homogenate aliquots containing about 20 AAN units per assay.

hydrolysis of the isopeptide bond of ε-*N*-methionyllysine. The organ location of the enzymatic activity responsible for hydrolyzing this isopeptide bond was determined (*see* Table IV). The proteolytic enzymes of the upper alimentary tract were not able to catalyze the hydrolysis of the isopeptide bond whereas the intestinal membrane-bound aminopeptidase (*64, 65, 66*) could hydrolyze this bond at a very high rate.

Digestion studies using homogenates of kidneys, intestine, and liver indicated a close relationship between the levels of aminopeptidase activity as measured with a specific synthetic substrate and that based on the enzymatic activity responsible for hydrolyzing the isopeptide bond (*see* Table V). Purified aminopeptidase N from the intestinal brush border of rabbit (*68*) and pig (*64*) hydrolyzed a dipeptide and the related isodipeptide with about the same efficiency (*see* Table VI). These results may explain the findings that methionine covalently bound

**Table VI. Kinetic Parameters of the Hydrolysis of a Dipeptide and Its Related Isodipeptide Catalyzed by Intestinal Aminopeptidase N**[a]

| | *α-Methionine–Lysine* | | *ε-Methionine–Lysine* | |
|---|---|---|---|---|
| *Enzymes* | $K_m$ *(mmol)* | $k_{cat}$ *($s^{-1}$)* | $K_m$ *(mmol)* | $k_{cat}$ *($s^{-1}$)* |
| Aminopeptidase N | | | | |
| rabbit | 0.35 | 45 | 0.17 | 22 |
| pig | 0.05 | 70 | 0.17 | 31 |

[a] Experiments were performed in 0.05*M* phosphate buffer at 37°C and pH 7.0.

to the ε-amino group of lysyl residues of casein was as available nutritionally as the free amino acid. Experiments are presently underway to study in detail the enzyme-catalyzed hydrolysis of isopeptide bonds in vitro and in vivo.

**The *N*-Carboxy-Amino Acid Anhydride Method.** Since the isopeptide bond of ε-methionyllysine was hydrolyzed readily by intestinal aminopeptidase and the released amino acid was biologically available, we decided to further increase the amount of covalently attached methionine through a polymerization reaction. The most suitable amino acid derivative for this approach is the *N*-carboxyanhydride or Leuchs' anhydride.

SYNTHESIS AND POLYMERIZATION OF LEUCHS' ANHYDRIDES. The *N*-carboxy-α-amino acid anhydrides, referred to as Leuchs' anhydrides or as NCAs, are synthesized either from *N*-alkoxycarbonyl derivatives of α-amino acids or from free amino acids. Cyclization of the amino acid derivative by $SOCl_2$ or similar reagents was described first by Leuchs in 1906, but direct synthesis involves treating the α-amino acid or its hydrochloride with phosgene as shown in Figure 4. This method, commonly termed the Fuchs–Farthing method, is the one that is used most frequently since the *N*-carboxyanhydride may be freed easily from the carbamyl chloride, isocyanate derivative, and hydrogen chloride by crystallization.

The stepwise addition of *N*-carboxy α-amino acid anhydrides to amino acids or peptides in organic solvents is well documented (for review *see* Refs. 69 and 70). More recently the method has been applied to peptide synthesis in aqueous solutions (*71, 72, 73*). Novel syntheses of *N*-substituted carboxyanhydrides of α-amino acids have been carried out (*74, 75*) and the crystal molecular structures of several of them have been determined (*76, 77, 78*).

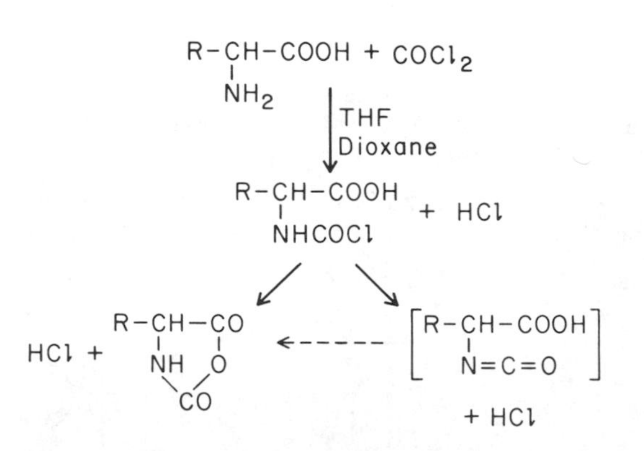


*Figure 4. Synthesis of* N*-carboxy-α-amino acid anhydrides (NCA). THF = tetrahydrofuran.*

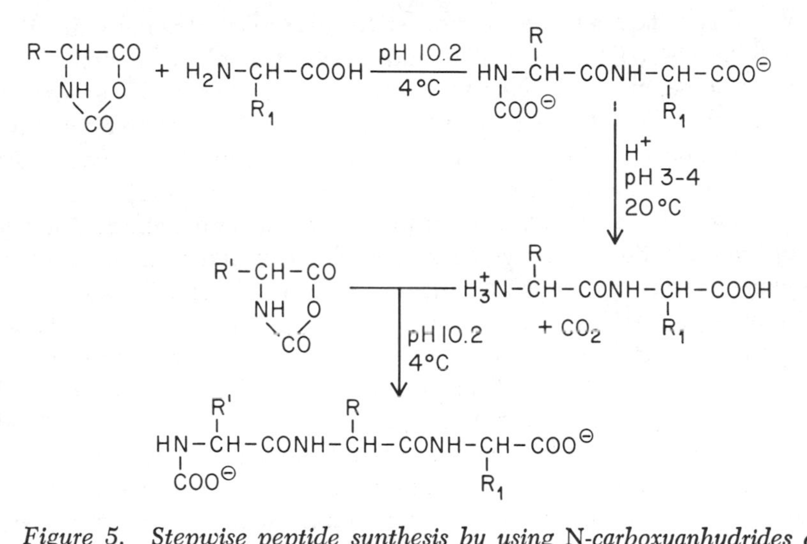


*Figure 5. Stepwise peptide synthesis by using* N-*carboxyanhydrides of amino acids*

As shown in Figure 5, the coupling reaction between an amino acid and an *N*-carboxy-α-amino acid anhydride is performed at alkaline pH but the resulting carbamate intermediate is decomposed at acidic pH in order to give the dipeptide and allow further reaction leading to polyamino acids. Polymer formation is performed most frequently at neutral pH since the half-life of the intermediate carbamate is very short (*see* Figure 6).

The kinetics and mechanism of polymerization of *N*-carboxy-α-amino acid anhydrides as well as the biological properties of poly-α-amino acids have been studied and reviewed extensively (*70, 79, 80*). The critical chain length for helix formation varies from one amino acid to another (*81, 82*). The critical chain length in L-methionine oligopeptides is the heptamer (*83*).

REACTION OF *N*-CARBOXYAMINO ACID ANHYDRIDES WITH PROTEINS. A number of proteins have been modified chemically with *N*-carboxy an-

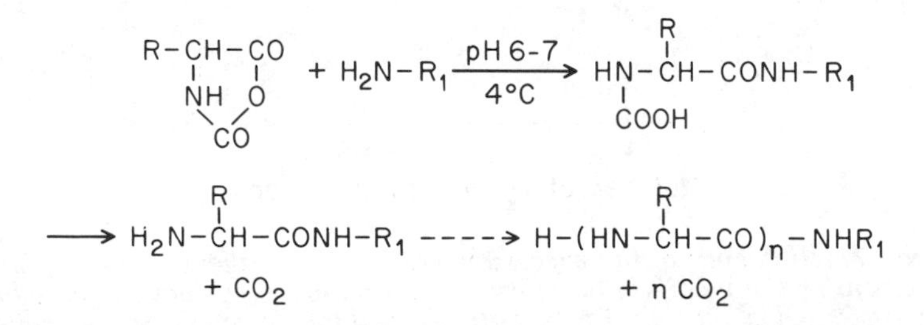


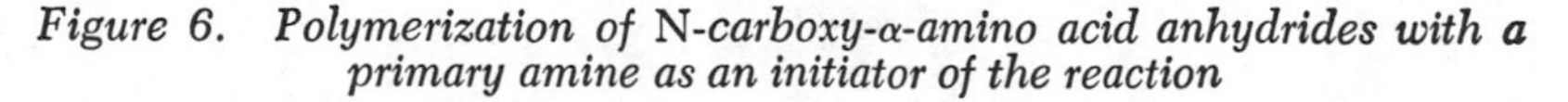


*Figure 6. Polymerization of* N-*carboxy-α-amino acid anhydrides with a primary amine as an initiator of the reaction*

hydrides of specific amino acids in order to study their structure–function relationships (*84, 85, 86*). Covalent introduction of phenylalanine, tyrosine, methionine, and isoleucine into whey proteins by the *N*-carboxyanhydride method has been accomplished also (*87*). The relative nutritive values, determined by the slope ratio assay with young rats, were 0.75 for whey proteins and 0.92 for the amino acid-enriched protein.

We have studied the influence of pH and casein concentration on the extent of modification of the lysyl residues of casein under single addition or polymerization conditions. As shown in Figure 7, at 4% casein concentration in a $NaHCO_3$ buffer, pH 10.2, using 1.5 mol of *N*-carboxymethionine anhydride per amino group allowed the introduction of 12 mol of methionine with an efficiency higher than 70%. An average of eight residues of lysine, of the 12 present in casein, were modified covalently. The weight of covalently bound methionine represented up to 25% of the weight of modified casein when five successive additions of the reagent were used (1:1 reagent:amino groups each time, *see* Figure 7). After the third addition, all of the lysyl residues were modified covalently; subsequent addition occurred on the amino groups of the methionine residues giving a high degree of polymerization of methionine on casein. On the basis of the data in Figure 7 the average chain length of the polymers is about four amino acids. Table VII shows that the efficiency of the covalent attachment of methionine to casein drops to 40–50% after the fifth addition of reagent mainly because of free amino

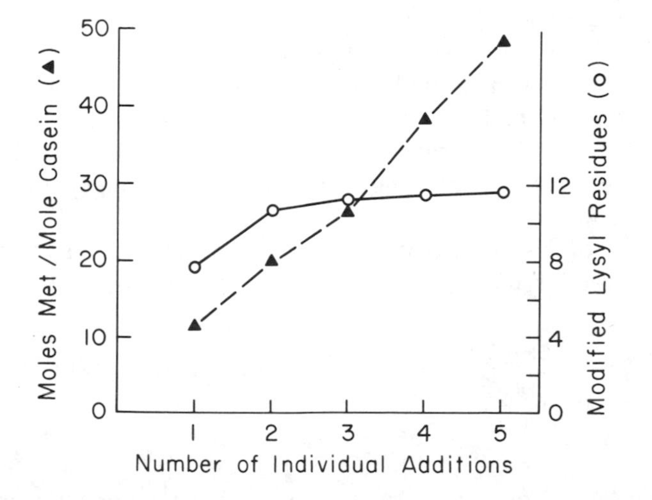


*Figure 7. Influence of the successive addition of reagent at pH 10.2 on the extent of the modification of casein and its lysyl residues. Each addition contained a ratio of 1.5* N*-carboxy methionine anhydride to amino groups.*

**Table VII. Casein Modification by *N*-Carboxymethionine Anhydride**

| *Type of Addition* | N-*Carboxy-methionine/ $NH_2$ Groups (molar ratio)* | *Efficiency of Reaction*[a] *(%)* | *Modified Lysyl Residues (%)* |
|---|---|---|---|
| Single addition at pH 6.5 | > 2 | 80–90 | 50 |
| Several additions[b] | 1.5 | 40–50 | 100 |

[a] Moles methionine covalently bound/moles methionine added to reaction (× 100).
[b] Four–five successive additions, each one being followed by an adjustment of pH from 10.2 down to 3–4 and then back to pH 10.2 for subsequent reaction.

acids in the medium. In contrast, the efficiency of the reaction is rather high (80–90%) under polymerization conditions. Table VIII shows that casein and lactoglobulin may be modified extensively by *N*-carboxymethionine anhydride as a result of copolymerization of the amino acid on lysyl residues.

The ability to covalently attach 50 (and probably more) residues of methionine (and other essential amino acids) to proteins has important, practical implications. For example, a 1:5 mixture of ε-methionyl casein (50 residues of methionine) and casein (5 residues of methionine), fed at 10% weight of the diet, would meet the methionine requirements of rats (*23*). Therefore the cost of modification and the carry over of products from side reactions would be reduced in the feed.

**Table VIII. Extent of Modification of Proteins**

| | *Amino Acid Content*[a] | | | *Free Amino Groups*[b] | |
|---|---|---|---|---|---|
| *Protein* | *Lysine* | *Methionine* | *Increase In Methionine* | *Number* | *Modification (%)* |
| Casein | 12 | 5 | 0 | 11.4 | 0 |
| Methionine–casein[c] | 12 | 67 | 62 | 4.9 | 58 |
| Methionine–casein[d] | 12 | 42 | 37 | traces | > 95 |
| β-Lactoglobulin | 15 | 4 | 0 | 14.6 | 0 |
| Methionine–lactoglobulin[e] | 15 | 57 | 53 | 6.1 | 58 |
| Methionine–lactoglobulin[f] | 15 | 54 | 50 | 0 | > 98 |

[a] Number of residues/mole of protein.
[b] Determined by protein guanidination with *O*-methylisourea.
[c] Modified by a single addition of *N*-carboxymethionine anhydride at pH 6.5. Molar ratio *N*-carboxymethionine anhydride:protein amino groups = 5:1.
[d] Modified by five successive additions of *N*-carboxymethionine anhydride at pH 10.2 with a 1.5:1.0 molar ratio each.
[e] As *c* with a molar ratio of 4:1.
[f] As in *d* but only four successive additions instead of five.

More details of the chemical modification of casein and other food proteins with carboxy-α-amino acid anhydrides will be published elsewhere. Depending on the distribution and length of copolymers of methionine covalently linked to proteins, distinct effects on functional and nutritional properties of the modified proteins are expected.

### *Some Functional Properties of Modified Casein*

Some physical and functional properties of casein modified by the covalent attachment of amino acids are given in Table IX. Despite extensive modification, the relative viscosities of 2% solutions of the modified proteins did not change significantly, with the exception of aspartyl casein which was more viscous. There was some decrease in the solubilities of aspartyl casein and tryptophyl casein as compared with the casein control. It is anticipated that adding some 11.4 tryptophyl residues per mole of casein would decrease the aqueous solubility of the modified protein. However the results with aspartyl casein are unexpected. The changes in viscosity, solubility, and fluorescence indicate that aspartyl casein is likely to be a more extended molecule than the casein control. There was a marked decrease in the fluorescence of aspartyl casein and tryptophyl casein (*see* Table IX). The ratios of the fluorescences of acetylmethionyl casein to methionyl casein and *t*-BOC-tryptophyl casein to tryptophan casein were 1.20 and 2.01, respectively, indicating the major effects that these acyl groups have on the structure of the casein.

**Table IX. Some Physical Properties of Casein Modified by Covalent Attachment of Some Amino Acids**[a]

| *Protein* | *% Modification*[b] | *Relative Viscosity*[c] | *Relative Solubility*[d] | *Relative Fluorescence*[e] |
|---|---|---|---|---|
| Casein (control) | 0 | 1.00 | 1.00 | 1.00 |
| Alanyl casein | 88 | 1.01 | 1.04 | 1.02 |
| Methionyl casein | 89 | — | — | 1.02 |
| Threonyl casein | ~80 | — | 1.02 | — |
| Asparaginyl casein | ~80 | — | 1.00 | — |
| Aspartyl casein | 83 | 1.06 | 0.85 | 0.68 |
| Tryptophyl casein | 92 | 0.99 | 0.47 | 0.58 |

[a] Adapted in part from Ref. *22*.
[b] Percentage of the 12 ε-amino groups of lysine residues modified as determined by TNBS titration and amino acid analysis.
[c] For 2% protein solution in 0.1*M* Tris · HCl buffer, pH 8.0, and 25°C. The value of $\eta$ foi casein control was 1.11 cSt.
[d] In 0.1*M* citrate buffer, pH 5.5, and 4°C. Solubility of casein control was 53 mg/mL.
[e] At maximum wavelength of fluorescence (330–340 nm) using 8.9 × $10^{-7}M$ solution of casein in 0.05*M* Tris · HCl buffer, pH 8.0, and 20°C.

## *Significance of Covalent Attachment of Amino Acids to Proteins*

The nutritive value of food proteins may be reduced during processing because of deteriorative changes of essential amino acids. For example, the methionine and cysteine contents of milk and fish products are known to decrease when they are sterilized or roller-dried at a high temperature (*87*). Covalent attachment of amino acids to proteins could reduce such detrimental effects. Moreover, if the added amino acid is a limiting one for the protein, modification also should increase the nutritional value. If these methods are to be used for protection and fortification, it is important to show that the newly formed isopeptide bond, e.g. $\epsilon$-methionyllysine, is hydrolyzed efficiently by intestinal aminopeptidase. This enzyme, which is responsible for the hydrolysis, is very abundant in the brush border of intestinal cells and it accounts for practically all of the peptidase activity of the absorptive membrane. The action of this protease helps to explain the good bioavailability of methionine covalently linked to the $\epsilon$-amino group of lysine. We still do not know whether this enzyme is also responsible for permitting the use of covalently bound *N*-acetyl-methionine (*23*). This amino acid derivative and its free form are equivalent sources of methionine in the rat (*88*) and human (*89*). Although several mammalian tissues contain enzymes capable of deacylating $\alpha$-*N*-acyl amino acids, they have been characterized primarily from kidneys (*90, 91*).

Another potential application of the covalent attachment of amino acids to proteins is the protection of ruminant feeds by reducing ruminal protein fermentation. Studies already have shown that chemical modification of proteins can be useful in decreasing the rumen fermentation of feed proteins (*92, 93*). The covalent attachment of methionine or its *N*-protected form to proteins in feeds could be of practical interest since this amino acid is considered to be the more limiting one for ruminants (*94, 95*) and to have faster intestinal uptake than nonessential amino acids (*96*). A hydroxy analog of methionine has been examined for its possible use instead of the natural amino acid (*97*). We suggest that feed fortification through covalent amino acid addition probably could be achieved by polymerizing methionine onto the feed proteins. The feasibility of multiple addition of essential amino acids to proteins is therefore important in this respect. However, the influence of chain length on functional properties and nutritional bioavailability of proteins is still unknown. The interactions of L-methionine oligopeptides in solution or covalently linked to the protein backbone may be completely different leading to different properties. For example, conjugation of poly-L-lysine to albumin (*98*) or poly D,L-alanyl peptides to asparaginases (*99*) enhanced the cellular uptake of proteins and prolonged the plasma clearance properties of the enzyme, respectively.

The currently available information indicates that the attachment or polymerization of amino acids to proteins should be a valuable procedure for modifying food proteins, but more research is needed to develop satisfactory processes and products. Procedures attuned to large-scale commercial processes need to be developed. Only a small amount (perhaps less than 1%) of the information available on these types of reactions with proteins deal with large-scale procedures, particularly from the economic standpoint.

Only a small amount of the information available on the properties of the products is useful for designing products for human use. Among the various biologically important areas that should be considered are the following: (1) the nutritional availabilities of the products with different types of covalent attachments; (2) the effects of long-term feedings of the different types of products; (3) the possible hazards due to side reactions, not only with different amino acid residues but also with other constituents sometimes present in food-grade proteins such as nucleic acids; and (4) the possibility that some individuals will respond differently to the modified products due to variations in their digestive systems which might lead to less efficient utilization or even to unfavorable physiological reactions.

Even less information is available on the functional properties of the different types of products that can be made. Naturally occurring proteins frequently have very different characteristics in foaming, whipping, gelling, coagulation, and dissolving properties. All of these characteristics should be amenable to changes by attaching amino acids. Among the various chemical substitutions that might be considered are the following: (1) different amino acids as substituents; (2) modification at different types of amino acid residues in the protein; (3) modification at different positions on the protein, e.g. at the COOH or $NH_2$ ends; and (4) different sizes of polymers of amino acids as substituents.

Although the above discussion might indicate that the attachment of amino acids to food proteins under commercial conditions is only remotely possible in the foreseeable future, many of the questions raised can be answered through carefully planned research. Some of the modifications may be useful, if not alone, perhaps in conjunction with enzymological methods. Therefore we encourage continued research in this potentially important area.

### *Literature Cited*

1. Bressani, R.; Elias, L. G.; Gomez Brenes, R. A. In "Protein and Amino Acid Functions"; Bigwood, E. J., Ed.; Pergamon: Oxford, 1972; p. 475.
2. Evans, R. J.; Bauer, D. H.; Sisak, K. A.; Ryan, P. A. *J. Agric. Food Chem.* **1974**, *22*, 130.
3. Liener, I. E. In "Protein Nutritional Quality of Foods and Feeds. Part 2"; Friedman, M., Ed.; Marcel Dekker: New York, 1975; p. 523.

4. Liener, I. E. *J. Am. Oil Chem. Soc.* **1979,** *56,* 121.
5. Ford, J. E.; Shorrock, C. *Br. J. Nutr.* **1971,** *26,* 311.
6. Carpenter, K. J.; Booth, V. H. *Nutr. Abstr. Rev.* **1973,** *43,* 424.
7. Cheftel, J. C. In "Food Proteins"; Whitaker, J. R.; Tannenbaum, S. R., Eds.; Avi: Westport, CT, 1977; p. 401.
8. Feeney, R. E. In "Chemical Deterioration of Proteins," *ACS Symp. Ser.* **1980,** *123,* 1.
9. Miller, R.; Groninger, H. S., Jr. *J. Food Sci.* **1976,** *41,* 268.
10. Richardson, T. In "Food Proteins: Improvement Through Chemical and Enzymatic Modification," *Adv. Chem. Ser.* **1977,** *160,* 185.
11. Whitaker, J. R. In "Food Proteins: Improvement Through Chemical and Enzymatic Modification," *Adv. Chem. Ser.* **1977,** *160,* 95.
12. Whitaker, J. R. *Food Technol.* **1978,** *32*(5), 175.
13. Beuchat, L. R. *Food Technol.* **1978,** *32,* 193.
14. Kinsella, J. E. *CRC Crit. Rev. Food Sci. Nutr.* **1978,** *10,* 147.
15. Jeunink, J.; Cheftel, J. C. *J. Food Sci.* **1979,** *44,* 1322.
16. Franzen, K. L.; Kinsella, J. E. *J. Agric. Food Chem.* **1976,** *24,* 914.
17. Barman, B. G.; Hansen, T. R.; Mossey, A. R. *J. Agric. Food Chem.* **1977,** *25,* 638.
18. Feeney, R. E. In "Food Proteins: Improvement Through Chemical and Enzymatic Modification," *Adv. Chem. Ser.* **1977,** *160,* 3.
19. Meyer, E. W.; Williams, L. D. In "Food Proteins: Improvement Through Chemical and Enzymatic Modification," *Adv. Chem. Ser.* **1977,** *160,* 52.
20. Fujimaki, M. *Ann. Nutr. Alim.* **1978,** *32,* 233.
21. Yamashita, M.; Arai, S.; Imaizumi, Y.; Amano, Y.; Fujimaki, M. *J. Agric. Food Chem.* **1979,** *27,* 52.
22. Puigserver, A. J.; Sen, L. C.; Gonzales-Flores, E.; Feeney, R. E.; Whitaker, J. R. *J. Agric. Food Chem.* **1979,** *27,* 1098.
23. Puigserver, A. J.; Sen, L. C.; Clifford, A. J.; Feeney, R. E.; Whitaker, J. R. *J. Agric. Food Chem.* **1979,** *27,* 1286.
24. Voutsinas, L. P.; Nakai, S. *J. Food Sci.* **1979,** *44,* 1205.
25. Altschul, A. M. *Nature (London)* **1974,** *248,* 643.
26. Harper, A. E.; Hegsted, D. M. In "Improvement of Protein Nutriture"; NAS-NRC: Washington, D.C., 1974; p. 184.
27. Benevenga, N. S.; Cieslak, D. G. In "Nutritional Improvement of Food and Feed Proteins," *Adv. Exp. Med. Biol.* **1978,** *105,* 379.
28. Sikka, K. C.; Johari, R. P. *J. Agric. Food Chem.* **1979,** *27,* 962.
29. Hernandez-Infante, M.; Herrador-Peña, G.; Sotelo-Lopez, A. *J. Agric. Food Chem.* **1979,** *27,* 965.
30. Matthews, D. M. In "Peptide Transport in Bacteria and Mammalian Gut"; Assoc. Sci. Publ.: Amsterdam, 1972; p. 71.
31. Ugolev, A. M.; Timofeeva, N. M.; Smirnova, L. F.; De Laey, P.; Gruzdkov, A. A.; Iezuitova, N. N.; Mityushova, N. R.; Roshchina, G. M.; Gurman, E. G.; Gusen, V. M.; Tsvetkova, V. A.; Shcherbakov, G. G. In "Peptide Transport and Hydrolysis," *Ciba Found. Symp.* **1977,** *50,* 221.
32. Kinsella, J. E. *J. Am. Oil Chem. Soc.* **1979,** *56,* 242.
33. Whitaker, J. R.; Fujimaki, M., Eds. In "Chemical Deterioration of Proteins," *ACS Symp. Ser.* **1980,** *123.*
34. Bergmann, M.; Fraenkel-Conrat, H. *J. Biol. Chem.* **1937,** *119,* 707.
35. Horowitz, J.; Haurowitz, F. *Biochim. Biophys. Acta* **1959,** *33,* 231.
36. Yamashita, M.; Arai, S.; Tsai, S. J.; Fujimaki, M. *J. Agric. Food Chem.* **1971,** *19,* 1151.
37. Fujimaki, M.; Arai, S.; Yamashita, M. In "Food Proteins: Improvement Through Chemical and Enzymatic Modification," *Adv. Chem. Ser.* **1977,** *160,* 156.
38. Monti, J. C.; Jost, R. *J. Agric. Food Chem.* **1979,** *27,* 1281.
39. Arai, S.; Yamashita, M.; Fujimaki, M. *Agric. Biol. Chem.* **1979,** *43,* 1069.

40. Morihara, K.; Oka, T. *Biochem. J.* **1977**, *163*, 531.
41. Widmer, F.; Johansen, J. T. *Carlsberg Res. Commun.* **1979**, *44*, 37.
42. Homandberg, G. A.; Mattis, J. A.; Laskowski, M., Jr. *Biochemistry* **1978**, *17*, 5220.
43. Kuhl, P.; Könnecke, A.; Döring, G.; Däumer, H.; Jakubke, H.-D. *Tetrahedron Lett.* **1980**, *21*, 893.
44. Sheehan, J. C.; Hess, G. P. *J. Am. Chem. Soc.* **1955**, *77*, 1067.
45. Sheehan, J. C.; Hlavka, J. J. *J. Org. Chem.* **1956**, *21*, 439.
46. Riehm, J. P.; Scheraga, H. A. *Biochemistry* **1965**, *4*, 772.
47. Hoare, D. G.; Koshland, D. E., Jr. *J. Am. Chem. Soc.* **1966**, *88*, 2057.
48. Wilchek, M.; Frensdorff, A.; Sela, M. *Biochemistry* **1967**, *6*, 247.
49. Carraway, K. L.; Koshland, D. E., Jr. *Methods Enzymol.* **1972**, *25*, 616.
50. Khorana, H. G. *Chem. Rev.* **1953**, *53*, 145.
51. Hoare, D. G.; Koshland, D. E., Jr. *J. Biol. Chem.* **1967**, *242*, 2447.
52. Li-Chan, E.; Helbig, N.; Holbek, E.; Chau, S.; Nakai, S. *J. Agric. Food Chem.* **1979**, *27*, 877.
53. Concon, J. M. *Anal. Biochem.* **1975**, *66*, 460.
54. Schwyzer, R.; Kappeler, H. *Helv. Chim. Acta* **1961**, *44*, 1991.
55. Zervas, L.; Borovas, R.; Gazis, E. *J. Am. Chem. Soc.* **1963**, *85*, 3660.
56. Carpino, L. A.; Han, G. Y. *J. Org. Chem.* **1972**, *37*, 3404.
57. Robinson, N. C.; Neurath, H.; Walsh, K. A. *Biochemistry* **1973**, *12*, 420.
58. Blumberg, S.; Vallee, B. L. *Biochemistry* **1975**, *14*, 2410.
59. Slotboom, A. J.; de Haas, G. H. *Biochemistry* **1975**, *14*, 5394.
60. Kowalski, D.; Laskowski, M., Jr. *Biochemistry* **1976**, *15*, 1309.
61. Wicker, C.; Puigserver, A., unpublished data.
62. Puigserver, A. J.; Sen, L. C.; Feeney, R. E.; Whitaker, J. R. *Ann. Biol. Anim. Biochim. Biophys.* **1979**, *19*, 749.
63. Puigserver, A. J.; Sen, L. C.; Clifford, A. J.; Feeney, R. E.; Whitaker, J. R. In "Nutritional Improvement of Food and Feed Proteins," *Adv. Exp. Med. Biol.* **1978**, *105*, 587.
64. Maroux, S.; Louvard, D.; Baratti, J. *Biochim. Biophys. Acta* **1973**, *321*, 282.
65. Takesue, Y. *J. Biochem. (Tokyo)* **1975**, *77*, 103.
66. Gray, G. M.; Santiago, N. A. *J. Biol. Chem.* **1977**, *252*, 4922.
67. Roncari, G.; Zuber, H. *Int. J. Protein Res.* **1969**, *1*, 45.
68. Feracci, H.; Maroux, S. *Biochim. Biophys. Acta* **1980**, *599*, 448.
69. Katchalski, E.; Sela, M. *Adv. Protein Chem.* **1958**, *13*, 243.
70. Bamford, C. H.; Block, H. In "High Polymers"; Frisch, K. C., Ed.; John Wiley & Sons: New York, 1972; Vol. 26, p. 687.
71. Hirschmann, R.; Schwam, H.; Strachan, R. G.; Schoenewaldt, E. F.; Barkemeyer, H.; Miller, S. M.; Conn, J. B.; Garsky, V.; Veber, D. F.; Denkewalter, R. G. *J. Am. Chem. Soc.* **1971**, *93*, 2746.
72. Pfaender, P.; Pratzel, H.; Blecher, H. U.S. Patent 3 951 741, 1976.
73. Kircher, K.; Berndt, H.; Zahn, H. *Justus Liebigs Ann. Chem.* **1980**, *2*, 275.
74. Akiyama, M.; Hasegawa, M.; Takeuchi, H.; Shimizu, K. *Tetrahedron Lett.* **1979**, *28*, 2599.
75. Halstrom, J.; Brunfeldt, K.; Kovacs, K. *Acta Chem. Scand.* **1979**, *B33*, 685.
76. Kanazawa, H.; Matsuura, Y.; Tanaka, N.; Kakudo, M.; Komoto, T.; Kawai, T. *Bull. Chem. Soc. Jpn.* **1976**, *49*, 954.
77. Kanazawa, H.; Kawai, T.; Ohashi, Y.; Sasada, Y. *Bull. Chem. Soc. Jpn.* **1978**, *51*, 2200.
78. Ibid., 2205.
79. Sela, M.; Katchalski, E. *Adv. Protein Chem.* **1959**, *14*, 391.
80. Szwarc, M. *Adv. Polym. Sci.* **1965**, *4*, 1.
81. Goodman, M.; Verdini, A. S.; Toniolo, C.; Phillips, W. D.; Bovey, F. A. *Proc. Natl. Acad. Sci. USA* **1969**, *64*, 444.
82. Goodman, M.; Naider, F.; Toniolo, C. *Biopolymers* **1971**, *10*, 1719.

83. Becker, J. M. *Biopolymers* **1974,** *13,* 1747.
84. Glazer, A. N.; Bar-Eli, A.; Katchalski, E. *J. Biol. Chem.* **1962,** *237,* 1832.
85. Virupakska, T. K.; Tarver, M. *Biochemistry* **1964,** *3,* 1507.
86. Kowalski, D.; Laskowski, M., Jr. *Biochemistry* **1976,** *15,* 1300.
87. Pieniazek, D.; Rakowska, M.; Szkilladziowa, W.; Grabarek, Z. *Br. J. Nutr.* **1975,** *34,* 175.
88. Boggs, R. W.; Rotruck, J. T.; Damico, R. A. *J. Nutr.* **1975,** *105,* 326.
89. Stegink, L. D.; Filer, L. J., Jr.; Baker, G. L. *J. Nutr.* **1980,** *110,* 42.
90. Endo, Y. *FEBS Lett.* **1978,** *95,* 281.
91. Endo, Y. *Biochim. Biophys. Acta* **1978,** *523,* 207.
92. Broderick, G. A. In "Protein Nutritional Quality of Foods and Feeds, Part 2"; Friedman, M., Ed.; Marcel Dekker: New York, 1975; p. 211.
93. Clark, J. H. In "Protein Nutritional Quality of Foods and Feeds, Part 2"; Friedman, M., Ed.; Marcel Dekker: New York, 1975; p. 261.
94. Champredon, C.; Pion, R.; Vérité, R. *Ann. Biol. Anim. Biochim. Biophys.* **1974,** *14,* 813.
95. Spires, H. R.; Clark, J. H.; Derrig, R. G.; Davis, C. L. *J. Nutr.* **1975,** *105,* 1111.
96. Bergen, W. G. *Fed. Proc., Fed. Am. Soc. Exp. Biol.* **1978,** *37,* 1223.
97. Champreden, C.; Pion, R.; Basson, W. D. *C. R. Hebd. Séances Acad. Sci., Ser. D* **1976,** *282,* 743.
98. Shen, W. C.; Ryser, H. J.-P. *Proc. Natl. Acad. Sci. USA* **1978,** *75,* 1872.
99. Uren, J. R.; Ragin, R. C. *Cancer Res.* **1979,** *39,* 1927.

RECEIVED November 14, 1980.

# 6

# Reversible Modification of Lysine: Separation of Proteins and Nucleic Acids in Yeast

JAYARAMA K. SHETTY and JOHN E. KINSELLA

Institute of Food Science, Cornell University, Stocking Hall, Ithaca, NY 14853

*Cyclic acid anhydride-modified ε-$NH_2$ groups of proteins destabilized the noncovalent interactions in yeast nucleoprotein complex facilitating the separation of nucleic acids from the modified protein at pH 4.0–4.2. Proteins modified using cyclic anhydride containing β-diene structure (cis double bond) were removed under acidic conditions rendering lysine nutritionally available. In this review, the application of reversible modifying reagents of the ε-$NH_2$ group of protein to separate nucleic acids and to isolate yeast proteins in an undenatured form is described.*

The quest for new sources of nutrients, especially proteins, must be continued despite localized surpluses and assurances that, statistically, there is a surplus of protein in the world. Research to expand productivity of conventional agriculture (crops, animals, and poultry), to develop new sources (marine, leaf, and microbial) of proteins, and to develop the basic knowledge base for successful exploitation of new sources must be continued because of the burgeoning population growth, the limitations of arable land, and the cost and shortage of petrochemical-based agricultural chemicals (fertilizers, pesticides, etc.) required for modern cropping systems.

Of the various new sources of nutrients, especially proteins, microbial biomass shows great potential. This applies particularly to yeasts which have a long tradition of successful use as foods, as producers of foods and beverages, and as food ingredients (*1, 2, 3, 4, 5*). The interest in and the supply of yeast and microbial biomass will increase rapidly with the growing activities in biomass conversion (alcohol production, etc.). Microbial biomass can become a significant source of food protein

0065-2393/82/0198-0169$07.50/0

in the near future if the appropriate technology for isolating and refining the protein is developed. The objective of research in our laboratory is to develop procedures that will facilitate the practical exploitation of yeast biomass as a source of nutrients and ingredients in foods (*1*).

In particular, we have concentrated on developing techniques for isolating protein; to attain this objective we have studied the practical use of chemical modification. The application of chemical modification to food proteins has been explored for several purposes: to block deteriorative interactions between reactive groups (e.g., $\epsilon$-$NH_2$ and reducing sugars); to improve functional properties (solubility, flavor, and thermal stability); to enhance nutritive value and digestibility; to facilitate the elucidation of interrelationship between structure and functional properties (*6, 7, 8, 9*); and, as discussed herein, to facilitate the preparation of protein isolates.

### *Chemical Modification*

Several comprehensive reviews concerning chemical modification of proteins (mostly enzymes) for elucidating structure functional relationships in biochemical reactions have been published. Means and Feeney (*9*) and Glazer et al. (*10*) discussed the chemistry of chemical modification of functional groups in proteins. Stark (*11*) and recently Delle (*28*) tabulated reagents according to their ability to alter charge and functionality, and Vallee and Riordan (*12*) presented a list of the common reagents that are used. Most of the derivatizing agents or compounds used could not be used for food proteins but these reviews and the current literature provide a source of ideas and information on protein modification. Fraenkel–Conrat et al. (*13*) acetylated ovomucoid (egg white trypsin inhibitor) and trypsin and studied their interactions. Riordan and Vallee (*14*) acylated carboxypeptidase with dicarboxylic anhydrides (succinic, glutaric, methylsuccinic, $\alpha,\alpha$-dimethylglutaric, and $\beta,\beta$-dimethylglutaric) and determined its peptidase and esterase activities. These activities then were related to factors such as modification of tyrosyl residues in the active center of the enzyme. Gounaris and Perlmann (*15*) succinylated pepsinogen and determined its amino acid composition, absorption spectrum, optical rotatory dispersion, the availability of COOH-terminal alanine to digestion with carboxypeptidase A, and the activation of S-pepsinogen. Later, Nakagawa and Perlmann (*16*) modified porcine pepsinogen with citraconic and maleic anhydrides and studied the properties of these derivatives in relation to the nature and extent of structural changes in the protein. For example, they followed the reversibility of the reactions and the amount of potential pepsin activity that could be recovered when activating the derivatives. From optical rotatory dispersion (ORD) and circular dichroism (CD) measurements, a conformational change occurred in all of these derivatives, and the reappearance of potential pepsin activity paralleled the change in optical rotatory properties. Succinylation is also an important tool in hybridization experiments that involve the reassociation of protein subunits with those that have been derivatized (*17*). Means and Feeney (*18*) reductively alkylated amino groups in proteins (lysozyme, insulin, ribonuclease, turkey ovomucoid, human serum transferrin, $\alpha$-chymotrypsin, and chymotrypsinogen) and observed changes in their physical properties using UV absorption spectra, sedimentation, and ORD measurements. Paik and Kim (*19*) investigated the effect of methylation on the susceptibility of pancreatic ribonuclease, polylysine, and calf thymus arginine-rich histone to proteolytic enzymes (trypsin and $\alpha$-chymotrypsin). These biochemical studies have provided us with chemical methods for protein modification and have described techniques for studying structural and functional alterations.

Of the numerous acylating agents, succinic anhydride has been used most often (*9*). Succinylated proteins usually show increases in intrinsic viscosity and concomitant decreases in sedimentation coefficient (*9, 21*). Succinylation of hemerythrin causes dissociation into subunits (*22*) and extensive succinylation of BSA increases its Stokes radius and the susceptibility of its disulfide bonds to reduction, and decreases its ability to precipitate anti-BSA immunoglobulins (*23*). The principal reaction of succinic anhydride with protein is through the $\epsilon$-amino group of lysine (*24*). Mühlrad et al. (*25*) found that a large molar excess of succinic anhydride was necessary to succinylate a significant percentage of the $\epsilon$-amino groups of lysine. Succinic anhydride also can react with other functional groups on the protein, i.e. tyrosine, histidine, and aliphatic hydroxy amino acids (*15, 20, 25, 26, 27*).

The chemical modification and characterization of food proteins have not been studied as extensively as have those of enzymes and other proteins of biochemical interest. Oppenheimer et al. (*29, 30*) observed that succinylated chicken myosin had increased viscosity only a molecular size similar to that of unmodified myosin. Egg white modified with 3,3-dimethylglutaric anhydride has increased heat stability (*31*) and *N*-succinylated egg yolk proteins are useful in making mayonnaise and salad dressings (*32*). Evans et al. (*32*) prepared acetyl, propionyl, *n*-butyryl, *n*-hexanoyl, *n*-octanoyl, and *n*-decanoyl derivatives of $\beta$-casein and studied their physiochemical properties, i.e., electrophoresis, sedimentation coefficients, partial specific volume, ORD, and nuclear magnetic resonance (NMR). Hoagland et al. (*33*) acetylated $\beta$-casein and investigated its response to tryptic hydrolysis and gel electrophoresis. He also studied the changes in selected properties of acetylated and succinylated $\beta$-casein, i.e., mobility during alkaline polyacrylamide gel electrophoresis (PAGE), elution from DEAE cellulose column, calcium ion sensitivity, and sedimentation patterns produced by removing positive charges and adding negative charges (*34*). These results then were used to evaluate the role of electrostatic interactions in the aggregation of $\beta$-casein.

Several workers have chemically modified milk proteins ($\kappa$-casein, casein) to determine the chemical basis for the stabilization of the casein micelle by $\kappa$-casein and to elucidate the mechanism of rennin action (*35, 36*). Creamer et al. (*37*) made several derivatives of sodium caseinate and showed their improved solubility. In preliminary feeding trials in which succinylated casein was the only protein source, supplementation with lysine was required for the normal growth of rats.

Groninger (*38*) succinylated fish myofibrillar protein and examined some of its chemical and functional properties. The fish myofibrillar proteins were succinylated at different levels and the degree of succinylation was related to a functional property such as emulsification capacity. The protein efficiency ratio for succinylated protein was lower than that of untreated fish protein. Grant (*39*) succinylated wheat flour proteins and analyzed their solubility, viscosity, and chromatographic behavior. The effects of acetylation and succinylation on the physicochemical and functional properties of several plant proteins was reviewed recently by Kinsella and Shetty (*6*).

The improvement in the nutritional quality of low-grade food proteins by covalent attachment of limiting essential amino acids has been demonstrated and discussed (*40, 41, 42*). Much of the available data indicates that chemical modification can be used to improve many desirable characteristics of food protein. However, in certain cases, the modifying agents may impair the digestibility of the modified protein or more generally the modified protein would not meet current regulatory specifications. However, because of the several practical benefits of modification, we need to explore the possibility of using reversible chemical modifying reagents for application to food proteins.

### *Reversible Modifying Reagents for Amino Groups in Proteins*

Several reversible blocking reagents for amino groups in proteins have been reported (*see* Table I). Succinyl derivatives cannot be removed easily from succinylated proteins under mild conditions; however, Braunitzer et al. (*43*) have shown that tetrafluorosuccinyl derivatives could be removed at pH 9.5 and 0°C (*see* Reaction 1):

$$\text{(P)—NH—}\overset{+}{\underset{\nwarrow\,^{-}OH}{\overset{O}{\overset{\|}{C}}}}\text{—CF}_2\text{—CF}_2\text{—COO}^- \xrightarrow[0°C]{pH\ 9.0} \text{(P)—}\ddot{N}H_2 + {}^{-}OOC\text{—CF}_2\text{—CF}_2\text{—COO}^- \quad (1)$$

This is because the large electron-withdrawing effects of the fluorine groups induce a larger positive charge at the carbonyl carbon than exists in the succinyl derivatives. This facilitates the nucleophilic attack by hydroxyl anions and results in hydrolysis of the peptide bond attached to the tetrafluorosuccinyl group. Nevertheless, the deacylation is very slow and only 60% of the tetrafluorosuccinylated lysine was deacylated in myoglobin (*44*) and 50% in lysozyme (*45*) after 4 d of incubation at pH 9.5 and 0°C.

Deacylation of derivatives formed by acid anhydrides (maleic, citraconic, and 2,3-dimethylmaleic) containing double bonds (cis) occurs at a much faster rate under mild conditions. Several mechanisms have been proposed for deacylating these derivatives formed from the double bond containing anhydrides (*9, 46, 47, 48, 49*). The most important structural factor in the deacylation process is the cis double bond which maintains the terminal carboxyl group in the spatial orientation that makes a nucleophilic attack on the amide carbon much more probable.

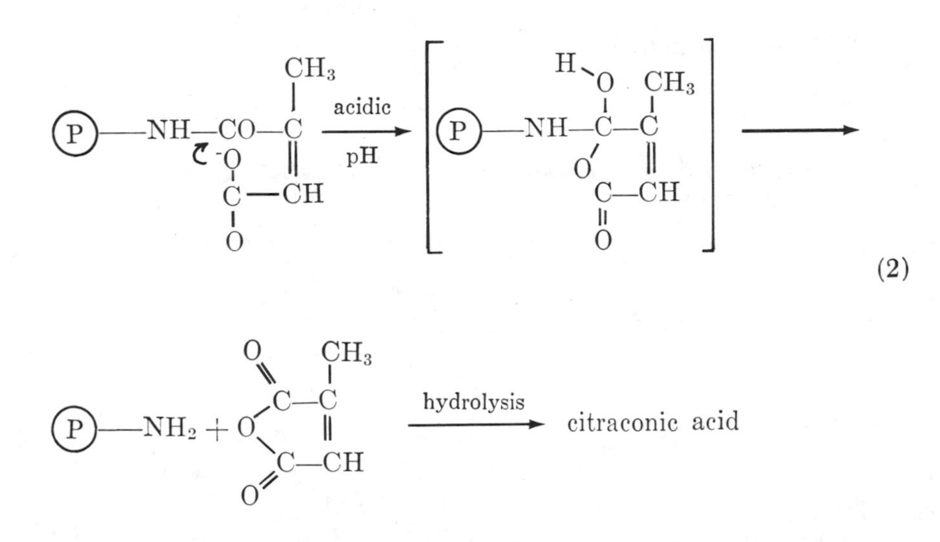


In succinyl derivatives, the freely rotating single bond between carbon 2 and 3 allows the terminal carboxyl group to assume many more orientations and as a result this drastically reduces the probability that it will stay in the proper conformation long enough for deacylation to occur. Kirby et al. (*46*) proposed an intramolecular catalysis of amide bond hydrolysis by a proton transfer from external general acids and also showed that in dilute acid the *O*-protonated amide is the reactive species that initiates the deacylation process (*see* Reaction 3).

Butler et al. (*59*) showed that methylating the unsaturated carbons in the maleic anhydride to give citraconic anhydride had a profound effect on the rate of deacylation. While deacylation of the maleyl amide

**Table I. Reversible Modifying Reagents**

| *Reagents* | |
|---|---|
| Reversible reductive alkylation | 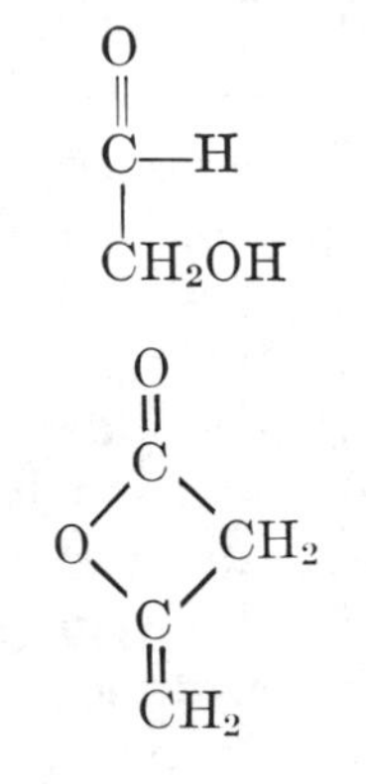 |
| Diketene | |
| Ethyl thiotrifluoroacetate | 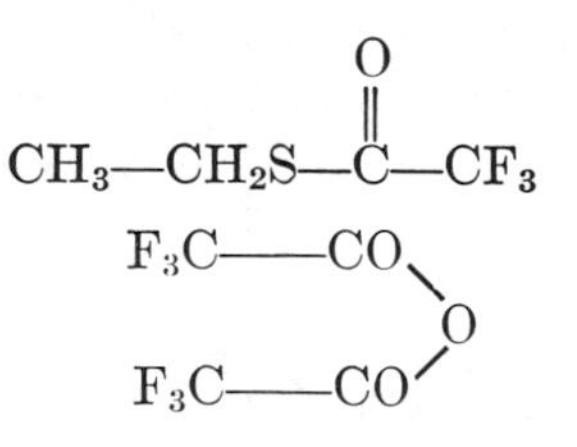 |
| Trifluoroacetic anhydride | |
| *exo-cis*-3,6-Endoxo-$\Delta^4$-tetrahydrophthalic anhydride | 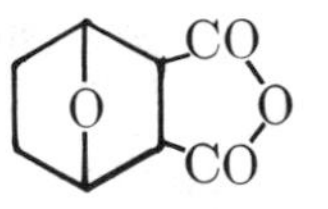 |
| *exo-cis*-3,6-Endoxo-$\Delta^4$-hexahydrophthalic anhydride | 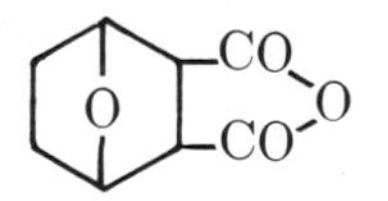 |
| Tetrafluorosuccinic anhydride | 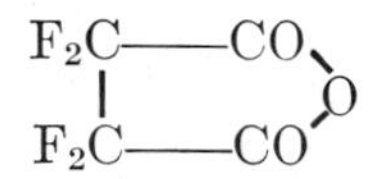 |
| Maleic anhydride | 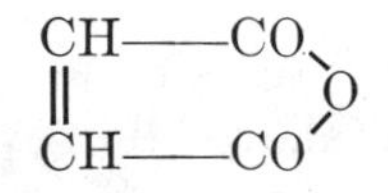 |
| Citraconic anhydride | 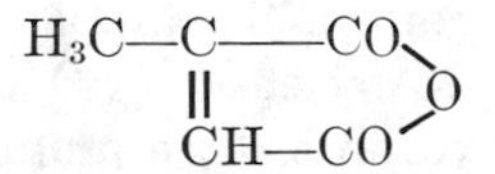 |
| 2,3-Dimethylmaleic anhydride | $H_3C—C—CO$ / $\Vert$ O / $H_3C—C—CO$ |

**for Amino Groups in Proteins**

| *Conditions for Reversal* | *References* |
| --- | --- |
| periodate treatment (10–20mM) at pH 7.5–9.3 for 30 min | *53* |
| threefold excess of hydroxylamine at pH 7, 25°C for 12 h | *54, 55* |
| 1*M* piperidine at 0°C | *56, 57* |
| 1*M* piperidine for 2 h at 0°C or carbonate buffer pH 10.7 at 25°C for 30 h | *58* |
| pH 3 for 4–5 h at 25°C | *50* |
| pH 3 for 4–5 h at 25°C | *50* |
| slow and incomplete, pH 9.5 for several days at 0°C | *43, 44, 45* |
| pH 3.5 for 11–12 h at 37°C | *49* |
| pH 3.5 for 3 h at 37°C | *51* |
| pH 3.5 for 5 min at 20°C | *45, 51* |

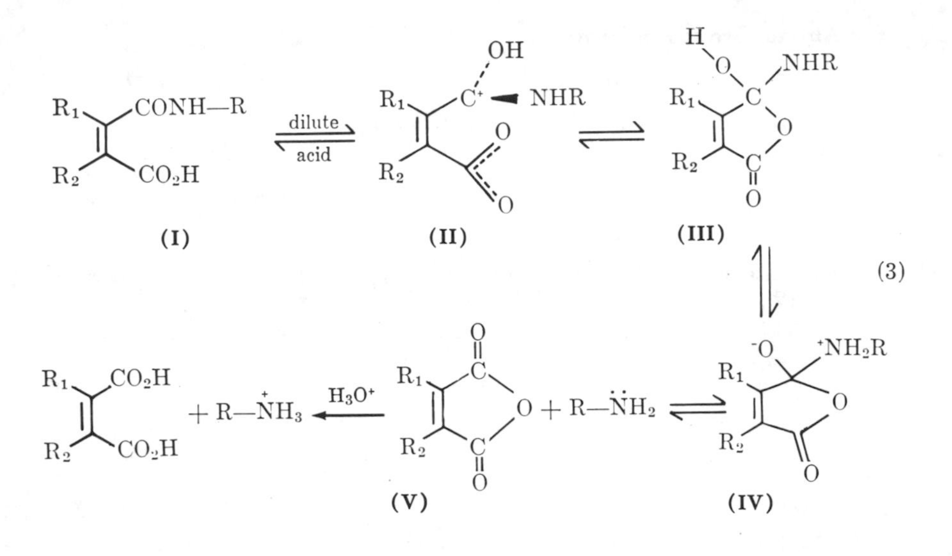


required a half-time ($t_{1/2}$) of 11–12 h at pH 3.5 and 37°C; the citraconyl (2-methylmaleic) amide was deacylated completely in 3 h (*45, 59*). The 2,3-dimethylmaleic derivative is extremely acid labile, deacylation being complete within 5 min at pH 3.5 and 20°C (*51*). Steric and electronic effects of the methyl substituents result in a net destabilization of the amide bond to acid-catalyzed ($H^+$) hydrolysis. This would explain why the 2,3-dimethylmaleic amide bond is the most labile of the three and the maleic amide bond is the strongest.

The reversibility of deacylation with these anhydrides has proved to be very useful in peptide chemistry. Citraconylation of insulin inhibited trypsin in hydrolysis at lysine but not at arginine (*51*). A method of diagonal paper electrophoresis was developed from trypsin digests of maleylated proteins which allowed the isolation of lysine and *N*-terminal peptides by deacylating them prior to electrophoresis in the second dimension (*59*). Although the extreme lability of 2,3-dimethylmaleic anhydride can pose problems, Puigserver and Desnuelle (*60*) found it very useful in the dissociation of carboxypeptidase A subunits which could be deacylated then to their native conformation by overnight dialysis at pH 6.0. By reversibly maleylating antibodies against negatively charged haptens, Freedman et al. (*61*) showed that positively charged lysine residues were required for binding of the anionic haptens.

The usefulness of a reversible reagent depends on the ease of removal of the blocking groups under conditions that will not lead to denaturation or impair recovery of critical biological properties. Studies with several reversible blocking reagents for amino groups in myoglobin and lysozyme showed that citraconic anhydride was the most satisfactory reagent, yielding homogeneous preparations identical in structure and function with the respective native proteins after deblocking (*45, 52*).

### *Application of Chemical Modification for Refining Proteins: Yeast Proteins*

**Potential of Yeast Proteins and Problems.** The current food shortages and continually rising prices necessitate exploring novel sources of food proteins (*1, 2, 3, 4, 5*). The possible use of yeast as a source of protein for human consumption has received much attention because of its ability to grow rapidly on a wide variety of substrates and its high content of good-quality proteins (*1, 62, 63*). In fact, yeasts and other microbial cells are expected to become significant sources of food-grade proteins in the future as the world population and demand for food exceeds the food supply from conventional sources.

The presence of cell wall materials and the high content of nucleic acids (NA) limit the full exploitation of yeast for human consumption. Cell wall material in unfractionated single-cell proteins (SCP) is undesirable because it reduces the bioavailability of proteins; it contains antigenic, allergenic agents, and factors causing nausea and gastrointestinal disturbances (*64, 65, 66*). For the above reasons and for many potential food applications of the protein, it is necessary to separate proteins from cell wall materials. Isolating proteins from yeast cells improves the digestibility by 50% compared with the broken cell which includes cell wall materials (*67*). However, the release of protein from microbial cells by such methods as autolysis, lytic enzymes, and chemical treatments causes considerable protein denaturation and degradation. This decreases yield, impairs digestibility, and reduces nutritional quality (*5, 68, 69, 70*).

The most significant problem associated with the utilization and consumption of yeast proteins is the high content of nucleic acids, mostly ribonucleic acid (RNA), the quantity of which may reach one-third of the total protein (*2, 67*). While the protein per se is of high nutritional value the presence of NA limits its use.

Normally dietary NA is hydrolyzed by pancreatic ribonucleases and absorbed in the small intestine. Guanine and adenine are metabolized further to uric acid before urinary excretion (*2*). Increased consumption of NA increased the uric acid levels above the urinary excretion rate, resulting in an increased plasma uric acid level. Uric acid has a p*K*a near 5.4. In the acidic urine, as much as half of the excreted compound may be in the form of the undissociated acid, which has a low solubility

(0.1 g/L of pure water at 38°C). The upper limit for the normal concentration of urate in the blood is about 0.35m*M* (60 μmol). The lower solubility of uric acid results in the precipitation of uric acid crystals that accumulate in the joints resulting in the disease called gout. This problem is alleviated in other mammals by the enzyme uricase which oxidizes uric acid to soluble and excretable allantoin; however, man lacks uricase and consequently consumption of conventionally prepared yeast protein could result in the development of gout (*71*). Scrimshaw (*64*) discussed several clinical studies concerned with the safe dietary level of NA for humans and recommended $< 2$ g/d for healthy adults. Thus the high content of NA in the conventionally prepared protein from yeast necessitates developing methods to reduce NA levels in the protein that is isolated for human consumption.

Different chemical and enzymatic methods for reducing the NA content in yeast proteins have been reported (*1, 68*). The most common method for obtaining protein with a low NA level consists of extracting the protein from mechanically disrupted yeast cells at an elevated temperature ($>$ 60°C) in concentrated alkali followed by precipitating the extracted protein at acidic pHs (*72, 73, 74, 75*). Treating food-grade proteins with alkali is undesirable because it causes denaturation and/or degradation of proteins which generally adversely affects the nutritional and functional properties. These changes include racemization of the amino acids, elimination and cross-linking of certain amino acids, and formation of potentially antinutritive compounds (*76, 77, 78, 79*). It has been known for many years that alkali treatment of proteins results in the decomposition of cystine, serine, phosphoserine, and threonine residues to form the corresponding unsaturated amino acids, i.e., dehydroalanine, etc. (*81, 82*). Unusual amino acids are formed by reacting dehydroalanine with the different nucleophiles in alkali-treated proteins (*see* Scheme I). Recently we observed that treating disrupted yeast cells with high concentrations of alkali at elevated temperatures ($>$ 60°C) to reduce the NA content in yeast protein caused the destruction of amino acids and the formation of lysinoalanine (*see* Table II). Alkali-treated proteins containing lysinoalanine show reduced in vivo and in vitro digestibility (*83, 84, 85*). Alkali treatment of soy proteins may induce nephrotoxic properties that are related to the lysinoalanine content (*86*). So while alkali processing reduces the NA level in the yeast protein isolate, it causes undesirable chemical changes that may adversely affect the functional properties, nutritive value, and safety of these products.

Endogeneous and exogeneous ribonucleases have been used also to reduce NA under mild conditions (*68*). However, extensive proteolysis during incubation for RNA hydrolysis and the cost of the enzymes present sizable drawbacks (*68, 87*).

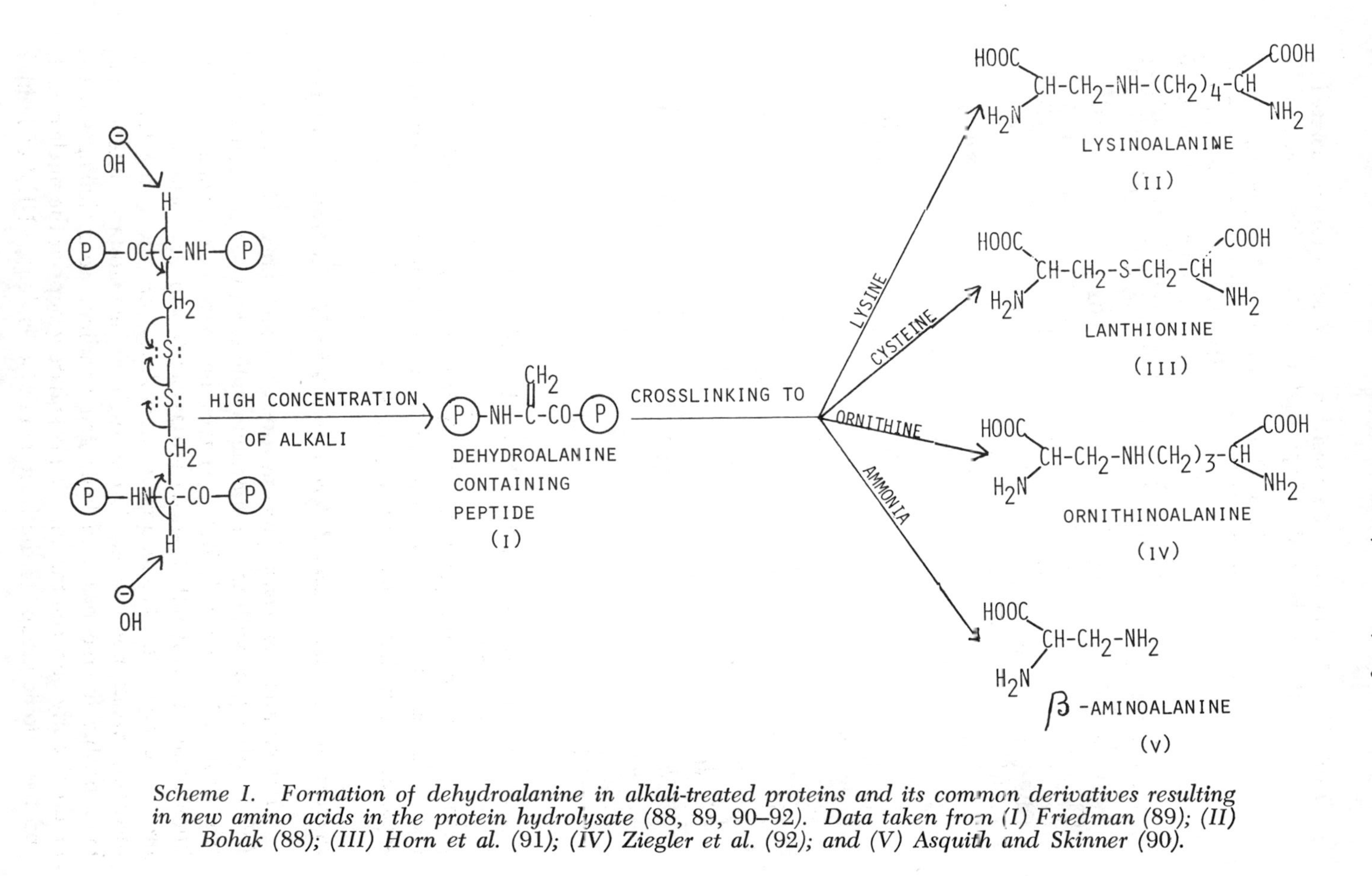


*Scheme I. Formation of dehydroalanine in alkali-treated proteins and its common derivatives resulting in new amino acids in the protein hydrolysate* (88, 89, 90–92). *Data taken from (I) Friedman* (89); *(II) Bohak* (88); *(III) Horn et al.* (91); *(IV) Ziegler et al.* (92); *and (V) Asquith and Skinner* (90).

**Table II. Amino Acid Composition of Yeast Protein Isolated by Different Methods (Expressed as g/16 g Nitrogen)**

| | *Yeast Protein Method of Preparation* | | | |
|---|---|---|---|---|
| *Amino Acid* | *Whole Yeast*[a] | *Succinylation Method*[a] | *High Temperature, Low Alkali, 85°C, pH 10.5, 2 h*[b] | *Low Temperature, High Alkali, 65°C, pH 12.5, 4 h*[b] |
| Aspartic acid | 7.98 | 10.14 | 8.02 | 9.18 |
| Threonine | 3.81 | 4.28 | 3.59 | 3.42 |
| Serine | 3.74 | 3.87 | 3.64 | 2.85 |
| Glutamic acid | 10.81 | 12.41 | 9.38 | 10.83 |
| Proline | 4.50 | 4.19 | 3.35 | 3.56 |
| Glycine | 3.08 | 3.47 | 2.95 | 3.33 |
| Alanine | 4.43 | 5.05 | 4.56 | 4.80 |
| Half cystine | — | 1.48 | 0 | 0 |
| Valine | 4.96 | 6.08 | 5.12 | 5.82 |
| Methionine | 1.82 | 2.09 | 1.15 | 1.43 |
| Isoleucine | 4.46 | 5.44 | 3.75 | 4.68 |
| Leucine | 6.59 | 7.94 | 6.18 | 7.36 |
| Tyrosine | 3.20 | 4.08 | 3.13 | 4.05 |
| Phenylalanine | 4.02 | 4.93 | 3.79 | 3.88 |
| Histidine | 1.94 | 2.42 | 1.59 | 1.83 |
| Lysine | 5.72 | 7.65 | 5.01 | 5.17 |
| Ammonia | 2.69 | 2.05 | 0.30 | 0.25 |
| Arginine | 4.17 | 5.27 | 3.94 | 3.90 |
| Lysinoalanine | — | — | 0.49 | 3.59 |

[a] Ref. *87*.
[b] Ref. *80*.

In summary the deleterious effects of alkali on the proteins and the large-scale impracticality of the heat–shock (endogeneous ribonuclease) process, plus the accompanying proteolysis and denaturation of proteins, clearly indicate the need for better methods to facilitate the large-scale separation of nucleic acids from yeast proteins.

### *Protein and Nucleic Acid Interactions: The Nucleoprotein Complex*

Precipitation of yeast proteins isoelectrically (pH 4.5), thermally (80°C, pH 6), or with salt (ammonium sulfate) causes the coprecipitation of nucleic acids (*93, 94, 99*) suggesting that nucleic acids in yeast homogenate(s) (disrupted yeast) are associated in the form of nucleoprotein complexes. Nucleoprotein complexes are held together by relatively weak noncovalent forces such as electrostatic interactions, hydrophobic interactions, and hydrogen bonding. Ionic linkages occur predominantly between the anionic phosphate groups of the nucleic acids and the cationic groups of the basic amino acids (e.g., $\epsilon$-$NH_3^+$ of lysine) of the proteins, i.e., a case of polycations combining with polyanions (*see* Scheme II). Because of the large number of such weak associations, the complex is quite stable under mild conditions. Disruption of these ionic interactions should facilitate separating proteins from the NAs. Modifying either the anionic groups of RNA or the $\epsilon$-$NH_2$ group of the protein may help achieve this objective.

### *Destabilization of Noncovalent Interactions: Lysine Modification*

There are several reagents that are used to modify the specific functional groups of proteins (*9, 10*). Using cyclic acid anhydrides to acylate free amino groups in proteins has provided valuable information concerning the physicochemical behavior and subunit interactions of various proteins (*9, 21, 95, 96, 97, 98*). Succinylation destabilizes oligomeric proteins and causes dissociation into subunits, particularly when the quaternary structure of the protein is stabilized by ionic interactions (*96, 100*). The increased electronegativity in an oligomeric protein after succinylation destabilizes the proteins due to repulsion between the subunits and results in dissociation. Based on this rationale we speculated that the ionic interactions that are responsible for the stability of the nucleoprotein complex could be destabilized also in an analogous manner.

### *Separation of Proteins from the Nucleoprotein Complex in Yeast by Succinylation Procedure*

In this method (*87, 101*) succinic anhydride was added continuously to the homogenized cells while the pH was maintained at 8.5 with 3*N* NaOH. Upon completing the reaction, the insoluble cell wall materials were separated. The pH of the supernatant was decreased then to 4.2 to precipitate the protein. The NA content of the protein progressively decreased with increasing succinylation (*see* Figure 1). More than 95% of the NAs remained in the solution at pH 4.2 when the protein was succinylated above 90%. Succinylation of proteins increased the net electronegativity and caused the electrostatic repulsions between the added carboxylic groups and the anionic phosphate groups of the nucleic acids which then destabilized the nucleoprotein complex (*see* Scheme II B). This procedure facilitated separating NAs and proteins at pH 4.2 and offered several significant advantages over existing methods. It increased the protein extractability producing a protein isolate containing 15–32% nitrogen, 1.8% NA, 7% carbohydrate, and 3.5% lipid. It eliminated proteolysis and there was no insolubilization of the protein as occurs during the heat–shock process for RNA hydrolysis (*87*). The amino acid composition of the protein that was isolated by the succinyla-

*Scheme II. Schematic indicating the manner in which cyclic acid anhydride derivatization of yeast protein by disrupting electrostatic inter-*

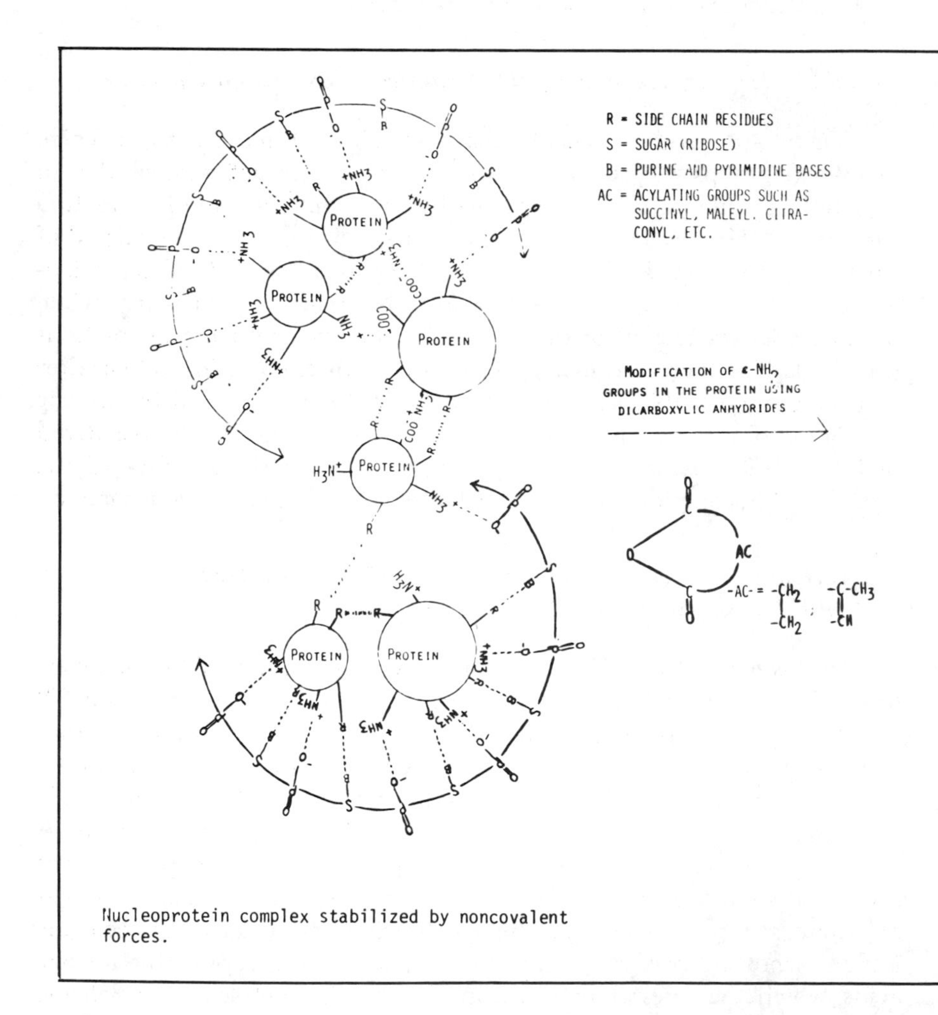


Nucleoprotein complex stabilized by noncovalent forces.

*actions between the ε-$NH_2$ group of protein and phosphate group of NAs facilitates the isolation of protein essentially free of RNA from microbial cells*

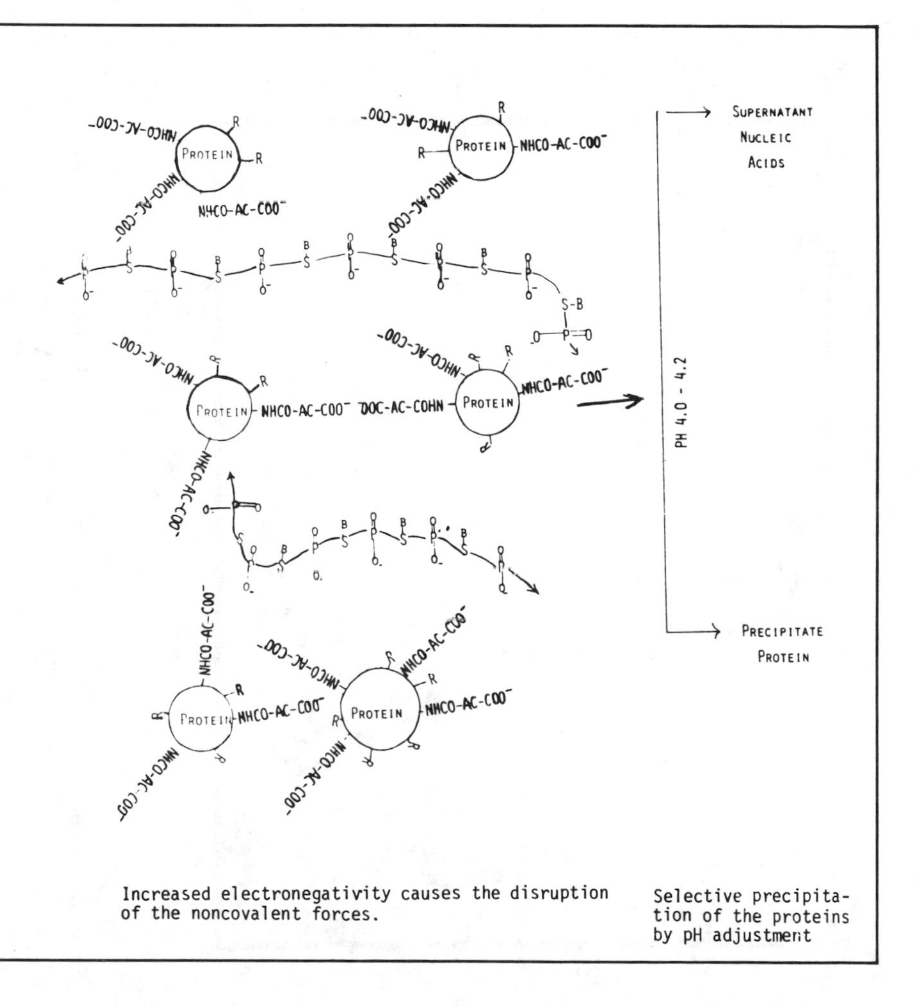

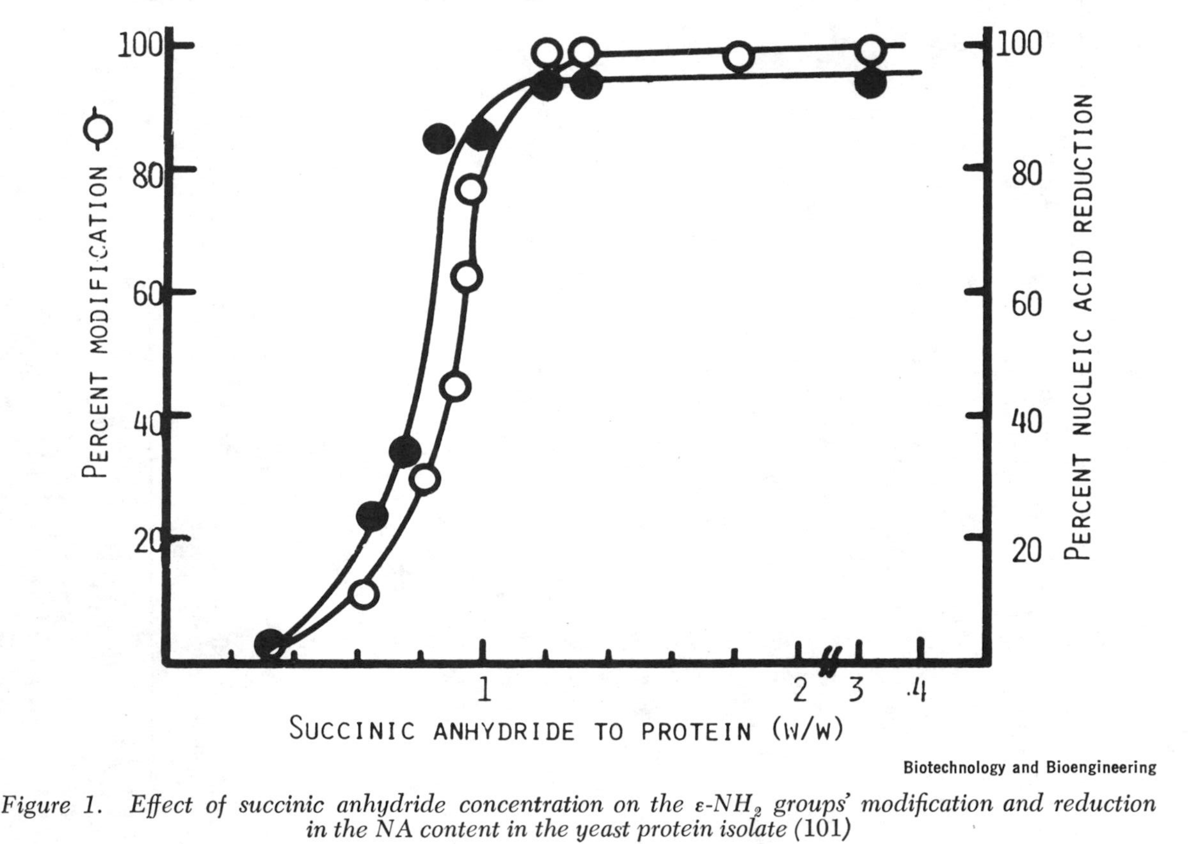


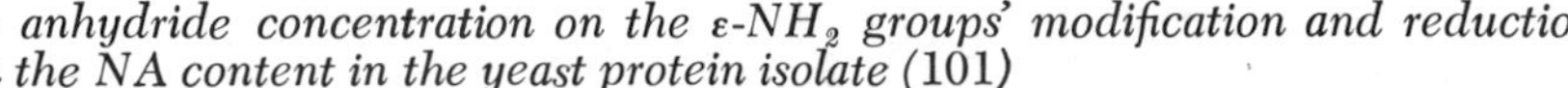


*Figure 1. Effect of succinic anhydride concentration on the ε-$NH_2$ groups' modification and reduction in the NA content in the yeast protein isolate* (101)

tion procedure showed an increased content of each amino acid compared with the protein that was isolated by the alkaline method (*see* Table II). The succinylated protein is white, bland, highly soluble, and possessed good functional properties (emulsion, stabilization, gelation, and solubility in the acidic pH range).

The only disadvantage of the succinylation procedure (which is practical and amenable to conventional cell disruption processes) is that the final product is a succinylated protein. Succinyl groups cannot be removed from the succinylated proteins under mild conditions. This could be a problem if succinylated yeast protein was a major source of dietary proteins. Therefore we explored the feasibility of using reversible modifying reagents (citraconic anhydride and maleic anhydride) to separate proteins from NAs and subsequently remove the modifying groups under mild acidic conditions.

## *Reversible Modification of Amino Groups of Lysine in Yeast Proteins Using Citraconic and Maleic Anhydrides*

**Isolation of Proteins with a Reduced Nucleic Acid Level.** The procedure is virtually identical to that described for succinylation of yeast proteins (*87*). In a typical experiment proteins, together with NA, were extracted from the disrupted yeast cells at pH 8.5–9.0 and centrifuged at 15,000 rpm for 30 min at 5°C. Citraconic anhydride then was added in small increments to the supernatant with constant stirring while the pH was maintained between 8.0–8.5 by adding 3.5*N* NaOH. After the stabilization of the pH, the pH of the solution was decreased to 4.2 to precipitate the proteins. Protein then was separated by centrifugation, dissolved in water (pH adjusted 8.5), dialyzed extensively against water (pH 8.5) at 5°C, and lyophilized.

The number of amino groups that were modified and the NA content in the protein precipitated at pH 4.2 then were determined by 2,4,6-trinitrobenzenesulfonic acid (*102*) and orcinol methods (*103*), respectively. Modifying the amino groups in yeast proteins increased with increasing concentration of the anhydride (*see* Figure 2). However, the extent of modifying the amino group of lysine in yeast proteins was greater with maleic anhydride (*see* Figure 2) compared with citraconic anhydride for the same molar ratio of anhydride to lysine. This indicated that the substitution of a methyl group in maleic anhydride decreased the reactivity of the anhydride. A maximum of 88% modification was achieved when citraconic anhydride was used whereas it was 95% with maleic anhydride.

The extent of NA reduction at pH 4.2 was proportional to the extent of lysine modification when citraconic anhydride was used (*see* Figure 3). The NA was separated more effectively in the citraconylated yeast proteins compared with maleylated yeast proteins. Comparing the 50% levels of modification, maleylation resulted in a 12% reduction in the NA whereas it was 50% following citraconylation.

Thus modifying the lysine amino group in yeast proteins using maleic or citraconic anhydride altered the NA–protein interactions and facilitated separating proteins from NAs at pH 4.2.

**Nutritional Availability of Acylated Lysine and Proteins.** The successful use of chemically derivatized proteins as food ingredients requires that they be digestible and nontoxic, and that the modified amino acid residues should be available nutritionally. Nutritional studies using modified food proteins are limited. The nutritional availability of several acylated lysines were studied by Bjarnason and Carpenter (*105*) and Mauron (*106*) and the results are summarized in Table III. The bioavailability of the acylated lysine varied significantly with the type of the acyl groups (*see* Table III). In addition acylated proteins (acetylated and succinylated) gave lower responses to the growth activity for the rats than equivalent supplements of unmodified proteins

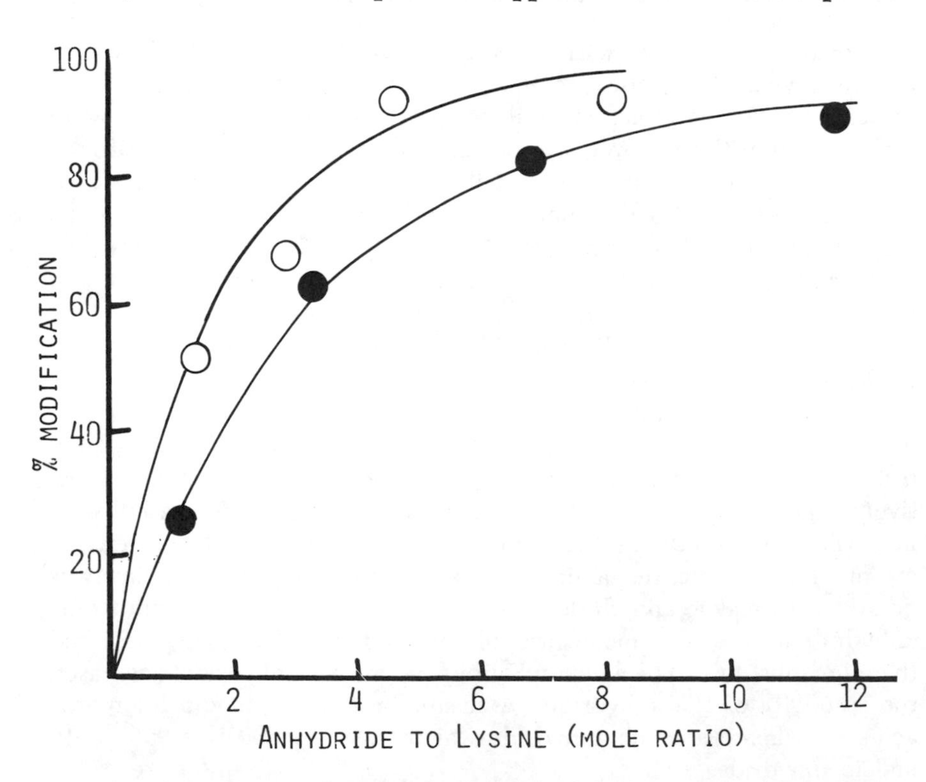


*Figure 2. Effect of cyclic acid anhydride concentration on the modification of ε-$NH_2$ groups in the yeast proteins: maleic anhydride (○); citraconic anhydride (●).*

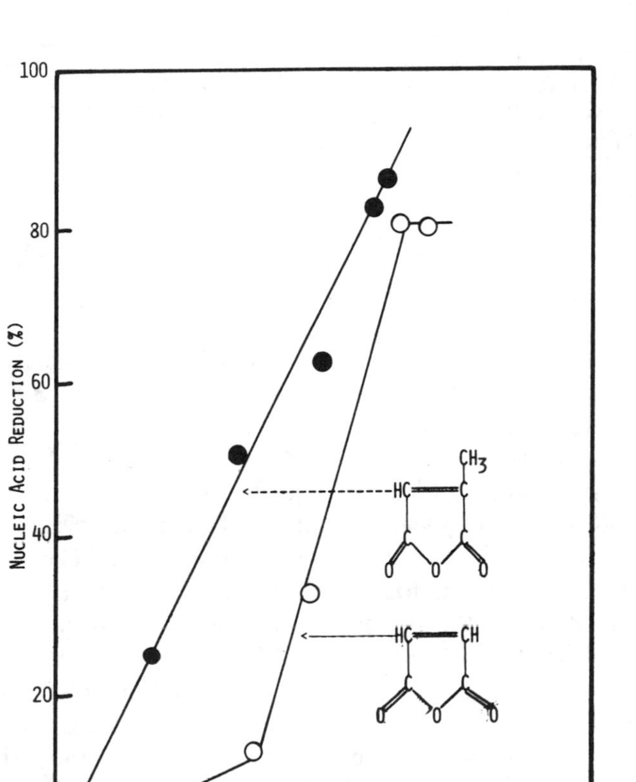


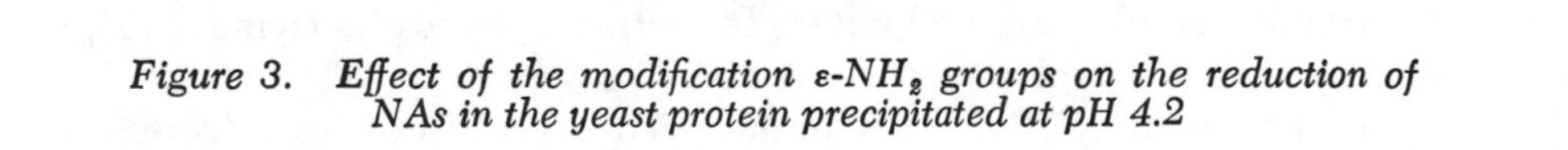

*Figure 3. Effect of the modification ε-$NH_2$ groups on the reduction of NAs in the yeast protein precipitated at pH 4.2*

**Table III. Nutritional Availability of Acylated Lysine Derivatives**[a]

| *Derivative* | *Percent Utilization as a Source of Lysine* | *Animal* |
|---|---|---|
| ε-*N*-Formyl-L-lysine | ~ 50 | rat |
| ε-*N*-Acetyl-L-lysine | ~ 50 | rat, chick |
| ε-*N*-(γ-Glutamyl)-L-lysine | ~ 100 | rat, chick |
| ε-*N*-(α-Glutamyl)-L-lysine | ~ 100 | rat |
| ε-*N*-Glycyl-L-lysine | ~ 80 | rat |
| ε-*N*-Glycyl-L-lysine | ~ 100 | rat |
| ε-*N*-(*N*-Acetylglycyl)-L-lysine | 0 | rat |
| ε-*N*-Propionyl-L-lysine | 0 | rat |
| ε-*N*-Propionyl-L-lysine | ~ 70 | chick |

[a] Ref. *108*.



(*37, 104*). Thus the successful application of acylated yeast proteins for human foods may, in certain cases, require removing modifying groups from the proteins after separating the NAs.

**Removal of Modifying Groups from the Modified Yeast Proteins.** Since citraconyl amide bond is acid labile, the citraconylated yeast proteins (90% modified) were incubated at different acidic pHs, i.e., pH 6, 5, 4, and 3 for different intervals of time at 30°C. The pH was controlled using an automatic pH stat during the incubation of the samples. After specified times, the pH of the samples was adjusted to 8.5 using 1*N* NaOH. The samples then were dialyzed and lyophilized. The rate at which the citraconyl groups were removed from the citraconylated proteins was determined by measuring the free ε-$NH_2$ group of lysine (*see* Figure 4). Deacylation increased by decreasing the pH from 6 to 4 and followed in the order of pH 3–4 > 5 > 6. Complete deacylation occurred at pH 3 and 4 within 3 h and it took 5 h at pH 5.0. However, at pH 6, the rate of deacylation was much slower and only 80% of the modified amino groups were deacylated even after a prolonged incubation. Similar results were reported for citraconylated lysozyme (*45*), citraconylated β-lactoglobulin (*107*), and 7S soy protein (*98*).

During the deacylation of the citraconylated yeast proteins a decrease in the pH of the system was observed in the acidic pHs mainly between 3 and 4.5. Above pH 5 no significant change in the pH occurred. The pH was maintained by adding the alkali. The maximum amount of alkali was consumed at pH 4. The pH was not altered when an acetate buffer of pH 4 was used during decitraconylation. At a low pH, the liberated amino groups are protonated fully whereas the liberated carboxylic groups are not. However the protonation of the amino groups does not

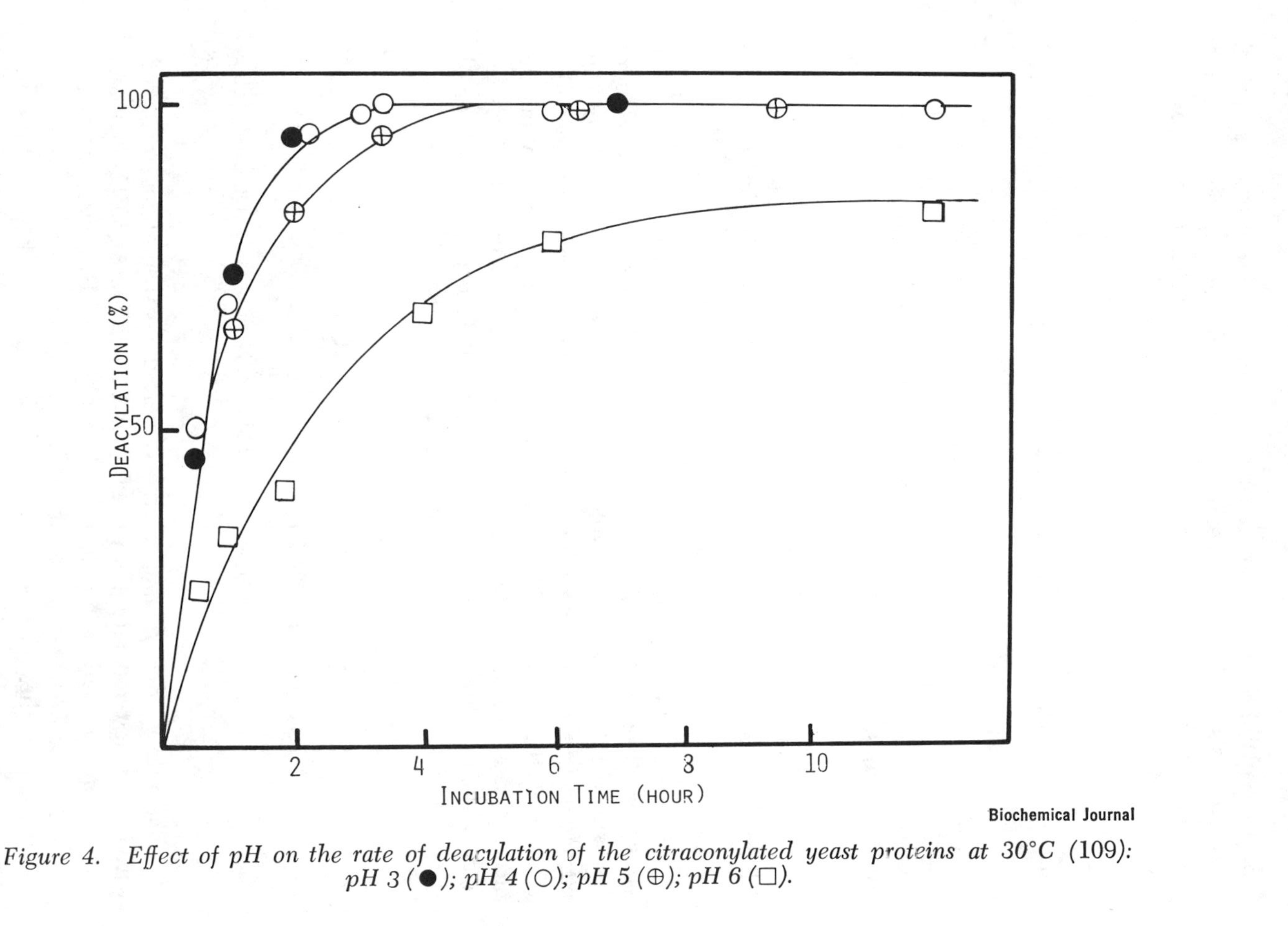


Biochemical Journal

*Figure 4. Effect of pH on the rate of deacylation of the citraconylated yeast proteins at 30°C* (109): *pH* 3 (●); *pH* 4 (○); *pH* 5 (⊕); *pH* 6 (□).

affect the pH of the medium because of the acid-catalyzed process during the hydrolysis of the amide bond formed by the anhydride containing $\beta$-diene structure (*46*). A simplified reaction scheme is shown in Reaction 4.

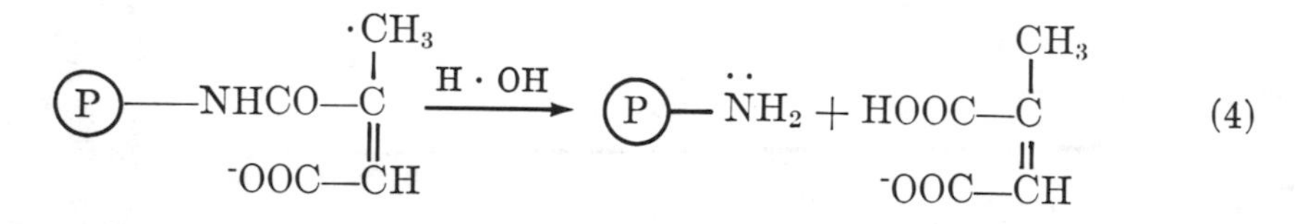


Then a proton from citraconic acid is released and protonation of the amino group occurs (Reaction 5).

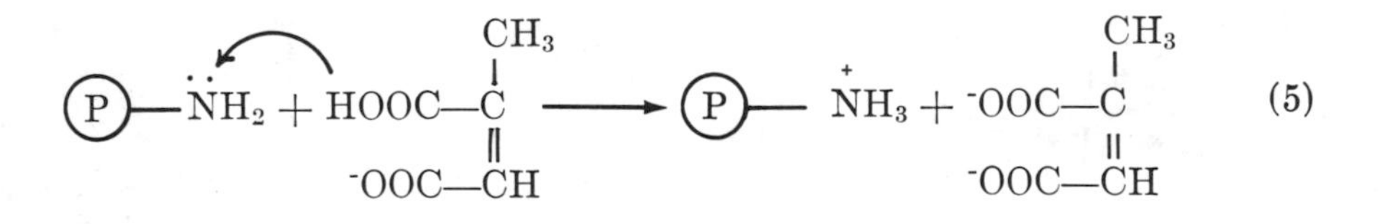


However, the observed decrease in the pH during decitraconylation between pH 3 and 4 can be explained by a decrease in the p*K*a of the citraconyl free carboxylic group in the citraconylated proteins upon the release of the proteins. The citraconyl free carboxylic group in the citraconylated yeast proteins might be protonated below pH 5. After deacylation, the p*K*a of these carboxylic groups decreased which caused the ionization and production of protons (*see* Reaction 6).

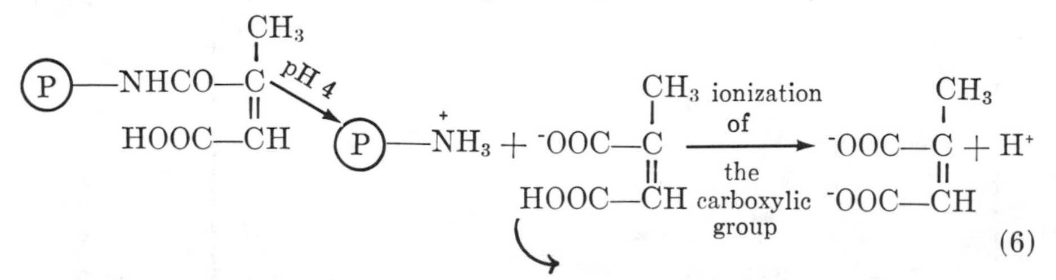


The rate of deacylation of citraconylated yeast proteins increased with increasing temperature at pH 4.0 (*see* Figure 5). At 50°C, complete deacylation occurred within 30 min of incubation. However, prolonged incubation ($>$ 12 h) was necessary at lower temperatures.

Based on these experimental data an integrated procedure for separating protein and NA is proposed. This procedure, which is outlined in Scheme III, is an extremely practical procedure for isolating and separating proteins from yeast nucleoprotein complexes. For their successful use in foods, the digestibility, the nutritional, and functional

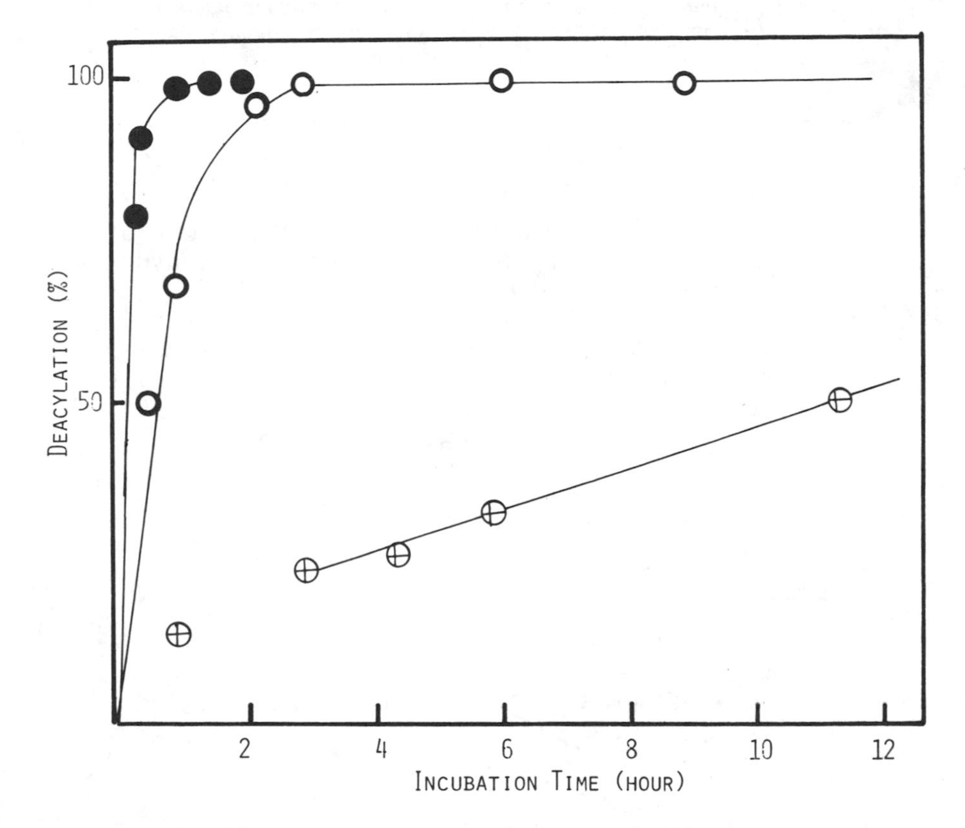


*Figure 5. Effect of temperature on the rate of deacylation of the citraconylated yeast proteins at pH 4: at 50°C (●); at 30°C (○); at 10°C (⊕).*

properties of decitraconylated yeast proteins must be determined. The possibility of the proteins being isolated in an undenatured form and thus having good functional properties is likely because, as reported (*45, 52*), decitraconylated proteins retain a high degree of their original structure.

### *Peptic Hydrolysis of Citraconylated and Decitraconylated Yeast Protein*

Acidic pH conditions caused the removal of citraconyl groups from the citraconylated yeast proteins, so the environmental conditions of the stomach, i.e., pH and temperature, may cause the deacylation to occur following consumption of citraconylated proteins. This then would make lysine nutritionally available. The peptic digestibility of the citraconylated and decitraconylated yeast proteins was determined; soy proteins and egg albumin were studied also for comparison. In a typical experi-

*Scheme III. Scheme for separating proteins and nucleic acids from a nucleoprotein complex using reversible modifying reagents of amino groups in the proteins*

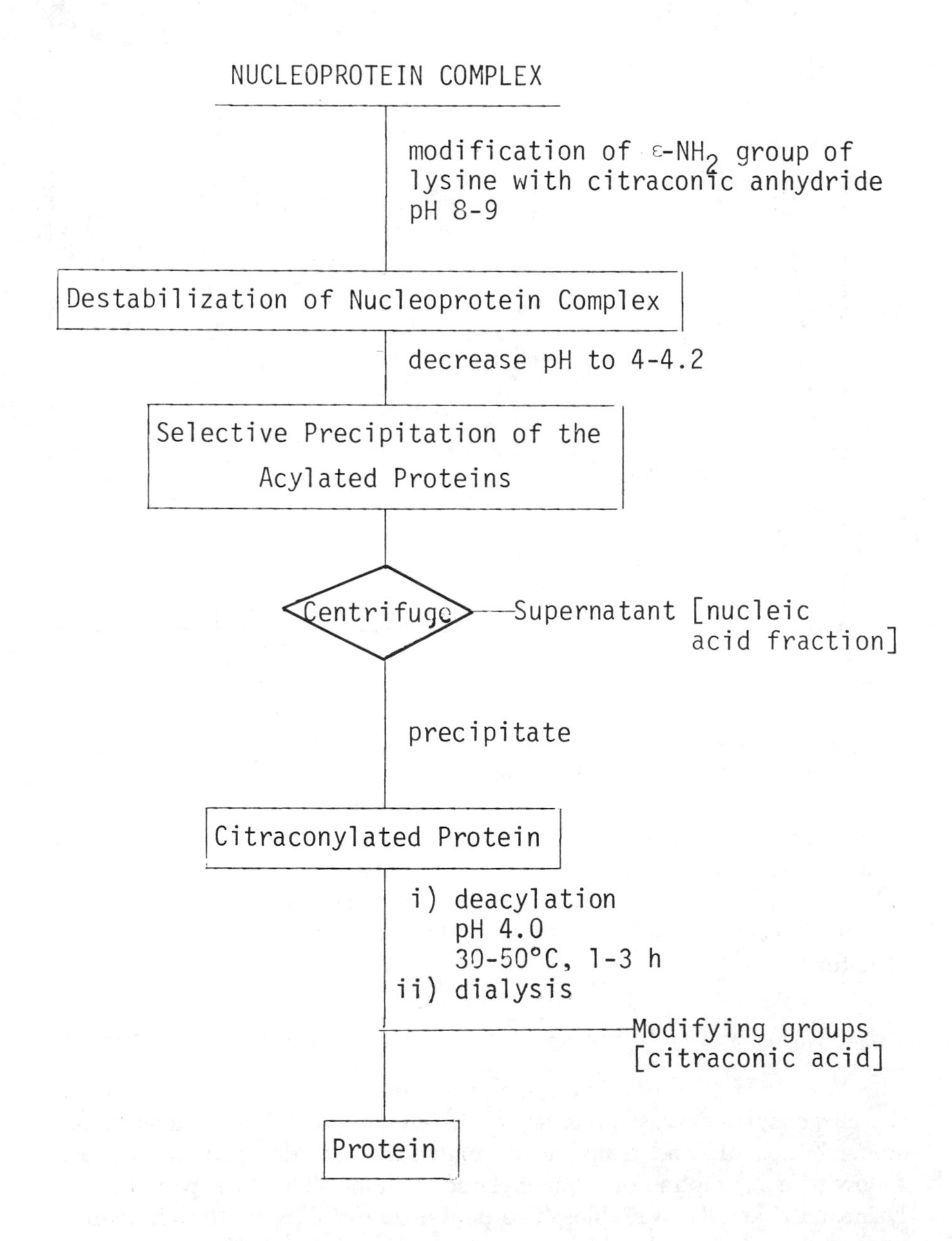

ment protein (1.0% concentration) was dissolved in HCl–KCl buffer, pH 1.6 and pepsin (Nutritional Biochemicals Corporation, Cleveland, OH) was added at a ratio of 1:500 (w/w) enzyme to protein and the mixture was incubated at 37°C. The degree of hydrolysis was determined at different intervals of time using the trichloroacetic acid method as reported by Yamashita et al. (*110*) and it was expressed as the trichloroacetic acid-soluble nitrogen/total nitrogen × 100 (%).

The rate of hydrolysis of egg albumin by pepsin exceeded that of soy proteins and decitraconylated yeast proteins (Figure 6). More than 90% egg albumin was hydrolyzed within 3 h of incubation whereas only 60% of the soy proteins and decitraconylated yeast proteins were hydrolyzed under the same conditions. Hydrolysis of egg albumin was not impaired by the presence of added citraconic acid (0.01*M* and 0.05*M*). So the hydrolysis of the citraconylated yeast proteins by pepsin was measured (*see* Figure 6). No differences occurred either in the rate or in the extent of hydrolysis compared with decitraconylated yeast proteins. More thorough nutritional studies and a safety evaluation of the citraconic acid must be conducted before either citraconylated or decitraconylated yeast proteins are used for human consumption.

For adoption as an ingredient in foods, the isolated protein should have appropriate functional properties (*111*). Currently we are examining some functional properties of decitraconylated yeast proteins.

## *Comments*

**Modification of Other Functional Groups of Proteins.** Determination of the sulfhydryl groups in the unmodified and modified yeast proteins using the method of Fernandez Diez et al. (*116*) indicated that the modification of the sulfhydryl groups occurred during citraconylation. Alkylation of protein thiol groups across the double bond of the citraconic anhydride was reported during citraconylation of the proteins (*98, 107, 112, 113, 114, 115*). Evidence for the involvement of sulfhydryl groups in the formation of higher-molecular-weight species after citraconylation of β-lactoglobulin was reported recently by Brinegar and Kinsella (*107*). Gibbons and Perham (*113*) reported the irreversible modification of sulfhydryl groups in aldolase by citraconic anhydride through alkylation involving the addition of -SH groups across the double bond of the citraconyl group. In our studies it was not clear whether the sulfhydryl groups were blocked irreversibly by citraconyl groups or citraconylated proteins.

Thus potential problems of the reversible citraconylation of proteins are the concomitant changes of the sulfhydryl groups. Thus, when modifying proteins with citraconic anhydride (especially when sulfhydryl groups

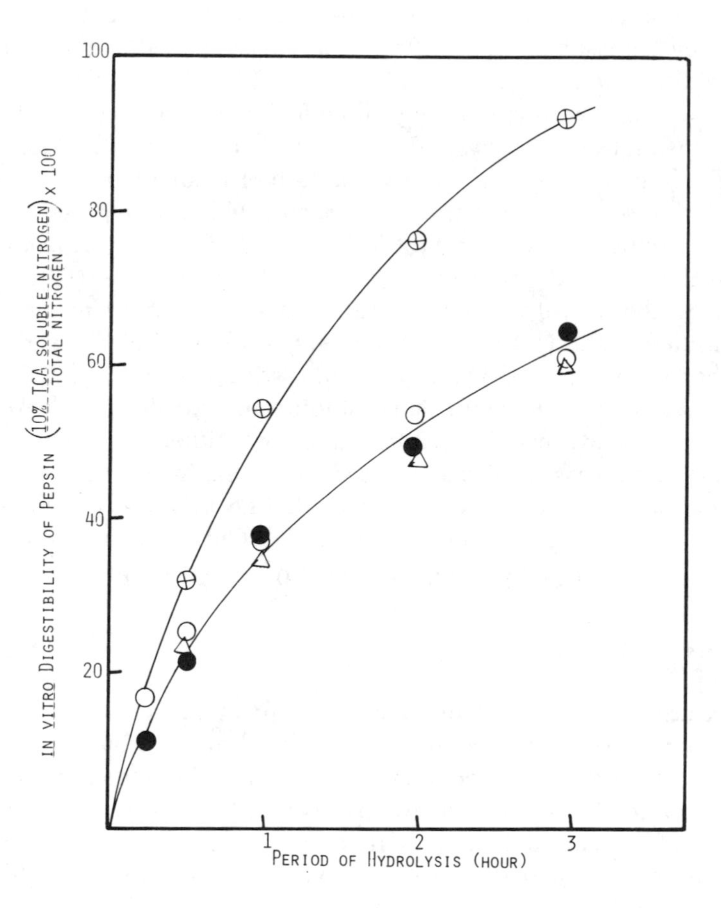


*Figure 6. Rate of hydrolysis of citraconylated and decitraconylated yeast proteins by pepsin: egg albumin (⊕); soy proteins (○); decitraconylated yeast proteins (●); citraconylated yeast proteins (△).*

are necessary in maintaining the stability of the protein or when disulfide exchange is induced easily) the irreversible blocking of sulfhydryl (groups) during lysine modification can be circumvented by protecting the -SH groups using reversible blocking reagents such as sodium tetrathionate prior to modification (*117*). Even though the modification of the phenolic groups of tyrosine is reported in many cases (*15, 26*), modification of tyrosine could not be detected by the hydroxylamine–$FeCl_3$ method in citraconylated yeast proteins (*118*).

### *Literature Cited*


1. Kinsella, J. E.; Shetty, J. K. "Nutritional Improvement of Food and Feed Proteins," *Adv. Exp. Med. Biol.* **1978**, *105*, 797.
2. Kihlberg, R. *Ann. Rev. Microbiol.* **1972**, *26*, 427.
3. Rose, A.; Morrison, J. S. In "The Yeasts"; Academic: New York, 1971; p. 1.
4. Kharatyan, S. G. *Ann. Rev. Microbiol.* **1978**, *32*, 30.
5. Johnson, J. C. In "Yeasts for Food and Other Purposes"; Noyes: Park Ridge, NJ, 1977.
6. Kinsella, J. E.; Shetty, J. K. In "Functionality and Protein Structure," *ACS Symp. Ser.* **1979**, *92*, 37.
7. Feeney, R. E. In "Food Proteins: Improvement Through Chemical and Enzymatic Modification," *Adv. Chem. Ser.* **1977**, *160*, 3.
8. Ryan, D. S. In "Food Proteins: Improvement Through Chemical and Enzymatic Modification," *Adv. Chem. Ser.* **1977**, *160*, 67.
9. Means, G. E.; Feeney, R. E. In "Chemical Modification of Proteins"; Holden-Day: San Francisco, 1971.
10. Glazer, A. N.; Delange, R. J.; Sigman, D. S. In "Chemical Modification of Proteins: Selected Methods and Analytical Procedures"; Wosk, T. S.; Wosk, E., Ed.; Am. Elsevier: New York, 1975.
11. Stark, G. R. *Adv. Protein Chem.* **1970**, *24*, 261.
12. Vallee, B. L.; Riordan, J. F. *Ann. Rev. Biochem.* **1969**, *38*, 733.
13. Fraenkel-Conrat, H.; Bean, R. S.; Lineweaver, H. *J. Biol. Chem.* **1949**, *177*, 385.
14. Riordan, J. F.; Vallee, B. L. *Biochemistry* **1964**, *3*, 1768.
15. Gounaris, A. D.; Perlmann, G. E. *J. Biol. Chem.* **1967**, *242*, 2739.
16. Nakagawa, Y.; Perlmann, G. E. *Arch. Biochim. Biophys.* **1972**, *149*, 476.
17. Klapper, M. H.; Klotz, I. M. In "Methods in Enzymology"; Hirs, C. H. W.; Timasheff, S. N., Eds.; Academic: New York, 1972; Vol. 25B, p. 530.
18. Means, G. E.; Feeney, R. E. *Biochemistry* **1968**, *7*, 2192.
19. Paik, W. K.; Kim, S. *Biochemistry* **1972**, *11*, 2589.
20. Habeeb, A. S. F. A.; Cassidy, H. G.; Singer, S. J. *Biochim. Biophys. Acta* **1958**, *29*, 587.
21. Shetty, J. K.; Rao, M. S. N. *Int. J. Pept. Protein Res.* **1978**, *11*, 305.
22. Klotz, I. M.; Keresztes-Nagy, S. *Biochemistry* **1963**, *2*, 445.
23. Habeeb, A. F. S. A. *Arch. Biochim. Biophys.* **1967**, *171*, 652.
24. Grant-Greene, M. C.; Friedberg, F. *Int. J. Protein Res.* **1970**, *2*, 235.
25. Mühlrad, A.; Corsi, A.; Granata, A. L. *Biochim. Biophys. Acta* **1968**, *162*, 435.
26. Freisheim, J. H.; Walsh, K. A.; Neurath, H. *Biochemistry* **1967**, *6*, 3010.
27. Freedman, M. H.; Grossberg, A. L.; Pressman, D. *Biochemistry* **1968**, *7*, 1941.
28. Delle, A. *Amino Acids Pept. Proteins* **1979**, *10P*, 108.
29. Oppenheimer, H.; Barany, K.; Hamoir, G.; Fenton, J. *Arch. Biochem. Biophys.* **1966**, *115*, 233.
30. Ibid., **1967**, *120*, 108.
31. Gandhi, S. K.; Schultz, J. R.; Boughey, F. W.; Forsythe, R. H. *J. Food Sci.* **1968**, *33*, 163.
32. Evans, M. T. A.; Irons, L. I. *Chem. Abstr.* **1970**, *73*, 3401d.
33. Hoagland, P. D.; Boswell, R. T.; Jones, S. B. *J. Dairy Sci.* **1971**, *54*, 1564.

34. Hoagland, P. D. *J. Dairy Sci.* **1966,** *49,* 783.
35. Nakai, S.; Wilson, H. K.; Herreid, E. O. *J. Dairy Sci.* **1964,** *47,* 1310.
36. Ibid., **1965,** *48,* 431.
37. Creamer, L.; Roeper, J.; Lohrey, E. *N. Z. J. Dairy Sci. Technol.* **1971,** 107.
38. Groninger, H. S. *J. Agric. Food Chem.* **1973,** *21,* 978.
39. Grant, D. R. *Cereal Chem.* **1973,** *50,* 417.
40. Puigserver, A. J.; Sen, L. C.; Clifford, A. J.; Feeney, R. E.; Whitaker, J. R. In "Nutritional Improvement of Food and Feed Proteins," *Adv. Exp. Med. Biol.* **1978,** *105,* 587.
41. Chan, E. L.; Helbig, N.; Holbek, E.; Chan, S.; Nakai, N. *J. Agric. Food Chem.* **1979,** *27,* 877.
42. Voutsinas, L. P.; Nakai, S. *J. Food Sci.* **1979,** *44,* 1205.
43. Braunitzer, G.; Beyreuther, K.; Fujiki, H.; Schrank, B. *Hoppe–Seyler's Z. Physiol. Chem.* **1968,** *349,* 265.
44. Singhal, R. P.; Atassi, M. Z. *Biochemistry* **1971,** *10,* 1756.
45. Habeeb, A. F. S. A.; Atassi, M. Z. *Biochemistry* **1970,** *9,* 4939.
46. Kirby, A. J.; McDonald, R. S.; Smith, C. R. *J. Chem. Soc. Perkin Trans. 2* **1974,** 1495.
47. Kirby, A. J.; Lancaster, P. W. *J. Chem. Soc. Perkin Trans. 2* **1972,** 1206.
48. Alderseley, M. F.; Kirby, A. J.; Lancaster, P. W.; McDonald, R. S.; Smith, C. R. *J. Chem. Soc. Perkin Trans. 2* **1971,** 1487.
49. Butler, P. J. G.; Harris, J. L.; Hartley, B. S.; Liberman, R. *Biochem. J.* **1967,** *103,* 78P.
50. Riley, M.; Perham, R. N. *Biochem. J.* **1970,** *118,* 733.
51. Dixon, H. B. F.; Perham, R. N. *Biochem. J.* **1968,** *109,* 312.
52. Atassi, M. Z.; Habeeb, A. F. S. A. In "Methods in Enzymology"; Hirs, C. H. W.; Timasheff, S. N., Eds.; Academic: New York, 1972; Vol. 25B, p. 546.
53. Geoghegan, K. F.; Ybarra, D. M.; Feeney, R. E. *Biochemistry* **1979,** *18,* 5392.
54. Marzotto, A.; Pajetta, P.; Scoffone, E. *Biochim. Biophys. Res. Commun.* **1967,** *26,* 517.
55. Marzotto, A.; Pajetta, P.; Galzigna, L.; Scoffone, E. *Biochim. Biophys. Acta* **1968,** *154,* 450.
56. Goldberger, R. F. In "Methods in Enzymology"; Hirs, C. H. W., Ed.; Academic: New York, 1967; Vol. 11, p. 317.
57. Schallenberg, E. E.; Calvin, M. *J. Am. Chem. Soc.* **1955,** *77,* 2779.
58. Goldberger, R. F.; Anfinsen, C. B. *Biochemistry* **1962,** *1,* 402.
59. Butler, P. J. G.; Harris, J. L.; Hartley, B. S.; Liberman, R. *Biochem. J.* **1969,** *112,* 679.
60. Puigserver, A.; Desnuelle, P. *Proc. Nat. Acad. Sci. USA* **1975,** *72,* 2442.
61. Freedman, M. H.; Grossberg, A. L.; Pressman, D. *Biochemistry* **1968,** *7,* 1941.
62. Mateles, R. T.; Tannenbaum, S. R., Eds. In "Single-Cell Proteins"; MIT Press: Cambridge, 1968.
63. Tannenbaum, S. R.; Wang, D. I. C., Eds. In "Single-Cell Proteins, II"; MIT Press: Cambridge, 1975.
64. Scrimshaw, N. S. In "Single-Cell Proteins II"; Tannenbaum, S. R., Wang, D. I. C., Eds.; MIT Press: Cambridge, 1975; p. 24.
65. Tannenbaum, S. R. In "Single-Cell Proteins"; Mateles, R. T.; Tannenbaum, S. R., Eds.; MIT Press: Cambridge, 1968; p. 343.
66. Young, V. R.; Scrimshaw, N. S.; Milner, M. S. *Chem. Ind.* **1976,** 588.
67. Worgan, J. T. In "Single-Cell Protein: Plant Foods For Man"; 1974, Vol. 1, p. 99.
68. Sinskey, A. J.; Tannenbaum, S. R. In "Single-Cell Proteins II"; Tannenbaum, S. R.; Wang, D. I. C., Eds.; MIT Press: Cambridge, 1975; p. 158.

69. Knorr, D.; Shetty, J. K.; Kinsella, J. E. *Biotech. Bioeng.* **1979,** *21,* 2011.
70. Shetty, J. K.; Kinsella, J. E. *Biotech. Bioeng.* **1978,** *20,* 755.
71. Young, V. R.; Scrimshaw, N. S. In "Single-Cell Proteins II"; Tannenbaum, S. R.; Wang, D. I. C., Eds.; MIT Press: Cambridge, 1975; p. 566.
72. Hedenskog, G.; Mogren, H. *Biotech. Bioeng.* **1973,** *15,* 129.
73. Vananuvat, P.; Kinsella, J. E. *J. Agric. Food Chem.* **1975,** *23,* 216.
74. Lindblom, M. *Biotech. Bioeng.* **1974,** *16,* 1495.
75. Newell, J. A.; Robbins, E. A.; Seeley, R. D. *U.S. Patent 3 867 555,* 1975.
76. Cheftel, J. C. In "Food Proteins"; Whitaker, J. R.; Tannebaum, S. R., Eds.; Avi: Westport, CT, 1977; p. 404.
77. Friedman, M., Ed. In "Protein Cross-linking: Nutritional and Medical Consequences"; Plenum: New York, 1977; 86B.
78. DeGroot, A. P.; Slump, P.; VanBeek, L.; Ferson, V. J. In "Evaluation of Proteins for Humans"; Bodwell, S., Ed.; Avi: Westport, CT, 1977; p. 270.
79. Masters, P. M.; Friedman, M. *J. Agric. Food Chem.* **1979,** *27,* 507.
80. Shetty, J. K.; Kinsella, J. E. *J. Agric. Food Chem.,* in press.
81. Patchornik, A.; Sokolovsky, M. *J. Am. Chem. Soc.* **1964,** *86,* 1206.
82. Nicolet, B. H.; Shinn, L. A.; Saidel, L. J. *J. Biol. Chem.* **1942,** *142,* 609.
83. DeGroot, A. J.; Slump, P. *J. Nutr.* **1969,** *98,* 45.
84. Finot, A. P.; Bujard, E.; Arnaud, M. In "Nutritional Improvements of Food and Feed Proteins," *Adv. Exp. Med. Biol.* **1978,** *105,* 549.
85. Provansal, M. M. P.; Cug, J-L. A.; Cheftel, J. C. *J. Agric. Food Chem.* **1975,** *23,* 938.
86. Woodard, J. C.; Short, D. D. *J. Nutr.* **1973,** *103,* 569.
87. Shetty, J. K.; Kinsella, J. E. *J. Food Sci.* **1979,** *44,* 1362.
88. Bohak, Z. *J. Biol. Chem.* **1964,** *239,* 2878.
89. Friedman, M. In "Protein-Crosslinking: Nutritional and Medical Consequences," Plenum: New York, 1977; 86B.
90. Asquith, R .S.; Skinner, J. D. *Texilveredlung* **1970,** *5,* 406.
91. Horn, M. J.; Jones, D. B.; Ringel, S. J. *J. Biol. Chem.* **1941,** *133,* 11.
92. Ziegler, K.; Melchert, I.; Lürken, C. *Nature (London)* **1967,** *214,* 404.
93. Snoke, E. R.; Klein, W. G. U.S. Patent 4 055 469, 1977.
94. Vananuvat, P., Ph.D. Thesis, Cornell Univ., Ithaca, NY, 1974.
95. Habeeb, A. F. S. A. *Can. J. Biochem. Physiol.* **1960,** *38,* 269.
96. Hass, L. F. *Biochemistry* **1964,** *3,* 535.
97. Klotz, I. M. In "Methods in Enzymology"; Hirs, C. H. W., Ed.; Academic: New York, 1967; Vol. 11, p. 567.
98. Brinegar, C.; Kinsella, J. E. *J. Agric. Food Chem.* **1980,** *28,* 818.
99. Shetty, J. K.; Kinsella, J. E., unpublished data.
100. Klotz, I. M.; Keresztes-Nagy, S. *Biochemistry* **1963,** *2,* 445.
101. Shetty, J. K.; Kinsella, J. E. *Biotech. Bioeng.* **1979,** *21,* 329.
102. Hall, R. J.; Trinder, N.; Givens, D. I. *Analyst* **1973,** *98,* 673.
103. Herbert, R.; Phipps, P. J.; Strange, R. E. In "Methods in Microbiology"; Norris, N. J.; Robbins, D. W., Eds.; Academic: New York, 1971; Vol. 5B, p. 244.
104. Groninger, H. S.; Miller, R. *J. Agric. Food Chem.* **1979,** *27,* 949.
105. Bjarnason, J.; Carpenter, K. J. *Br. J. Nutr.* **1969,** *23,* 859.
106. Mauron, J. In "Protein and Amino Acid Functions"; Bigwood, E. J., Ed.; Pergamon: Oxford, 1972.
107. Brinegar, C.; Kinsella, J. E. *Int. J. Pept. Protein Res.,* in press.
108. Cheftel, J. C. In "Food Proteins: Improvement Through Chemical and Enzymatic Modification," *Adv. Chem. Ser.* **1977,** *160,* 401.
109. Shetty, J. K.; Kinsella, J. E. *Biochem. J.* **1980,** *189,* 363.
110. Yamashita, M.; Arai, A.; Matsuyama, J.; Gonda, M.; Kato, M.; Fujimaki, M. *Agric. Biol. Chem. (Tokyo)* **1970,** *34,* 1484.
111. Kinsella, J. E. *Crit. Rev. Food Sci. Nutr.* **1976,** *7,* 219.

112. Butler, P. J. G.; Hartley, B. S. In "Methods in Enzymology"; Hirs, C. H. W.; Timasheff, N. S., Eds.; Academic: New York, 1972; Vol. 25B, p. 191.
113. Gibbons, I.; Perham, R. N. *Biochem. J.* **1970,** *116,* 843.
114. Calam, D. H.; Waley, S. G. *Biochem. J.* **1963,** *86,* 226.
115. King, L.; Perham, R. N. *Biochemistry* **1971,** *10,* 981.
116. Fernandez Diez, M. J.; Osuga, D. T.; Feeney, R. E. *Arch. Biochem. Biophys.* **1964,** *107,* 449.
117. Liu, T-Y. *J. Biol. Chem.* **1967,** *242,* 4029.
118. Habeeb, A. F. S. A.; Atassi, M. Z. *Immunochemistry* **1969,** *6,* 555.

RECEIVED October 14, 1980.

# Proteinaceous Surfactants Prepared by Covalent Attachment of L-Leucine *n*-Alkyl Esters to Food Proteins by Modification with Papain

MICHIKO WATANABE and SOICHI ARAI

Department of Agricultural Chemistry, University of Tokyo,
Bunkyo-ku, Tokyo 113, Japan

*This chapter aims at maximizing the functionality of food proteins with the smallest possible modification to their structure. For this purpose we used a papain-catalyzed unusual reaction that proceeded efficiently under intentionally abnormal conditions. The reaction was defined as a type of aminolysis of a protein by an amino acid ester, leading to the formation of a product to which the amino acid ester was attached covalently. This reaction, when applied to a mixture of a hydrophilic protein and a lipophilic* L*-leucine* n*-alkyl ester, produced a proteinaceous surfactant. The attachment of a* $C_4$*–*$C_6$ *alkyl ester of leucine led to greatly improved whippability, while a potent emulsifying activity resulted when a longer-chain alkyl ester of leucine was attached. The products were characterized by their highly amphiphilic function. The possibility of using these surfactants as food ingredients is discussed.*

The importance of studying protein functionality is stressed as worldwide demand for functional proteins usable as food ingredients has increased in recent years. Hammonds and Call (*1*) estimated the maximum market potential for protein ingredients at approximately 3.1 billion pounds annually. Kinsella (*2*) stated that about 80% of this quantity has to possess a high degree of functionality.

Much is known of protein functionality in terms of water solubility and dispersibility, the heat coagulation (or heat-setting) property, water

0065-2393/82/0198-0199$05.75/0

and/or fat absorbability, oil-emulsifying activity, foamability and whippability, gelling property, and the spinning property, etc. To increase the use of proteins for food, the improvement and, if possible, the maximization of their functionality will become more and more important. Recent advances in research in these areas and further developments have been reviewed by many authors (*3, 4, 5, 6*).

The functionality of food proteins can be modified by using enzymes, especially proteases. Several of the enzymatic modification processes could be used practically even on an industrial scale. Our laboratory has contributed to the knowledge on nutritional improvement of food proteins through modification with proteases (*7*) with the development of a novel one-step process that permits covalent incorporation of methionine and other amino acids directly into, for example, soy protein with the aid of papain (*8, 9, 10, 11*). From an overall point of view this process is estimated to be primarily a type of aminolysis of protein (substrate) by an amino acid ester (nucleophile) added to the reaction system (*see* Figure 1). It follows that the amino acid ester is attached covalently to the C-terminal of the peptide that has been produced from the protein substrate (*9*). Such an unusual reaction involving papain can proceed efficiently only by creating intentionally the following unconventional conditions: (1) the use of an alkaline medium adjusted to a pH value of 9–10 and (2) increasing the concentration of the protein

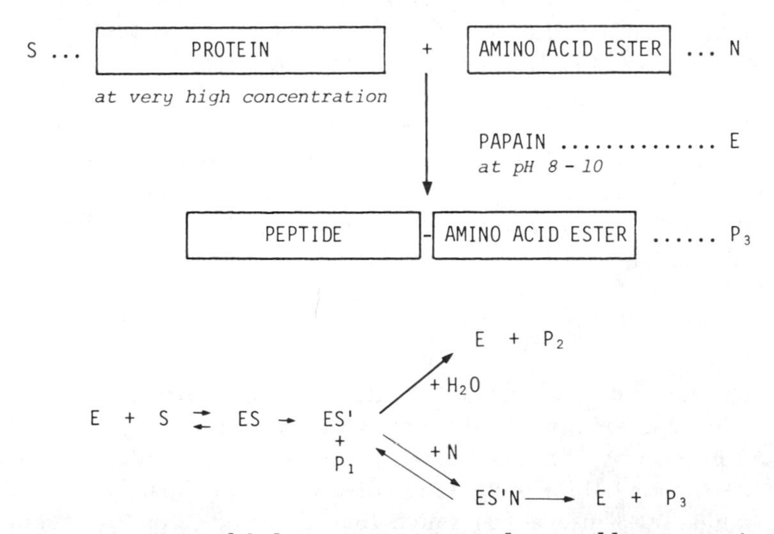


*Figure 1. A simplified representation of the possible process for the papain-catalyzed hydrolysis and aminolysis: E, enzyme (papain); S, substrate (protein); ES, Michaelis complex; ES', peptidyl enzyme; N, nucleophile (amino acid ester); $P_1$ and $P_2$, products formed from S by hydrolysis; $P_3$ (or SN'), product formed from ES' by aminolysis with N.*

substrate in this medium up to 20% (w/w) or higher. Only when both of these requirements are satisfied does such an otherwise limited reaction as the covalent incorporation of amino acid ester occur and can be utilized in practice for nutritional improvement of food proteins (*7, 8*).

This enzymatic process may be applied to modifying protein functionality as well. Provided that a hydrophilic protein (substrate) and a highly hydrophobic or lipophilic amino acid ester (nucleophile) are used, it would be possible to obtain a product ($P_3$ in Figure 1) with a structure such that hydrophilic and lipophilic regions in the molecule are localized from each other. It is expected, as a consequence, that a proteinaceous surfactant with an adequately amphiphilic function would be produced.

The present study deals first with the papain-catalyzed attachment of lipophilic L-norleucine *n*-dodecyl ester to a hydrophilic protein. To determine more clearly the mode of reaction, well-defined $\alpha_{s1}$-casein was selected as the protein for this purpose.

## *Covalent Attachment of a Lipophilic Nucleophile to a Hydrophilic Protein with the Use of Papain*

**Primary Mode of Reaction and Consequences** (*9*). Bovine $\alpha_{s1}$-casein (Variant B) was prepared from fresh milk (*12*) and immediately succinylated to increase water dispersibility (*13*). This succinylated $\alpha_{s1}$-casein preparation, molecular weight of ca. 25,000 daltons, was used as the hydrophilic protein substrate. The compound L-norleucine 1-$^{13}$C-dodecyl ester, prepared from $K^{13}CN$ and 1-bromoundecane through four steps (*14, 15, 16*), was used as the lipophilic nucleophile. C-13 NMR measurements showed that this sample gave only one signal at a distance of 65.2 ppm from the signal of tetramethylsilane (TMS).

The enzymatic process was carried out under the following conditions: medium, 20% (v/v) acetone in 1*M* carbonate (pH 9) containing 10m*M* 2-mercaptoethanol; concentration of succinylated $\alpha_{s1}$-casein in medium, 20% (w/w); concentration of L-norleucine 1-$^{13}$C-dodecyl ester in medium, 0.25*M;* concentration of papain (recrystallized) in medium, 0.02% (w/w); incubation temperature, 37°C; and incubation time up to 60 min. For further details refer to the published papers (*7, 8*).

We used Sephadex G-15 chromatography to remove low-molecular-weight species including unreacted L-norleucine 1-$^{13}$C-dodecyl ester from each of the samples resulting from the incubations for 1 min and 30 min. The high-molecular-weight fraction, representing about 95% of the original sample on a dry-matter basis, was subjected to C-13 NMR measurement. The result is shown in Figure 2; a 65-ppm signal assignable to 1-methylene C-13 is observed for the high-molecular-weight fraction from the 1-min incubation mixture and much more clearly for that from

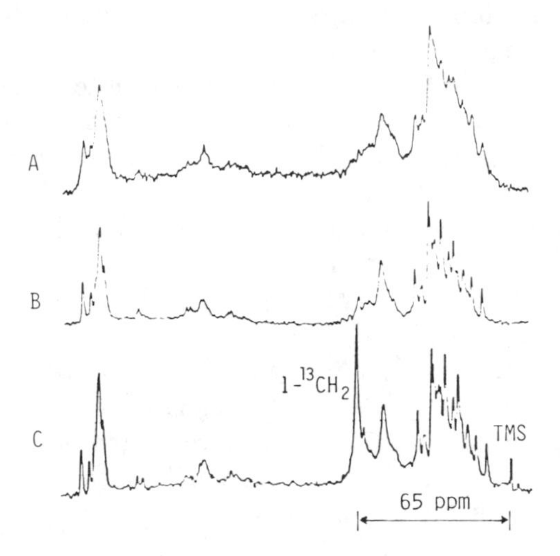


*Figure 2. 100-MHz C-13 NMR spectra: A, succinylated $\alpha_{s1}$-casein before incubation; B and C, products resulting from, respectively, 1-min and 30-min incubation of this protein with papain in the presence of L-norleucine 1-$^{13}$C-dodecyl ester. Each spectrum resulted from 45,000 transients obtained by using the 8K-frequency domain data points (spectral width, 5,000 Hz; recycle time, 0.69 sec; and flip angle, 38°).*

the 30-min incubation mixture. For further confirmation, a sufficient amount of unlabeled L-norleucine *n*-dodecyl ester was added to the high-molecular-weight sample from the 30-min incubation product and the resulting mixture was treated likewise by Sephadex G-15 prior to C-13 NMR measurement. The C-13 NMR spectrum observed for the high-molecular-weight fraction obtained was very similar to that shown in Figure 2. Since no C-13 dilution effect was found in this experiment, it was concluded that during the enzymatic process L-norleucine 1-$^{13}$C-dodecyl ester was attached covalently to the high-molecular-weight fraction.

Analysis by polyacrylamide gel electrophoresis (PAGE) in the presence of sodium dodecyl sulfate (SDS) (*17*) demonstrated that during the enzymatic process the substrate, succinylated $\alpha_{s1}$-casein, underwent degradation to an approximately 20,000-dalton polypeptide (*see* Figure 3). This polypeptide was isolated from the 30-min incubation mixture by Sephadex G-75 chromatography and investigated for its C-13 spectrum, dodecanol and norleucine contents, N- and C-terminal structures, and emulsifying activity. Prior to quantification of the dodecanol by gas chromatography (GC) the polypeptide sample was treated with 1*N* NaOH. Terminal structures were investigated with leucine aminopeptidase (*18*) and with carboxypeptidase A (*19*). For measuring the emulsifying activity, the method proposed by Franzen and Kinsella (*20*) was used and the activity represented in terms of a cream-phase volume divided by a total volume. Table I summarizes the results obtained from these measurements with the 20,000-dalton product.

Another experiment was carried out with the 5-min, 15-min, and 30-min incubation mixtures in order to determine the correlation between their content of the 20,000-dalton product and the emulsifying activity. The result was that the emulsifying activity increased in accordance with the formation of this polypeptide (*see* Table II). No increase in the emulsifying activity results when succinylated $\alpha_{s1}$-casein is incubated with papain in the absence of L-norleucine *n*-dodecyl ester.

All of the data from Table II indicate that, under the unconventional conditions intentionally set in the present study, papain can catalyze the aminolysis of the ES′ intermediate that probably occurs from succinylated $\alpha_{s1}$-casein (substrate) and papain (enzyme) by L-norleucine *n*-dodecyl ester (nucleophile), with formation of a surface-active 20,000-dalton product to which this lipophilic nucleophile is attached covalently as illustrated in Figure 1. The observed amphiphilic function of the 20,000-dalton product is probably a result of the formation of a localized hydro-

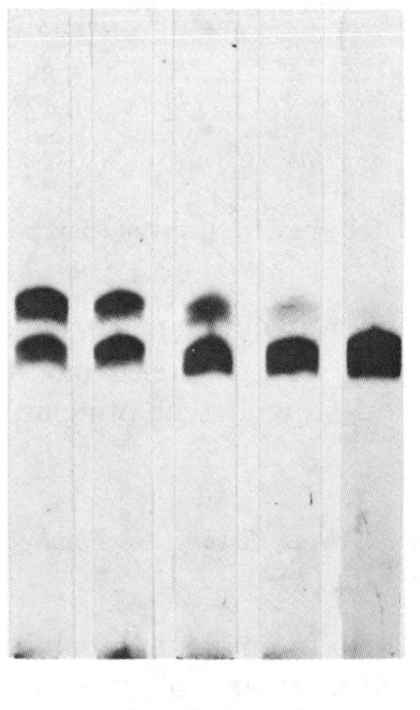


*Figure 3. SDS–PAGE of a high-molecular-weight fraction of the reaction mixture sampled at different times of incubation*

**Table I. Properties of a 20,000-Dalton Polypeptide Formed from Succinylated $\alpha_{s1}$-Casein by Treatment with Papain in the Presence of L-Norleucine 1-$^{13}$C-Dodecyl Ester**

| *Investigated Item* | *Result* |
|---|---|
| C-13 NMR spectrum | a clear 65-ppm signal observed[a] |
| 1-Dodecanol content | 0.88 mol/mol polypeptide |
| Norleucine content | 0.85 mol/mol polypeptide |
| C-terminal amino acid | norleucine (dodecyl ester form) |
| N-terminal amino acid | arginine masked by succinylation[b] |
| Emulsifying activity[c] | 0.63[d] |

[a] As in Figure 1.
[b] Concluded from the fact that this polypeptide shows no ninhydrin response.
[c] *See* text.
[d] A 20,000-dalton polypeptide was formed also when a similar papain treatment was conducted in the absence of L-norleucine dodecyl ester. This polypeptide, however, showed an emulsifying activity of only 0.40. Also *see* Table II.

**Table II. Relationship Between the Amount of a 20,000-Dalton Product Formed in the Incubation Mixture and Its Emulsifying Activity**

| *Incubation Time (Min)* | *Amount of a 20,000-Dalton Product Formed[a] (%)* | *Emulsifying Activity[b]* |
|---|---|---|
| 0 | 0 | 0.39 |
| 5 | 9 | 0.42 |
| 15 | 33 | 0.50 |
| 30 | 49 | 0.59 |

[a] Roughly estimated from SDS–PAGE followed by densitometry with stained zones.
[b] *See* Table I.

philic–lipophilic structure (*see* Figure 4). This chapter demonstrates the fact that small changes in the chemical structures of proteins result in large changes in their functionality.

### *Covalent Attachment of* L-*Leucine* n-*Alkyl Esters to Food Proteins for Improving Their Functionality* (*10, 11*)

The function of a surfactant depends on its hydrophilicity–lipophilicity balance (HLB). Efficient emulsification of oil generally requires a low HLB, while the whipping characteristic arises at a larger HLB. This chapter is an attempt to prepare proteinaceous surfactants with different HLBs by the enzymatic attachment of amino acid esters with different lipophilicity. For this purpose L-leucine *n*-alkyl esters (Leu–$OC_i$), the alkyl chain length, $i$, varying from 2 to 12, were used. As

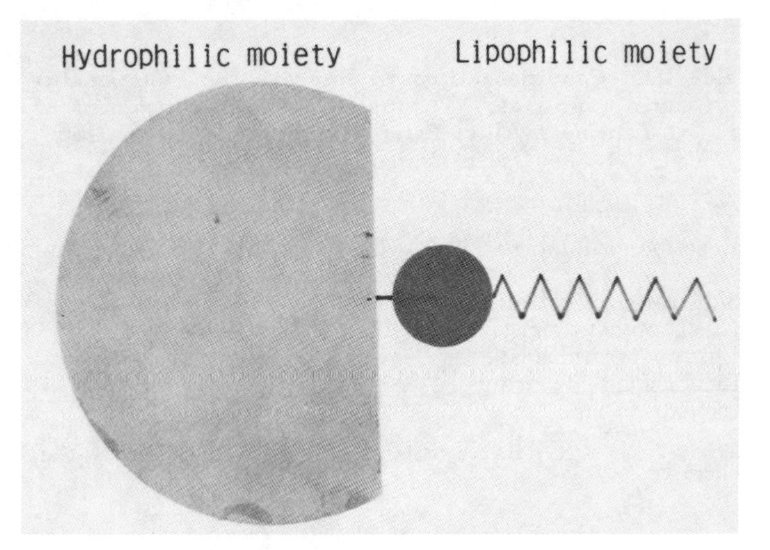


*Figure 4. Proposed amphiphilic structure for a proteinaceous surfactant molecule*

hydrophiles, commercially available food proteins such as gelatin, fish protein concentrate (FPC), soy protein isolate (SPI), whole casein preparation, and ovalbumin were selected. Proteins other than gelatin were succinylated prior to being used in order to enhance their hydrophilicity.

**Gelatin.** Gelatin as a food ingredient has long been in use. Commercially available preparations of gelatin commonly are used as jellying agents, bodying (or binding) agents, emulsion stabilizers, etc. (*21*). However, gelatin is characterized by its highly hydrophilic nature, which causes poor performance with regard to its whipping and emulsifying functions. Our idea is that the addition of a proper degree of lipophilicity to the molecule of gelatin would give rise to an adequately amphiphilic function.

A commercial preparation of gelatin (Kanto Kagaku Co.) was subjected to the enzymatic reaction as described in Table III. The conditions primarily follow those required for the one-step process with papain (*7, 8*). The reaction was stopped by acidifying to pH 1 with 1*N* HCl and the acidified mixture was dialyzed in running water. Lyophilization of the nondiffusible fraction gave a proteinaceous product in powder form, which was purified further by treating it with hot acetone to remove the low-molecular-weight species that might still remain. Each of the purified products was investigated for leucine and alkanol content and mobility in SDS-PAGE, and also was evaluated for three kinds of

**Table III. Conditions Used to Improve the Functionality of Gelatin (Substrate) Through Covalent Attachment of L-Leucine n-Alkyl Ester (Nucleophile) with the Use of Papain (Enzyme)**

| *Item* | *Condition* |
|---|---|
| Reaction medium | 1*M* carbonate (pH 9) containing 2m*M* 2-mercaptoethanol |
| Substrate concentration | 33% (w/w) |
| Nucleophile vs. substrate | 0.1 mol/100 g |
| Enzyme[a] vs. substrate | 1 g/100 g |
| Incubation temperature | 37°C |
| Incubation time | 15 min |

[a] Activity: $1.16 \times 10^{-2}$ BAPA units (BAPA refers to *N*-benzoyl-L-arginine *p*-nitroanilide).

**Table IV. Chemical and Functional Properties of Products Prepared from Gelatin[a] by Papain-Catalyzed Attachment of L-Leucine n-Alkyl Esters**

| *Chain Length of Alkyl Moiety* | *Average Molecular Weight*[b] | *Leucine Increment*[c] *(mol/ 7,500 g)* | *Alkanol Content*[d] *(mol/ 7,500 g)* | *Whippability*[e] | *Foam Stability*[e] | *Emulsifying Activity*[f] |
|---|---|---|---|---|---|---|
| $C_0$[g] | 12,000 | 0.0 | 0.0 | 1.8 | 0.95 | 0.00 |
| $C_2$ | 8,300 | 1.4 | 1.1 | 3.4 | 0.95 | 0.55 |
| $C_4$ | 8,200 | 1.5 | 1.2 | 4.3 | 0.60 | 0.60 |
| $C_5$ | | | | 4.4 | 0.60 | 0.62 |
| $C_6$ | 7,500 | 1.5 | 1.1 | 4.5 | 0.55 | 0.66 |
| $C_7$ | | | | 4.6 | 0.65 | 0.68 |
| $C_8$ | 7,300 | 1.5 | 1.2 | 4.7 | 0.70 | 0.69 |
| $C_9$ | | | | 4.7 | 0.85 | 0.71 |
| $C_{10}$ | 7,300 | 1.5 | 1.2 | 2.5 | 0.90 | 0.72 |
| $C_{11}$ | | | | 1.9 | 1.00 | 0.73 |
| $C_{12}$ | 7,300 | 1.4 | 1.1 | 1.3 | 1.00 | 0.74 |

[a] The gelatin sample before the papain treatment gave the following data: whippability, 1.70; foam stability, 0.90; and emulsifying activity, 0.00.

[b] Stained zones were scanned densitometrically. The average molecular weights were roughly estimated from the peak positions. The molecular-weight distribution of each product ranged from 2,000–40,000 daltons.

[c] Difference between the leucine content of each sample and that of the gelatin hydrolysate.

[d] Columns: 5% polyethylene glycol 20M on Chrom-P (3 mm × 2 m) for $C_2$ and $C_4$ alkanols and 3% silicone OV-17 on Chrom-G (3 mm × 3 m) for $C_5$–$C_{12}$ alkanols.

[e] *See* text.

[f] *See* Table I.

[g] Referring to the gelatin hydrolysate resulting from the papain treatment in the absence of any L-leucine *n*-alkyl ester.

functionality: whippability, foam stability, and emulsifying activity. To measure the whippability, a 1% dispersion (20 mL) of each sample in 0.01*M* phosphate (pH 7) was homogenized at 10,000 rpm for 3 min. The homogenate was transferred immediately into a glass cylinder and after 0.5 min was measured for total and drainage volumes. The whippability was represented by (total volume − drainage volume)/initial volume (20 mL).

Subsequently, the whipped sample was allowed to stand at 25°C for another 30 min and then the drainage volume was measured. With this volume the foam stability was represented by the ratio of drainage volume/initial volume (20 mL).

To measure the emulsifying activity, each 200-mg sample dispersed in 0.01*M* phosphate, pH 7 (20 mL), was homogenized with 20 mL corn oil at 10,000 rpm for 3 min, and the resulting emulsion was centrifuged at 500 rpm for 5 min to separate a cream layer. The emulsifying activity was represented by the ratio of cream layer volume/total volume.

Table IV collates the data obtained, showing that in the first place any of the products has a molecular-weight distribution of 2,000–40,000 daltons, with an average ca. 7,500 daltons. Table IV also shows that, for the products other than gelatin hydrolysate, the observed leucine increments and alkanol contents are controlled well in a narrow range of 1.4–1.5 mol/7,500 g and 1.1–1.2 mol/7,500 g, respectively. Regardless of these similarities among the products, their functionality varies greatly. The highest whippability and foam stability resulted from the attachment of L-leucine $C_8$ and $C_6$ alkyl esters, respectively, whereas the emulsifying activity tended to increase gradually with chain length of the alkyl moiety (*see* Table IV).

**Succinylated Proteins.** Most food protein systems are primarily hydrophobic in nature. Though some proteins are often soluble in isolated form, in most cases it is due to the native structure where the hydrophilic groups are on the outside. Common treatments for the denaturation of proteins often make their buried hydrophobic regions exposed to the surrounding water. Consequently, denatured proteins tend to aggregate with a decrease in their water solubility. Effective solubilization of denatured proteins may call for chemical modifications such as glycosylation, phosphorylation, sulfonylation, carboxylation, acylation, etc. One of the convenient processes widely used for this purpose is succinylation, which blocks protein amino groups replacing them with carboxyl anions. A number of investigators (*6, 13, 22, 23*) have applied succinylation to food proteins, with a fair success in improving their water solubility and/or dispersibility. The attachment of anionic succinyl moieties to a protein results in an increase in its molecular surface area. This is apparently due to a contribution of the electric repulsion force occurring between the identical charges. The resulting increase in surface area is an important requirement for a protein to gain surface activity.

We succinylated FPC, SPI, casein, and ovalbumin as well as gelatin, with the result that their whippability, foam stability, and emulsifying activity were improved to a certain extent (*see* Table V).

Further studies were conducted to improve to a greater extent the functionality of succinylated proteins by the covalent attachment of L-leucine *n*-alkyl esters. Each of the above proteins was succinylated (*13*) and then subjected to papain treatment under conditions similar to those shown in Table III. The incubation mixture was dialyzed and then purified by treatment with hot acetone as described previously. The products were measured for whippability, foam stability, and emulsifying activity. Table VI shows the results. In most cases the highest whippability results from the enzymatic process with incorporation of $C_4$–$C_8$ of L-leucine. To improve the foam stability, the use of $C_4$–$C_6$ esters of L-leucine is preferable. The use of longer-chain alkyl esters of L-leucine was required to give a very high degree of emulsifying activity to succinylated FPC, casein, and ovalbumin. Somewhat different data were obtained with succinylated SPI, which may be poorer in hydrophilicity than the other proteins used. In comparison with the above,

**Table V. Changes in Functional Properties of Proteins Due to Succinylation[a]**

| *Protein* | *Whippability[b]* | *Foam Stability[b]* | *Emulsifying Activity[b]* |
|---|---|---|---|
| Fish protein concentrate (FPC) | | | |
| untreated | 1.6 | 0.80 | 0.00 |
| succinylated | 2.4 | 0.31 | 0.57 |
| Soy protein isolate (SPI) | | | |
| untreated | 1.9 | 0.85 | 0.47 |
| succinylated | 2.8 | 0.64 | 0.53 |
| Casein | | | |
| untreated | 2.5 | 0.50 | 0.39 |
| succinylated | 3.4 | 1.00 | 0.47 |
| Ovalbumin | | | |
| untreated | 1.2 | 0.90 | 0.00 |
| succinylated | 2.4 | 0.76 | 0.50 |
| Gelatin | | | |
| untreated | 1.7 | 0.90 | 0.00 |
| succinylated | 2.4 | 1.00 | 0.23 |

[a] Each of the proteins was dispersed in an alkaline medium (pH 7–9) and then treated with succinic anhydride amounting to 10 times the concentration level (molar basis) of amino groups present in the protein. The degrees of succinylation measured by a usual method (*13*) were as follows: 98.0% for FPC, 95.2% for SPI, 94.0% for casein, 97.7% for ovalbumin, and 98.8% for gelatin.

[b] *See* text.

**Table VI. Functionality of Products Prepared from Succinylated Proteins[a] by Papain-Catalyzed Attachment of L-Leucine *n*-Alkyl Esters**

| *Product* | *Whippability* | *Foam Stability* | *Emulsifying Activity* |
|---|---|---|---|
| Suc-FPC hydrolysate (30 min)[b] | 2.8 | 0.68 | 0.51 |
| Suc-FPC–Leu–$OC_2$ (30 min)[c] | 3.8 | 0.35 | 0.63 |
| –$OC_4$ (30 min) | 4.4 | 0.12 | 0.68 |
| –$OC_6$ (30 min) | 4.4 | 0.35 | 0.69 |
| –$OC_8$ (30 min) | 4.5 | 0.85 | 0.69 |
| –$OC_{10}$ (30 min) | 2.7 | 0.90 | 0.70 |
| –$OC_{12}$ (30 min) | 2.4 | 0.97 | 0.70 |
| Suc-SPI hydrolysate (30 min) | 2.0 | 1.00 | 0.49 |
| Suc-SPI–Leu–$OC_2$ (30 min) | 3.7 | 0.40 | 0.61 |
| –$OC_4$ (30 min) | 3.9 | 0.20 | 0.64 |
| –$OC_6$ (30 min) | 4.3 | 0.27 | 0.65 |
| –$OC_8$ (30 min) | 3.9 | 0.80 | 0.66 |
| –$OC_{10}$ (30 min) | 2.2 | 0.98 | 0.64 |
| –$OC_{12}$ (30 min) | 2.0 | 1.00 | 0.64 |
| Suc-casein hydrolysate (30 min) | 2.4 | 0.80 | 0.41 |
| Suc-casein–Leu–$OC_2$ (30 min) | 3.9 | 0.90 | 0.56 |
| –$OC_4$ (30 min) | 4.0 | 0.45 | 0.57 |
| –$OC_6$ (30 min) | 4.0 | 0.18 | 0.58 |
| –$OC_8$ (30 min) | 4.1 | 0.32 | 0.59 |
| –$OC_{10}$ (30 min) | 3.6 | 0.71 | 0.61 |
| –$OC_{12}$ (30 min) | 3.2 | 0.75 | 0.64 |
| Suc-ovalbumin hydrolysate (30 min) | 2.0 | 0.68 | 0.41 |
| Suc-ovalbumin–Leu–$OC_2$ (30 min) | 3.4 | 0.36 | 0.62 |
| –$OC_4$ (30 min) | 3.5 | 0.20 | 0.63 |
| –$OC_6$ (30 min) | 3.4 | 0.21 | 0.63 |
| –$OC_8$ (30 min) | 3.4 | 0.24 | 0.64 |
| –$OC_{10}$ (30 min) | 3.4 | 0.60 | 0.72 |
| –$OC_{12}$ (30 min) | 3.4 | 0.72 | 0.83 |
| Suc-gelatin hydrolysate (15 min) | 3.2 | 1.00 | 0.00 |
| Suc-gelatin–Leu–$OC_2$ (15 min) | 3.3 | 1.00 | 0.34 |
| –$OC_4$ (15 min) | 3.4 | 1.00 | 0.35 |
| –$OC_6$ (15 min) | 3.7 | 1.00 | 0.39 |
| –$OC_8$ (15 min) | 4.0 | 0.85 | 0.50 |
| –$OC_{10}$ (15 min) | 3.1 | 0.85 | 0.51 |
| –$OC_{12}$ (15 min) | 2.9 | 0.87 | 0.53 |

[a] For the degrees of succinylation, *see* Table V.

[b] Succinylated FPC treated with papain for 30 min in the absence of L-leucine *n*-alkyl ester. Similar notations are used for the hydrolysates of other succinylated proteins.

[c] Product prepared from FPC by treatment with papain for 30 min in the presence of L-leucine ethyl ester (Leu–$OC_2$). Similar notations are used for other products resulting from the papain treatment of succinylated proteins in the presence of L-leucine *n*-alkyl esters.

Table VII. Physicochemical Properties

| *Surfactant* | *Fluorescence Intensity*[a] *(recorder response)* |
|---|---|
| Gelatin hydrolysate (15 min) | 28 |
| Gelatin–Leu–$OC_6$ (15 min) | 60 |
| Gelatin–Leu–$OC_{12}$ (15 min) | 381 |
| Suc-FPC hydrolysate (30 min) | 131 |
| Suc-FPC–Leu–$OC_4$ (30 min) | 146 |
| Suc-FPC–Leu–$OC_{12}$ (30 min) | 696 |

[a] Apparatus: Hitachi model 204 fluorescence spectrometer. Conditions: excitation at 365 nm and measurement at 450 nm.
[b] Apparatus: Plaxis model PR-1005 pulsed NMR analyzer. Temperature for measurement: 25°C.
[c] Not measured because no stable foam was formed.
[d] Each of the data obtained with a foam phase separated.

the enzymatic process with succinylated gelatin gave distinctly different data; the product resulting from the attachment of the $C_4$ or $C_6$ alkyl ester of L-leucine did not make a stable foam, while the products with longer-chain alkyl esters of L-leucine showed high whippability and foam stability. This may reflect the fact that succinylated gelatin is too hydrophilic compared with the other succinylated proteins.

### *Parameters Related to the Whipping and Emulsifying Functions of Surfactants Prepared from Proteins by Enzymatic Modification (24)*

According to the above-mentioned performance (*see* Tables IV and VI), we selected gelatin–Leu–$OC_6$ (15 min) and suc-FPC–Leu–$OC_4$ (30 min) as whipping surfactants, and gelatin–Leu–$OC_{12}$ (15 min) and suc-FPC–Leu–$OC_{12}$ (30 min) as emulsifying surfactants; the abbreviated designation of these surfactants is explained in detail in Tables IV and VI. The hydrolysates of gelatin and suc-FPC were used as controls. All of these samples were characterized for their physicochemical properties in terms of hydrophobicity (or lipophilicity) and their ability to interact with water and oil. The degree of interaction with water and oil was estimated from pulsed NMR measurement of their spin–spin relaxation times ($T_2$) and also from ESR measurement of rotational correlation time ($\tau_c$) observed for an added free radical.

**Hydrophobicity Measurement.** A conventional method (25) of using 1-anilinonaphthalene-8-sulfonate (ANS) as a semiquantitative hydrophobic probe was applied. To $8 \times 10^{-5}M$ ANS in $0.1M$ Tris–HCl (pH 8.5) was added the sample at a concentration of 0.05% (w/w) and the resulting solution measured for the intensity of its emission spectrum

**of Proteinaceous Surfactants**

| $T_2$ (ms)[b] | | $T_2$ (ms)[e] | | |
|---|---|---|---|---|
| *Before Whipping* | *After Whipping* | *Before Emulsification* | *After Emulsification* | $\tau_c$[g] ($\times 10^{-9}$ s) |
| 126 | —[c] | — | —[f] | 1.6 |
| 85 | 43[d] | 81 | 78 | 9.3 |
| 148 | —[c] | 145 | 75 | 10.3 |
| 114 | —[c] | — | —[f] | 1.7 |
| 83 | 49[d] | 74 | 68 | 9.4 |
| 118 | —[c] | 140 | 68 | 9.9 |

[e] Apparatus: Plaxis model PR-1005 pulsed NMR analyzer. Temperature for measurement: 0.5°C.
[f] Not measured because no stable emulsion was formed.
[g] Apparatus: JEOL JES–PE–3X ESR spectrometer. Each ESR spectrum was recorded at 9.2–9.3 GHz by using a modulation frequency of 100 KHz and a 10-mV incident microwave power.

at 450 nm. For the samples, gelatin hydrolysate, gelatin–Leu–$OC_6$, and gelatin–Leu–$OC_{12}$, the observed spectral intensities tended to increase in this order (*see* Table VII). This result indicates that the emulsifying surfactant, gelatin–Leu–$OC_{12}$, is endowed with a much higher degree of hydrophobicity (or lipophilicity) than the whipping surfactant, gelatin–Leu–$OC_6$. Similar relationships were found with the other samples, suc-FPC hydrolysate, suc-FPC–Leu–$OC_4$, and suc-FPC–Leu–$OC_{12}$ (*see* Table VII).

**Pulsed NMR Measurements.** In a stable foam or emulsion system, surfactant molecules are arranged at an interface to interact with an air or oil particle. The interaction takes place in the form of an air–surfactant–water or oil–surfactant–water complex. In both cases the rotational freedom of water molecules is restricted to a greater or lesser extent. The average rotational correlation time of a water molecule is know to correlate well with its $T_2$ value measured by pulsed NMR spectrometry (*26, 27*). In short, $T_2$ can be considered as an index for average mobility of water molecules. According to the present knowledge, $T_2$ for free water is approximately 2 s while that for bound water is smaller depending on the intensity of binding (*26*).

We dispersed each sample (1.5 g) in 10 mL of 0.01*M* phosphate (pH 7) and submitted the dispersion to pulsed NMR measurement. $T_2$ was obtained from the following equation

$$A_{echo} = A_0 \exp(-2\tau/T_2)$$

where $A_{echo}$ is the echo amplitude in the spectrometry, $\tau$ is the decay time, and $A_0$ is the echo amplitude at $\tau = 0$. When the whipping surfactant, gelatin–Leu–$OC_6$, was used as a sample, a clear decrease in $T_2$

was observed (*see* Table VII). Also, on whipping this dispersion a further decrease in $T_2$ resulted. A similar tendency was found with the other whipping surfactant, suc-FPC–Leu–$OC_4$. For the dispersions containing the emulsifiable surfactants, gelatin–Leu–$OC_{12}$ and suc-FPC–Leu–$OC_{12}$, the observed $T_2$ values were rather large. This indicates that, unless oil is present, the molecules of these surfactants do not have a great ability to prevent water molecules from tumbling.

$T_2$ studies with an oil-in-water emulsion system would provide information on the interaction between an emulsifying surfactant and water. In this experiment we used hydrogenated coconut oil which had a high freezing temperature. First, a dispersion of each surfactant (0.5 g) in 10 mL of 0.01*M* phosphate (pH 7) was measured for $T_2$ at 0.5°C, below the freezing temperature of the hydrogenated coconut oil used. Subsequently, this oil (5 mL) and the above-mentioned dispersion (5 mL) were mixed with each other and the mixture was well homogenized at 60°C for 3 min. Immediately an entire homogenate was transferred into an NMR sample tube (10 mm in diameter), cooled at 0.5°C for 30 min, and then submitted to $T_2$ measurement. At this temperature the hydrogenated coconut oil was mostly frozen, making a negligible contribution to the echo amplitude at the millisecond level. Contributory species must be only the protons of water and those of unfreezable oil. However, $T_2$ due to the latter never exceeds 50 ms. Any larger $T_2$ value to be observed, therefore, can be ascribed exclusively to that of water. As Table VII shows, it was actually observed that, on emulsification with oil, both gelatin–Leu–$OC_{12}$ and suc-FPC–Leu–$OC_{12}$ gave appreciably decreased $T_2$ values. For example, $T_2$ observed for the former was 145 ms in the absence of oil and 75 ms as a result of its emulsification with oil. It is considered that very loosely bound water generally gives a $T_2$ value ranging from 50 to 100 ms (*26*).

**ESR Measurement.** The behavior of a stable free radical added to a medium reflects an average state of molecular rotation of the medium. Oil-soluble free radicals are used commonly to probe the rotation of oil molecules. The mobility of any oil-soluble probe is represented by its rotational correlation time ($\tau_c$) as calculated from the following equation (*28*):

$$\tau_o = [(h_0/h_1)^{1/2} + (h_0/h_{-1})^{1/2} - 2]\left[\frac{(3.06 \times 10^8)^2}{4\pi\sqrt{3}w_0}\right]^{-1} \text{(sec)}$$

The symbols involved in the equation are given in Figure 5. The probes used most widely in the field of food science are of the nitroxide class (*29, 30*). We selected 2-(10-carboxydecyl)-2-hexyl-4,4-dimethyl-3-oxazolodinyloxy methyl ester (Structure I) as an oil-soluble probe. A $\tau_c$

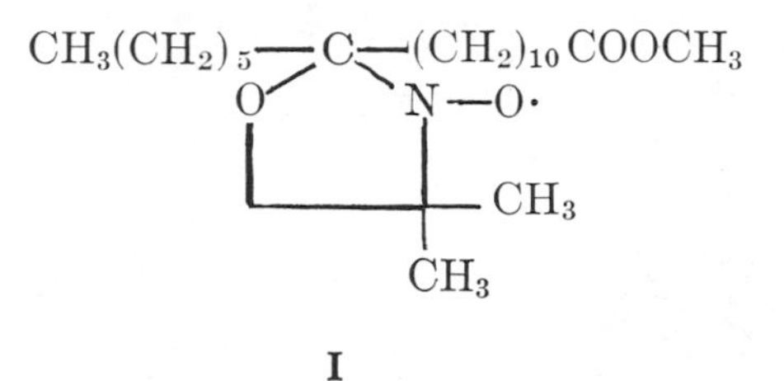

I

value of $1.6 \times 10^{-9}$ s was obtained with this probe ($4 \times 10^{-4}M$) in corn oil at 10°C. Next, a mixture of surfactant (0.5 g) dispersed in 5 mL 0.01*M* phosphate buffer, pH 7.0 and 5 mL corn oil containing the probe was well homogenized to a fine emulsion, which was submitted immediately to ESR measurement under simulated conditions. As Table VII shows, gelatin–Leu–$OC_{12}$ gave a $\tau_c$ value of $10.3 \times 10^{-9}$ s which was distinctly larger than $\tau_c$ ($1.6 \times 10^{-9}$ s) obtained in the absence of any surfactant. A similar oil-immobilizing function was found in suc-FPC–Leu–$OC_{12}$. Schenouda and Pigott (*30*) have reported that a one-order increase in $\tau_c$ results when a lipid is bound tightly to a protein. Our observation of the $\tau_c$ increase (*see* Table VII) may indicate the occurrence of a similar surfactant–oil interaction. Since the interaction involving a surfactant must take place at the site of an alkyl moiety, its chain length determines its characteristics as a surfactant.

**Discussion.** According to the result of the experiment using ANS, the attachment of a proper degree of hydrophobicity (or lipophilicity) to a hydrophilic proteinaceous moiety can maximize whippability, while a potent emulsifying activity results when a much higher degree of hydrophobicity is attached. All of the data obtained from pulsed NMR and ESR measurements lead us to speculate that the whipping–surfactant molecules can rearrange at an air–water interface with their hydrophilic proteinaceous parts exposed to the aqueous environment and their lipophilic alkyl ends to the air (*see* Figure 6A). Thus a ternary air–surfactant–water complex may be formed, where the water molecules are

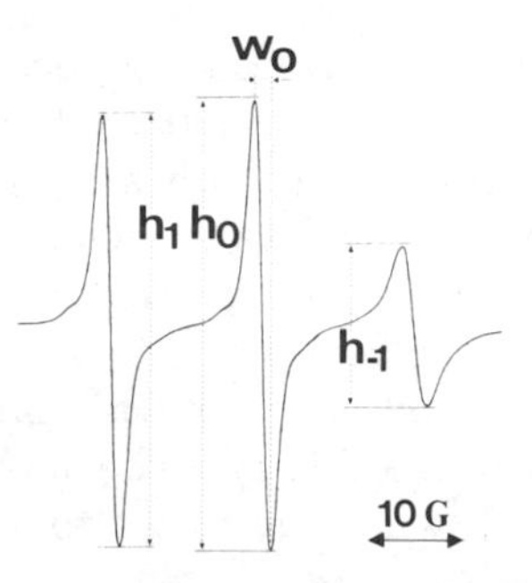


*Figure 5. A typical ESR spectrum of 2-(10-carboxydecyl)-2-hexyl-4,4-dimethyl-3-oxazolodinyloxy methyl ester in corn oil and parameters for obtaining its rotational correlation time ($\tau_c$)*

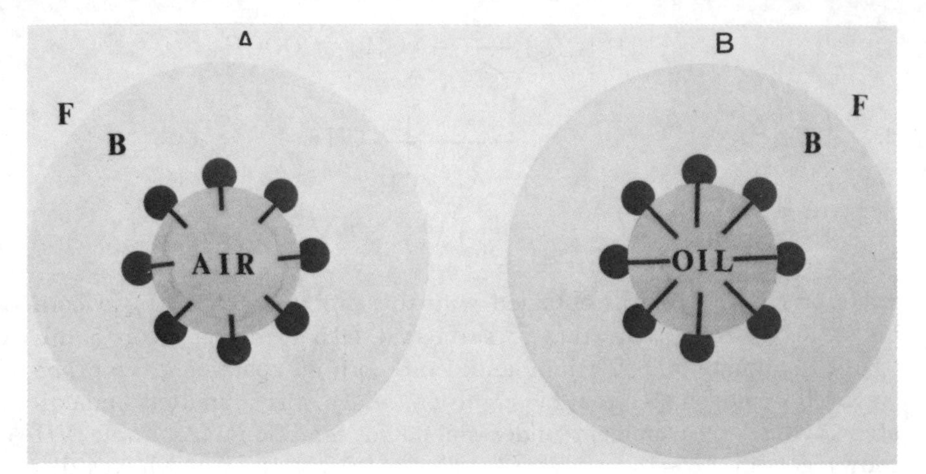


*Figure 6. Schematics of proteinaceous surfactant molecules arranged at an air–water interface (A) and at an oil–water interface (B): B, bound water; F, free water.*

bound loosely. A barrier made up of this kind of water molecules around air particles may contribute to foam stability.

A similar mechanism could apply to the emulsifying surfactants; their molecules may restrict the rotation of water as well as oil molecules through the formation of a ternary oil–surfactant–water complex (*see* Figure 6B). The resulting water barrier surrounding the oil particles would be able to hinder their coalescence and consequently maintain a stable emulsion.

### *Use of Proteinaceous Surfactants as Food Ingredients* (31)

With a series of applications to well-known food products, we tested the suitability of the proteinaceous surfactants prepared from gelatin by the papain-catalyzed attachment of L-leucine *n*-alkyl esters. Using each of these surfactants we tried to prepare snow jelly, ice cream, mayonnaise-like food, and bread.

**Snow Jelly.** Snow jelly was selected as a typical whipped food. This generally is prepared by whipping a sugar solution containing a surfactant and subsequently setting the foam system with a mixture of gelatin and milk at low temperature. The final ingredient composition we used was as follows: 40 g granulated sugar; 11.6 g dried milk; 7 g gelatin; 121.4 g water; and 20 g surfactant dispersion. Gelatin–Leu–$OC_2$, gelatin–Leu–$OC_4$, and gelatin–Leu–$OC_6$ were selected as whipping surfactants according to the performance shown in Table IV. The surfactant dispersion was made by adding 1 g of each whipping agent to 19 g of water. Dried egg white and fresh cream also were used as controls. To prepare a surfactant dispersion, 2.5 g dried egg white was added to 17.5 g water, while 20 g fresh cream was used without any treatment. All of the snow jelly samples were molded in an ice bath and stored at 20°C for 1 h, prior to texturometric measurements for hardness, adhesiveness, and cohesiveness (32). Any of the surfactants prepared from gelatin by the covalent attachment of the L-leucine *n*-alkyl esters were well suited to the preparation of snow jelly. In particular, gelatin–Leu–$OC_6$ resulted in giving high degrees of hardness and cohesiveness which were comparable with those of the snow jelly sample made from fresh cream (*see* Table VIII).

**Ice Cream.** Since ice cream is a colloidal system with air and oil particles, the whipping and the emulsifying functions of an added surfactant will contribute to giving a fine structure. From that point of view, gelatin–Leu–$OC_4$ (15 min) and gelatin–Leu–$OC_{12}$ (15 min) were selected as whipping and emulsifying surfactants, respectively (*see* Table IV). The most widely used surfactants, monostearin and monoolein, were used as controls (33). The final adopted ingredient composition was as follows: 20 g dried milk; 30 g granulated sugar; 10 g hydrogenated coconut oil; 10 g millet jelly powder; 0.4 g carboxymethylcellulose; 0.6 g surfactant; and 129 g water. The overrun, a typical parameter of the performance, was represented by:

$$\frac{\text{specific gravity of ice cream mix (g/mL)} - \text{specific gravity of soft cream (g/mL)}}{\text{specific gravity of soft cream (g/mL)}} \times 100$$

where soft cream is the product prepared from an ice cream mix by whipping. The performance is shown in Table IX, indicating that either of the two proteinaceous surfactants can give a higher degree of overrun by mixing (whipping) in a shorter time than the lipid-type surfactants. Also, gelatin–Leu–$OC_{12}$ is better than gelatin–Leu–$OC_4$ as a surfactant for ice cream use. This result may reflect the fact that the emulsifying function of a surfactant makes a great contribution to the overrun in soft cream.

**Mayonnaise.** Mayonnaise is an oil-in-water emulsion prepared generally by using either whole egg or egg yolk. Instead of these, gelatin–Leu–$OC_{12}$ (15 min) (*see* Table IV) was tried as an emulsifying surfactant to prepare a mayonnaise-like food. Tween-60 as well as whole egg and egg yolk were used as controls. The final adopted ingredient composition was as follows: 54 g soybean oil; 0.9 g table salt; ~ 10 g surfactant dispersion. To prepare surfactant dispersions of two different concentra-

**Table VIII. Texturometric Measurements[a] of Snow Jelly**

| *Surfactant* | *Hardness (kg) Average ± Statistical Error* |
|---|---|
| Gelatin–Leu–$OC_2$ (15 min) | 1.09 ± 0.03 |
| Gelatin–Leu–$OC_4$ (15 min) | 1.05 ± 0.06 |
| Gelatin–Leu–$OC_6$ (15 min) | 1.17 ± 0.03 |
| Egg white | 0.64 ± 0.04 |
| Fresh cream | 1.19 ± 0.09 |

[a] Conditions: apparatus, Zenken Model GTX-2 texturometer; sample size, 28 mm in diameter and 10 mm in thickness; lucite plunger, 30 mm in diameter; clearance, 2 mm for measurement of hardness and adhesiveness and 3 mm for measurement of cohesiveness; temperature at the time of measurement, 20°C.

**Table IX. Effects of Different Surfactants on the Overrun as**

| | *Mixing Time (min)* | |
|---|---|---|
| *Surfactant* | *0* *Average ± Statistical Error* | *1* *Average ± Statistical Error* |
| Gelatin–Leu–$OC_4$ (15 min) | 20.1 ± 0.5 | 54.4 ± 1.2 |
| Gelatin–Leu–$OC_{12}$ (15 min) | 14.9 ± 0.9 | 47.3 ± 1.1 |
| Monostearin | 8.7 ± 0.4 | 33.6 ± 1.3 |
| Monoolein | 2.9 ± 0.9 | 39.3 ± 0.8 |

tion levels, 0.32 or 0.64 g of gelatin–Leu–$OC_{12}$ was added to 9 g of 3.5% acetic acid. Five grams of raw egg and raw egg yolk were dispersed separately in an equal amount (5 g) of 3.5% acetic acid. A Tween-60 dispersion was prepared by adding 0.32 g of this surfactant to 9 g of 3.5% acetic acid. All of the mayonnaise samples were stored at 20°C for 1 h prior to texturometric and sensory evaluations. A test panel rated highly the texture of the mayonnaise-like food prepared by using gelatin–Leu–$OC_{12}$ even at a 0.5% concentration level (*see* Table X). The texturometric data also showed that, with respect to both hardness and adhesiveness, gelatin–Leu–$OC_{12}$ gave results between whole egg and egg yolk (*see* Table X).

**Bread.** The use of surfactants with lipid-emulsifying activity is essential in bread making. One of the examples is monostearin (*34*). We used gelatin–Leu–$OC_{12}$ (15 min) and compared the effect of this proteinaceous surfactant with that of monostearin. The ingredient composition we used was as follows: 50 g strong wheat flour; 1.25 g dry yeast powder; 1 mg potassium bromate; 1 g table salt; 2 g granulated sugar; 2.5 g hydrogenated coconut oil; 0.4 g surfactant; and 30 g water. Baked samples were allowed to stand for 1 h at room temperature and then were

**Samples Prepared by Using Different Surfactants**

| Adhesiveness (TM unit[b]) Average ± Statistical Error | Cohesiveness (TM unit[b]) Average ± Statistical Error |
|---|---|
| 0.17 ± 0.04 | 0.81 ± 0.01 |
| 0.12 ± 0.05 | 0.90 ± 0.02 |
| 0.12 ± 0.05 | 0.87 ± 0.07 |
| 0.08 ± 0.04 | 0.50 ± 0.08 |
| 0.13 ± 0.04 | 0.91 ± 0.03 |

[b] Referring to texturometer unit.

**a Function of Mixing Time for Preparing Soft Cream Samples**

| Mixing Time (min) | | | |
|---|---|---|---|
| *2* Average ± Statistical Error | *4* Average ± Statistical Error | *6* Average ± Statistical Error | *8* Average ± Statistical Error |
| 77.5 ± 1.7 | 108.9 ± 1.0 | 140.3 ± 1.4 | 153.3 ± 1.3 |
| 75.7 ± 1.4 | 110.1 ± 0.8 | 150.0 ± 1.1 | 163.2 ± 0.8 |
| 68.2 ± 1.2 | 87.4 ± 1.4 | 89.1 ± 1.1 | 89.4 ± 1.0 |
| 50.7 ± 0.6 | 46.7 ± 0.1 | 45.6 ± 0.5 | 42.9 ± 1.0 |

measured for loaf volume. Subsequently, the samples were stored at 20°C and $a_w$ 0.93 for up to 4 d. Crumb pieces were sampled every 24 h and their hardness measured with a texturometer. Gelatin–Leu–$OC_{12}$ was very effective in giving a large loaf volume of bread; the effect was comparable to that of monostearin (*see* Table XI). Bread made without using any surfactant gave a significantly smaller loaf volume. In the bread sample prepared with the proteinaceous surfactant, the crumb remained softer over a 4-d storage period under our conditions (*see* Table XI).

### *Conclusion*

The present study is an attempt to maximize the functionality of food proteins through the smallest possible modification. This object can be attained partly by applying a papain-catalyzed unusual reaction that proceeds efficiently under intentionally abnormal conditions. These conditions were characterized by at least two requirements: extremely high substrate concentration and alkaline pH (*see* Table III). The reaction taking place is defined primarily as a type of aminolysis of a protein (sub-

**Table X. Texturometric Measurement[a] and Sensory Evaluation of Different**

| *Surfactant* | *Surfactant Concentration* |
|---|---|
| Gelatin–Leu–$OC_{12}$ (15 min) | 0.5 |
| Gelatin–Leu–$OC_{12}$ (15 min) | 1.0 |
| Whole egg | — |
| Egg yolk | — |
| Tween-60 | 0.5 |

[a] Conditions: apparatus, Zenken model GTX-2 texturometer; sample size 70 mm in diameter and 6 mm in thickness; nickel plunger, 50 mm in diameter; clearance, 1 mm; temperature at the time of measurement, 20°C.

**Table XI. Effects of Surfactants on the Loaf Volume of Bread During**

| *Surfactant* | *Loaf Volume (mL) Average ± Statistical Error* |
|---|---|
| Gelatin–Leu–$OC_{12}$ (15 min) | 80.0 ± 1.8 |
| Monostearin | 79.3 ± 2.0 |
| None | 74.6 ± 2.0 |

[a] Conditions for measurement: apparatus, Zenken model GTX-2 texturometer; sample size, 1 cm × 1 cm × 1 cm; lucite plunger, 18 mm in diameter; clearance, 1 mm; temperature at the time of measurement, 20°C.

strate) by an added amino acid ester (nucleophile), leading to the formation of a polypeptide product to which the amino acid ester is attached covalently (*see* Figure 1).

This reaction, when applied to a mixture of a hydrophilic protein (substrate) and a lipophilic L-leucine *n*-alkyl ester (nucleophile), can produce a proteinaceous surfactant with proper amphiphilic functions (*see* Figure 4).

The surfactant characteristics of the product vary depending on its lipophilicity, which is related to the chain length of the *n*-alkyl moiety attached. The attachment of a $C_4$–$C_6$ *n*-alkyl ester of L-leucine gives rise to greatly improved whippability as well as to very stable foams, while a potent oil-emulsifying activity results when a longer-chain *n*-alkyl ester, especially *n*-dodecyl ester, of L-leucine is attached. This relationship was found with several substrates used (*see* Table VI).

**Mayonnaise and Mayonnaise-Like Samples Prepared by Using Surfactants**

| *Texturometric Measurement* | | |
|---|---|---|
| *Hardness (kg) Average ± Statistical Error* | *Adhesiveness (TM Unit[b]) Average ± Statistical Error* | *Sensory Evaluation* |
| 0.87 ± 0.01 | 1.77 ± 0.02 | smooth and soft |
| 1.06 ± 0.02 | 2.26 ± 0.02 | smooth and hard |
| 0.75 ± 0.01 | 1.68 ± 0.02 | smooth and soft |
| 1.49 ± 0.01 | 3.31 ± 0.02 | smooth and very hard |
| 0.26 ± 0.01 | 0.50 ± 0.01 | oily and very soft |

[b] Referring to texturometer unit.

**Samples and the Changes in Hardness of Their Crumb Storage**

| *Hardness[a] (kg)* | | | |
|---|---|---|---|
| *Day 1 Average ± Statistical Error* | *Day 2 Average ± Statistical Error* | *Day 3 Average ± Statistical Error* | *Day 4 Average ± Statistical Error* |
| 2.4 ± 0.1 | 2.8 ± 0.1 | 3.4 ± 0.3 | 4.3 ± 0.1 |
| 3.0 ± 0.2 | 3.4 ± 0.3 | 4.2 ± 0.2 | 4.6 ± 0.1 |
| 3.3 ± 0.1 | 3.8 ± 0.2 | 4.3 ± 0.2 | 4.9 ± 0.3 |

Supporting data were obtained by instrumental analyses for the highly whippable surfactants such as gelatin–Leu–$OC_6$ and suc-FPC–Leu–$OC_4$ and the highly emulsifiable surfactants such as gelatin–Leu–$OC_{12}$ and suc-FPC–Leu–$OC_{12}$ (*see* Table VI for the abbreviated notations). According to pulsed NMR measurements of the spin–spin relaxation time ($T_2$) of the water proton (*see* Table VII), it is estimated that the molecules of the whipping surfactants are arranged well at an air–water interface, with formation of a ternary air–surfactant–water complex (*see* Figure 6).

An ESR study using an oil-soluble free radical probe to observe rotational correlation time, $\tau_c$, of oil (*see* Table VII), and pulsed NMR measurements indicate that the molecules of the emulsifying surfactants surround an oil particle to form a ternary oil–surfactant–water complex (*see* Figure 6).

Several of the proteinaceous surfactants prepared in the present work can be used in food applications. In particular, gelatin–Leu–$OC_6$ and gelatin–Leu–$OC_{12}$ were useful as ingredients for selected food items: the former for snow jelly (*see* Table VIII) and the latter for ice cream (*see* Table IX), mayonnaise (*see* Table X), and bread (*see* Table XI).

The present study actually shows examples of how small changes in the chemical structures of food proteins can result in large changes in their functionality.

### *Literature Cited*


1. Hammonds, T. M.; Call, D. L. *Chem. Technol.* **1972**, *2*, 156.
2. Kinsella, J. E. *CRC Crit. Rev. Food Sci. Nutr.* **1976**, *7*, 219.
3. Feeney, R. E. In "Food Proteins: Improvement Through Chemical and Enzymatic Modification," *Adv. Chem. Ser.* **1977**, *160*, 3.
4. Whitaker, J. R. In "Food Proteins: Improvement Through Chemical and Enzymatic Modification," *Adv. Chem. Ser.* **1977**, *160*, 95.
5. Pour-El, A. In "Toxic Chemical and Explosives Facilities," *ACS Symp. Ser.* **1979**, *92*, ix.
6. Franzen, K. L.; Kinsella, J. E. *J. Agric. Food Chem.* **1976**, *24*, 788.
7. Yamashita, M.; Arai, S.; Imaizumi, Y.; Amano, Y.; Fijimaki, M. *J. Agric. Food Chem.* **1979**, *27*, 52.
8. Yamashita, M.; Arai, S.; Amano, Y.; Fijimaki, M. *Agric. Biol. Chem. (Tokyo)* **1979**, *43*, 1065.
9. Arai, S.; Watanabe, M., presented at the *Ann. Meet. Jpn. Agric. Chem. Soc., Fukuoka, April, 1980.*
10. Watanabe, M.; Toyokawa, H.; Shimada, A.; Arai, S., presented at the *Ann. Meet. Jpn. Agric. Chem. Soc., Fukuoka, April, 1980.*
11. Watanabe, M.; Shimada, A.; Arai, S., unpublished data.
12. Zittle, C. A.; Custer, J. H. *J. Dairy Sci.* **1963**, *46*, 1069.
13. Hoagland, P. *J. Dairy Sci.* **1966**, *49*, 783.
14. Stark, C. M.; Liotta, C. In "Phase Transfer Catalysis"; Academic: New York-San Francisco-London, 1978; p. 111.
15. Fieser, L. F.; Fieser, M. In "Reagents for Organic Synthesis"; John Wiley & Sons: New York-London-Sydney, 1967; p. 584.
16. Kato, T.; Makizumi, K.; Ohno, S.; Izumiya, N. *J. Jpn. Chem. Soc.* **1962**, *83*, 1151.
17. Weber, K.; Osborn, M. *J. Biol. Chem.* **1969**, *244*, 4406.
18. Hill, R. L.; Smith, E. L. *J. Biol. Chem.* **1957**, *228*, 557.
19. Fraenkel-Conrat, H.; Harris, J. I.; Levy, A. L. In "Methods of Biochemical Analysis"; Glick, D., Ed.; Interscience: New York, 1955; Vol. 2, p. 359.
20. Franzen, K. L.; Kinsella, J. E. *J. Agric. Food Chem.* **1976**, *24*, 914.
21. Jones, N. R. In "The Science and Technology of Gelatin"; Ward, A. G.; Courts, A., Eds.; Academic: London, 1977; p. 366.
22. Beuchat, L. R. *J. Agric. Food Chem.* **1977**, *25*, 258.
23. Peace, K. N.; Kinsella, J. E. *J. Agric. Food Chem.* **1978**, *26*, 716.
24. Watanabe, M.; Shimada, A.; Arai, S., unpublished data.
25. McClure, W. O.; Edelman, G. M. *Biochemistry* **1966**, *5*, 1908.
26. Kuntz, I. D.; Kauzmann, W. *Adv. Protein Chem.* **1974**, *28*, 239.
27. Fennema, O. In "Food Proteins"; Whitaker, J. R.; Tannenbaum, S. R., Eds.; Avi: Westport, CT, 1977; p. 50.
28. Waggoner, A. S.; Griffith, O. H.; Christensen, C. R. *Proc. Natl. Acad. Sci.* **1967**, *57*, 1198.

29. Berliner, L. J. In "Spin Labeling, Theory and Application"; Academic: London-New York-San Francisco, 1976; p. 592.
30. Schenouda, S. Y. K.; Pigott, G. M. *J. Agric. Food Chem.* **1976,** *24,* 11.
31. Watanabe, M.; Shimada, A.; Yazawa, E.; Kato, T.; Arai, S., presented at the *Ann. Meet. Jpn. Agric. Chem. Soc., Fukuoka, April, 1980.*
32. Friedman, H. H.; Whitney, J. E.; Szczesniak, A. S. *J. Food Sci.* **1963,** *28,* 390.
33. Berger, K. G. In "Food Emulsions"; Friberg, S., Ed.; Marcel Dekker: New York, 1976; p. 141.
34. Van Haften, J. L. *J. Am. Oil Chem. Soc.* **1979,** *56,* 831A.

RECEIVED October 10, 1980.

8

# Modification of Proteins with Proteolytic Enzymes from the Marine Environment

N. F. HAARD, L. A. W. FELTHAM, N. HELBIG, and E. J. SQUIRES

Department of Biochemistry, Memorial University of Newfoundland, St. John's, Newfoundland, Canada A1B 3X9

*Pepsin (E.C. 3.4.4.1) was isolated from the stomach lining of three marine fish ranging in habitat from arctic to temperate temperatures in the northwest Atlantic (arctic cod, Greenland cod, and American smelt). The fish pepsins had a more alkaline pH optimum than mammalian pepsins characterized thus far and pH optima were dependent on assay temperature. Fish pepsins exhibited activation energies for the hydrolysis of hemoglobin (pH 1.9) ranging from 4.1 to 8.8 kcal/mol in contrast to 11.2 kcal/mol for porcine pepsin. Activation energies were affected markedly by the assay pH for fish pepsins and only slightly affected for porcine pepsin. Temperature optima for fish pepsins were 15°–20°C lower than for porcine pepsin and the* $K'_m$ *for hemoglobin was substantially higher for fish pepsins than for porcine pepsin. The* $K'_m$ *for fish pepsins was variable with different assay temperatures. This chapter discusses low-temperature-adapted digestive enzymes and their utility as food-processing aids and the general importance of proteolysis in marine food products.*

Understanding proteolysis is pertinent to the science and technology of marine food products for a variety of reasons. Proteases that are interesting and important to the seafood technologists include intracellular enzymes, primarily those associated with muscle tissue, extracellular digestive enzymes, and extracellular enzymes originating from microorganisms or parasites. Since most marine organisms are poikilothermic and reside in widely differing environmental temperatures, we would expect that thermodynamic and kinetic properties of such proteolytic enzymes differ considerably. Hence the effect of temperature on physio-

0065-2393/82/0198-0223$05.50/0

logical disorders and microbial deterioration of marine products that are linked to proteolytic action would be expected to vary considerably. Moreover, proteolytic enzymes isolated from marine organisms may find special utility as food-processing aids because of their unique properties such as cold stability, high-molecular activity at low temperature, thermal instability, etc. The intent of this treatise is to comment briefly on the importance of endogenous and exogenous proteolytic enzymes in marine products and to present recently obtained data describing the properties of pepsins isolated from certain low-temperature-adapted marine poikilotherms.

### *Importance of Proteolytic Enzymes in Marine Food Products*

Some examples relating the importance of proteolytic enzymes to the quality, quality assessment, and processing of fish, fishery wastes, and other marine products are summarized in Tables I and II.

**Quality.** The presence of endogenous intracellular and extracellular proteases has been linked to the rapid deterioration of certain fishery products during cold storage (*1*) and heating (*2*) and to specific disorders such as belly burst in capelin (*3*), the himidori phenomenon (loss of elasticity) in Kamaboko preparation (*4*), and excessive tissue softening, such as milky hake caused by proteolytic enzymes produced by the protozoa *Chloromyscium thyrsites* where the infected tissue has the consistency of toothpaste (*5*). There is also evidence relating endogenous proteases (cathepsins) to desirable attributes of fish flesh such as flavor and to seasonal variations in quality. Moreover, exogenous sources of protease, such as subtilisin, have been used to monitor the ecological impact of effluent on fish quality (*6*).

**Table I. Relevance of Proteolytic Enzymes to the Quality of Seafood**

| | *Source of Proteolytic Activity* | | |
|---|---|---|---|
| *Example* | *Intra-cellular* | *Diges-tive* | *Exoge-nous* |
| Deterioration of krill | − | + | + |
| Belly burst, e.g. capelin | − | + | − |
| Facilitating bacterial enumeration | + | + | + |
| Texture softening | + | + | + |
| Himidori phenomenon in Kamaboko | + | − | − |
| Milky hake, etc. | − | − | + |
| Seasonal variation in flavor | + | − | − |
| Quality assessment | − | − | + |

**Table II. Proteolytic Enzymes as Food-Processing Aids**

| *Example* | *Intracellular* | *Digestive* | *Exogenous* |
|---|---|---|---|
| Fish sauces | + | + | + |
| Facilitate salting | + | − | − |
| Ripening fish preserves | + | + | + |
| Skinning | + | − | + |
| Fish silage | + | + | ? |
| Fish protein isolates | − | + | + |
| Modifying FPC | − | − | + |

**Food-Processing Aids.** Various traditional products made from fish such as sauces, preserves, cheese, and salt fish are dependent on or are influenced by proteases associated with the flesh. Also, exogenous sources of proteolytic activity have been used to accelerate or improve these traditional practices. For example, bromelain treatment of mackerel flesh solubilizes 75% of the protein nitrogen and gives a product having similar characteristics to oriental fish sauce (*7*). Other sources of proteolytic activity such as fungal extracts (*8*) and 0.5% pronase (*9*) also have been used to accelerate sauce preparation from various species of fish. Enzyme preparations from the organs of herring, mackerel, sardine, and sprats have been used to refine the taste of mackerel preserves (*10*), and exogenous enzymes have been used to accelerate the ripening of herring fillets (*11*). Endogenous protease activity also has been related to the suitability of fish for salting (*12*). Proteolytic enzymes from *Bacillus subtilus* (subtilo peptidases) can facilitate complete removal of skins and residual intestines from the ventral cavity of certain fish (*13*) and a Swedish patent (*14*) describes skin removal by adding acetic or citric acid to fish and presumably involves activating endogenous intracellular proteases.

Exogenous and endogenous sources of proteases have been used also to process inedible fish or plant-processing wastes to produce oils, meals, fish solubles, flavoring components, and other by-products. The primary emphasis has been to produce biological fish meals or fish protein isolates for animal feed or human consumption. Hale (*15*) recommends pepsin, papain, or pancreatin for preparing fish protein isolates. Others recommend proteases from *Aspergillus* species to eliminate fishy taste and off-odor (*16*); a combination of alkaline proteases and acidic proteases to obtain a fish protein isolate of excellent organoleptic and nutritional quality (*17*); papain to obtain a fish hydrolysate having no bitter taste for use as a beverage additive (*18*); ficin to produce a spray-dried fish protein concentrate in the SINTEF process (*19*); bromelain hydrolysis to obtain a freeze-dried protein concentrate from cod or haddock (*20*); neutrase to obtain a spray-dried product from krill (*21*); protosubtilin to aid autolysis of fish wastes (*22*); alcalase, followed by pancreatin treatment, to yield a low-molecular-weight fish protein hydrolysate from cod offal (*23*); bacterial proteases to prepare a meal from fish wastes (*24*); and visceral enzymes to prepare products useful for animal feed and human consumption (*25*). The traditional fish silage is prepared from offal by acidifying with mineral or organic acids, preferably formic acid, and allowing the material to digest to a stable autolysis product which is useful as animal feed (*26*). Fish silage appears to be gaining recognition in North America although it has been used for several decades in certain European countries.

Finally, numerous attempts have been made to treat fish protein concentrate (FPC) prepared by heating and/or solvent extraction with proteolytic enzymes to improve the functional properties of the protein components. Hevia et al. (*27*) observed that cysteine-activated ficin and pronase were most effective in solubilizing FPC although the hydrolysate had a bitter taste. Bhumiratama (*28*) used trypsin to solubilize 85% of the initial solids of the FPC and used a continuous-flow reactor to minimize product inhibition. Das et al. (*29*) observed that papain or pepsin treatment of catfish FPC had improved foaming and dispersion and Spinelli et al. (*30*) incorporated enzyme-modified fish proteins into bakery products, mayonnaise, desserts, and beverages. Miller and Groninger (*31*) showed that hydrolysis of acylated fish proteins with bromelain yielded a product with increased aeration capacity and foam stability but lowered emulsifying, gel, and water absorption properties.

These examples illustrate the broad manner in which protease activity can impact the quality and processing of marine products. Another facet of this topic is that proteolytic enzymes from fishery wastes may find general use as aids in the processing of other foods and in nonfood applications. Millions of metric tons of fish offal, rich in digestive enzymes, presently are not used and present a serious, growing disposal problem in areas where fish meal plants are not located. The new northern cod fishery off the coast of Newfoundland will in itself provide an estimated 500,000,000 lb of offal by 1985.

### *Proteolytic Enzymes as Food-Processing Aids*

Proteolytic enzymes from animal, plant, and microbial sources currently are used extensively as food-processing aids. Generally, a given proteolytic enzyme has been chosen for a specific process or application based on empirical trials. In more recent years, with our better understanding of the properties of such enzymes, the choice has been made by more thoughtful approaches. The criteria for the choice of a given protease in an application have been discussed elsewhere (32). The choice is made easier with a knowledge of specificity and other properties such as pH optimum, temperature optimum, response to inhibitors and activators, and stability. The specificity has been the only consideration in many applications. A protease with a narrow specificity, such as chymosin, is preferred in cheese manufacture since the actions of other proteases do not lead necessarily to curd formation or give rise to low yields and undesirable texture or flavor. Alternatively, producing protein hydrolysates for flavorants or for removing protein from some bound or intractable form, as in leaf protein concentrate, requires proteases of very broad specificity. The pH at which an enzyme is active also will affect its usefulness in a given process operation. For chillproofing of beer an enzyme must be active at or near pH 4.5. Other applications, such as tenderization of meat or the recovery of flesh from skeletal wastes, also may be limited by the pH range of the raw material. The presence of inhibitors in the raw material also may affect the choice of protease for a given application. Fungal enzymes are more useful than trypsin or plant proteases in dough-mixing and -baking operations because wheat flour contains protease inhibitors and oxidizing agents.

The temperature optimum of catalysis by a protease also can limit the utility of the enzyme in certain applications. Advantage has been taken of proteases that remain active during cooking operations such as papain as a meat tenderizer or proteases as gluten modifiers during baking operations. Prior emphasis has been on exploiting the use of enzymes active at high temperatures rather than at low temperatures, as we propose. The utility of proteases is also dependent on their thermal stability. Often there is an advantage to thermal stability; however, thermal stability also can give rise to problems in certain processes. For example, a heat-stable protease such as papain can survive the thermoprocess of canned hams with resulting continued hydrolysis and undesirable texture softening. Excessive proteolysis also gives rise to bitterness in cheese or in biological FPCs. Accordingly, there are instances where thermal instability may be viewed as an asset for a given application.

### *Low-Temperature Adaptation of Enzymes*

Hultin (33) has suggested that using low-temperature processing with enzymes could provide various benefits including lower thermal energy requirements, protection of the substrate or products of the reaction from degradation, minimization of side reactions, prevention of the destruction of other substances associated with the raw material, and minimization of microbial enumeration under certain circumstances.

The Northwest Atlantic fishery is a source of poikilothermic organisms adapted to low temperatures and is accordingly a promising source of low-temperature-adapted digestive enzymes. There is relatively little available information on the properties of digestive enzymes of fish from a low habitat temperature. It is, however, clear that organisms are capable of adaptation to low temperatures. The arctic cod fish (*Boreogadus saida*) remains active at environmental temperatures between −1.5° and 2.8°C (*34*) while fish from temperate waters undergo general anesthesia at such low temperatures (*35*). It is clear that many factors are involved in the adaption of an organism to low temperature (*36*). It is important to recognize that empirical changes in the kinetic and thermodynamic parameters of enzymes are associated with low-temperature adaption (*37*). Affected enzymic properties include substrate affinity, molecular activity, cold stability, activation energy for catalysis, sensitivity to modulators, isoenzyme pattern, time-dependent interconversions, and specificity (*33*). Not all of these changes are associated necessarily with temperature-adapted enzymes. The paucity of research in this area makes it difficult to know how general these modifications are and how many factors operate to modify the activity of a single enzyme.

**Molecular Activity.** Generally enzymes from low-temperature-adapted organisms have a higher molecular activity at cold temperatures (*38*). For example, $V_{max}$ for glyceraldehyde-3-phosphate dehydrogenase decreases thirtyfold for the rabbit enzyme when the assay temperature is reduced from 35° to 5°C while the $V_{max}$ for the lobster enzyme decreases by only tenfold over the same temperature range. Similarly, the $V_{max}$ for rabbit glycogen phosphorylase drops seventy-fivefold when the temperature is reduced from 30° to 0°C as compared with fifteenfold for the lobster enzyme. It has been suggested that high-molecular activity under conditions of low thermal energy is due to a more flexible protein conformation (*39*). It appears that certain fish contain proteolytic enzymes in the alimentary tract which are highly active at low temperatures since burnt belly spoilage syndrome caused by proteolysis can occur in less than 1 d in susceptible species held on ice (*40*).

**Thermal Stability.** Increased catalytic efficiency at low temperatures is associated often with a decrease in thermal stability (*41*). Proteolytic enzymes from the pyloric caeca of cod, herring, and mackerel are more heat labile than bovine trypsin (*42*); the thyroid protease of burbot from cold waters had lower thermal stability than that from burbot in a warmer habitat (*43*). The half-life of myofibrillar ATPase at 37°C averages 1 min for the enzyme from Antarctic Ocean fish species, 70 min for the enzyme from fish species in the Indian Ocean, and 600 min for the enzyme from East African hot spring fish (*33*).

**Substrate Affinity.** Enzyme–substrate affinity, as estimated by the Michaelis constant ($K_m$), is affected also by habitat temperature (*37*). In many cases, the temperature range at which the Michaelis constant is minimal coincides closely with the temperature of the organism's native environment. Wiggs (*43*) observed that the $K_m$ of thyroid protease from burbot is lower for winter fish than for summer fish. Hofer et al. (*44*) also showed that the $K_m$ of trypsin from various species, including fish, correlated with the temperature preferendum of each species. The response of $K_m$ to temperature appears to relate to the balance of electrostatic and hydrophobic interactions of the active site with substrate and intramolecular bonds within the enzyme (*38, 39, 45, 46, 47*).

**Activation Energy.** The above observations are consistent with additional findings that the activation energy ($E_a$) for catalysis generally relates to the habitat temperature of the organism. Enzymes from organisms adapted to low temperatures have lower $E_a$s than those from fish in a warmer environment. The $E_a$ (calories/mole) for fish myofibrillar ATPase ranges from 7,400 for *Nothena rossii* (0–2°C habitat) to 33,000 for *Tipalia grahami* (35°–38°C habitat) (*33*).

**Specificity.** Differences in enzyme specificity also have been related to habitat. Proteolytic activity appears to be more nonspecific in pelagic species of fish (*48*) although more study is needed to generalize on this point. It is quite significant that trypsin from agastric Cyprinid species of fish hydrolyze native proteins much more effectively than trypsin from gastric fish or mammals. It would appear that gastric species possess a trypsin that requires prior denaturation of proteins by acid (*49, 50, 51*). These observations deserve further study since one can anticipate an advantage in using a protease that more efficiently hydrolyzes native protein structures.

### *Characterization of Pepsins from Low-Temperature-Adapted Fish*

We recently have isolated and characterized pepsin(s) (E.C. 3.4.4.1) from three species of fish which range in habitat from arctic to subarctic regions in the northwest Atlantic Ocean. *Boreogadus saida,* commonly known as arctic cod or polar cod, is primarily an arctic fish and has been maintained in tanks at sub-zero temperatures (*34*). Greenland cod (rock cod, *Gadus ogac*) ranges from arctic waters to the Gulf of St. Lawrence in coastal waters and has a habitat temperature that is often below 0°C. American smelt (*Osmerus mordax*) is also a coastal fish which ranges from Lake Melville, Labrador to temperate waters in Virginia. We have isolated pepsin from smelt collected in Lake Melville, the northern most range of the species, and also from Gambo Pond, Newfoundland. Specimens of Greenland cod were obtained from Lake Melville and arctic cod were captured by trawlers in arctic waters north of Labrador. Pepsinogens were isolated from the stomach linings and purified by techniques described elsewhere (*52*). The arctic cod pepsin was purified partially while the Greenland cod and smelt pepsins appeared to be pure as judged by polyacrylamide gel electrophoresis (PAGE). Porcine pepsin was obtained from Sigma Chemical Co., St. Louis, MO. The results presented represent the early stages of our survey of pepsins from marine organisms in the northwest Atlantic. We will discuss the following properties of pepsins: pH optima; thermal stability; $E_a$ for catalysis; and substrate affinity. In all cases the substrate used was crystalline hemoglobin (Sigma, St. Louis, MO) purified of trichloroacetic acid (TCA)-soluble components by extensive dialysis against 0.06*N* HCl.

**pH Optima.** Porcine pepsin is active between pH 1 and 4 with a sharp maximum for several proteins at about pH 1.8 to 2. For synthetic substrates, the optimum pH is between 2 and 4 (*32*). We observed no influence of temperature on the pH optimum of porcine pepsin (PP) with hemoglobin. However, the pepsins isolated from marine poikilotherms from a low-temperature habitat had pH optima approximately one unit higher than the optimum pH 1.9 for porcine pepsin at 30°C (*see* Figures 1, 2, and 3.)

Pepsins isolated from the arctic to subarctic species, arctic cod and Greenland cod, also differed from PP in exhibiting a pronounced acid shift in pH optimum at lower assay temperatures (*see* Figures 1, 2, and 3). Studies with synthetic substrates and PP showed that substrates having free carboxylic acid groups exhibit a lower pH optimum than their esters (*54*). Hence, the present observations may reflect differences in ionization of certain groups in the active-site region of pepsins from

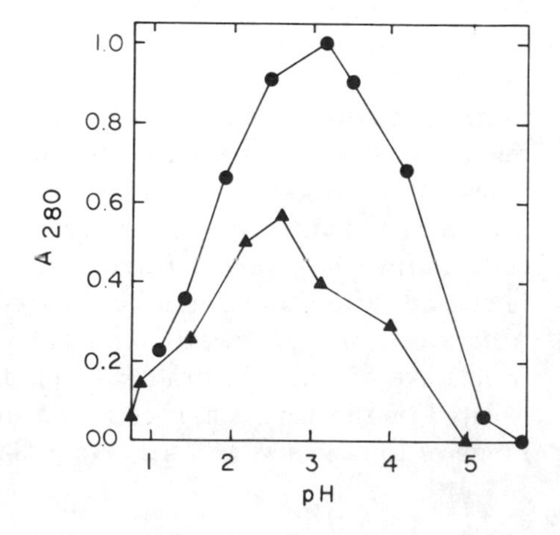


*Figure 1. Optimum pH of pepsin partially purified from the stomach lining of arctic cod: 30°C (●); 5°C (▲); A 280 = absorbancy of TCA solubles at 280 nm. Pepsin was assayed with acid-denatured hemoglobin (2%) at the indicated temperatures and activity was monitored by measurements of TCA-soluble products* (*53*).

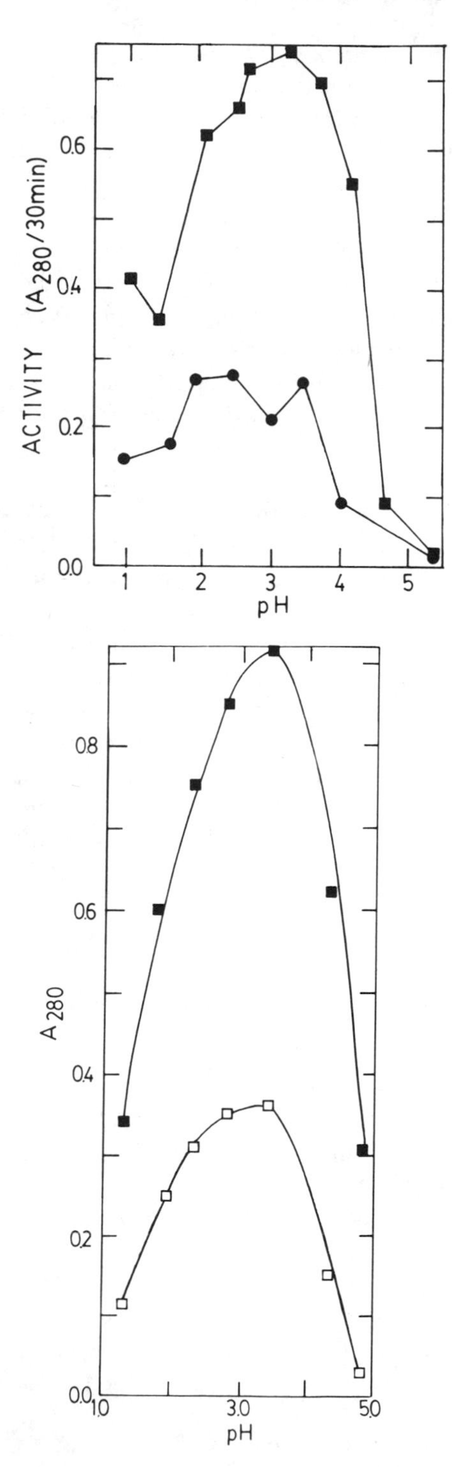


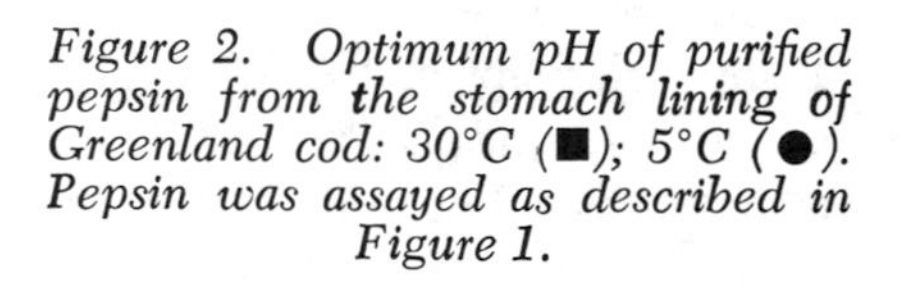
*Figure 2. Optimum pH of purified pepsin from the stomach lining of Greenland cod: 30°C (■); 5°C (●). Pepsin was assayed as described in Figure 1.*

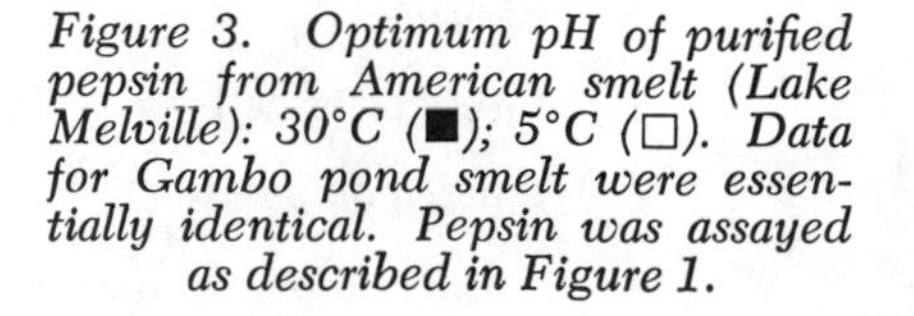
*Figure 3. Optimum pH of purified pepsin from American smelt (Lake Melville): 30°C (■); 5°C (□). Data for Gambo pond smelt were essentially identical. Pepsin was assayed as described in Figure 1.*

low-temperature-adapted organisms (*55*). Further studies with synthetic peptides should provide useful information on this point.

**Thermal Stability.** The temperature optimum for arctic cod pepsin is approximately 32°C (*see* Table III) and the enzyme is unstable when incubated at temperatures above 37°C (*see* Figure 4) in contrast to porcine pepsin which has a temperature optimum of approximately 47°C and is unstable at temperatures above 50°C (*see* Figure 4). Accordingly, there is a difference in thermal stability of approximately 13°C for the two enzymes. Similar differences in temperature optima were observed when Greenland cod and American smelt were compared with PP (*see* Table III). These data are consistent with previous reports for intracellular enzymes and for crude preparations of pyloric caeca enzymes from other low-temperature-adapted organisms (*42*).

**Energy of Activation.** The $E_a$ for catalysis of hemoglobin hydrolysis by fish pepsins (FP) also differed considerably from the porcine enzyme. At pH 1.9 the Arrhenius plots were generally linear from 0°C to 30°C (*see* Figures 5, 6, 7, and 8) and $E_a$s range from 4.1 kcal/mol for arctic cod pepsin to 8.8 kcal/mol for American smelt pepsin obtained from Gambo pond (*see* Table IV). In contrast, porcine pepsin has an $E_a$ of 11.8 kcal/mol under identical assay conditions. However, at pH 3, near the optimum for the fish enzymes at higher temperatures, the Arrhenius plots are more complex and exhibit deviations from linearity over the temperature range of 0° to 30°C (*see* Figures 9, 10, 11, and 12, Table IV). Moreover, the state of purity of the Greenland cod enzyme (*see* Figures 9 and 10, Table IV) also influenced the slopes and break points as did the source of the smelt (*see* Figures 11 and 12, Table IV). Accordingly, caution is necessary when interpreting such data with respect to biological adaption. For example, at pH 3 the $E_a$s for hemoglobin hydrolysis are nearly identical for porcine, Greenland cod, and smelt (Gambo) enzymes; at pH 1.9 there appears to be adaption with respect to habitat temperatures and $E_a$s (*see* Table IV). The differing results may be accounted for partly by the influence of temperature on pH

**Table III. Temperature Optimum of Pepsins from Different Organisms**[a]

| *Pepsin Source* | *Optimum Temperature (°C)* |
|---|---|
| Porcine | 47 |
| Arctic cod | 32 |
| Greenland cod | 30 |
| American smelt | 30 |

[a] Assay procedure as described elsewhere (*53*) at pH 1.9.

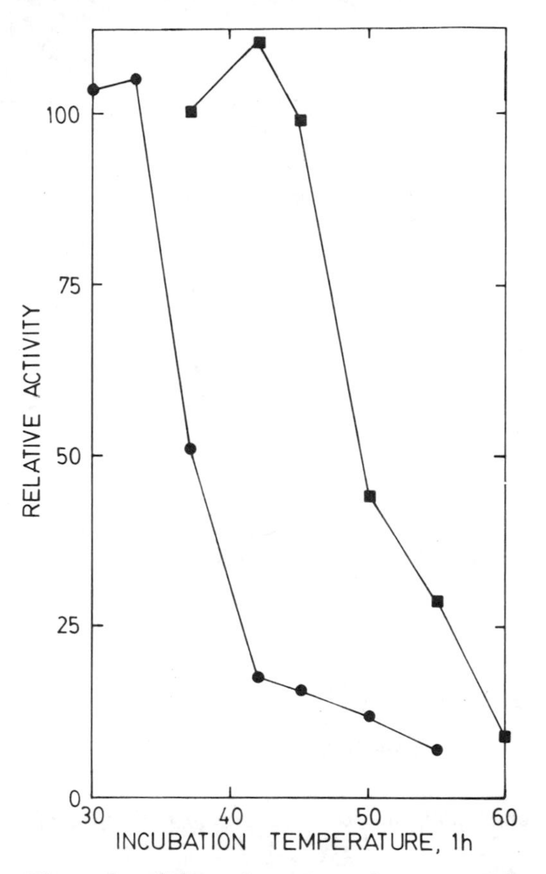


*Figure 4. Thermal stability of arctic cod pepsin and porcine pepsin. Enzymes were incubated at pH 1.9 without substrate for 1 h and subsequently assayed at 30°C: arctic cod (●); pig (■).*

optimum because the porcine enzyme is identical at 5°, 17°, 30°, and 37°C. However pepsin from smelt exhibited a modest acid shift in pH optimum at reduced assay temperatures (*see* Figure 3) and a modest change (slope, linearity) in the Arrhenius plots at reduced assay pH. Similarly Greenland cod pepsin, whose pH optimum shifts from 3.2 to 2 when the temperature is reduced from 30°C to 5°C, exhibits a most pronounced change in $E_a$ at low assay temperatures. Further study is, of course, necessary to clarify the interrelationship of pH and $E_a$ but it would appear from these data that ionizable group(s) may be involved intricately with the adaption of pepsin(s) to low temperatures. Interestingly within the same species (smelt), fish captured from lower-temperature waters (Lake Melville) had an $E_a$ which was approximately 400 cal/mol lower than smelt captured in Gambo pond (*see* Table IV).

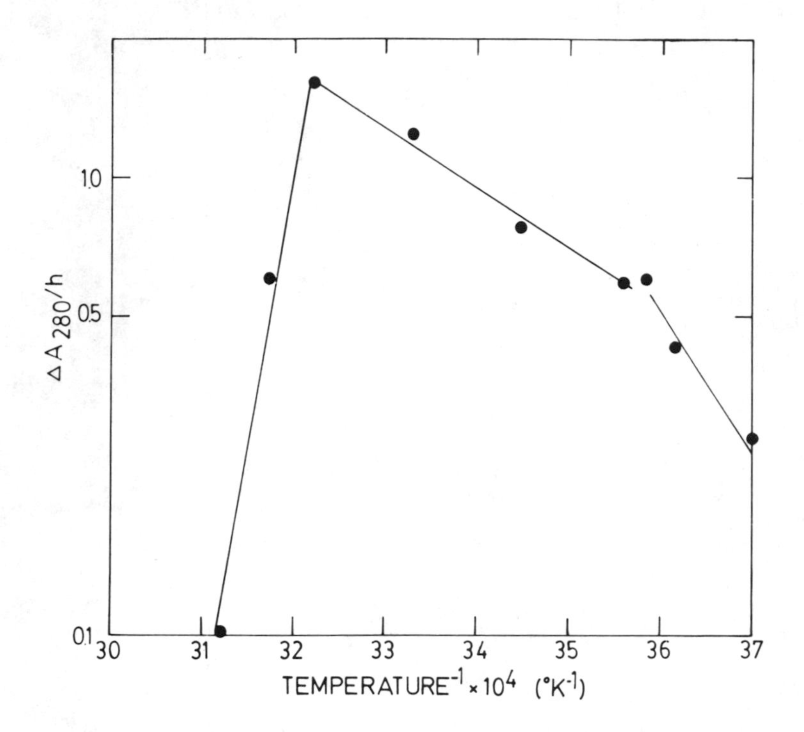


*Figure 5. Arrhenius plot for arctic cod pepsin hydrolysis of hemoglobin at pH 1.9 (53).*

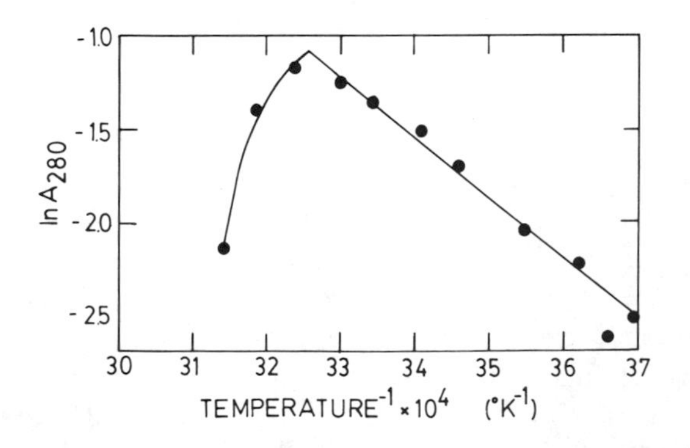


*Figure 6. Arrhenius plot for Greenland cod pepsin-catalyzed hydrolysis of hemoglobin at pH 1.9 (53).*

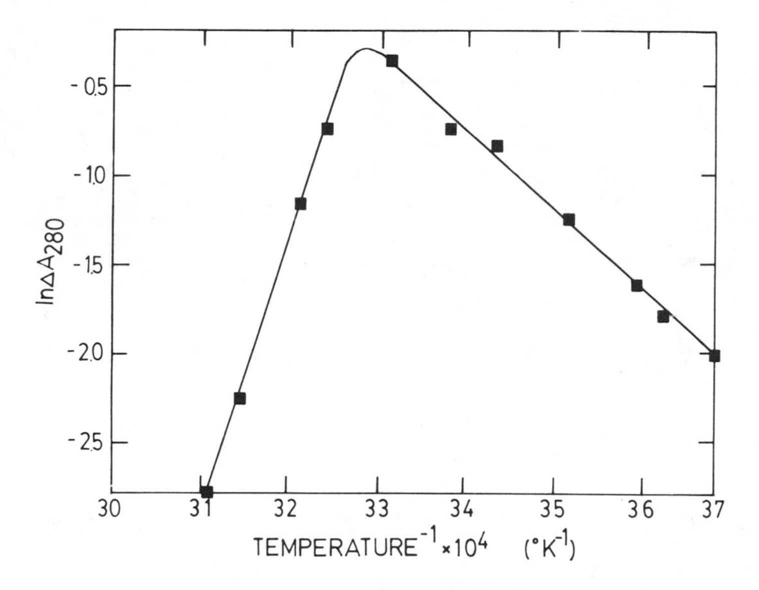


*Figure 7. Arrhenius plot for American smelt (Lake Melville) pepsin-catalyzed hydrolysis of hemoglobin at pH 1.9 (53).*

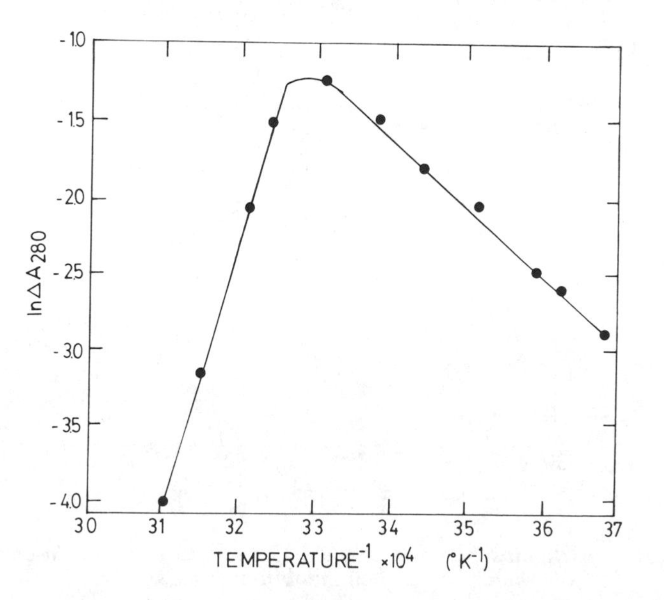


*Figure 8. Arrhenius plot for American smelt (Gambo Pond)-catalyzed hydrolysis of hemoglobin at pH 1.9 (53).*

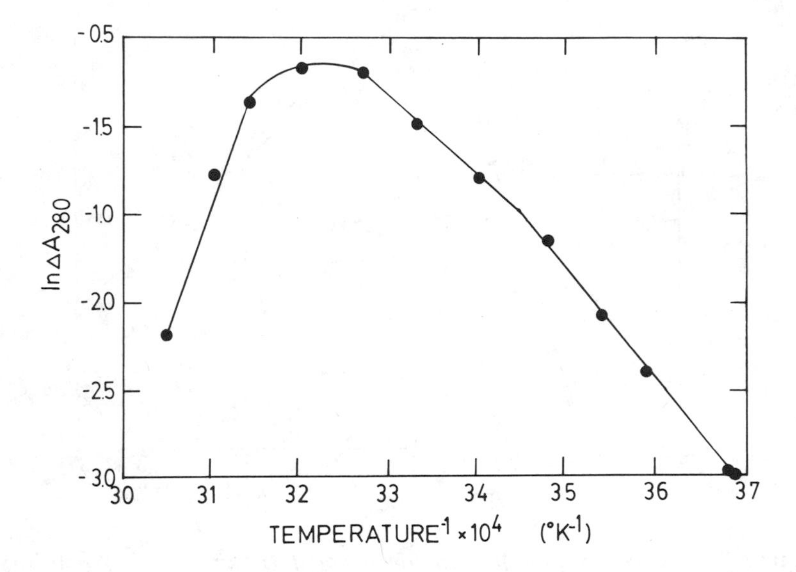


*Figure 9. Arrhenius plot for crude Greenland cod pepsin-catalyzed hydrolysis of hemoglobin at pH 3.0.*

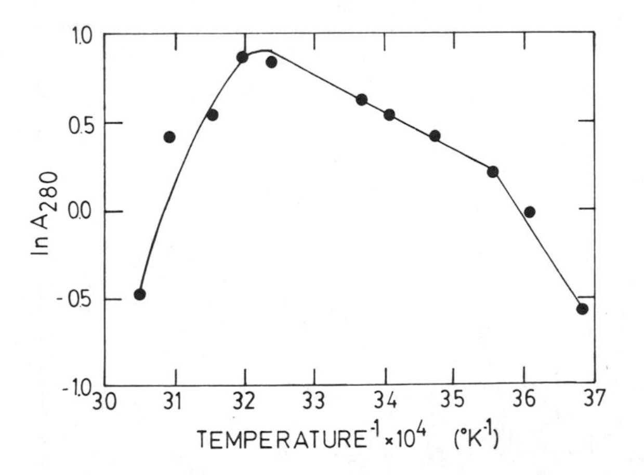


*Figure 10. Arrhenius plot for purified Greenland cod pepsin-catalyzed hydrolysis of hemoglobin at pH 3.0.*

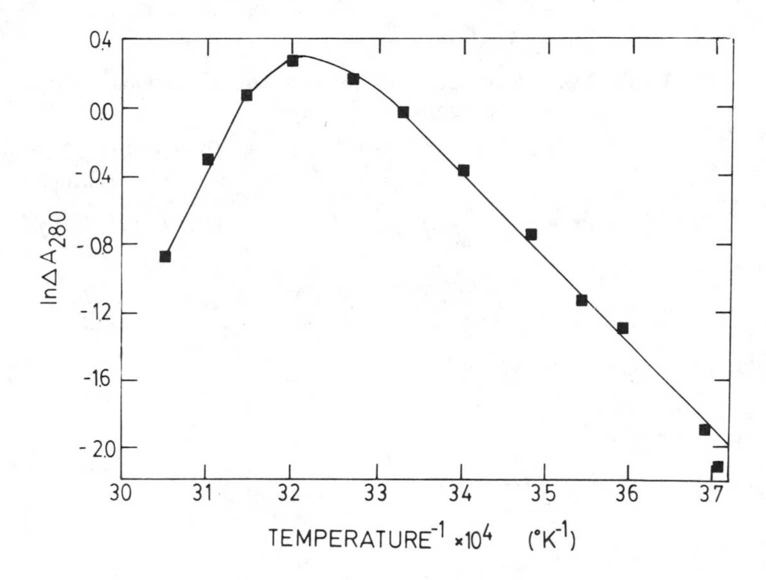


*Figure 11. Arrhenius plot for American smelt (Lake Melville) pepsin-catalyzed hydrolysis of hemoglobin at pH 3.0.*

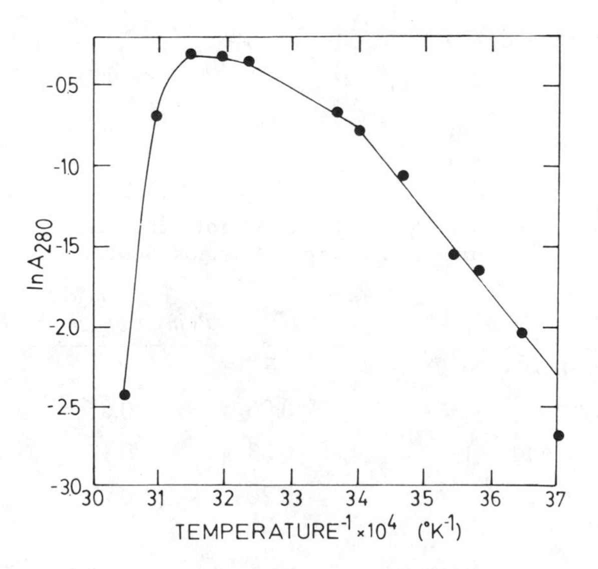


*Figure 12. Arrhenius plot for American smelt (Gambo Pond)-catalyzed hydrolysis of hemoglobin at pH 3.0.*

**Table IV. Activation Energies for Catalysis of Hemoglobin Hydrolysis**

| *Pepsin Source* | *Temperature (°C)* | *Energy of Activation (kcal/mol)* | | |
|---|---|---|---|---|
| | | *pH 1.9* | *pH 3.0* | *pH 4.0* |
| Porcine | 0–50 | 11.8 | | |
| | 0–25 | | 13.7 | |
| | 25–50 | | 4.2 | |
| Greenland cod (crude homogenate) | −3–20 | — | 9.9 | — |
| | 20–40 | — | 4.7 | — |
| Greenland cod (pure) | −1–10 | 6.6 | 12.3 | — |
| | 10–35 | 6.6 | 4.2 | — |
| | −3–25 | — | — | 12.3 |
| | 25–40 | — | — | 4.9 |
| Arctic cod | −1–30 | 4.1 | — | — |
| Smelt (Gambo) | 0–20 | — | 12.9 | — |
| | 20–30 | — | 8.5 | — |
| | −1.5–39 | 8.8 | — | — |
| Smelt (Lake Melville) | −3–30 | — | 9.9 | — |
| | −1–30 | 8.4 | — | — |

**Table V. Apparent Michaelis Constants for Hemoglobin and Pepsin Isolated from Various Sources**

| *Pepsin Source* | $K_m'$ (*m*M) *Temperature (°C)* | | |
|---|---|---|---|
| | *5* | *17* | *37* |
| Porcine | 0.05 | 0.05 | 0.05 |
| Greenland cod | 0.53 | 0.47 | 1.14 |
| Arctic cod | 0.20 | 0.27 | 0.40 |

***Substrate Affinity***

The apparent Michaelis constants for pepsin from low-temperature-adapted fish appear to be approximately one order of magnitude higher than for PP with hemoglobin as the substrate (*see* Table V). Other studies in our laboratory indicate a similar trend when bovine serum albumin (BSA) is the substrate (porcine pepsin $K_m' = 0.03\text{m}M$, Greenland cod pepsin $K_m' > 5\text{m}M$ at 5°, 17°, 37°C). The striking feature of these data is the relatively high $K_m'$ for the FP in contrast to the values reported previously for mammalian pepsins. While the reaction with PP is saturated at approximately 0.15m$M$ hemoglobin (unpublished data), the FP are not saturated at 1m$M$ concentration of substrate (*see* Figures 13, 14, and 15). The data presented in Table V, representing $K_m'$ at three temperatures with respect to dramatic adaptation of these enzymes with respect to substrate affinity and temperature. The data indicate that the $K_m'$ for the fish enzymes is temperature dependent and that $K_m'$ is lower at 5°C than at 37°C. A more detailed study of the relationship between $K_m'$ and temperature with American smelt pepsins is summarized in Figure 16. First, the influence of temperature on $K_m'$ differs for pepsins isolated from smelt recovered from Lake Melville and Gambo pond. Second, the $K_m'$ for pepsin from Lake Melville decreases as the assay temperature is reduced and is minimal at the habitat temperature of −2°C. The pattern of change in $K_m'$ for the pepsin from Gambo pond

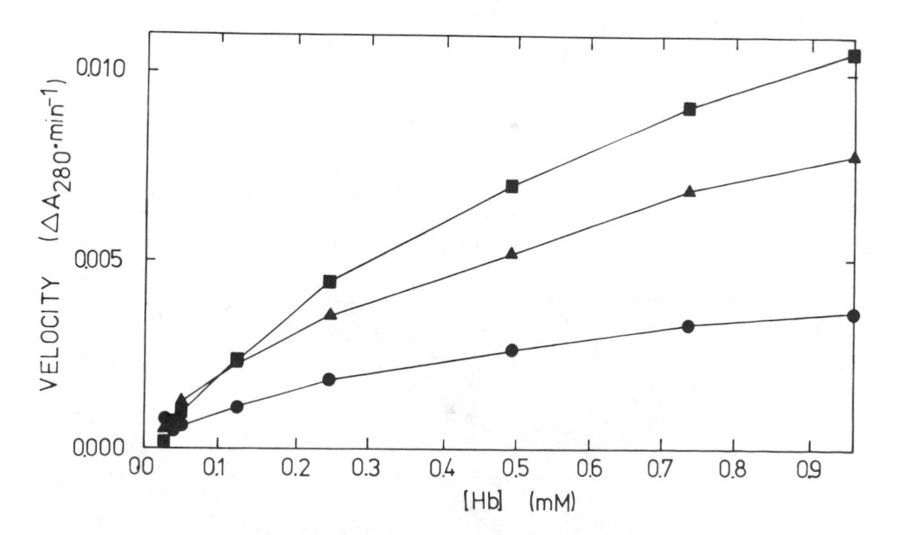


*Figure 13. Substrate velocity curve influence of hemoglobin concentration on its rate of hydrolysis catalyzed by Greenland cod pepsin at 37°C (■), 17°C (▲), and 5°C (●).*

fish is more complex and shows minimum $K'_m$ at 17°C and −2°C. We do not know whether these fish represent different subspecies of *Osmerus mordax*. These data are certainly interesting with respect to the relationship between habitat temperature and low-temperature adaption within a species. The data again support the view that substrate affinity is optimal at the habitat temperature of the organism (*37*). The relatively high $K'_m$s observed in this study may be reflective of the carnivorous nature of these species in contrast to the hog whose diet contains relatively lower levels of protein. Indeed, it is somewhat difficult to rationalize the apparent relationshop between the $K'_m$ and the temperature for pepsin, since unlike intracellular enzymes, we would not expect that a high substrate affinity for a digestive enzyme would have an adaptive function. One would not anticipate that substrate concentration in the stomach would generally be limiting.

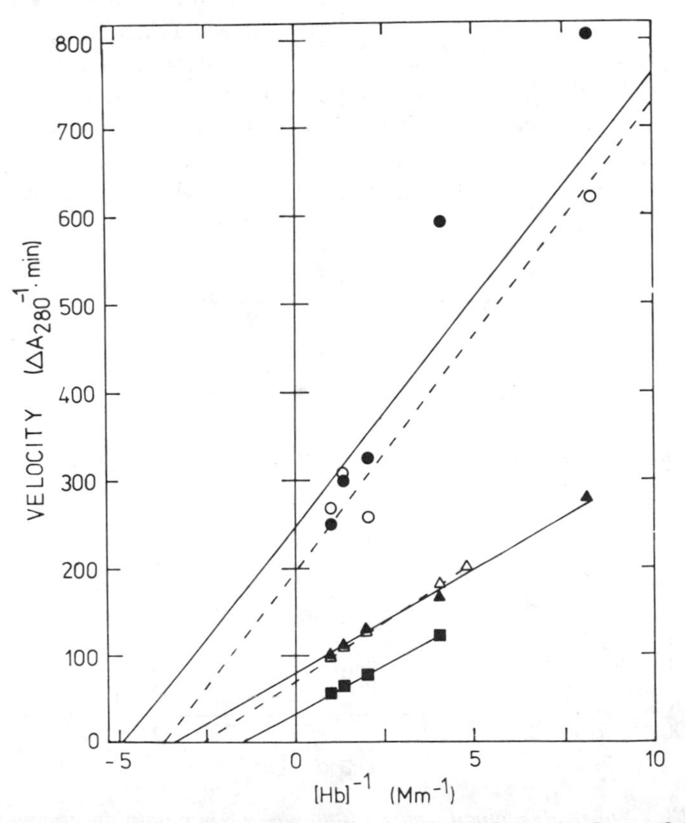


*Figure 14. Lineweaver–Burke plot for hydrolysis of hemoglobin by Greenland cod pepsin at 37°C (■), 17°C (▲), and 5°C (●).*

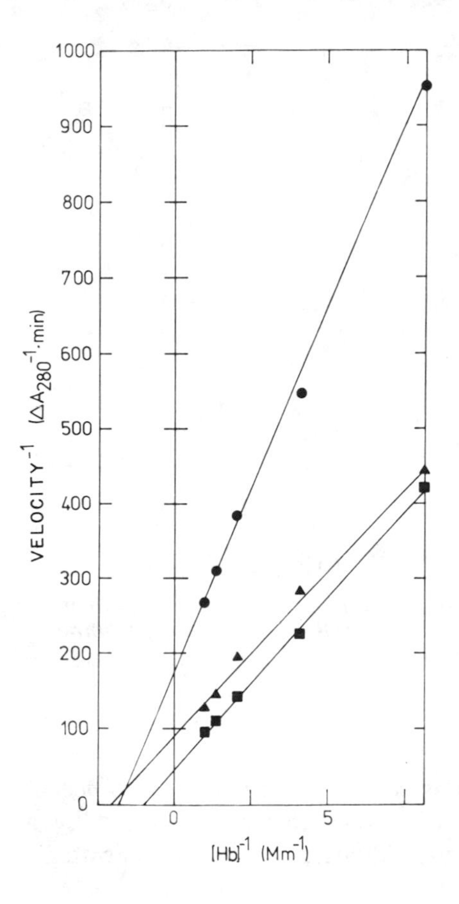


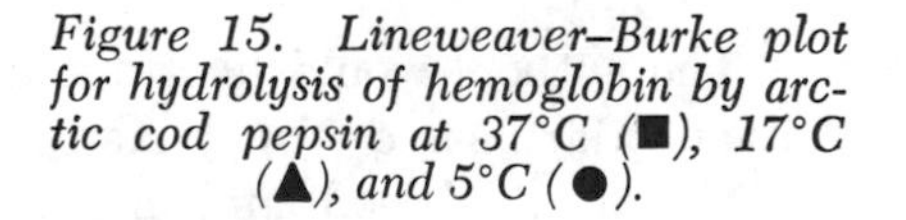
*Figure 15. Lineweaver–Burke plot for hydrolysis of hemoglobin by arctic cod pepsin at 37°C (■), 17°C (▲), and 5°C (●).*

## *Conclusions*

The present study indicates that the extracellular enzyme, pepsin, exhibits striking differences from its mammalian homologue with respect to optimum pH, $E_a$ for catalysis, thermal stability, and substrate affinity. These data are interesting from the viewpoint of biological adaption at low temperatures, but they also provide some substance to our contention that enzymes from fish plant wastes can have sufficiently unique properties to justify their use over conventional sources of enzymes used as food-processing aids. The relatively low $E_a$s for protein hydrolysis by fish pepsins indicate they may be especially useful for protein modifications at low temperatures. Alternatively, the poor thermal stability of the fish pepsins studied indicate that the enzymes can be inactivated by relatively mild blanching temperatures. The reality of this concept will have to await studies where the pepsins are used as food-processing aids. Such studies are currently underway in our laboratory.

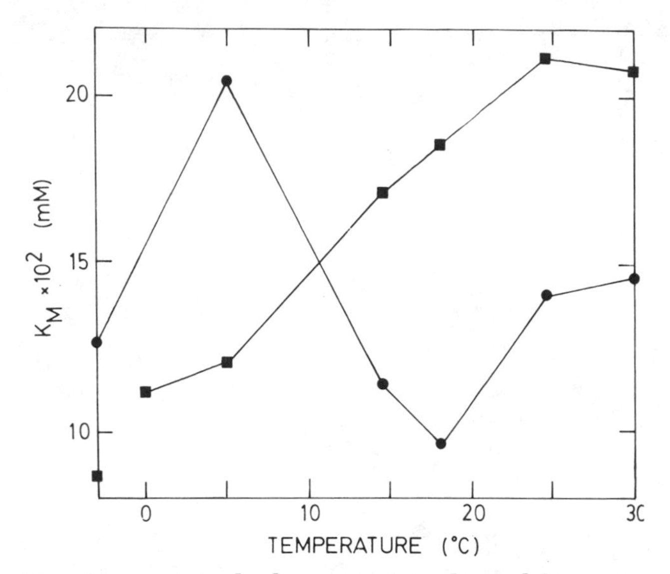


*Figure 16. Apparent Michaelis constant for hemoglobin and American smelt pepsin at different assay temperatures: Lake Melville (■); Gambo Pond (●).*

### *Acknowledgments*

The authors would like to acknowledge the technical assistance of Antonio Martin and Nancy Kariel. This research was supported by a grant from the Food and Agriculture Division of NSERC Strategic Grants, Canada.

### *Literature Cited*

1. Murakami, K.; Kimoto, K. *Int. Congr. Food Sci. Technol. Abstr.* **1978,** 234.
2. Deng, J. C. *Int. Congr. Food Sci. Technol. Abstr.* **1978,** 183.
3. Gildberg, A. *J. Food Technol.* **1978,** *13,* 409.
4. Iwata, K.; Kobashe, K.; Hase, J. *Bull. Jpn. Soc. Sci. Fish.* **1979,** *45*(2), 157.
5. Connell, J. J. "Control of Fish Quality"; Fishing News Ltd.: Surrey, England, 1975; p. 12.
6. Weis, H.; Smerat, S. *Z. Lebensm.–Unters.–Forsch.* **1976,** *162*(4), 373.
7. Beddows, C. G.; Ismail, M.; Steinkraus, K. H. *J. Food Technol.* **1976,** *11*(4), 379.
8. Murayama, S.; Calvey, D. L.; Nitayachin, P. *Bull. Tokai Reg. Fish. Res. Lab.* **1962,** *32,* 155.
9. Myazawa, K.; van Le, C.; Ito, K.; Matsumoto, F. *J. Fac. Appl. Biol. Sci., Hiroshima Univ.* **1979,** *18*(1), 55.
10. Shenderyuk, V. I.; Lisovaya, V. P. *Rybn. Khoz.* **1978,** 9, 69.
11. Henning, W. *Allg. Fischwirtschaftztg.* **1976,** *28*(17), 9.
12. Zhellosvskoya, A. M.; Kamilov, F. Izv. Vyssh. Uchebn. Zaved., Pishch. *Tekhnol.* **1977,** *2,* 74.

13. Rayumovskaya, R. G.; Chernogortstev, A. P. *Izv. Vyssh. Uchebn. Zaved., Pishch. Tekhnol.* **1976**, *5*, 92.
14. Haraldsson, R.; Lindblod, S.; Stahlberg, W. Swedish Patent 393514, 1977.
15. Hale, M. B. *Food Technol.* **1969**, *23*(10), 107.
16. Jeffreys, G. A.; Krell, A. J. U.S. Patent 3170794, 1965.
17. Ikeda, I.; Takasaki, T. U.S. Patent 4036993, 1977.
18. Prabhu, P. V.; Radhakrishan, A. G.; Arul. James, M. *India Fish. Technol.* **1975**, *12*(2), 127.
19. Mohr, V. *SIK Rapport* **1975**, *317*, K1.
20. Makie, J. M. "Fishery Products"; Kreuyer, R., Ed.; Fishing News Ltd.: Surrey, England, 1974; p. 136.
21. Flechlemmacher, W.; Wanke, W. *Inform. Fischwirtschaft* **1978**, *25*(6), 196.
22. Razumovskaya, R. G.; Chernogortsev, A. P. *Izv. Vyssh, Uchebn. Zaved., Pishch. Tekhnol.* **1976**, *4*, 76.
23. Lalasidis, G.; Bostroe, S.; Sjoeerg, L. *J. Agric. Food Chem.* **1978**, *3*, 751.
24. Kinumaki, T.; Iseki, S.; Yuguchi, H. *Bull. Tokai Reg. Fish. Res. Lab.* **1974**, *77*, 55.
25. Belhomme, P. French Patent, 2352498, 1977.
26. Patterson, I.; Windsor, M. *J. Sci. Food Agric.* **1974**, *25*, 369.
27. Hevia, P.; Whitaker, J. R.; Olcott, H. S. *J. Agric. Food Chem.* **1976**, *24*(2), 383.
28. Bhumiratama, S. *Diss. Abstr. Int.* **1976**, *36*(11), 5701; Order No. 76-2464.
29. Das, K.; Shukri, N. A.; Al-Nasira, S. K. *Indian J. Food Sci. Technol.* **1979**, *16*(2), 58.
30. Spinelli, J.; Groninger, H. S.; Koury, B. "Fishery Products"; Kreuyer, R., Ed.; Fishing News Ltd.: Surrey, England, 1974; p. 283.
31. Miller, R.; Groninger, H. S. *J. Food Sci.* **1976**, *41* (2), 268.
32. Yamamoto, A. "Enzymes in Food Processing"; Reed, G., Ed.; Academic: New York, 1975; p. 123.
33. Hultin, H. O. "Study of a Proteolytic Enzyme from Poikiolothermic Organisms Adapted to Low Temperature," Univ. Massachusetts Marine Station, Gloucester, 1978.
34. Pesiss, C.; Field, *J. Biol. Bull.* **1950**, *98*, 213.
35. Parker, G. H. *Proc. Soc. Exp. Biol. Med.* **1939**, *42*, 186.
36. Johnson, I.; Eyring, H.; Stover, B. "The Theory of Rate Processes in Biology and Medicine"; John Wiley & Sons: New York, 1974; p. 703.
37. Hazel, J. R.; Prosser, C. L. *Physiol. Rev.* **1974**, *54*, 620.
38. Somero, G. *J. Exp. Zool.* **1975**, *194*, 175.
39. Low, P.; Somero, G. *Comp. Biochem. Physiol.* **1974**, *49B*, 307.
40. Haard, N. F.; Martins, I.; Newburg, R.; Botta, R. *J. Can. Inst. Food Sci. Technol.* **1979**, *12*, 84.
41. Thompson, T. L.; Militzer, W.; Georgi, C. *J. Bacteriol.* **1958**, *76*, 337.
42. Rossebo, L.; Underdal, B. *Arch. Fischereiwiss.* **1972**, *23*, 128.
43. Wiggs, A. *Can. J. Zool.* **1974**, *52*, 1071.
44. Hofer, R.; Ladurner, H.; Gattrigir, A.; Wieser, W. *J. Comp. Physiol.* **1975**, *99*, 345.
45. Hochachka, P.; Storey, K.; Baldwin, J. *Comp. Biochem. Physiol.* **1975**, *52B*, 13.
46. Hochachka, P.; Norberg, C.; Baldwin, J.; Fields, J. *Nature* **1976**, *260*, 648.
47. Low, P.; Soamero, G. *J. Exp. Zool.* **1976**, *198*, 1.
48. Chesley, L. *Biol. Bull.* **1934**, *66*, 133.
49. Pfleiderer, G.; Zwilling, R. *Naturwissenschaften* **1972**, *59*, 3969.
50. Pfleiderer, G.; Zwilling, R. *Biochem. Z.* **1966**, *344*, 127.
51. Jany, K., Ph.D. Dissertation, Univ. of Heidelberg, 1972.
52. Haard, N. F.; Feltham, L.; Helbig, N.; Squires, J., unpublished data.

53. Anson, M. L. *J. Gen. Physiol.* **1938**, *22*, 79.
54. Ihove, K.; Voynick, I. M.; Delprerre, G. R.; Fruton, J. S. *Biochemistry* **1966**, *5*, 2473.
55. Fruton, J. S. "The Enzymes"; Boyer, P., Ed.; Academic: New York, 1971; Vol. 3, p. 154.

RECEIVED November 6, 1980.

# 9

# The Role of Lime in the Alkaline Treatment of Corn for Tortilla Preparation

AUGUSTO TREJO-GONZALEZ, ALEJANDRO FERIA-MORALES, and CARLOS WILD-ALTAMIRANO

Department of Food Research, Center of Research and Teaching for Agricultural and Food Engineering, University of Guanajuato, Irapuato, Guanajuato, Mexico

*Several variables were studied during the lime treatment of corn (nixtamalization) for preparing tortillas: time, temperature, and lime concentration; incorporation of water and calcium by the corn grain; chemical analyses of the resultant flours; and SEM observations of corn grain sections at various stages of alkaline treatment. Lime treatment improved the nutritive value of corn by increasing the availability of lysine in the glutelin fraction of the protein. The gelatinization of starch was the most important physical–chemical event occurring during the period of lime treatment. The calcium taken up by the corn grain during treatment appears to be bound to the starch grains. Starch isolated from lime-treated corn took up approximately three times more calcium than starch isolated from untreated corn. Calcium can be replaced by the divalent cation magnesium. Substitution of corn by other starch-containing foods (wheat, broadbean, garbanzo, potato, and rice) was studied also. The combination of these foods with corn resulted in tortillas with better nutritional and organoleptic qualities, reaching protein and lysine contents of 14.4% and 4.04 g/100 g protein, respectively.*

From Pre-Colombian times to the present, the development of Mexico has been linked intimately with the cultivation and consumption of corn. The domestication of corn terminated nomadism in Mesoamerica and resulted in the establishment of the first human settlements, which

0065-2393/82/0198-0245$05.00/0

evolved eventually to achieve a level of cultural and scientific development whose excellence amazed its own and foreign peoples. Corn was so important to the Mesoamerican societies that it was deified, Cinteotl, being the Aztec goddess of maize and Yum Kaax, the Mayan god of maize and vegetation.

A wild progenitor of corn, known to have existed 7000–8000 years ago, was cultivated over 5000 years ago (*1*). The intensification of corn-based agriculture was linked closely to the development of the following innovation in corn-processing technology: the alkaline treatment, specifically lime treatment [called nixtamalización in Spanish, derived from the Nahuatl word for lime-treated corn prepared as dough, nixtamal (*2*)]. The lime treatment of corn was so important that it is considered to have differentiated the indigenous societies and have influenced their respective development (*3*). Figure 1 demonstrates that the centers of civilization that reached a high level of organization and achievement coincided with those in which lime treatment of corn was practiced.

Corn is recognized generally as deficient in the essential amino acids lysine and tryptophan, and in niacin, the pellagra-preventing B vitamin, synthesized from tryptophan. Lime treatment results in denaturation of the corn protein, particularly the glutelins, making them more readily digestible (*4*). Since the major proportion of the lysine and tryptophan residues are found in the glutelin fraction, lime treatment enhances the availability of these two essential amino acids. Therefore, the alkaline treatment with lime results in a product of greater biological value than corn prepared by other cooking methods. For that reason, lime treatment of corn is considered to have influenced the development of the Mesoamerican societies because it permitted dietary dependence on a single cultivated crop (*3*). Other corn-consuming societies not practicing lime treatment were required to greatly supplement the corn diet to prevent malnutrition.

Today corn remains the most important source of calories and protein for many people of Mexico and Central America: in the rural areas it supplies approximately 70% of the caloric intake. In terms of overall yearly consumption, about 186 kg of corn per capita are consumed in Mexico (*5*). Stated in another way, of the 16.7 million hectares of cultivated arable land in Mexico, over 7.2 million hectares are dedicated to cultivating corn (*6*).

Much work has been carried out on the nutritional benefits derived from the lime treatment of corn (*7,8*). Although this aspect is very important, an understanding of the overall physical–chemical changes occurring in the corn, which determine the mechanical and sensory properties of the resultant food products, is also essential. Such information can lead to modification of the traditional lime treatment process to

*Figure 1. Localization of pre-Colombian corn-consuming societies (3). Corn treatment: no alkali treatment (○); alkali treatment with lye or wood ashes (□); alkali treatment with lime (●).*

improve the quality and nutritional value of the derived foods. The work presented here strives to further our understanding of lime treatment as a traditional food-processing technique.

### *The Lime Treatment Process: Nixtamalization*

**Lime Treatment of Corn for Tortilla Preparation.** The lime treatment of corn for tortilla preparation consists of the following (*see* Figure 2): a 3:1 mixture of lime water (1.3%, w/v) and dry corn grains is

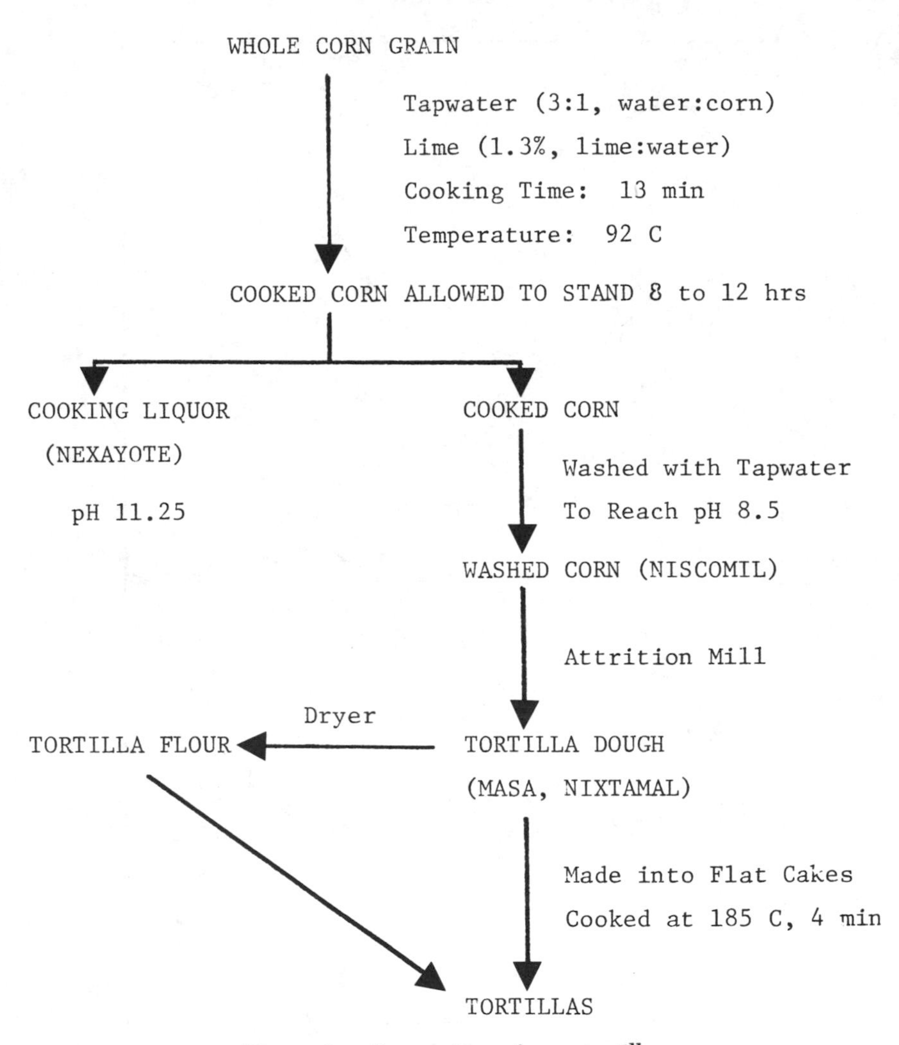


*Figure 2. Preparation of corn tortillas.*

heated to boiling until the pericarp is loosened. The cooked corn (Nixtamal) then is allowed to stand at room temperature for 8 to 12 h, after which the cooking liquor (Nexayote) is removed and the corn is washed with water until the pH of the rinse water decreases to 8.5. The washed corn then is milled forming the dough (Masa). The basic tortilla is made from a quantity of dough sufficient to form a disc about 10 cm in diameter and 0.3 cm in thickness. This is cooked on a hot plate at 185°C for 15 s on one side, 30 s on the other, and then returned to the original side to repeat the cooking until the steam produced within the tortilla is sufficient to inflate it. In commercial tortilla preparation the dough may be dried and ground to produce a flour that subsequently is rehydrated for tortilla elaboration.

**Parameters of the Lime Treatment Process.** Parameters of lime treatment that can be varied and greatly affect subsequent tortilla quality are: (1) the corn grain, in particular the endosperm type and pericarp thickness and composition; (2) temperature and time of cooking; and (3) the concentration of the lime water. Corns of the same variety with different endosperm types required varying periods of time to achieve the same degree of cooking (loosening of the pericarp) as shown in Table I. The soft endosperm type, Opaque-2, requires much less cooking at the same temperature than the hard endosperm type. The hardness of the endosperm, determined by the ratio of amylopectin to amylose in the corn starch (9), greatly influences the plasticity of the doughs prepared from the lime-treated corn (*see* Table I). The time required for cooking the corn with lime water, however, apparently depends on characteristics other than endosperm type—most probably composition and thickness of the pericarp: distinct lime treatment conditions are required for different corn varieties, as illustrated in Tables IIa and b. In addition, for a given corn variety (normal or dent corn), it is possible to carry out the lime treatment process under widely differing conditions of time and temperature: from 70° to 92°C, with cooking periods from 13 to 70 min, and with or without a steeping period. The criteria applied to determine the adequacy of the lime treatment conditions were the mechanical and sensory properties of the resultant tortillas, including extensibility, plasticity, flexibility, softness, inflatability, smoothness, flavor, and odor.

Different alkali processes have been and are applied to corn, including the use of lye in hominy preparation and the use of wood ashes for preparing arepas. However, lime, as a source of alkali, is the most important technique currently used for the processing of corn for human

**Table I. Characteristics of the Lime Treatment of Corns with Different Endosperm Types[a]**

| *Corn Type*[b] | *Treatment Time (min)* | *Temperature (°C)* | *Residual Water (mL)* | *Dough Moisture (%)* | *Dough Plasticity* |
|---|---|---|---|---|---|
| Opaque-2 | 30 | 80 | 530 | 56.6 | Poor |
| Modified | 45 | 80 | 670 | 46.0 | Intermediate |
| Normal | 55 | 80 | 620 | 42.0 | Good |

[a] Preparation for the lime treatment: 500 g corn + 1000 mL water + 21 g lime.
[b] Normal, modified, and opaque corns of the variety Veracruz 181 X Old Group 2.

**Table IIa. Characteristics of Doughs Prepared from Lime-Treated Corns**[a]

| *Corn Variety* | *Endosperm Type* | *Treatment Time (min)* | *Dough Yield (g)* | *Dough pH* | *Rinse Water pH* |
|---|---|---|---|---|---|
| Cacahuacintle | soft | 11 | 280 | 6.4 | 8.5 |
| Chalco | normal | 24 | 250 | 5.1 | 8.7 |
| Palomero Reventador Toluqueño | hard | 13 | 261 | 6.4 | 9.0 |

[a] Conditions for the lime treatment: 200 g corn + 400 mL water + 6 g lime, heated to 92°C. Initial pH of limewater: 12.4.

**Table IIb. Characteristics of the Cooking Liquor from Lime-Treated Corns**[a]

| *Corn Variety* | *Volume Cooking Liquor (mL)* | *pH* | *Soluble Solids (%)* | *Calcium (mg)* | *Hemicelluloses (%)* |
|---|---|---|---|---|---|
| Cacahuacintle | 250 | 10.5 | 5.7 | 1075 | 6.4 |
| Chalco | 180 | 11.1 | 8.0 | 774 | 7.2 |
| Palomero Reventador Toluqueño | 260 | 11.2 | 5.9 | 1120 | 7.5 |

[a] Conditions for the lime treatment: 200 g corn + 400 mL water + 6 g lime, heated to 92°C. Initial pH of limewater: 12.4.

**Table III. Characteristics of the Cooking Liquor (Nexayote)**[a]

| *Characteristic* | *Amount or Value* |
|---|---|
| pH | 11.0 |
| Total solids, ppm | 28,000 |
| Soluble solids, ppm | 17,000 |
| Ash solids, ppm | 8,200 |
| Soluble organic nitrogen, ppm | 75 |
| Calcium, ppm | 9,600 |
| BOD, 5-day, ppm | 6,415 |
| Total sugars, ppm | 14,500 |
| Total hemicelluloses, ppm | 8,460 |
| Turbidity, Jacks units | 201 |
| Color, units | 2,450 |

[a] From the lime treatment of Palomero Reventador Toluqueño.

consumption. The chemical characteristics of lime water are such that the concentration of the lime always exceeds the saturation point of calcium hydroxide, which is 0.15% under ambient conditions. The lime concentrations used for tortilla preparation may range from 0.8% (7) to 2.1%, as reported here, with an initial pH value of 12.4. After cooking, the pH of the residual lime water is approximately 11.0. To remove excess lime after cooking, the corn must be rinsed adequately with water until the pH of the rinse water approaches 8.5; otherwise the resulting dough and tortillas would be alkaline in taste. Typical characteristics of the residual lime water, or cooking liquor, are shown in Table III. This cooking liquor is a rich organic medium and a very contaminating effluent from the tortilla factories.

### *Physical and Chemical Changes Occurring in the Corn Grain as a Result of Lime Treatment*

**General Changes in Corn Grain Composition.** Samples of 18 varieties of corn, including hybrids and native varieties from distinct geographical areas of Mexico, were studied (*see* Figure 3 and Table IV). From Tables V and VI the most notable differences between the untreated and lime-treated corns are demonstrated: an increase in mineral content and decreases in total protein, crude fiber, and ether extract.

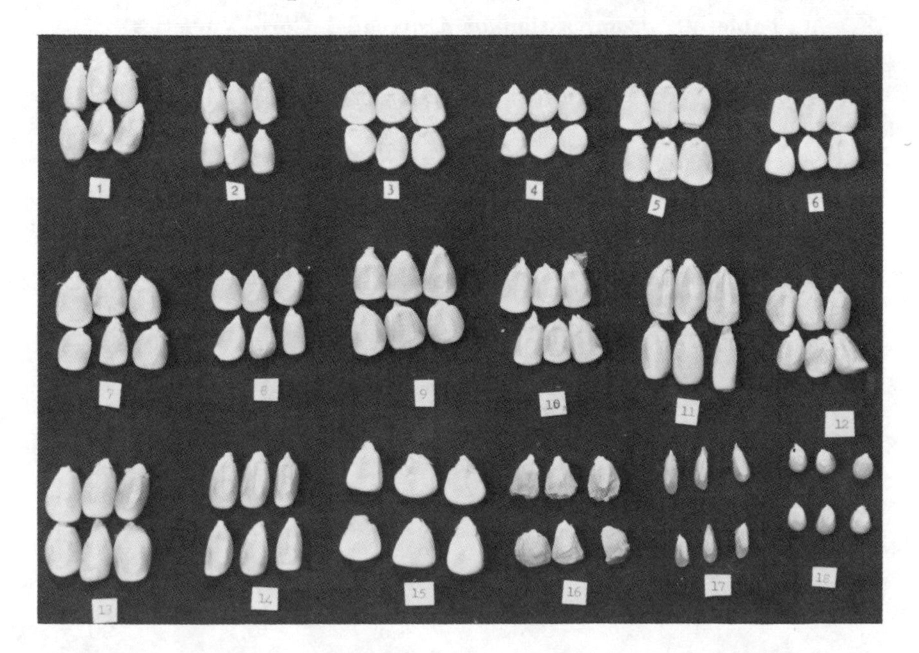


*Figure 3. The 18 corn varieties used to study the nixtamalization process. Variety names correspond to those indicated in Table IV.*

**Table IV. Moisture Content and Endosperm Type of Corn Varieties**

| *Variety Number* | *Variety Name* | *Zone of Cultivation* | *Endo-sperm Type* | *% Mois-ture* |
|---|---|---|---|---|
| 1 | V-15 (Tol. 68) | humid temperate | hard | 6.7 |
| 2 | VS11.F2. (68) | humid temperate | hard | 6.2 |
| 3 | Zapalote Chico (69APL) | humid temperate | hard | 7.1 |
| 4 | Zapalote Grande (69APL) | humid temperate | hard | 7.5 |
| 5 | Native of Copital | dry tropics | hard | 7.3 |
| 6 | H-507.F2. | dry tropics | hard | 7.1 |
| 7 | C. de Villagran (65) | Bajio, rainy season | hard | 6.8 |
| 8 | H-220.F2. (65) | Bajio, rainy season | hard | 7.1 |
| 9 | Celaya 11.F2 (66R) | Bajio, rainy season | hard | 5.7 |
| 10 | H-366.F2 (66R) | Bajio, rainy season | hard | 6.8 |
| 11 | Chalco, small (58–68) | central highlands | hard | 7.7 |
| 12 | H-28 (67) | central highlands | hard | 7.6 |
| 13 | Chalco (68) | central highlands | hard | 7.5 |
| 14 | H-129 (68) | central highlands | hard | 7.8 |
| 15 | Cacahuacintle | central highlands | soft | 6.8 |
| 16 | Sweet corn | central highlands | — | 7.5 |
| 17 | Palomero Reventador Toluqueño | humid temperate | hard | 6.6 |
| 18 | Commercial popcorn | humid temperate | hard | 7.5 |

**Table V. Composition of Untreated Corn Varieties**

| *Variety Number* | *Total Protein (%)* | *Crude Fiber (%)* | *Ether Extract (%)* | *Ash (%)* | *Calcium (mg/kg)* |
|---|---|---|---|---|---|
| 1 | 10.2 | 2.2 | 5.0 | 1.7 | 80 |
| 2 | 10.1 | 2.6 | 5.7 | 1.7 | 80 |
| 3 | 11.5 | 1.7 | 5.9 | 1.8 | 70 |
| 4 | 11.8 | 1.8 | 6.6 | 1.9 | 70 |
| 5 | 10.2 | 1.9 | 3.9 | 1.7 | 70 |
| 6 | 8.7 | 2.5 | 2.5 | 2.0 | 70 |
| 7 | 11.5 | 2.2 | 5.4 | 1.7 | 80 |
| 8 | 10.0 | 2.4 | 5.8 | 1.7 | 80 |
| 9 | 11.2 | 2.2 | 5.0 | 1.5 | 70 |
| 10 | 11.0 | 2.4 | 5.5 | 1.5 | 70 |
| 11 | 11.5 | 2.2 | 5.5 | 1.5 | 80 |
| 12 | 11.4 | 2.2 | 5.7 | 1.6 | 80 |
| 13 | 10.9 | 2.3 | 5.4 | 1.5 | 80 |
| 14 | 9.4 | 2.5 | 5.8 | 1.6 | 80 |
| 15 | 9.1 | 1.8 | 2.2 | 2.9 | 80 |
| 16 | 12.9 | 2.9 | 3.9 | 1.5 | 70 |
| 17 | 14.0 | 2.2 | 6.8 | 1.7 | 60 |
| 18 | 13.3 | 2.8 | 4.6 | 1.6 | 90 |
| Average | 11.0 | 2.3 | 5.1 | 1.7 | 76 |

**Table VI. Composition of Lime-Treated Corn Varieties**

| *Variety Number* | *Total Protein (%)* | *Crude Fiber (%)* | *Ether Extract (%)* | *Ash (%)* | *Calcium (mg/kg)* |
|---|---|---|---|---|---|
| 1 | 9.9 | 1.9 | 4.8 | 2.1 | 900 |
| 2 | 9.8 | 2.1 | 4.9 | 2.1 | 900 |
| 3 | 11.3 | 1.5 | 5.0 | 2.3 | 1140 |
| 4 | 10.5 | 1.6 | 5.5 | 2.3 | 1090 |
| 5 | 9.2 | 1.8 | 4.5 | 2.3 | 1130 |
| 6 | 8.5 | 1.9 | 4.9 | 2.3 | 1280 |
| 7 | 11.4 | 1.9 | 5.0 | 2.1 | 980 |
| 8 | 9.8 | 2.1 | 5.2 | 2.1 | 1240 |
| 9 | 10.8 | 1.9 | 4.5 | 2.5 | 1690 |
| 10 | 10.4 | 2.2 | 4.6 | 2.3 | 1630 |
| 11 | 10.2 | 1.9 | 4.4 | 2.3 | 1030 |
| 12 | 11.3 | 2.0 | 4.8 | 2.3 | 1030 |
| 13 | 10.5 | 2.2 | 5.3 | 2.2 | 1260 |
| 14 | 9.3 | 2.2 | 5.2 | 2.3 | 1160 |
| 15 | 9.0 | 1.5 | 0.9 | 2.5 | 1430 |
| 16 | 12.7 | 2.6 | 3.0 | 2.9 | 1600 |
| 17 | 13.4 | 2.0 | 5.0 | 2.2 | 1400 |
| 18 | 12.8 | 2.1 | 3.7 | 2.3 | 1280 |
| Average | 10.6 | 2.0 | 4.5 | 2.3 | 1230 |

**Absorption of Water and Calcium by the Corn Grain.** The absorption of water and calcium by corn of the variety Palomero Reventador Toluqueño with a heating period of 13 min at 92°C was studied. The absorption of water as a function of time indicated a very rapid initial uptake (during the first 30 min) (*see* Figure 4). The corn continued to absorb water until by 135 min after heating, the absorbed water was 105% of the initial weight of the corn. The absorption of calcium by the corn grain follows a similar pattern, although it occurred over a much longer period of time (*see* Figure 5). Calcium was measured by atomic absorption spectrometry in corn grains after 12 h of steeping. The calcium content of the lime-treated corn grain was about 4.5 times that of the untreated grains (640 and 140 mg/kg, respectively).

**Scanning Electron Microscope Observations of Corn Grain Sections.** Scanning electron micrographs (SEMs) of corn grain sections fractured perpendicularly to the long axis were prepared. These illustrate changes in the structure of the endosperm during the nixtamalization process. Figure 6 shows a close-up view of the untreated corn grain, indicating the arrangement of the starch grains enveloped by glutelins. Observations of the same area in corn grains with 4 and 8 h of steeping at room temperature are shown in the micrographs (*see* Figure 6). It is evident from these micrographs that the main structural changes are occurring in

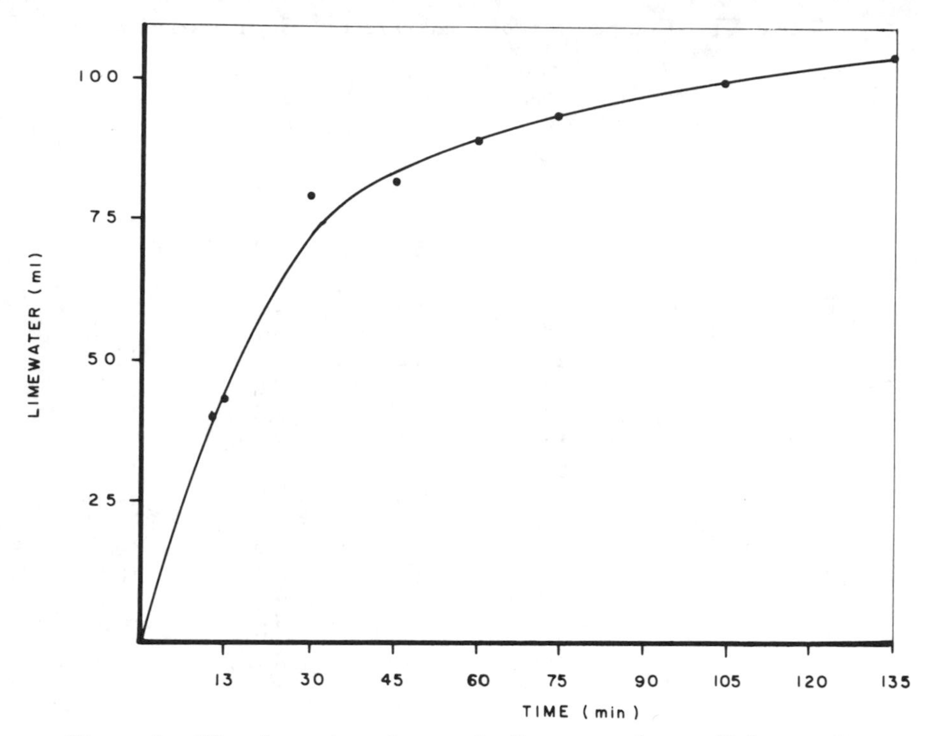


*Figure 4. The absorption of water by lime-treated corn (Palomero Reventador Toluqueño).*

the starch granules as a result of gelatinization. In the lime treatment process starch gelatinization occurs in the presence of calcium ions in alkaline medium at room temperature.

**Calcium Content of Starch Isolated from Lime-Treated Corn.** Starch is the major component of the corn grain and derived products (*see* Table VII). To explore in more detail the distribution of calcium after the nixtamalization process, calcium was measured in starch isolated from untreated and lime-treated corn. These analyses (*see* Table VIII) indicated that calcium was fixed or bound in some way to the starch. There was approximately 2.9 times more calcium in starches isolated from lime-treated corns than in starches isolated from untreated grains. Figure 7 shows micrographs of isolated starches.

**The Requirement of a Divalent Alkaline Earth Metal.** Lime-treated corn produces a dough whose functional properties depend on the pasting characteristics of starch. The hydroxides of different alkaline earth metals were tested to determine whether they could replace calcium in the nixtamalization process. It was not possible to prepare tortillas with adequate sensory or mechanical properties from corn treated with

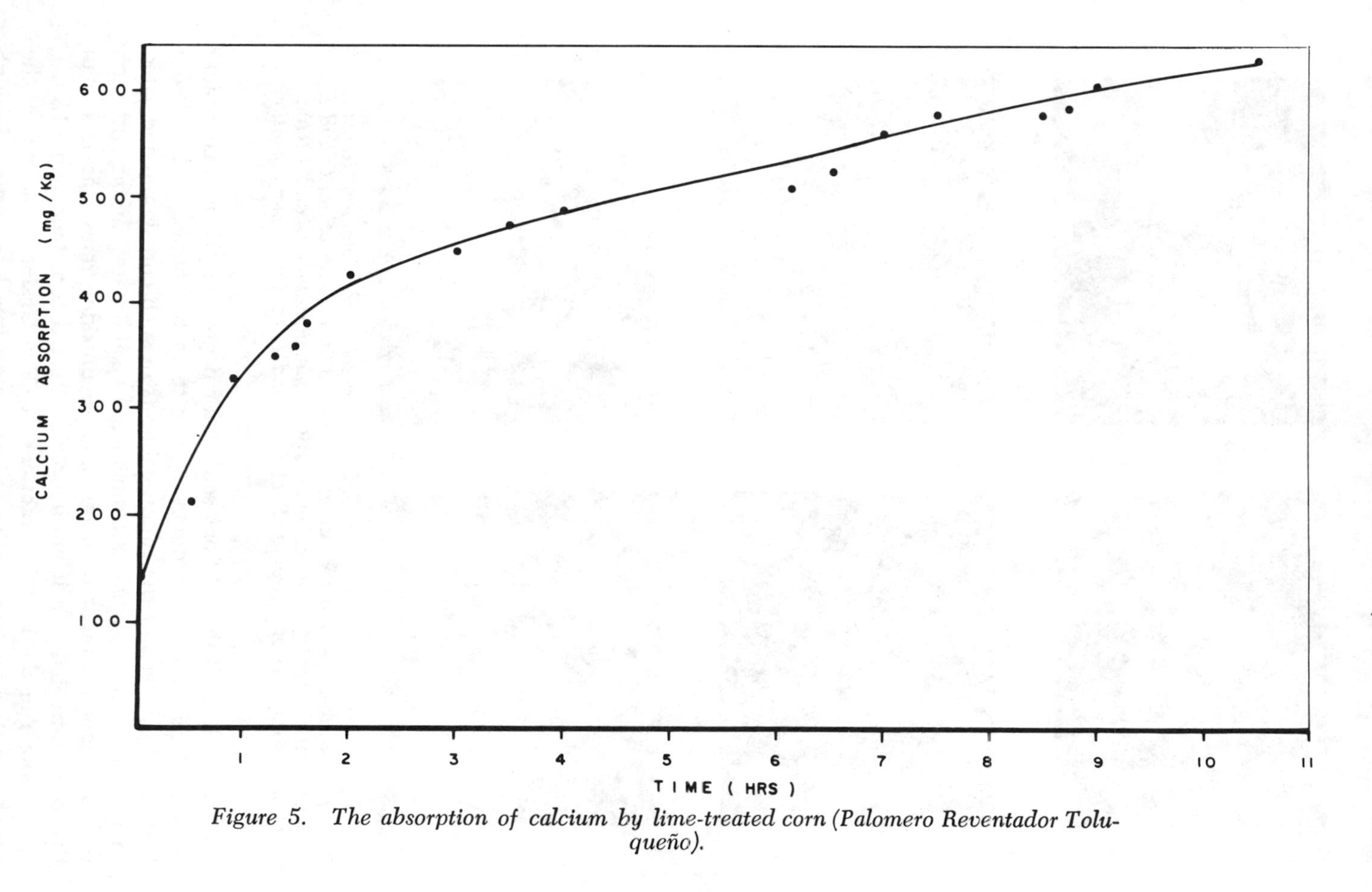


*Figure 5. The absorption of calcium by lime-treated corn (Palomero Reventador Toluqueño).*

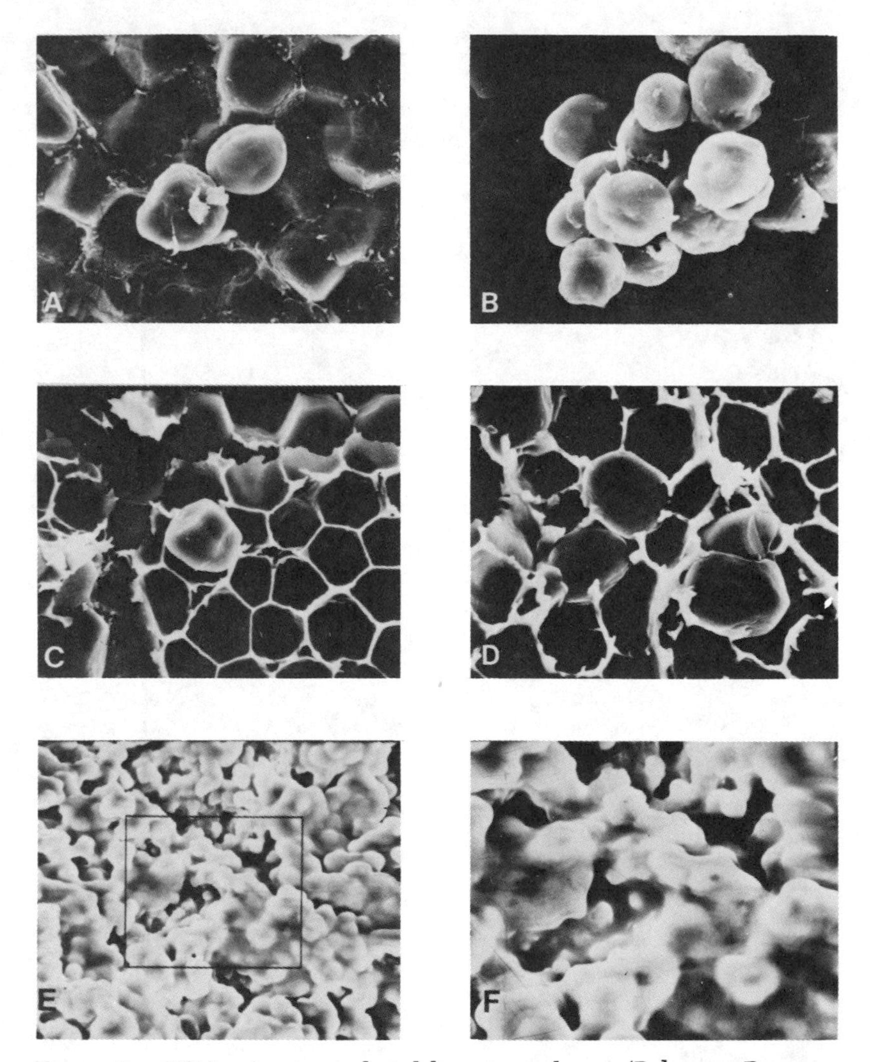


*Figure 6. SEMs of untreated and lime-treated corn (Palomero Reventador Toluqueño): A and B, untreated corn, 130× magnification; C and D, lime-treated corn, 4 h of steeping, 130× magnification E and F, lime-treated corn, 8 h of steeping, E at 60× and F at 130× magnification.*

the monovalent alkaline cations sodium or potassium. Using magnesium hydroxide resulted in a product similar to that obtained with calcium hydroxide. Analysis of the cation contents of such corn grains indicated that magnesium displaced calcium in the alkali-treated grains, although the total cation content of magnesium hydroxide grains was less than that of lime-treated or calcium hydroxide-treated corn (*see* Table IX).

**Changes in the Protein Fractions of Corn Grains.** Changes in the starch component determine the principal mechanical properties of dough

**Table VII. Average Composition of Dough Prepared from Lime-Treated Palomero Reventador Toluqueño Corn[a]**

| *Component* | *% of Dough* |
|---|---|
| Moisture | 9.0 |
| Ether extract | 5.0 |
| Crude fiber | 2.0 |
| Protein (N $\times$ 6.25) | 10.2 |
| Ash | 2.2 |
| Nitrogen free extract | 70.6 |
| Calcium (milligrams/killigram corn) | 640 |
| Available lysine (grams/100 g protein) | 2.9 |
| Total free sugars (as glucose) | 2.6 |

[a] Dry weight basis.

**Table VIII. Calcium Content of Starch Isolated from Different Corns[a]**

| *Corn Type* | *Untreated Corn* | *Lime-Treated Corn* |
|---|---|---|
| | *(mg Ca/kg dry weight starch)* | |
| Palomero Reventador Toluqueño (hard endosperm) | 50 | 150 |
| Opaque-2 (soft endosperm) | 56 | 164 |
| Cacahuacintle (soft endosperm) | 66 | 182 |

[a] Prepared with method from Ref. *10*.

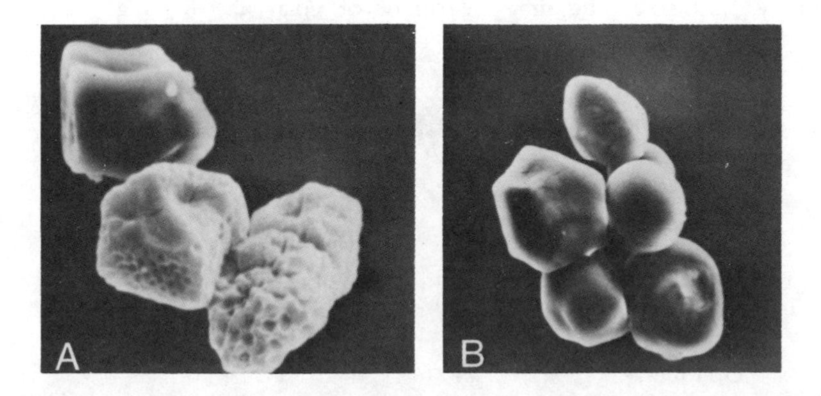


*Figure 7. SEMs of starch isolated from untreated and lime-treated corn (Palomero Reventador Toluqueño): A, starch from untreated corn, 110× magnification; B, starch from lime-treated corn, 75× magnification.*

**Table IX. Calcium and Magnesium Content of Corn[a]**

| *Corn Treatment* | *Calcium* | *Magnesium* |
|---|---|---|
| | *(mg/kg)* | |
| Untreated | 225 | 935 |
| Lime treated[b] | 1000 | 1000 |
| $Mg(OH)_2$ treated[b] | 200 | 1510 |
| $Ca(OH)_2$ treated[b] | 1065 | 800 |

[a] Chalco corn, normal endosperm.
[b] 1.3%, w/v.

from lime-treated corn. Changes in the protein component of lime-treated corn determine the principal nutritional benefits derived from the process. Protein fractions from untreated and lime-treated corns were determined by modifying the classic grain protein extraction procedure (*see* Table X). It is apparent that the nixtamalization process results in a decrease of protein that is extractable as either albumins, globulins, zein, or glutelins. The decreases in the globulin and glutelin fractions were most notable. Correspondingly, protein that wasn't extractable in one of these fractions —residue—increased notably in the lime-treated corns. These results corroborate data indicating a decrease in total grain protein, but an increased availability of lysine as a consequence of the lime treatment process (*see* Table XI) (lysine values for the opaque corns are high and may indicate the presence of an interfering compound in the opaque corns (*12*).

There has been some concern about the effect of alkaline conditions on the formation of toxic amino acids in corn, particularly lysinoalanine (*13*). However, since lysinoalanine was found in very small amounts in foods prepared under more rigorous conditions than those applied traditionally, and since the lime treatment of corn has been a processing

**Table X. The Distribution of Nitrogen in**

| *Corn Type* | *Protein Fraction (mg N/g)* | | |
|---|---|---|---|
| | *Albumins* | *Globulins* | *Zein-I* |
| *Untreated* | | | |
| Opaque-2 | 3.10 | 2.53 | 1.47 |
| Modified | 1.55 | 2.91 | 1.18 |
| Normal | 1.90 | 1.55 | 2.51 |
| *Lime Treated* | | | |
| Opaque-2 | 2.72 | 0.26 | 1.25 |
| Modified | 2.46 | 0.59 | 0.30 |
| Normal | 2.05 | 0.40 | 0.58 |

[a] Method from Ref. *11*.

**Table XI. Whole Endosperm Protein and Available Lysine in Untreated and Lime-Treated Corns[a]**

| *Corn Type* | *Whole Endosperm (mg N/g)* | *Available Lysine (g/100 g protein)* |
|---|---|---|
| Untreated | | |
| Opaque-2 | 14.24 | 6.50 |
| Modified | 16.48 | 2.60 |
| Normal | 17.60 | 1.40 |
| *Lime Treated* | | |
| Opaque-2 | 13.60 | 9.01 |
| Modified | 15.52 | 3.25 |
| Normal | 15.68 | 4.50 |

[a] Method for available lysine from Ref. *12*.

technique used for hundreds of years in Mesoamerica, and since lysinoalanine is specifically toxic to rats but not to other mammals, it is likely that the small amounts of lysinoalanine and lanthionine resulting from the traditional lime treatment process of corn have no important physiological consequences in humans (*14, 15*).

**Lime Treatment of Other Grains, Legumes, and Rootcrops.** Because changes in the starch component determine the principal mechanical properties of dough prepared from lime-treated corn, it seemed reasonable that knowledge derived from the study of lime-treated corn could be extended to other starch-containing foods, such as cereals, legumes, roots, and tubers. The potato can be taken as an example, and divalent cation incorporation (*see* Table XII) and SEMs (*see* Figure 8) indicate that similar changes occur in this starchy food as occur in corn as a result of the lime treatment process.

**Endosperm Protein Fractions of Corn Flour[a]**

| Protein Fraction (mg N/g) | | | | |
|---|---|---|---|---|
| *Zein-II* | *Glutelins* | *Total* | *Residue* | *Whole Endosperm* |
| 0.14 | 5.26 | 12.51 | 1.73 | 14.24 |
| 0.64 | 3.39 | 9.68 | 6.80 | 16.48 |
| 1.76 | 4.29 | 12.02 | 5.58 | 17.60 |
| 0.00 | 4.82 | 9.06 | 4.54 | 13.60 |
| 0.32 | 1.13 | 4.91 | 10.61 | 15.52 |
| 1.60 | 1.12 | 5.86 | 9.82 | 15.68 |

**Table XII. Calcium and Magnesium Content of Potato Tubers**

| *Potato Treatment* | *Calcium* (mg/kg) | *Magnesium* (mg/kg) |
|---|---|---|
| Untreated | 250 | 500 |
| Lime treated[a] | 1630 | 490 |
| $Mg(OH)_2$ treated[a] | 350 | 1560 |
| $Ca(OH)_2$ treated[a] | 2550 | 540 |

[a] 1.3%, w/v.

Tortillas prepared from mixtures of corn with wheat, broadbean, rice, chickpea, and/or potato were studied. These mixes resulted in tortillas of very similar mechanical and sensory properties as those prepared from corn alone. However, the contents of protein and available lysine were in many cases higher (*see* Table XIII). Such mixtures provide an efficient way to improve the quantity and quality of protein in tortillas.

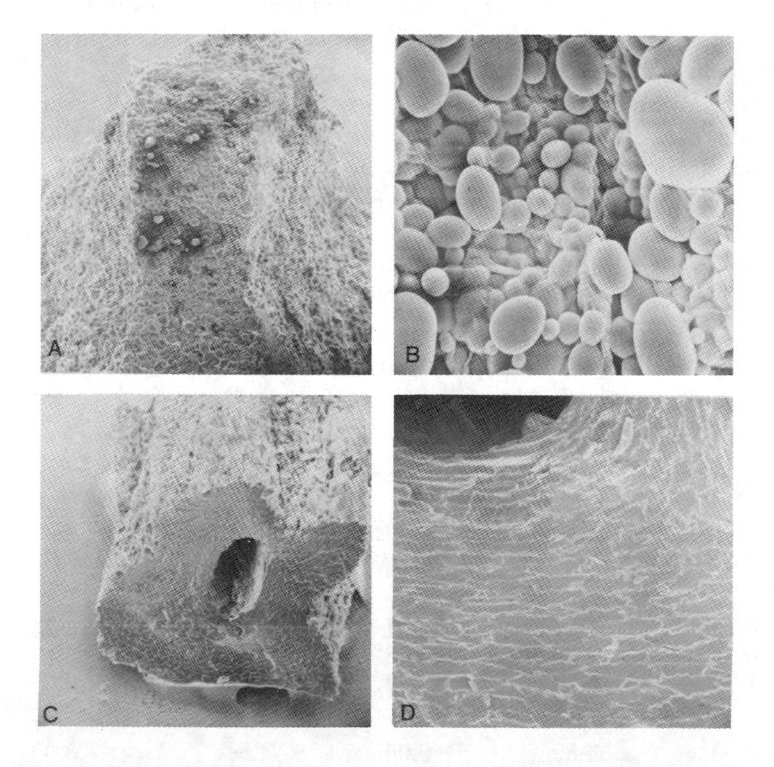


*Figure 8. SEMs of untreated and lime-treated potato tuber: A and B, untreated potato tuber, A at 30× and B at 300× magnification; C and D, lime-treated potato tuber, C at 12× and D at 55× magnification.*

**Table XIII. Protein and Available Lysine in Tortillas Prepared from Different Mixtures of Corn, Other Grains, and Potatoes[a]**

| | *Mixture* | *Protein (%)* | *Available Lysine (%)* |
|---|---|---|---|
| 1. | Corn (Palomero Reventador Toluqueño) | 10.2 | 2.87 |
| 2. | Corn–wheat | 11.1 | 2.82 |
| 3. | Corn–broadbean | 11.2 | 3.41 |
| 4. | Corn–chickpea | 11.6 | 3.91 |
| 5. | Corn–rice | 8.3 | 3.00 |
| 6. | Corn–potato | 9.0 | 2.89 |
| 7. | Corn–wheat–broadbean | 13.6 | 3.92 |
| 8. | Corn–wheat–chickpea | 12.3 | 3.48 |
| 9. | Corn–wheat–broadbean–potato | 12.9 | 3.51 |
| 10. | Corn–wheat–chickpea–potato | 14.4 | 3.85 |
| 11. | Corn–wheat–broadbean–rice–potato | 13.5 | 4.04 |

[a] Lime treatment conditions and relative proportions of starchy foods not given, patent pending.

The protein level achieved with these mixtures of starchy foods is comparable with that reported for corn and soybean mixtures (*see* Table XIV). An additional advantage of the lime treatment of mixtures of starch-containing foods is the possibility of substituting more abundant for less abundant foods when necessary.

### *Distinctive Features of the Nixtamalization Process: The Role of the Alkaline Environment*

The concentration of lime used in the nixtamalization process is ten times in excess of the saturation point. After cooking and steeping, about 40% of the original calcium ion remains in the lime water suspension. Of the 60% of calcium removed, about one-third is associated with the

**Table XIV. Protein Content of Tortillas Prepared from Corn and Mixtures of Other Grains and Potato[a]**

| | *Mixture* | *Protein (%)* |
|---|---|---|
| 1. | Corn (Palomero Reventador Toluqueño) | 9.5 |
| 2. | Corn–soybean (10%) | 12.9[b] |
| 3. | Corn–soybean (15%) | 14.4[c] |
| 4. | Corn–wheat–broadbean | 13.6 |
| 5. | Corn–wheat–broadbean–rice–potato | 13.5 |

[a] Lime treatment conditions and relative proportions of starchy foods not given, patent pending.
[b] Ref. *16*.
[c] Ref. *17*.

discarded pericarp, and about two-thirds is bound to the washed corn grains. The lime, as the alkaline component in the process, serves two distinct purposes: (1) it maintains the pH of the grain–lime water mixture at 12.4, which is the level of alkalinity necessary to hydrolyze the hemicelluloses of the pericarp (*18*); and (2) it provides a sufficiently alkaline medium to favor the changes occurring in the grain endosperm. One change occurring as a result of lime treatment is a change in protein conformation, giving rise to an increased availability of lysine in the glutelin fraction (*3, 4*). This change bears important nutritional consequences for those peoples dependent on corn as a staple food.

Another change occurring in the corn grain as a result of nixtamalization is that undergone by the starch component, the chemical gelatinization of starch. Some salts lower the temperature at which gelatinization is initiated and the rate at which it proceeds (*19*). In particular, calcium salts increase the number of starch granules gelatinized as a function of time (*20*). The same observations have been made by other authors (*21, 22*), and although the mechanism of this chemical gelatinization is not known, some speculations have been made. At pH values of 11.0 and higher, the amylose molecule carries negative charges, indicating that the hydroxyl groups dissociate in that pH range permitting interaction with calcium ions (*23*). Also the formation of an adduct between glucose, maltose, and amylose with calcium hydroxide and magnesium hydroxide molecules by electrostatic interaction has been suggested (*24*).

We have observed that the same mechanical properties of doughs prepared from lime or calcium hydroxide-treated corn were achieved when corn was treated with magnesium hydroxide. It seems reasonable to consider that the elaboration of products from nixtamalized corn is based mostly on the physical–chemical changes occurring in the starch component of the corn grain. This is substantiated further by the fact that other starchy foods can be lime treated to yield products similar to those obtained from corn.

A further beneficial feature provided by the alkaline medium during the nixtamalization process is the destruction of aflatoxins in corn contaminated by *Aspergillus flavus* (*25*).

### *Acknowledgment*

We gratefully acknowledge the collaboration of Dr. Richard Falk, from the Botany Department at the University of California, Davis, in preparing the SEMs.

### *Literature Cited*

1. MacNeish, R. In "The Prehistory of the Tehuacan Valley"; Byers, D., Ed.; Univ. of Texas Press: Austin, Texas, 1967; Vol. 1, pp. 3, 114, 290.

2. Leander, B. "Herencia Cultural del Mundo Nahuatl"; Secretaría de Educación Pública, México, D.F., 1972.
3. Katz, S. H.; Hediger, M. L.; Valleroy, L. A. *Science* **1974,** *184,* 765.
4. Paulis, J. W.; James, C.; Wall, J. S. *J. Agric. Food Chem.* **1969,** *17,* 1301.
5. Anon. "Mexico 1976: Facts, Figures, Trends;" Banco Nacional de Comerico Exterior, México, D.F., 1976.
6. Wellhausen, E. J. *Sci. Am.* **1976,** *235,* 129.
7. Cravioto, R. O.; Anderson, R. K.; Lockhard, E. E.; Miranda, F. de P.; Harris, R. S. *Science* **1945,** *102,* 91.
8. Bressani, R.; Scrimshaw, N. S. *J. Agric. Food Chem.* **1958,** *6,* 774.
9. Cluskey, J. E.; Knutson, C. .; Inglett, G. E. *Stärke* **1980,** *32,* 105.
10. Anon. "Standard Analytical Methods of the Member Companies of the Corn Industries Research Foundation;" Ames, Iowa, 1972, B-64-1.
11. Landry, J.; Moureaux, T.; Bacle, M. *Bull. Soc. Chim. Biol.* **1971,** *52,* 1021.
12. Tsai, C.-Y.; Hansel, L. W.; Nelson, O. E. *Cereal Chem.* **1972,** *49,* 572.
13. Sternberg, M.; Kim, C. Y.; Schwende, F. J. *Science* **1975,** *190,* 992.
14. De Groot, A. P.; Slump, P.; Feron, V. J.; Van Beek, L. *J. Nutr.* **1976,** *106,* 1527.
15. Sanderson, J.; Wall, J. S.; Donaldson, G. L.; Cavins, J. F. *Cereal Chem.* **1978,** *55,* 204.
16. Del Valle, F.; Montemayor, E.; Bourges, H. *J. Food Sci.* **1976,** *41,* 349.
17. Bressani, R.; Murllo, B.; Elias, L. G. *J. Food Sci.* **1974,** *39,* 577.
18. Wolf, M. J.; MacMasters, M. M.; Cannon, J. A.; Rosewall, E. C.; Rist, C.E. *Cereal Chem.* **1953,** *30,* 451.
19. Katz, J. R.; Muschler, F. J. F. *Biochem. Z.* **1933,** *257,* 385.
20. Sandstedt, R. M.; Kempf, W.; Abbott, R. C. *Stärke* **1960,** *12,* 333.
21. Mangels, C. E.; Bailey, C. H. *J. Am. Chem. Soc.* **1933,** *55,* 1981.
22. Guy, E. J.; Vettel, H. E.; Pallansch, M. J. *Cereal Sci. Today* **1967,** *12,* 200.
23. Doppert, H. L.; Staverman, A. J. *J. Polym. Sci.* **1966,** *4,* 2367.
24. Moulik, S. P.; Khan, D. P. *Carbohydr. Res.* **1975,** *41,* 93.
25. Ulloa-Sosa, M.; Schroeder, H. N. *Cereal Chem.* **1969,** *46,* 397.

RECEIVED November 14, 1980.

# PHARMACOLOGICAL ASPECTS

# 10

# Design of Site-Specific Pharmacologic Reagents

## Illustration of Some Alternative Approaches by Reagents Directed Towards Steroid–Hormone-Specific Targets

WILLIAM F. BENISEK, JOHN R. OGEZ, and STEPHEN B. SMITH[1]

Department of Biological Chemistry, School of Medicine,
University of California, Davis, CA 95616

*Current approaches to the site-specific affinity labeling of steroid-binding sites on enzymes and nonenzymatic proteins from bacterial and mammalian species are reviewed. Equations describing the kinetics of affinity labeling derived from equilibrium and material balance relationships and the application and limitation of these equations to the elucidation of labeling mechanism are discussed. Also, the importance of the kinetic parameter,* $k_2/K_R$, *the second-order rate constant for the reaction between reagent and target protein, and the reagent concentration in maximizing the specificity of the labeling reaction is emphasized. Examples of classical affinity reagents, photoaffinity reagents, and* $k_{cat}$ *or mechanism-based affinity reagents based on the steroid structure and targeted towards various enzymes and binding proteins are presented. Emphasis is given to studies reported during the past 5 years.*

This chapter reviews some of the progress which has been made during the preceding 10 years in the development and application of reagents that irreversibly react in a site-selective manner with proteins capable of binding steroids wih moderate to high affinity. Important reviews of this field have been presented by Warren et al. (*1*) and Katzenellenbogen (*2*).

[1] Current address: Department of Pharmacology, SJ-30, Laboratory of Molecular Pharmacology, Howard Hughes Medical Institute, University of Washington School of Medicine, Seattle, WA 98195.

0065-2393/82/0198-0267$11.75/0

Unlike earlier coverage of this subject, this review is organized according to reagent type rather than to the macromolecular species selected as the reagent target. We hoped that this organizational format will facilitate the reader's efforts to design new steroid-binding-site-specific reagents best suited to his particular system under investigation.

As a preamble to our admittedly selective review of steroidal affinity reagents, their intended target molecules, and their mutual chemical interaction, we will present a summary of the essential concepts related to affinity labeling in general. It is against this general background that the specific examples from the steroid hormone field will be displayed. The examples chosen to illustrate the various reagent design approaches are not intended to provide a fully comprehensive catalog of steroidal affinity reagents. Rather, our aim is to provide a few examples of each reagent type, illustrating the strengths and weaknesses of the various approaches. The reader may note that many other equally suitable examples could have been selected instead of those which follow. We can only reply that it was necessary to maintain a certain degree of arbitrariness in selecting examples so that we could establish reasonable limits to the length of this chapter and to the workload its preparation would entail. We apologize to those investigators whose excellent work was omitted in our efforts to achieve brevity without too much sacrifice of breadth. Coverage of the literature extends through August, 1980.

One of the goals of many workers in this field is the development of drugs that specifically can interrupt steps in steroid hormone synthesis or action. A few of the affinity reagents to be discussed have been tested for their physiological effects on hormone-dependent processes in mammalian systems. The results of these in vivo applications will be indicated for those systems where they are available.

### *Basic Principles of Affinity Labeling*

**Qualitative Aspects.** Affinity labeling is a way of selectively chemically modifying functional groups that are part of or proximal to a ligand-binding site on a protein. The conditions of the modifying reaction are optimized for maximal reaction of the reagent with binding-site residues with minimal reaction at residues remote from the binding site. This specificity is achieved by designing the reagent's molecular structure in such a way that reaction with residues in or near the ligand-binding site occurs much more rapidly than do reactions at points well removed from the target site. Thus the desired selectivity is relative and not absolute. Factors controlling the selectivity of the modification reaction are discussed in the following section.

In all methods of affinity labeling the necessary rate enhancement needed for site-specific reaction is a direct consequence of the formation of a noncovalent complex between the protein and the reagent (or its inactive proreagent form) (as shown later) with the reagent binding in the ligand-binding site to be modified. In such a complex, protein functional groups in the site "see" a reagent concentration that is often many orders of magnitude higher than do any other functional groups external to the site. Then, if the detailed stereochemistry of the complex permits, a much more rapid reaction will occur with residues lying near the reactive part of the reagent than with similar residues lying at sites remote from the reagent. In order to achieve complex formation between the reagent and the target protein, the structure of the reagent must be such that it results in the formation of noncovalent bonds between it and one or more protein functional groups. The natural, in vivo ligands with which the protein interacts do just that via the same sorts of noncovalent bonds. Consequently, researchers have closely patterned the structures of candidate affinity reagents after the structures of the natural ligands. In the case of steroid-binding proteins and enzymes it is clear that the prime candidates for site-selective reagents will be steroid derivatives bearing some sort of group that is chemically reactive (or can be made chemically reactive) towards the usual protein functional groups. Figure 1 illustrates this concept.

Three distinct classes of affinity reagents have been used in affinity-labeling experiments. These are the classical affinity reagents (*3, 4*), the photoaffinity reagents (*5, 6*), and the $k_{cat}$ or mechanism-based affinity reagents (*7*). The essential difference between the latter two reagent types and classical reagents is that classical reagents possess a reactive group that is preformed during chemical synthesis whereas the other types contain a masked reactive group that undergoes activation via light absorption (photoaffinity reagents) or enzymic catalysis ($k_{cat}$ reagents). Thus, photoaffinity reagents and $k_{cat}$ reagents might be termed proreagents. The special characteristics of proreagents will be discussed later.

Members of all three reagent types have been developed for studies of steroid-binding proteins. The relationship between affinity-labeling reagents and proreagents in their interactions with target proteins is shown by the reactions in Figure 2. Both reagents and proreagents form reversible complexes with the target protein (*see* Reaction 1). At this point the reaction pathways diverge. Classical reagents may proceed to execute Reaction 2, an irreversible covalent reaction with the protein in which all or part of the reagent molecule covalently bonds to the protein. In the case of photoaffinity labeling the proreagent is designed to be unreactive towards the protein. However, once the protein–proreagent complex is formed, the experimenter may choose to initiate the covalent reaction part of the whole process by irradiating the system with light of a wavelength known to activate the proreagent (*see* Reaction 3)

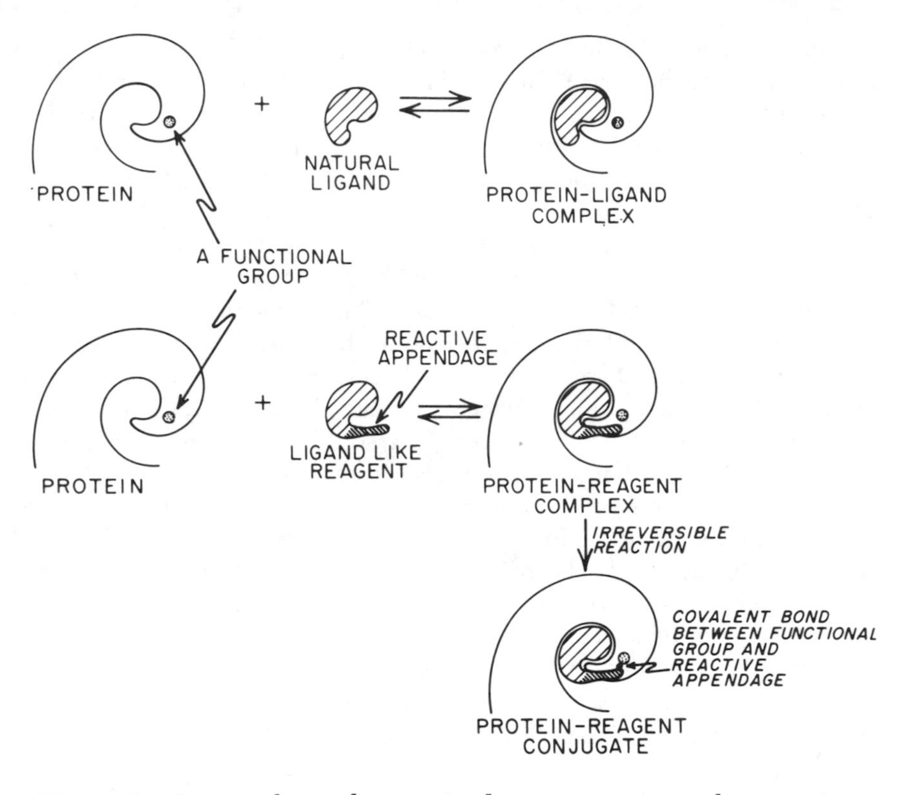


*Figure 1. Structural complementarity between reagent and target site usually is required for site selection reaction.*

yielding the reagent itself now in the binding site. The reagent now may proceed to react with the target protein as in Reaction 4. In the systems described to date, photoreagents may be short-lived electronically excited states or highly reactive species produced by chemical decomposition of excited states. Carbenes and nitrenes, by far the most widely utilized photoreagents in other systems, are good examples of reactive species produced by photodecomposition of proreagents such as diazo and azo derivatives, respectively. If the target protein is an enzyme, the possibility of labeling via a $k_{cat}$ reagent exists. In this approach the proreagent is designed to be a substrate of the target enzyme (*see* Reaction 5). The enzyme-generated product, unlike the substrate, is a reactive reagent which, hopefully, reacts with the enzyme (*see* Reaction 6) before dissociation of the E·R complex occurs (*see* Reaction 7).

Photoaffinity-labeling reagents and $k_{cat}$ reagents possess special features which may recommend them as reagents of choice in certain applications. The built-in requirement for light absorption by photoactivated proreagents in order to initiate the labeling process temporally decouples

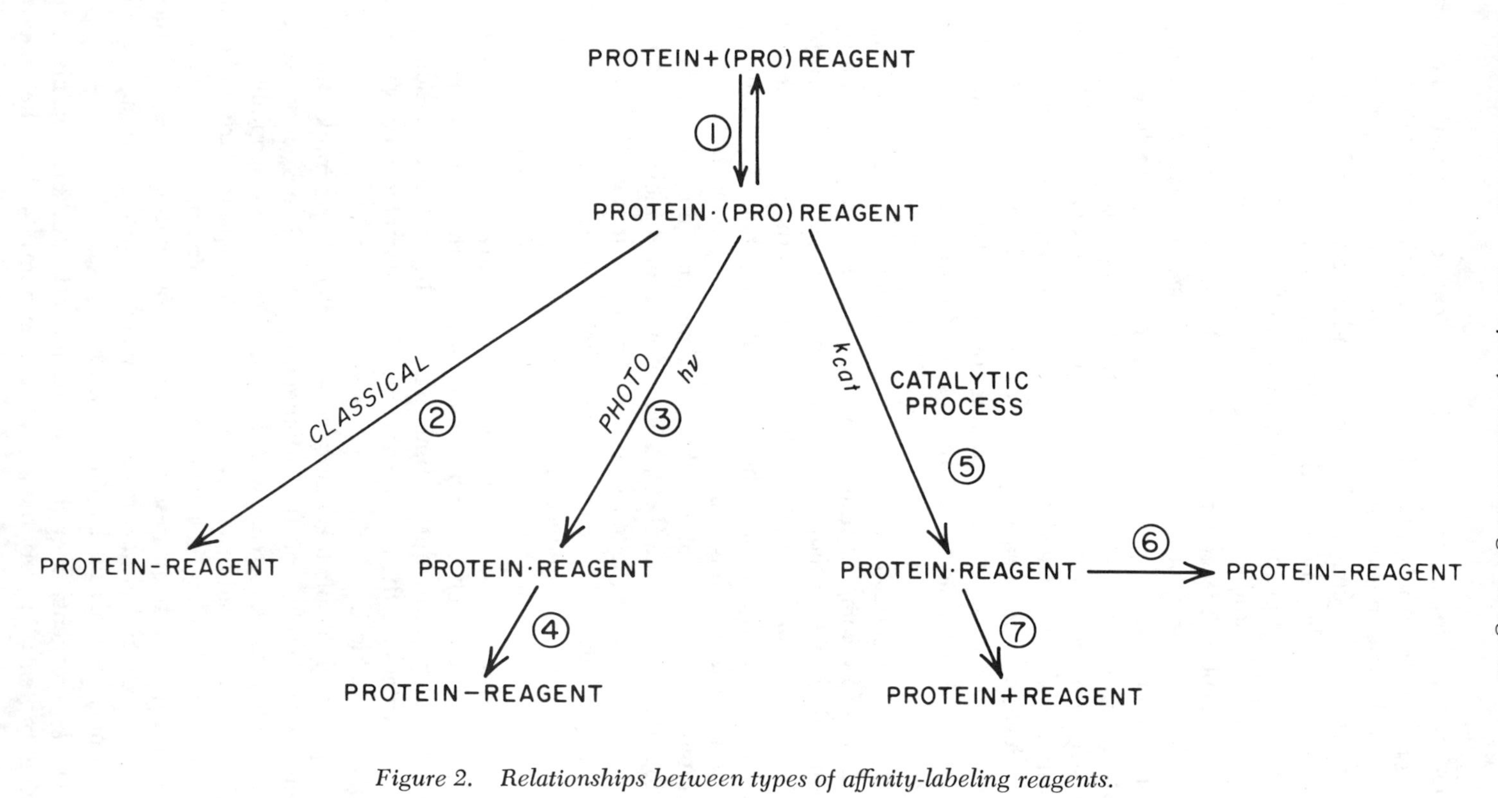


*Figure 2. Relationships between types of affinity-labeling reagents.*

the noncovalent binding step from the covalent reaction step. This feature can be used to advantage in systems in which the target protein undergoes changes in conformation, quaternary structure, location within the cell, etc., subsequent to the initial binding event. Westheimer and his co-workers, who invented the photoaffinity-labeling approach (*8*), were also the first to exploit the separability of the binding and labeling reactions in their studies on the photoaffinity labeling of chymotrypsin via formation of diazoacetylchymotrypsin. Thus, the experimenter simply may wait until the protein–proreagent complex has assumed the particular state of interest (conformational, geographical, or whatever) and initiate the labeling reaction with light. A comparison of the labeling chemistries obtained with the protein in different states would provide interesting and important data on the nature of differences between the states. Such an experimental approach would not be possible with classical affinity reagents. Another feature of some types of photoaffinity reagents is the potential for enhanced reaction specificity vis à vis classical reagents. Systems in which the rate of proreagent dissociation is slow (e.g., intracellular steroid receptor proteins) permit the experimenter to allow complex formation to occur between protein and proreagent and then to remove excess unbound proreagent by adsorption, gel filtration, or some other suitable separate method, all in the dark. Then the sample is irradiated before significant amounts of bound proreagent dissociate. Thus labeling is performed under nonequilibrium conditions. This protocol can help to minimize nonspecific labeling reactions due to unbound photoactivated proreagent molecules. Another source of enhanced reaction specificity in photoaffinity labeling is the rather short solution lifetimes of some photogenerated reactive species. This property serves to reduce nonspecific reactions since unbound photoactivated species will decay to unreactive products before they can collide and react with a protein; bound photoactivated species, already having passed through the diffusional phase of the process, can react intramolecularly with protein functionalities before deactivation. Approximate values for the lifetime of some photogenerated reactive species are given in Table I.

A property of photolabeling reagents that is desirable for some systems is the ability of the photogenerated carbenes, nitrenes, and ketone-excited states to react with hydrocarbon and nucleophilic groups (*5, 12, 13*). This broad reactivity spectrum may be particularly important for success in labeling nonenzymatic steroid-binding proteins, which might be expected to possess nucleophile-poor binding sites. Classical affinity reagents and all $k_{cat}$ reagents of which we are aware are usually reactive only towards nucleophilic groups.

The $k_{cat}$ reagents offer the greatest potential for site specificity for labeling proteins in very heterogeneous systems. Since the reactive

**Table I. Approximate Lifetimes at 25°C of Light-Generated Reactive Species**

| *Species* | *Lifetime (s)* | *Ref.* |
|---|---|---|
| Carbenes | $< 10^{-3}$ | *5* |
| Nitrenes | $\sim 10^{-3}$ | *9* |
| Nitrophenyl ether-excited state (triplet) | $10^{-7}$–$10^{-10}$ | *10* |
| Ketone-excited state | | |
| (singlet) | $10^{-10}$–$10^{-8}$ | *11* |
| (triplet) | $10^{-5}$–$10^{-6}$ | *11* |

species is generated at the active site of the target enzyme, the functional groups of that site "see" the reactive moiety before any groups outside of the site or any group on a nontarget protein. Thus the chances for their reacting preferentially with the reagent are enhanced relative to what they would be if the reactive form of the $k_{cat}$ reagent were generated outside of the target site. However, in many instances the reactive species dissociates from the target site before it reacts with that site. In this event $k_{cat}$ reagents are not expected to be any more specific than the reactive product would be when used as a classical reagent.

**Quantitative Aspects.** In this section we shall summarize the characterisitic kinetic properties of affinity-labeling reactions. Our purpose in doing this is to point out those kinetic features of an affinity-labeling reaction which distinguish it from nonspecific reactions. Upon recognizing the kinetic distinctions between site-specific and nonspecific reactions, we then shall be able to identify the parameters that determine the degree of specificity of any labeling reaction. Suitable manipulation of some of these parameters by the experimenter can optimize the specificity of the modification reaction. It is often important to determine whether a particular modification reaction occurs via a true affinity-labeling mechanism (Reactions 1 and 2 in Figure 2 for classical reagents) or whether an ordinary bimolecular chemical modification of a ligand-binding-site group occurs. The experimental criteria that permit this distinction of mechanism to be made are kinetic in nature. It is for this reason that a presentation of the theoretical aspects of affinity-labeling kinetics is made here. Other discussions of affinity-labeling kinetics have been given by Kitz and Wilson (*14*), Meloche (*15*), Baker (*16*), and Farney and Gold (*17*). Baker's discussion of alternative mechanisms that exhibit affinity-labeling-type kinetics (i.e., show a rate saturation effect) is very useful and aspects of his treatment will be incorporated into the following discussion. For simplicity we shall assume that the reagent is of the classical type.

Let P represent the protein to be labeled via reaction with the affinity-labeling reagent, R. Let G* represent the functional group with which R reacts with is in or very near to the ligand-binding site. Let $G_m$ represent the set of $m$ functional groups outside of the target site with which R also may react nonspecifically. Thus, the unmodified protein may be represented by $G_mPG^*$. Let L represent a nonreactive ligand that binds to the target site. The dissociation constant of the P·L complex will be $K_L$. Noncovalent complexes will be indicated by a · between the species participating in the complex. Covalent bonds between R and groups G* or G will be indicated by a —.

**True Affinity Labeling.** EFFECT OF REAGENT CONCENTRATION. The reactions to be considered are given by Equation 1.

$$G_mPG^* + R \underset{k_{-1}}{\overset{k_1}{\rightleftarrows}} G_mPG^* \cdot R \xrightarrow{k_2} G_mPG^* - R \tag{1}$$

Assume that $G_mPG^*$ and $G_mPG^* \cdot R$ are fully active in the assay system because of a large dilution factor which, by mass action, results in the complete dissociation of $G_mPG^* \cdot R$. Then the rate of inactivation of $G_mPG^*$ will be

$$v_2 = -d\,\{[G_mPG^*] + [G_mPG^* \cdot R]\}/dt = k_2\,[G_mPG^* \cdot R] \tag{2}$$

We shall further assume at all times during the reaction that $G_mPG^* \cdot R$ is in rapid equilibrium with $G_mPG^* + R$, i.e. that $k_{-1} >> k_2$. This is a valid approximation since for many of the steroid-binding enzymes which have been subjected to classical affinity labeling by a variety of steroidal alkylating agents $k_2$ has ranged between 0 and 0.014 $sec^{-1}$. Although the dissociation constant, $K_R$, of the most stable protein · steroid complexes is of the order of $10^{-9}M$, for most of the steroid-metabolizing enzymes which have been investigated by affinity labeling $K_R \approx 10^{-5}M$. Taking the latter value as one typical of $K_R$ and assuming that the association constant, $k_1$, is approximately $10^7M^{-1}$ $sec^{-1}$ (*18*), we may write that $k_{-1}/10^7 = 10^{-5}$ from whence $k^{-1} = 100$ $sec^{-1}$ which is far in excess of any observed value for $k_2$.

The equilibrium constant for the dissociation of the protein · reagent complex is given by Equation 3.

$$\frac{k_{-1}}{k_1} = K_R = \frac{[G_mPG^*][R]}{[G_mPG^* \cdot R]} \tag{3}$$

The material balance equation, Equation 4, applies.

$$P_t = [G_mPG^*] + [G_mPG^* \cdot R] + [G_mPG^* - R] \tag{4}$$

$P_t$ is the total concentration of the target protein.

Combining Equations 3 and 4 we obtain Expression 5 for the instantaneous concentration of $G_mPG^* \cdot R$.

$$[G_mPG^* \cdot R] = \frac{\{P_t - [G_mPG^* - R]\}[R]}{K_R + [R]} \quad (5)$$

Inserting this expression for $[G_mPG^* \cdot R]$ into Equation 2 we produce

$$-\frac{d\,\{[G_mPG^*] + [G_mPG^* \cdot R]\}}{dt} = \frac{k_2\,\{P_t - [G_mPG^* - R]\}\,[R]}{K_R + [R]} \quad (6)$$

We can simplify this unwieldy equation by making two variable changes. If we represent the total concentration of active species (as assayed after substantial dilution) by $P_a$, then we may write that

$$P_a = [G_mPG^*] + [G_mPG^* \cdot R] = P_t - [G_mPG^* - R]$$

and Equation 6 becomes

$$-\frac{d\,P_a}{dt} = \frac{k_2\,P_a\,[R]}{K_R + [R]} \quad (7)$$

Rearranging Equation 7 and integrating

$$\int_{P_t}^{P_a} \frac{d\,P_a}{P_a} = \int_0^t \frac{k_2[R]dt}{K_R + [R]}$$

we obtain

$$\ln P_a - \ln P_t = -\frac{k_2[R]\,t}{K_R + [R]}$$

$$\ln \frac{P_a}{P_t} = -\frac{k_2\,[R]\,t}{K_R + [R]} = -k_{obs}t \quad (8)$$

Inspection of Equation 8 reveals that the kinetics of true affinity labeling are first order as long as [R] does not change significantly during the reaction. Furthermore, the observed first-order rate constant, $k_{obs}$, is a hyperbolic function of [R], i.e.

$$k_{obs} = \frac{k_2\,[R]}{K_R + [R]} \quad (9)$$

It is this latter characteristic which distinguishes affinity-labeling kinetics from the kinetics of simple bimolecular reactions which do not involve prior formation of a reagent–protein complex. Thus, plots of $k_{obs}$ vs. [R] will be hyperbolic, with the value of $k_{obs}$ approaching that of $k_2$ as [R] becomes large relative to $K_R$. Obedience of experimentally determined values of $k_{obs}$ to Equation 9 is a necessary, but not sufficient condition to demonstrate that a true-affinity-labeling mechanism is followed.

Simple, graphical methods for testing the fit of rate data to Equation 9 and for estimating the kinetic parameters $k_2$ and $K_R$ involve using linearized forms of Equation 9. By far the most widely used linear form is that of Kitz and Wilson (*14*). Taking reciprocals of both sides of Equation 9, we obtain

$$\frac{1}{k_{obs}} = \frac{K_R}{k_2} \cdot \frac{1}{[R]} + \frac{1}{k_2} \qquad (10)$$

A plot of $1/k_{obs}$ vs. $1/[R]$ will be linear if the reaction obeys Equation 9. From the extrapolated intercepts and slope of such a Kitz–Wilson plot one readily can obtain values for $k_2$ and $K_R$. Figure 3 shows a theoretical Kitz–Wilson plot that indicates the values of slope and intercepts from which $k_2$ and $K_R$ can be calculated.

Enzymologists will recognize the complete equivalence of the Kitz–Wilson and Lineweaver–Burk plots. This equivalence extends to the practical statistical problems characteristic of the Lineweaver–Burk plot. [These have been discussed by Segel (*19*) and Cornish–Bowden (*20*).]

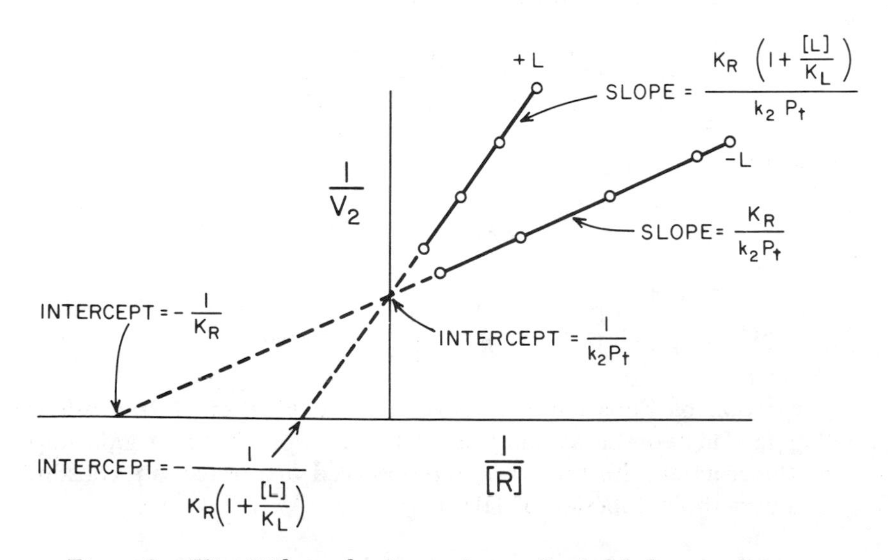


*Figure 3. Kitz–Wilson plot for verifying affinity-labeling mechanism.*

They center on the effect of errors in measuring $k_{obs}$ at low values of $k_{obs}$; these errors are amplified in the value of $1/k_{obs}$. However, it is points farthest out on the Kitz–Wilson (or Lineweaver–Burk) plot—those corresponding to the lowest values of $k_{obs}$—that most strongly influence the slope selected for the straight line that best fits the data. Thus, error in the estimation of $K_R$ is introduced.

This effect can be reduced if affinity-labeling kinetic data are analyzed by other types of linear plots used by enzyme kineticists. Two which appear to be useful improvements over the Kitz–Wilson plot are analogs of the Eadie–Hofstee and the Eisenthal–Cornish–Bowden plots (*20, 21*).

To obtain the analog of the Eadie–Hofstee plot we can multiply both sides of Equation 10 by $k_{obs}$ $k_2$ and we obtain

$$k_2 = K_R \frac{k_{obs}}{[R]} + k_{obs}$$

which upon rearrangement yields the desired Equation 11.

$$k_{obs} = k_2 - K_R \cdot \frac{k_{obs}}{[R]} \tag{11}$$

A plot of $k_{obs}$ vs. $k_{obs}/[R]$ will be linear with a slope of $-K_R$ and intercepts on the $k_{obs}$ and $k_{obs}/[R]$ axes of $k_2$ and $k_2/K_R$, respectively. This plot is less sensitive to errors in $k_{obs}$ than is the reciprocal plot because $k_{obs}$ appears in the numerators of the independent and dependent variables.

Eisenthal and Cornish–Bowden (*21*) and Cornish–Bowden (*20*) have described a rather different type of enzyme kinetics plot that should be useful in analyzing affinity-labeling kinetics. The equation that forms the basis of the direct linear plot (*21*) is obtained from Equation 11 by rearrangement of terms to give Equation 12

$$k_2 = k_{obs} + K_R \cdot \frac{k_{obs}}{[R]} \tag{12}$$

From Cornish–Bowden's point of view one regards $k_2$ and $K_R$ as variables and $k_{obs}$ and $[R]$ as constants. Seen in this way Equation 12 is the equation of a straight line plotted with respect to axes labeled $k_2$ and $K_R$. The line will have a slope of $k_{obs}/[R]$ and intercepts of $k_{obs}$ on the $k_2$ axis and $-[R]$ on the $K_R$ axis. Thus, a single experimental measurement of reaction velocity at a particular reagent concentration gives a pair of values, $(k_{obs})_1$ and $-[R]_1$. These are the axial intercepts of the straight line described above; the points of this line correspond to each of the infinite number of pairs of values that $k_2$ and $K_R$ might have which would result in the experimentals $(k_{obs})_1$ and $[R]_1$. This is shown graphically in Figure 4A. Only one of the points of this line corresponds to the true values of $k_2$ and $K_R$. This point can be identified by including kinetic data obtained at a different reagent concentration, $[R]_2$, resulting in a different modification rate constant, $(k_{obs})_2$. The data pair $[R]_2$ and $(k_{obs})_2$ will define a second straight line, all points of which represent the possible values of $k_2$ and $K_R$ which would result in the particular $(k_{obs})_2$ observed at reagent concentration $[R]_2$. The two lines will intersect at Point A (*see* Figure 4B) which specifies the unique values for $k_2$ and $K_R$ (denoted in Figure 4B as true $k_2$ and true $K_R$) which simultaneously accounts for both sets of data. In fact, several sets of data at various [R]s resulting in various $k_{obs}$s will specify several straight lines that will intersect in a common zone (rather than a point due to the inevitable experimental errors) as shown in Figure 4C. From the scatter of the points of intersection of several lines, the best estimate for true $k_2$ and true $K_R$ for the system can be selected. The statistical advantages of the direct linear plot vis-à-vis other linear plots have been discussed by Cornish–Bowden (*20*).

Accurate estimation of $K_R$ is important since the value obtained can be important in deciding whether or not the reagent is inactivating at the site targeted or by an indirect mechanism. $K_R$ values obtained from affinity-labeling kinetics should agree with $K_R$ values obtained independently if site inactivation occurs by Equation 1 at the site which becomes inactivated. If the site to be labeled is a catalytic site, then R should behave as a competitive inhibitor of the enzyme, with $K_I = K_R$. In some cases R may be a substrate of the enzyme. If $K_m$ is a true dissociation constant, it should equal $K_R$. If such equalities are not found, one must suspect that inactivation occurs by reaction at some second site whose integrity is required for the targeted site to be active.

EFFECT OF NONREACTIVE LIGAND CONCENTRATION: PROTECTION EFFECTS. Since true affinity labeling depends on the formation of a complex between target protein and reagent, anything that reduces the equilibrium concentration of that complex will decrease the rate of the labeling reaction. Such an effect will be produced by the presence of a nonreactive ligand, L, of the protein, which from other studies is known to bind to the target site. In the case of enzymes, L may be a substrate or competitive inhibitor; in the case of noncatalytic binding proteins, L may be the physiological ligand or a chemical relative of it. The characteristics of the perturbation of a labeling reaction by L constitute an important, widely used test of the target-site specificity of the labeling reaction.

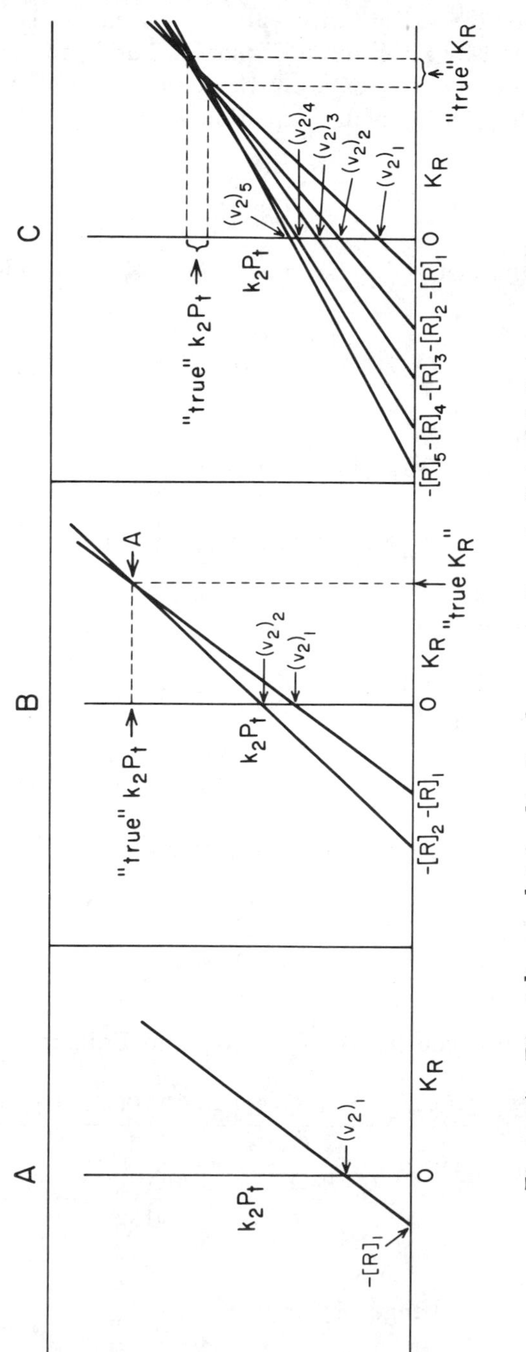

*Figure 4. Direct linear plot for kinetic characterization of affinity-labeling reactions.*

The basis for the quantitative effects of L on the labeling kinetics may be understood by returning to Equation 3 and proceeding with the derivation of the rate law as before with the inclusion of the equilibrium expression for the binding of L (Equation 13),

$$K_L = \frac{[G_mPG^*][L]}{[G_mPG^* \cdot L]} \tag{13}$$

and modifying the material balance equation (4) to include $G_mPG^* \cdot L$. Thus,

$$P_t = [G_mPG^*] + [G_mPG^* \cdot R] + [G_mPG^* - R] + [G_mPG^* \cdot L] \tag{14}$$

Combining Equations 13 and 14 one obtains

$$P_t = [G_mPG^*] + [G_mPG^* \cdot R] + [G_mPG^* - R] + \frac{[G_mPG^*][L]}{K_L}$$

and inserting Equation 3

$$P_t = \frac{K_R[G_mPG^* \cdot R]}{[R]} + [G_mPG^* \cdot R] + [G_mPG^* - R] + \frac{[L]}{K_L} \cdot \frac{K_R[G_mPG^* \cdot R]}{[R]}$$

Solving for $[G_mPG^* \cdot R]$ we derive Equation 15

$$[G_mPG^* \cdot R] = \frac{\{P_t - [G_mPG^* - R]\}[R]}{K_R\left(1 + \frac{[L]}{K_L}\right) + [R]} \tag{15}$$

Inserting this expression for $[G_mPG^* \cdot R]$ into Equation 2 we produce

$$-\frac{d\{[G_mPG^*] + [G_mPG^* \cdot R] + [G_mPG^* \cdot L]\}}{dt} = \frac{k_2\{P_t - [G_mPG^* - R]\}[R]}{K_R\left(1 + \frac{[L]}{K_L}\right) + [R]} \tag{16}$$

and introducing the change of variable

$$P_a = [G_mPG^*] + [G_mPG^* \cdot R] + [G_mPG^* \cdot L] = P_t - [G_mPG^* - R]$$

we obtain Equation 17.

$$-\frac{dP_a}{dt} = \frac{k_2 P_a [R]}{K_R\left(1 + \frac{[L]}{K_L}\right) + [R]} \tag{17}$$

After rearranging and forming the integrals

$$\int_{P_t}^{P_a} \frac{d\,P_a}{P_a} = -\int_o^t \frac{k_2 [R]\, dt}{K_R\left(1 + \frac{[L]}{K_L}\right) + [R]}$$

we may integrate both sides to give Equation 18:

$$\ln \frac{P_a}{P_t} = -\frac{k_2 [R]\, t}{K_R\left(1 + \frac{[L]}{K_L}\right) + [R]} = -k_{obs}\, t \tag{18}$$

Just as in the absence of L, first-order kinetics are expected for site inactivation. The observed first-order rate constant, $k_{obs}$, is a function of [R] and [L], i.e.

$$k_{obs} = \frac{k_2 [R]}{K_R\left(1 + \frac{[L]}{K_L}\right) + [R]} \tag{19}$$

Comparing Equations 9 and 19 emphasizes that the nonreactive ligand does not change qualitatively the form of the dependence of $k_{obs}$ on [R], but only increases the apparent dissociation constant of the protein–reagent complex by the factor $1 + [L]/K_L$. Since this factor is always greater than unity for $[L] > 0$, it follows that L always will decrease $k_{obs}$ at any given [R]. One can say that L protects the protein from affinity labeling by R.

The effects of L on the appearance of the Kitz–Wilson plot is exactly analogous to the effect of an enzyme competitive inhibitor on a Lineweaver–Burk plot.

Taking reciprocals of both sides of Equation 19 we then have

$$\frac{1}{k_{obs}} = \frac{K_R\left(1 + \frac{[L]}{K_L}\right)}{k_2} \cdot \frac{1}{[R]} + \frac{1}{k_2} \tag{20}$$

The plot of $1/k_{obs}$ vs. $1/[R]$ will be now increased in slope by the factor $1 + [L]/K_L$. The intercept on the $1/[R]$ axis will be $-1/[K_R(1 + [L]/K_L)]$ while the intercept on the $1/k_{obs}$ axis will be unchanged compared with a Kitz–Wilson plot for $[L] = 0$. From the increase in slope and/or change in intercept and the known [L] one may calculate

$K_L$. Figure 3 compares Kitz–Wilson plots obtained plus and minus L. Of course the effects of L on affinity-labeling kinetics could also be analyzed quantitatively by Eadie–Hofstee or direct linear-type plots.

Estimated in this way from labeling kinetics, $K_L$ should be compared with independent measures of $K_L$ such as those obtained from enzyme kinetics, equilibrium dialysis, or the various spectroscopic techniques. The agreement of $K_L$s obtained from affinity-labeling kinetics and enzyme kinetics or physical techniques provides important evidence that R reacts at the L binding site.

**Bimolecular Labeling with Self Protection.** Baker (*16*) has pointed out an interesting alternative labeling mechanism that can be confused with true affinity labeling (refer to previous section on true affinity labeling) if investigation of the kinetics of the reaction is limited to the kinds of studies outlined in the previous section. In bimolecular labeling with self protection the reagent forms a complex with the protein at the proper target site, but for some reason, such as improper positioning of the reactive part of the reagent with respect to any reactive group, G*, in the site, it fails to react covalently with the protein. On the other hand, G* still may be modifiable by R via a simple bimolecular collision mechanism that does not involve complex formation. If this situation obtains, the kinetics of the labeling reaction will follow a rate law that is indistinguishable in mathematical form from that for true affinity labeling. To see why this is so, we must derive the rate equation for the self-protection mechanism.

Effect of Reagent Concentration. The equilibria and reactions involved are as follows:

$$G_mPG^* + R \overset{K_R}{\rightleftarrows} G_mPG^* \cdot R \tag{21}$$

$$G_mPG^* + R \overset{k_3}{\rightarrow} G_mPG^* - R \tag{22}$$

The rate of inactivation of the target site is given by Equation 23.

$$\frac{d\,\{[G_mPG^*] + [G_mPG^* \cdot R]\}}{dt} = -\,k_3\,[G_mPG^*]\,[R] \tag{23}$$

As in the previous section on true affinity labeling we have the material balance equation

$$P_t = [G_mPG^*] + [G_mPG^* \cdot R] + [G_mPG^* - R] \tag{24}$$

from which the concentration of assayable active species is given by $P_a$

$$P_a = P_t - [G_mPG^* - R] = [G_mPG^*] + [G_mPG^* \cdot R] \quad (25)$$

applying the equilibrium expression, Equation 3, we find that

$$P_a = [G_mPG^*]\left(1 + \frac{[R]}{K_R}\right)$$

from whence

$$[G_mPG^*] = \frac{P_a}{1 + \frac{[R]}{K_R}} \quad (26)$$

Combining Equations 25 and 26 with the rate equation (23) we obtain Equation 27

$$\frac{d\,P_a}{dt} = -\frac{k_3\,K_R\,P_a\,[R]}{K_R + [R]} \quad (27)$$

Rearranging Equation 27 and forming integrals on both sides

$$\int_{P_t}^{P_a} \frac{d\,P_a}{P_a} = -\int_o^t \frac{k_3\,K_R\,[R]\,dt}{K_R + [R]}$$

and we obtain Equation 28 after integration

$$\ln\frac{P_a}{P_t} = -\frac{k_3\,K_R\,[R]\,t}{K_R + [R]} = -k_{obs}\,t \quad (28)$$

From Equation 28 we note that

$$k_{obs} = \frac{k_3\,K_R\,[R]}{K_R + [R]} \quad (29)$$

Comparing Equation 29 with Equation 9 shows that the two expressions for $k_{obs}$ differ only in the constants in the numerators of the right-hand sides. Both mechanisms predict first-order kinetics for the loss of site activity and identical dependence of the observed first-order rate constant, $k_{obs}$, on the [R]. The similarity of Equations 9 and 29 demonstrates that the documentation of saturation kinetics as evidenced by linear Kitz–Wilson or Eadie–Hofstee plots or by the critria of the direct linear plot does not prove that true affinity labeling is involved necessarily in a site-inactivating reaction.

EFFECT OF THE NONREACTIVE LIGAND. To analyze the effect of L on self-protection kinetics one shall consider Equations 3, 13, and 14. Combining these equations and expressing $P_a$ in terms of $[G_mPG^*]$, we find that

$$P_a = P_t - [G_mPG^* - R] = [G_mPG^*] \quad 1 + \frac{[R]}{K_R} + \frac{[L]}{K_L}$$

solving for $[G_mPG^*]$ and inserting it into Equation 23 we obtain the following rate law:

$$\frac{d\,P_a}{dt} = - \frac{k_3\,P_a\,[R]}{1 + \frac{[R]}{K_R} + \frac{[L]}{K_L}} \tag{30}$$

$$= - \frac{k_3\,K_R\,P_a\,[R]}{K_R\left(1 + \frac{[L]}{K_L}\right) + [R]}$$

Rearranging Equation 30 and forming integrals on both sides,

$$\int_{P_t}^{P_a} \frac{d\,P_a}{P_a} = - \int_{o}^{t} \frac{k_3\,K_R\,[R]\,dt}{K_R\left(1 + \frac{[L]}{K_L}\right) + [R]}$$

and we integrate both sides which gives

$$\ln \frac{P_a}{P_t} = \frac{-\,k_3\,K_R\,[R]\,t}{K_R\left(1 + \frac{[L]}{K_L}\right) + [R]} = -\,k_{obs}\,t \tag{31}$$

Therefore, $k_{obs}$ is a function of both [R] and [L], i.e.,

$$k_{obs} = \frac{k_3\,K_R\,[R]}{K_R\left(1 + \frac{[L]}{K_L}\right) + [R]} \tag{32}$$

This expression for $k_{obs}$ is of the same mathematical form as the analogous expression for the effects of [R] and [L] in true affinity labeling (*see* Equation 19). In other words, protection by [L] is quantitatively the same for both mechanisms and thus cannot differentiate between them.

**Criteria for Distinguishing Between True Affinity Labeling and Bimolecular Self-Protection Mechanisms.** In the case of a labeling reaction that exhibits saturation kinetics and the appropriate protection effects (i.e., follows integrated rate equations of the form of Equations 18 and 31, it may be critically important to distinguish between true affinity and bimolecular self-protection labeling mechanisms. For example, the purpose of the affinity-labeling attempt may be to obtain information about the relative spatial positions of site functional groups and the functional groups of ligands that bind in the site. Such proximity relationships may be inferred from the position of the reactive group on the affinity reagent and the identity of the protein functional group with which the reaction occurs. In favorable cases several similar reagents are used, each having the reactive group positioned at a different part of the reagent structure. Such a library of isomeric reagents can be used to roughly map out the functional group format of the ligand-binding site. The assumption usually is made that each of the affinity reagents bind more or less congruently with the binding geometry of the physiologically operant ligand. Good examples of this type of study from the steroid field are found in the work of Warren, Sweet, and their collaborators (*1*) and that of Crastes de Paulet and co-workers (*22*). However, if labeling actually occurs via the self-protection mechanism, the assumption of congruency is moot since labeling occurs via a bimolecular collision and not from the congrent reagent · protein complex. Consequently, no conclusions about the structural characteristics of the ligand-binding site can be drawn from the data obtained.

Fortunately, tests exist that may permit a determination of the labeling mechanism. These are based on considerations of the degree of site specificity of the protein's reaction with the reagent and on the effect of reagent structure on the labelability of the target site. If the reaction in question proceeds by self protection, the magnitude of the second-order rate constant for the covalent reaction, $k_3$ in Equation 22, could well be about the same as the second-order rate constants for the reaction of R with many of the $m$ reactive groups, G, in $G_m PG^*$ that are not part of the ligand-binding site. In other words, the intrinsic chemical reactivities of G and G* may be very similar. Chemical modification studies using simple reagents that do not mimic the structure of ligands that bind to the target site should show if G* is hyperreactive, of normal reactivity, or hyporeactive. If G* is not hyperreactive, then a reaction proceeding by the self-protection mechanism will be accompanied by a low degree of specificity for G*. Using radiolabeled R, determination of the stoichiometry of label incorporation into protein should reveal low specificity by a high ratio of R conjugated to protein at any fractional extent of site modification. Unfortunately, lack of specificity coupled with normal G* reactivity, though suggestive of self protection, is not a rigorous proof since low specificity is not an exclusive characteristic of self protection. For example, it is not difficult to imagine a ligand-binding site containing a hyporeactive group that could be labeled by a true affinity labeling mechanism at a rate that was comparable with the rate of reaction of the affinity reagent with extrasite groups, a situation characterized by poor specificity.

To make matters more ambiguous, recall that there are many examples of ligand-binding sites, often enzyme active sites, that contain groups that are hyperreactive. Directed towards such a target, true affinity labeling reagents as well as self-protection reagents will react with high site specificity. Consequently, good site specificity cannot be accepted as a proof of true affinity labeling. Again, studies with nonaffinity reagents should reveal the hyperreactive condition which characterizes this circumstance.

One condition permits an unambiguous distinction between a true affinity labeling mechanism and self protection. If hyperreactivity is not present in the target site then good labeling specificity by a reagent that binds in the site (as evidenced by a linear Kitz–Wilson plot) argues for true affinity labeling as the dominant reaction mechanism. If hyperreactivity is a characteristic of the target site, no distinction between the two mechanisms can be made on a specificity basis.

Another approach to distinguish between true affinity labeling and self protection is to assess how variation of reagent structure affects the rate of labeling. Thus, a true affinity labeling reagent should react with a target site faster than a nonbinding reagent bearing an identical reactive group. In contrast, self-protection reagents would react at rates that are similar to those nonbinding reagents. One also can characterize a labeling mechanism by comparing the rates of reaction of pairs of candidate affinity reagents which differ in the stereochemistry of attachment of the chemically reactive grouping. In a situation where both members of a stereoisomeric pair bind noncovalently to the target site, a large difference in reactivity would be expected if true affinity labeling were the actual mechanism for both reagents, whereas little difference in reactivity would be observed for self-protection reagents. This difference in kinetic behavior is expected since true affinity reagents react covalently from a complex of relatively well-defined geometry; changes in that geometry due to changes in reagent stereochemistry will have a large effect on rate. Self-protection reagents react via a stereochemically much less well-defined collisional complex in which rate differences between the members of a stereoisomeric pair of reagents usually would be small, but if significant, should be similar in magnitude to the intrinsic chemical reactivity difference between the reagents. The latter difference can be determined by reactivity studies using low-molecular-weight target molecules.

**Specificity of Affinity Labeling.** In this section we shall analyze those factors that influence the specificity of classical affinity reagents.

We already have indicated briefly additional specificity-influencing factors that are special to photoaffinity and $k_{cat}$ reagents.

Equation 9 is the expression for the observed first-order rate constant for site modification by a true affinity reagent, R. Let us rewrite Equation 9 as in Equation 33 to indicate that reaction is occurring with a specific group, G*, in the target site. Thus

$$k^*_{obs} = \frac{k_2 [R]}{K_R + [R]} \tag{33}$$

In a case in which a pure protein is the target of the reagent we also must consider the rate of reaction of R with all nontarget site functional groups that, under the conditions used, can react at a significant rate. We shall assume that each of these nonspecifically reacting groups reacts via a second-order bimolecular reaction with R, with the identical second-order rate constant, $k_4$. For $G_mPG^*$ we shall recognize $m$ such groups, G. Since $[R] >> P_t$ we are in the condition of pseudo-first-order reaction kinetics. Hence the total observed pseudo-first-order rate constant for all nonspecific reactions will be given by Equation 34.

$$k_{obs} = m\, k_4\, [R] \tag{34}$$

The degree of target-site specificity of the reaction of R with $G_mPG^*$ can be characterized by the enhancement, $E$, which we will define as the ratio of the first-order rate constants for the specific and the bulk aggregate of the nonspecific reactions. That is,

$$E = \frac{k^*_{obs}}{k_{obs}} = \frac{k_2}{mk_4\, (K_R + [R])} \tag{35}$$

This equation summarizes in a convenient form the parameters that determine the specificity of a labeling reaction. These factors are:

$k_2$: an increase in the value of the first-order rate constant for covalent bond formation will increase $E$. The term $k_2$ can be increased by using a more reactive group on R. However this approach is not likely to affect much change in $E$ since $k_4$ will be increased by about the same amount. A better approach would be to manipulate $k_2$ by varying the position or orientation of the reactive group on R in order to achieve an optional orientation of the reactive groups on the reagent and the target site. Thus $k_2$ might be maximized with little effect on $k_4$. Prior knowledge of the structure of the target site clearly is of great assistance in this optimization. Lacking this kind of knowledge, one is forced to rely on a trial and error protocol. However, reagent structures that maximize $k_2$ may not bind as well to the target site, i.e., $K_R$ may be increased, partially or completely cancelling the increase in $k_2$. Here again, structural information may help to guide the selection of the reagents most likely to have high $k_2$s along with minimal $K_R$s.

$k_4$: minimizing the intrinsic reactivity of R will minimize $k_4$ and therefore increase $E$. But, as pointed out earlier, $k_2$ will be diminished by the same factor as $k_4$ and no net effect on $E$ will be obtained.

$m$: $m$ is a fixed property of the target protein and is not easily susceptible to manipulation. In principle, it should be possible to reduce $m$ by using a reversible blocking group of G. G* could be protected from blockade by a site ligand, L. Then the G-blocked protein could be labeled with R more specifically than the native protein. As a third step the blocking groups on G could be removed.

$K_R$: the more stable the complex between reagent and target protein, the more specific will be the labeling reaction, other things being equal. Detailed studies of structure–function relationships for ligands that bind to the protein will be very helpful in the design of Rs that have a minimal $K_R$. Unfortunately, other things usually don't remain equal because structural changes that affect $K_R$ may affect $k_2$ as well. The effect may or may not be advantageous.

[R]: minimizing the reagent concentration will increase $E$. The effect will be limited by the magnitude of $K_R$ since $K_R +$ [R] is the denominator in Equation 35. Thus, decreasing [R] much below $K_R$ will have only a minor effect on $E$. The most important thing about [R] is to avoid needlessly high values, which will result in a substantial increase in $k_{obs}$ with relatively little increase in $k^*_{obs}$. The benefits of minimizing [R] are realized best by using a reagent with minimal $K_R$.

$k_2/K_R$: under the conditions of minimal [R], which tends to increase $E$, Equation 35 approaches the limiting form of Equation 36.

$$E = \frac{k_2/K_R}{m\, k_4} \tag{36}$$

This expression for $E$ emphasizes the fact that the most specific reagents will have the largest values for $k_2/K_R$. The expression $k_2/K_R$ is the parameter which is the most fundamental determinant of labeling specificity. It has a very simple kinetic significance; reference to Equation 7 shows that it is the second-order rate constant for the inactivation reaction which is obtained at low [R].

The special problems associated with labeling a specific target site in a heterogeneous mixture of proteins such as that which exists in cells can be examined in terms of the specificity equation (*see* Equation 35). In terms of specificity of labeling, the essential difference between labeling a pure protein and the same protein as a minor constituent in a complex mixture is the value of $m$. If the reactions of R with groups on nontarget proteins are bimolecular, then we may regard all such reactions as contributors to $k_{obs}$ in Equation 34. This is equavalent to $m$ becoming very large compared with its value for a pure protein. Clearly this will result in reduced specificity. If one regards $K_R$ as the most important determinant of $E$, then to achieve a particular value of $E$, $K_R$ will have to be lower when labeling a complex mixture than when labeling a pure protein. A simple calculation using Equation 35 can be made which illustrates the implications of the large increases in $m$ associated with going to heterogeneous systems.

Assume that $E = 1$ for labeling a particular pure protein of mol wt 50,000 for which $m = 50$. The condition, $E = 1$, is regarded as a minimum acceptable level of specificity—a sort of break-even level. Now, consider labeling the same protein when it represents $10^{-5}$ mole fraction of the total protein present. Assume that the other proteins each possess 50 groups reactive towards R. Thus $m$ increases by a factor of $10^5$. Consideration of Equation 35 shows that E now has a value of $10^{-5}$, meaning that most of the reagent which reacts with protein reacts nonspecifically. Such a low level of specificity means that that reagent would not be useful in the heterogeneous system. Thus, in order to compensate for the large value of $m$, the reagent would have to be redesigned to have a much lower value of $K_R$. If the original reagent had a $K_R$ of $10^{-4}M$ (typical of many of the steroid enzyme reagents to be discussed in later sections), the redesigned reagent would have to have a $K_R$ of $\sim 10^{-9}M$ in order to achieve the break-even condition. This is about as stable a complex as reasonably can be expected, considering that the intracellular steroid receptor–steroid complexes, among the most stable known, exhibit $K_d$ values of ca. $10^{-9}M$. The take-home lesson of this exercise is that reagents which are useful in labeling trace-level proteins in complex, heterogeneous mixtures must bind much more strongly to their targets than do reagents that are to label the same protein in a homogeneous state.

### *Representative Examples of Affinity Reagents Directed Towards Steroid-Binding Sites*

Reagents that have been designed to react preferentially with steroid-binding site functional groups have been developed in each of the three reagent-type categories: classical affinity, photoaffinity, and $k_{cat}$ reagents. The target proteins at which these reagents have been aimed fall into two classes: steroid-metabalozing enzymes—particularly those from certain bacterial species—and steroid-binding proteins from extra- and intracellular vertebrates.

The selection of target proteins for affinity labeling has been determined by a blend of independent factors. Primary among these are availability of the target in quantities that are adequate for detailed chemical characterization of the labeling reactions and the biological importance of the target protein in hormone-dependent physiology. Unfortunately, many of the most interesting proteins from the standpoint of hormone biochemistry and pharmacology are among the least abundant proteins of mammalian cells or extracellular fluids.

In the following sections we shall present selected examples of labeling in which the three categories of reagents have been directed towards various steroidal enyzmes and binding proteins. The selection of examples is meant to be illustrative rather than exhaustive or encyclopedic, reflecting the bias of the authors that illustration will suffice to stimulate the imaginations of future investigators.

**Classical Affinity Reagents.** HALOGEN-CONTAINING STEROIDS. A large number of steroidal affinity reagents have been synthesized during the previous 10 years which bear electrophilic reactive groups. These reagents have the potential, therefore, of reacting with protein nucleophiles via nucleophilic displacement reactions. They are adaptations of the early work of Elliott Shaw (*23*) who introduced chloromethyl ketone derivatives of *N*-blocked amino acids as affinity alkylating agents for serine proteases. Following the example of this germinal study, one notes that the steroid tetracyclic ring system provides a multiplicity of sites that can be made reactive as alkylating centers by attachment of a good leaving group such as bromine for nucleophilic displacement reactions. Alternatively, one may prepare reagents in which bromoacetoxy or bromoacetamido moieties are attached at one of several positions on the steroid nucleus. The advantages of this type of alkylating function have been discussed by Naider, Becker, and Wilchek (*24*).

A good example of a reagent in which the alkylating carbon atom is one of the ring carbons is found in the work of Chin and Warren (*25*). These workers prepared three isomeric bromoprogesterones for use in site-specific alkylation of the catalytic site of the $3\alpha,20\beta$-hydroxysteroid dehydrogenase from the bacterium *Streptomyces hydrogenans* (E.C.1.1.1.53). A reaction that is catalyzed by this enzyme is the reversible reduction (by NADH) of the 20-keto group of the steroid substrate to the $20\beta$-alcohol. An example of the reaction is shown in Figure 5. Recent studies have shown that the pure enzyme also has dehydrogenase activity towards $3\alpha$-hydroxysteroids (*26*) and that this $3\alpha$ activity is very likely a property of the $20\beta$ catalytic site (*27*). The kinetic mechanism of the enzyme has been studied in some detail by Szymanski and Furfine (*28*) who found that the addition of substrates and release of products followed a random rapid equilibrium mechanism.

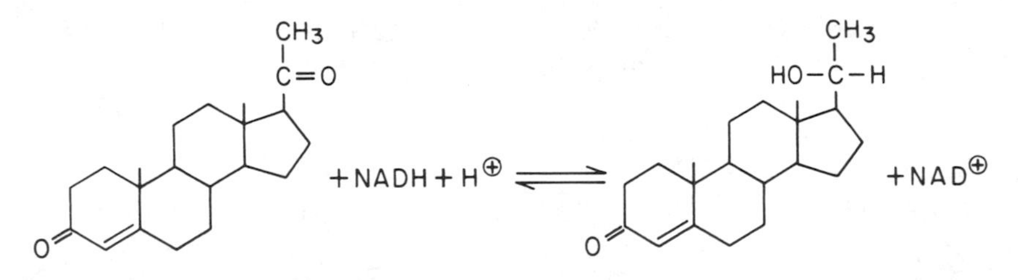


*Figure 5. Reaction catalyzed by 3α,20β-hydroxysteroid dehydrogenase.*

This means that this dehydrogenase can form binary complexes with steroid substrates, binary complexes with pyridine nucleotide, and ternary complexes with both substrates. This behavior contrasts with that usually observed for NAD-linked dehydrogenases in which the ketone or aldehyde substrate can bind only to the NADH–enzyme binary complex but not to free enzyme (*29*).

The physical properties of 3α,20β-hydroxysteroid dehydrogenase have not been investigated as extensively as have those of other dehydrogenases. Blomquist (*30*) found that the enzyme had an apparent molecular weight of about 110,000 as determined by agarose gel filtration and density gradient centrifugation. In SDS gel electrophoresis only a single species, migrating with a molecular weight of 27,000, was detected. These data suggest that the dehydrogenase is a tetramer of subunits of very similar if not identical chain length.

Chin and Warren (*25*) investigated the interaction of 2α-bromo-, 6α-bromo-, and 6β-bromoprogesterones as candidate affinity reagents for the dehydrogenase. The chemical reactivity of each bromosteroid towards protein nucleophiles was tested in reactions with free amino acids and glutathione at pH 7.0. Only cysteine and glutathione reacted and did so with relative rates of 1:3:15, respectively, when glutathione was the nucleophile. All three bromosteroids were shown to be good substrates of the enzyme with $K_m$ values of $4 \mu M$, $14.3 \mu M$, and $8.2 \mu M$, respectively. This shows that these compounds bind at the catalytic site. Szymanski and Furfine (*28*) have shown that for at least some substrates $K_m$ values closely approximate the true dissociation constants of enzyme steroid binary complexes. Assuming this to be true for the bromosteroids, they bind with high affinity to the target enzyme active site. When the dehydrogenase was treated with each bromosteroid, the 2α isomer did not inactivate the enzyme whereas the 6α and 6β isomers inactivated it via first-order processes although at very different rates. The kinetic constants associated with the inactivation reactions are given in Table II. Subsequent experiments in which enzyme inactivated with $^3$H-labeled 6β-bromoprogesterone was characterized chemically, showed that inactivation was paralleled by incorporating 1 mol of steroid/mole of enzyme

**Table II. Inactivation of 3α,20β-Hydroxysteroid Dehydrogenase by Bromoprogesterones**

| *Bromoprogesterone* | $k_{obs}$ [a] $(h^{-1})$ | $k_2$ [b] $(h^{-1})$ | *Relative* $k_2$ | *Relative Reactivity* [a] *with Glutathione* |
|---|---|---|---|---|
| 2α | 0 | 0 | 0 | 0.33 |
| 6α | 0.0144 | 0.222 | 1 | 1 |
| 6β | 0.347 | 3.19 | 14.36 | 5 |

[a] Calculated from the data of Chin and Warren (*25*).
[b] First-order rate constant for conversion of enzyme–steroid complex to enzyme–steroid covalent conjugate as defined in Equation 1. This was calculated from $k_{obs}$ and from the $K_m$ values of the steroids assuming that $K_m = K_R$.

(mol wt ~ 100,000) and that the site of covalent attachment was a cysteine sulfur. It is instructive to note from the data in Table II that the reactivity ratio for 6β/6α, though somewhat larger than the intrinsic reactivity ratio of the reagents towards the thiol nucleophile, glutathione, is really not enormously different from the model reaction. On the other hand, the lack of reactivity of the 2α isomer towards the dehydrogenase is completely at variance with its intrinsic reactivity. Presumably this is the result of a lack of juxtaposition of an SH group to the No. 2 carbon atom of the steroid in the enzyme–reagent complex. One must conclude from these data that orientational as well as reactivity factors operate to determine the actual rate of the labeling reaction and that reactive reagents can form complexes with the target enzyme (e.g., 2α has the lowest $K_m$ of the three bromosteroids) yet not be effective in site inactivation. A similar failure to inactivate this dehydrogenase was obtained with 21-iododeoxycortisone by Ganguly and Warren (*31*), a reagent that reacts with thiols substantially more rapidly than iodoacetate and which is also a substrate of the dehydrogenase. Presumably the positioning of the iodoacetyl group and enzyme nucleophiles prevents close approach of the reactive species.

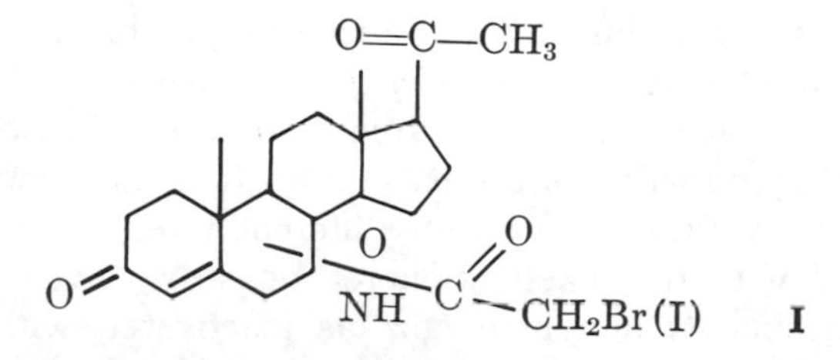


A variant of the reagents of Chin and Warren is the class of reagents represented by Compound **I.**

In this type of reagent a haloacetoxy (usually bromo) or haloacetamide moiety is a substituent of the steroid nucleus. This places the reactive part of the reagent 2–4 Å from the ring system. Thus protein groups that are not in actual contact with the normal bound substrate may be sites of covalent attachment of the successful affinity reagent. This will tend to blur the image of the nucleophile distribution in the binding site which is deduced from the pattern of group labeling by a series of isomeric haloacetoxy steroids. However, for many purposes specific labeling per se rather than the details of the structure of the steroid-binding site is sufficient for the purposes of the investigator. In addition, identification of the attachment residue is facilitated by using haloacetoxy steroids rather than halosteroids. This is because total acid hydrolysis of the haloacetoxy steroid-labeled protein results in hydrolysis of the reagent ester linkage leading to the accumulation of a carboxymethyl amino acid (or of glycolic acid if a carboxylate oxygen is the nucleophilic reactant). Since these carboxymethyl derivatives elute at well-known characteristic times from the analytical columns of an amino acid analyzer, the extent of modification of residues susceptible to alkylation can be obtained in a straightforward manner. The elution volumes for amino acid carboxymethyl derivatives have been tabulated by Gurd (*32*). Another advantage of haloacetyl steroids is their ready synthesis from the corresponding hydroxysteroid in radiolabeled form, since tritium or C-14 labels can be incorporated easily into the reagent via the [$^{3}$H]-bromoacetyl or [$^{14}$C]bromoacetyl group. Synthetic procedures and analytical methods associated with the characterization of haloacetoxysteroid-labeled proteins have been discussed in detail by Warren, Arias, and Sweet (*1*).

Warren, Sweet, and their colleagues have reported (*1, 31, 33–36*) on the reactions of a number of haloacetoxy progesterones and related compounds with $3\alpha,20\beta$ hydroxysteroid dehydrogenase. Table III summarizes the kinetic results obtained with bromoacetyl derivatives which inactivated the enzyme. All of these inactivation reactions appeared to be active-site-specific by the criterion of a significant protection effect by nonreactive steroid substrates of the enzyme. The question of whether these inactivations proceed by a true affinity labeling mechanism or by the self-protection mechanism can be answered by the following observations. First, the enzyme is not inactivated by simple alkylating agents such as iodoacetate either alone or in the presence of steroid substrate (*31*) under conditions used with the steroidal reagents. In addition, no carboxymethyl derivatives of amino acids are found in acid hydrolysates of iodoacetate-treated enzyme. Thus, the active site does not contain groups that are hyperreactive towards alkylating reagents of the haloacetyl type. Second, the chemistry of the alkylation reaction is affected by the position of the alkylating function on the steroid nucleus. Third, the reactions proceed with good chemical specificity. Taken together and bearing in mind the criteria discussed in the previous section on criteria for distinguishing between true affinity and bimolecular self-protective labeling mechanisms, it is clear that these reactions follow the true affinity labeling mechanism.

The kinetic data for the set of steroidal reagents tabulated in Table III provides an interesting opportunity to discuss criteria for the selection of reagents that might be suitable for some pharmacological or other in vivo application in which high reaction specificity was desired. Since all of these reagents have essentially the same reactive group, their nonspecific reaction rates with cellular nucleophiles are probably very similar. Thus, if the target protein happened to be this $3\alpha,20\beta$ hydroxysteroid dehydrogenase, the reagent that exhibited the greatest reactivity towards this enzyme would be the most specific reagent. We might choose as the most reactive and specific reagent that whose $k_2$ value was largest. Thus, the reactivity/specificity order would be $6\beta > 16\alpha > 17 > 2\alpha \approx 21 > 11\alpha$. However, in order to maximize specificity we have seen that one should use the lowest possible concentration of reagent. Under these conditions Equation 36 gives a good approximation to $E$.

Consequently, under maximum specificity conditions the reagent that reacts the fastest will be that for which the value of $k_2/K_R$ is the largest. From the data for $k_2/K_m$ in Table III, which we take as a good approximation to $k_2/K_R$ (*28*), we now order the reagents as $6\beta > 2\alpha > 16\alpha > 11\alpha > 17 > 21$, a rather different result. Note also that the reac-

**Table III. Inactivation[a] of $3\alpha,20\beta$-Hydroxysteroid Dehydrogenase (*S. hydrogenans*) by Bromoacetoxyprogesterone and Bromoacetamidoprogesterone**

| *Progesterone Derivative* | $K_m$ *or* $K_I$ ($\mu$M) | $k_{obs}$[b] ($h^{-1}$) | $k_2$ ($h^{-1}$) | $\frac{k_2}{K_m}$ ($\mu M^{-1}\ h^{-1}$) | *Residue Alkylated* | *Ref.* |
|---|---|---|---|---|---|---|
| $2\alpha$-Bromoacetoxy | 19 | 0.151 | 0.20 | 0.0105 | Met | *35* |
| $6\beta$-Bromoacetoxy | 20 | 1.93 | 2.57 | 0.129 | CySH | *34* |
| $11\alpha$-Bromoacetoxy | 25 | 0.0578 | 0.082 | 0.00328 | Met | *34, 35* |
| $16\alpha$-Bromoacetoxy | 145 | 0.347 | 1.18 | 0.00814 | His | *33* |
| 17-Bromoacetoxy | 1830 | 0.0277 | 0.87 | 0.000475 | COOH | *27* |
| 21-Bromoacetamido | 770 | 0.0144 | 0.20 | 0.000260 | His | *36* |

[a] Inactivation reactions were carried out with $60\mu M$ reagent in $0.05M$ phosphate buffer, pH 7.0 at 25°C.

[b] The $k_{obs}$ was decreased significantly in all cases when reactions were conducted in the presence of substrate.

tivity range when $k_2/K_R$ is taken as the appropriate measure of reactivity spans a factor of 500 whereas the reactivity range with $k_2$ as the measure is only 31. The correct choice of the reactivity parameter is obviously of critical importance in evaluating a group of reagents targeted at the same site.

Some interesting in vivo effects of bromoprogesterones and bromoacetoxyprogesterones on the pregnant rat have been reported by Warren's group. Clark, Sweet, and Warren (*37*) investigated the effects of several reactive steroids on the course of pregnancy in rats. The derivatives tested were 21-bromoacetoxyprogesterone, 16α-bromoacetoxyprogesterone, 11α-bromoacetoxyprogesterone, 2α-bromoacetoxyprogesterone, 6β-bromoprogesterone, and 2α-bromoprogesterone. The steroids were adminstered via steroid-impregnated silicone polymer gels implanted in the right cornua of the uterus. The left cornua served as an internal control. The clearest effects were seen in the case of 16α-bromoacetoxyprogesterone. This derivative produced resorption of fetuses, the effect being a local one affecting most strongly those fetuses closest to the implanted gel. Similar but statistically less clear-cut effects were observed with 11α-bromoacetoxyprogesterone and 6β-bromoprogesterone. The effect of 16α-bromoacetoxyprogesterone may be at the level of the uterine progesterone receptor protein. This hypothesized site of action is suggested by the observation that [$^3$H]progesterone uptake by uterine cytosol was diminished significantly by 16α-bromoacetoxyprogesterone treatment. Significantly, 16α-acetoxyprogesterone, a structurally similar nonreactive steroid, did not stimulate fetal resorption.

In more recent studies Sweet and others (*38, 39*) have synthesized reactive derivatives of the synthetic progestin, medroxyprogesterone acetate. The effects of 16α-bromoacetoxymedroxyprogesterone and medroxyprogesterone-17-bromoacetate on the pregnant rat were investigated. The former compound produced effects similar to those of 16α-bromoacetoxyprogesterone which was discussed previously. The latter compound did not stimulate resorption but, rather, exhibited a prolonged progestational activity in castrated pregnant animals.

EPOXYSTEROIDS. The epoxy group is a weak alkylating agent which has been incorporated into a number of protein ligands for affinity labeling (*40, 41*). Pollack and his co-workers, who have synthesized and tested a large number of epoxysteroids as potential site-specific reagents, recently have taken this approach. The target protein in these studies has been the $\Delta^5$-3-ketosteroid isomerase from the bacterium *Pseudomonas testosteroni* (E.C.5.3.3.1).

The structure and properties of this enzyme have been reviewed thoroughly by Talalay and his co-workers (*42*). Isomerase activities have been observed in all mammalian steroidogenic tissues. However, these mammalian isomerases have been difficult to study due to their membrane localization and instability during purification attempts. In contrast, the bacterial enzyme is a very stable water-soluble enzyme that can be obtained readily in 100-mg quantities in a pure state.

The isomerase catalyzes the allylic isomerization of $\Delta^5$-3-ketosteroids to their $\Delta^4$ isomers. This reaction is shown in Figure 6. A wide variety of substituents, R, are tolerated on C-17. The enzyme is inhibited competitively by a very large number of steroids including hormones such as testosterone, estradiol, and progesterone. Thus, it constitutes a useful test system for development and chemical studies of the reactions of many steroidal affinity reagents. The isomerase has a molecular weight of 27,000 and consists of two, probably identical, subunits each of mol wt 13,500. The complete primary structure of the polypeptide is known as a result of the work of Benson, Jarabak, and Talalay (*43*), and the *N*-terminal part of their structure has been largely confirmed independently by Ogez, Tivol, and Benisek (*44*), though some discrepancies in amide assignments still exist. The enzyme contains two independent steroid binding sites per dimer (*45, 46*) that have similar or identical affinities for steroids. Thus, binding of steroids to the isomerase can be characterized by a single dissociation constant. No evidence for the presence of dissociable or nondissociable cofactors of the enzyme has been reported.

The catalytic mechanism is thought to involve the concerted action of at least one general base (nucleophile) and one general acid functional group. The pH dependency of $V_{max}$ (*47*) and chemical modification data (some of which will be discussed in the section on photoactivated reagents) is suggestive of the possible importance of an aspartate carboxylate, a histidine imidazole, and an unidentified group with a $pK_a'$ of 9.3 (in the Michaelis complex) as participants in the transfer of a hydrogen from the $4\beta$ to the $6\beta$ position on the steroid. This hydrogen is thought to be transferred as a proton since there is partial loss of deuterium into the solvent proton pool when a $4\beta[^2H]$-$\Delta^5$-3-ketosteroid is the substrate (*48*). The generally accepted mechanism for the isomer-

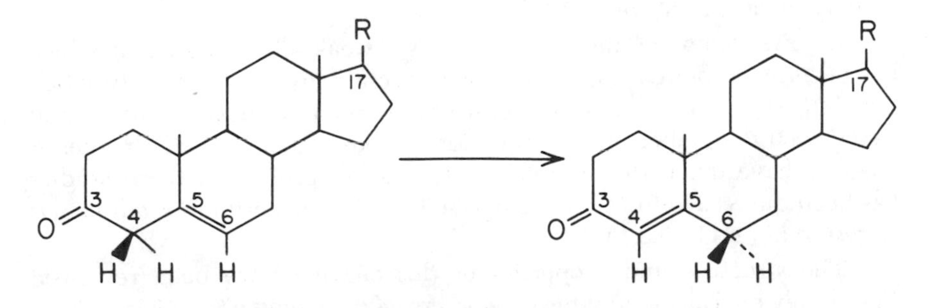


*Figure 6. Steroid isomerase reaction.*

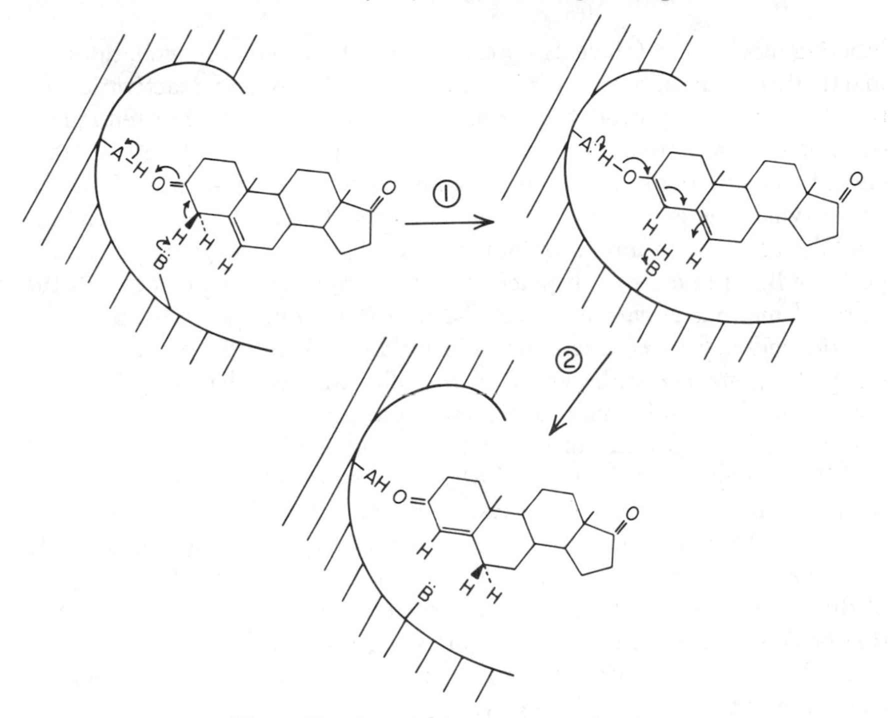

*Figure 7. Steroid isomerase mechanism.*

ase reaction is presented in Figure 7. In Step 1 of Figure 7 the enzyme's acidic group, AH, donates a proton to the substrates' 3-keto oxygen thus facilitating the removal of the 4$\beta$ proton by the enzyme's basic group, B. This generates a dieneol intermediate that goes to the $\Delta^4$ product by reprotonation from BH at the 6$\beta$ position and transfer of the dienolic proton back to A, thus regenerating the catalytic groups AH and B.

Pollack and his co-workers recently have reported (*49*) the effects of the $\alpha$ and $\beta$ isomers of spiro-3-oxiranyl-5$\alpha$-androstan-17$\beta$-ol on the activity of steroid isomerase. The results of the study are summarized in Figure 8 which shows the structures of the epoxy steroids. The 3$\alpha$ isomer, though a good competitive inhibitor of the enzyme, did not irreversibly inactivate it under the conditions used (0.033*M* phosphate buffer, pH 7.0, 21°C). In contrast, the 3$\beta$ isomer, also a good competitive inhibitor, inactivated the enzyme. Kitz–Wilson kinetics were observed from which a $K_R$ of 19$\mu M$ was calculated. This value was in good agreement with the value of $K_I$ determined from reversible inhibition of the enzymatic reaction by the 3$\beta$ epoxide. Furthermore, the nonreactive competitive inhibitor, 19-nortestosterone, protected the enzyme from 3$\beta$-epoxide inactivation. From Kitz–Wilson plots its $K_L$ was found to be 6.5$\mu M$, close to the kinetic $K_I$ of 9.1$\mu M$. These results very strongly support an active site labeling reaction by the 3$\beta$-epoxide which follows the true affinity labeling mechanism (since 3α-epoxide, which binds equally well, does not inactivate). The authors suggest that the mechanism of inactivation by the 3β-epoxide is a catalytic-site-promoted reaction that is essentially a diversion of the normal catalytic mechanism (*see* Figure 7) onto a dead-end irreversible path. In this hypothetical mechanism shown in Figure 9, the enzyme's general acid, AH, protonates the 3β-oxirane oxygen, making it a better leaving group (oxonium ion), so that some enzyme nucleophile perhaps B?) rapidly can displace it by nucleophilic displacement on the oxiranyl methylene carbon or on C-3 of the steroid. Knowledge of the pH dependency of $k_2$ might provide evidence bearing on the involvement of an enzyme acidic group in the alkylation mechanism. The supposition of the requirement for general acid catalysis by AH rests on the documented sluggishness of epoxide reactions with nucleophiles (*50*).

Were they available, it would be interesting to compare the crystal structures of the 3α- and 3β-epoxysteroids with the known (*51*) structure of the excellent isomerase substrate androst-5-ene-3,17-dione in order to see which isomer's epoxy oxygen lies closest in space to the 3-keto oxygen of the substrate when the steroid ring systems are superimposed. Inspection of Dreiding models by the authors suggests that the 3β-epoxy oxygen might be very close to the 3-keto oxygen's position, while the 3α-epoxy oxygen might be about 1.7 Å away.

In a subsequent study Bevins and co-workers (*52*) have synthesized several 17-epoxysteroids in both α and β isomeric forms. None of the α isomers inactivated the enzyme, including one which was an excellent competitive inhibitor of the enzyme. In contrast, all of the 17β-epoxy

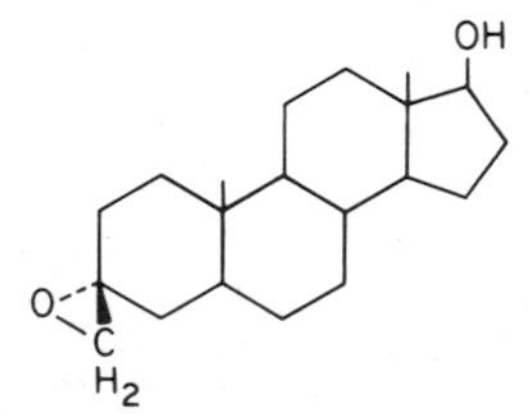


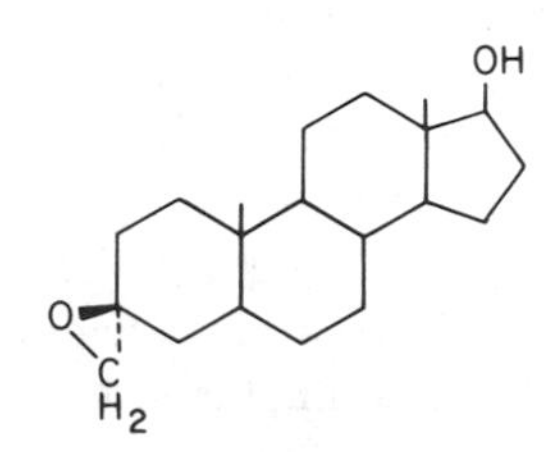


*Figure 8. The 3-epoxysteroids as isomerase affinity reagents.*

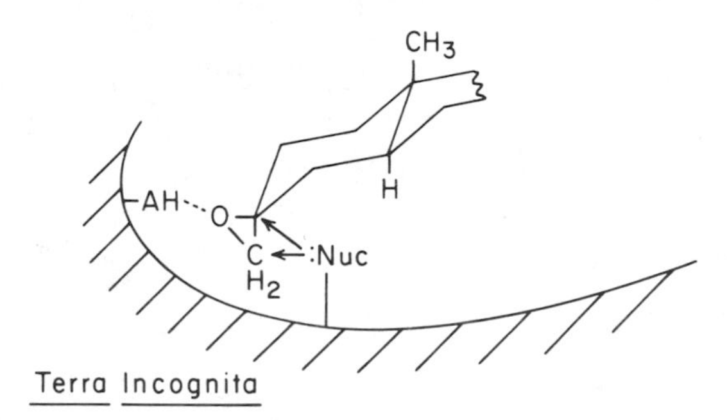


*Figure 9. Possible stereochemical mechanism for inactivation by 3β-epoxysteroid.*

steroids inactivated the enzyme, the kinetics obeying linear Kitz–Wilson plots, and all showing appropriate protection effects by 19-nortestosterone. The data are summarized in Figure 10. In view of the hypothesis of enzyme-mediated general acid catalysis for efficient inactivation by steroidal epoxides (*49*), it is remarkable that three of the four 17β-epoxysteroids exhibit limiting first-order rate constants, $k_2$ in Figure 10, which are substantially greater than that of the 3β-epoxysteroid for which the data from Figure 8 is reproduced in Figure 10. To account for this fact the authors make the interesting suggestion that 17β-epoxysteroids can bind in a backwards mode to the enzyme active site such that Ring D occupies the Ring A subsite and Ring C occupies the Ring B subsite. In such a mode, the 17β-oxygen might occupy the same locus as would the 3β-oxygen of a 3β-epoxysteroid binding in the frontwards mode. Data have not been presented yet by Pollack and co-workers concerning the stoichiometry of the presumed covalent attachment of epoxysteroids to isomerase. No doubt the stoichiometry and identification of the site(s) of alkylation are being pursued by these investigators. In connection with the backwards binding hypothesis, it will be very interesting to learn whether or not 3β- and 17β-epoxysteroids react with the same or different nucleophiles. Also, it will be interesting to learn if the methylene carbon or C-3 is the site of nucleophile attack since protonated epoxides often undergo nucleophilic attack at the more highly substituted carbon (*53*).

Bevins and co-workers (*52*) also have examined several other epoxy steroids: among them 4,5α; 4,5β; 5,6α; 5,6β; and 5,10β epoxides. However none of these behaved as irreversible inhibitors of the isomerase.

**Light-Generated Reagents.** NITRENES AND CARBENES. Steroidal nitrenes and carbenes, generated by photolysis of steroidal azides and diazo compounds, respectively, have been used by several groups in attempts to photoaffinity label various proteins, notably nonenzymatic steroid binding proteins. The particular properties of these highly reactive species which recommend them for site-specific labeling have been discussed earlier. Achievements using these reagents have been reviewed

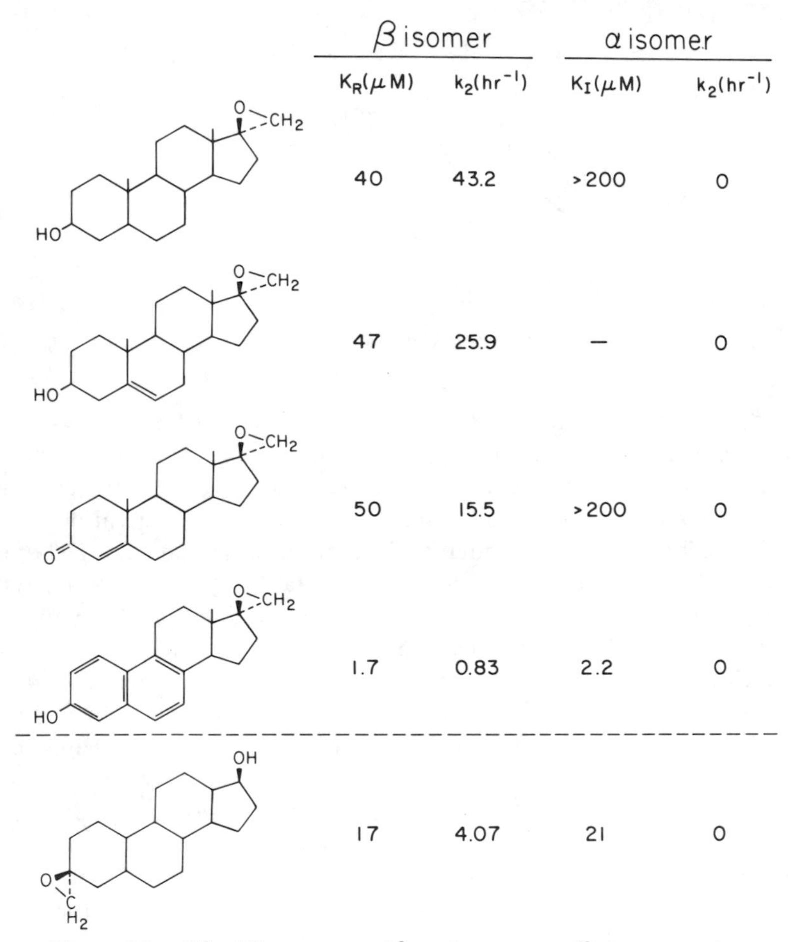


*Figure 10. The 17-epoxysteroids as isomerase affinity reagents.*

recently by Katzenellenbogen (*2*) and Chowdry and Westheimer (*54*). These excellent sources can be consulted for detailed and comprehensive coverage of applications. Only illustrative examples will be presented.

Katzenellenbogen and his co-workers have made a very systematic investigation of a large number of photoactivatable estrogen derivatives and hexestrol derivatives that are designed to label the estrogen receptor present in the uterine cytosol. This work has been reviewed thoroughly by Katzenellenbogen (*2*). Among the most successful photoreagents are derivatives of *meso*-hexestrol (Compound **II**) (dihydrodiethylstilbesterol), a compound that has potent estrogenic activity due to its ability to bind to the estrogen receptor.

Katzenellenbogen (*55*) has investigated hexestrol derivatives **III** and **IV**. Compound **III**, hexestrol diazoketopropyl ether, is a proreagent

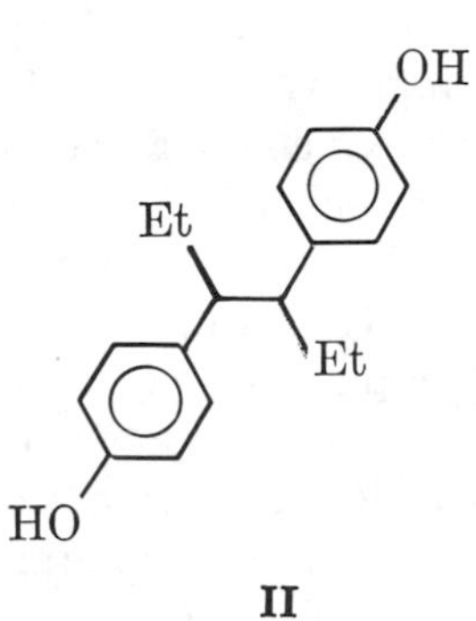


**II**

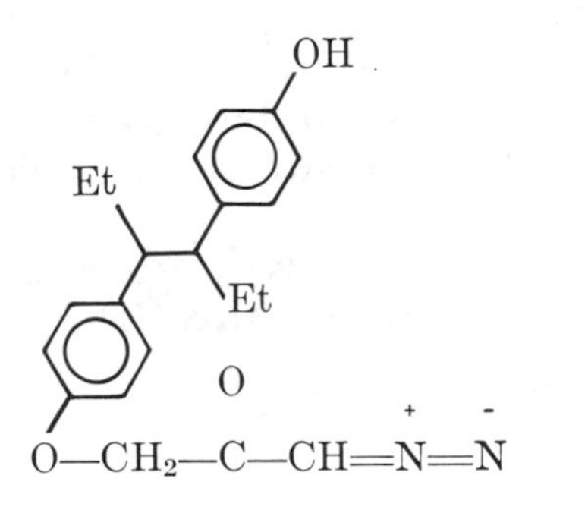

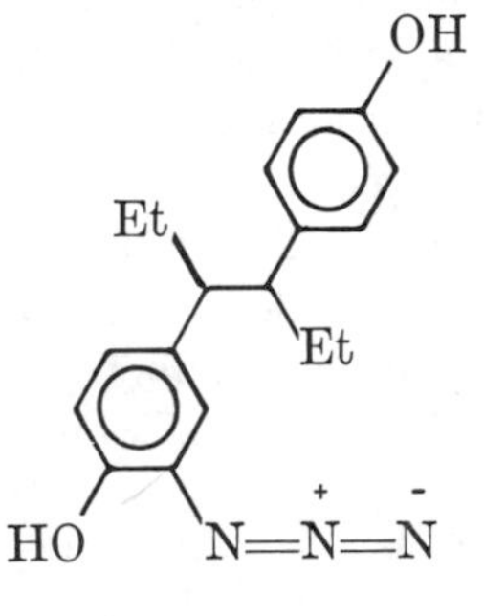


**IV**

that loses $N_2$ when irradiated at 254 nm and gives rise to a reactive carbene. Compound **IV,** hexestrol-3-azide, is a proreagent that yields an aromatic nitrene plus $N_2$ when photolyzed. Equilibrium binding studies showed that hexestrol, bound to a partially purified rat uterine receptor, has an association constant 18.5 times that of the natural ligand, 17$\beta$-estradiol. The binding constant for Compound **III** was 0.041 times $K_a$ (estradiol) and that for Compound **IV** was 4.1 times $K_a$ (estradiol). Roughly similar binding affinities towards partially purified lamb uterine receptor preparations were found.

Partially purified receptor preparations from rat and lamb uteri were saturated with (as a control) Compound **II, III,** or **IV** and then subjected to photolysis following the experimental procedures developed earlier (*56*) for use with these and a number of other photoactivatable proreagents. Following photolysis at 254 nm for a time period sufficient to photolyze all of the proreagent, the percentage of estradiol-binding sites inativated by the photolysis procedure was measured. This percentage is termed the inactivation efficiency. Additional photolyses using tritium-labeled proreagents were conducted which were designed to measure the amount of covalent coupling of Compounds **III** and **IV** to receptor protein. This was accomplished by radioactivity analysis of sliced SDS gel electropherograms. The results obtained are summarized in Table IV. Experiments in which the receptor was loaded with estradiol prior to photolysis in the presence of Compounds **III** and **IV** showed the estradiol-protected receptor from inactivation (*56*) and also from photoattachment to Compound **IV** (*55*), thus indicating binding-site specificity for the photoreactions. The most significant aspect of the data of Table IV is that, unlike alkylating agents, photoinactivation via carbenes does not result necessarily in covalent attachment of the photoreagent moiety even though the target site has undergone a covalent change. This is apparently the situation in the case of Compound **III.** On the other hand, the nitrene-generating reagent, Compound **IV,** inactivates largely if not totally via a covalent attachment reaction. The chemistries involved in these interesting reactions have not been elucidated yet due to the crudeness of the receptor preparations used and the extremely small quantities of receptor obtainable.

Human corticosteroid binding globulin (CBG) has been the target for photoaffinity labeling by the carbene precursor, [6,7-$^3$H]-21-diazo-21-deoxycorticosterone (Compound **V**)

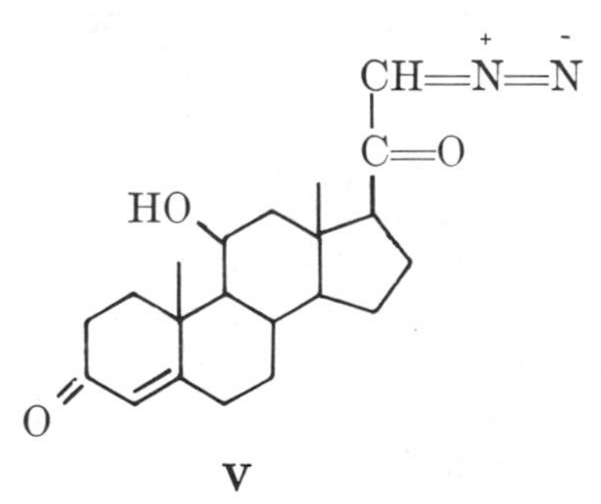


**V**

**Table IV. Photoinactivation and Photocovalent Attachment of Hexestrol Derivatives to Uterine Estrogen Receptor[a]**

| | *Rat* | | *Lamb* | |
|---|---|---|---|---|
| *Compound* | *Inactivation Efficiency (%)* | *Covalent Attachment* | *Inactivation Efficiency (%)* | *Covalent Attachment (as percent of sites labeled)* |
| **II** | 0 | — | 0 | 0 |
| **III** | 15 | — | 10 | 0 |
| **IV** | 15 | — | 20 | 15–20 |

[a] Data taken from Katzenellenbogen et al. (*55*).

by Marver et al. (57). These researchers found that purified CBG photolyzed by 254 nm light in the presence of Compound **V** incorporated tritium covalently, presumably by covalent coupling of the light-generated carbene (or some reactive species generated from the carbene by rearrangement) to the protein. The labeling reaction appeared to be site specific since the natural ligands, corticosterone and cortisol, protected the protein from labeling whereas aldosterone, which does not bind to CBG, did not afford significant protection. Further evidence for site specificity was provided by a labeling experiment conducted in the presence of the carbene scavenger, Tris. This substance did not decrease the labeling of CBG when it was included in the photolysis reaction mixtures. This indicates that carbene generated from unbound Compound **V** does not lead to labeled CBG since such carbenes would be scavenged by Tris. In addition, the failure of Tris to suppress the labeling reaction demonstrates that reactive species generated from bound Compound **V** that dissociate before reaction with CBG do not subsequently label CBG either specifically or nonspecifically. Further reports on the chemistry of the labeling reaction have not appeared yet.

A recent report from Katzenellenbogen's laboratory (58) reveals some of the complexities that can be encountered in photoaffinity labeling work. In this study two estrogen-derived photoreagents were evaluated as site-specific reagents using rat $\alpha$-fetoprotein as the target. This serum protein binds 17$\beta$-estradiol with high affinity. The reagents used were 16-diazoestrone, Compound **VI,** and 4-azidoestradiol, Compound **VII.** Both reagents bind reversibly to $\alpha$-fetoprotein.

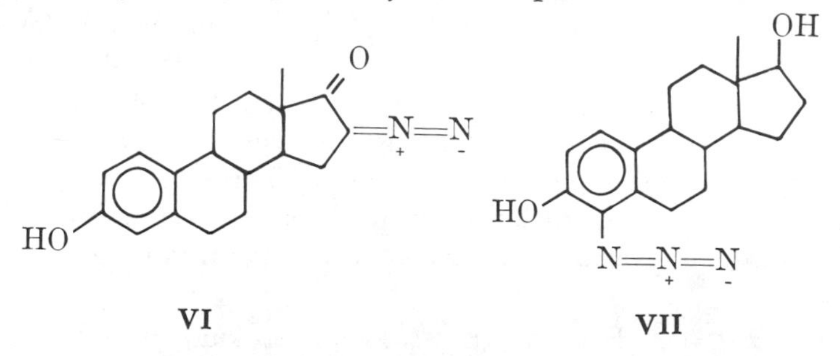


Irradiation of $\alpha$-fetoprotein in the presence of Compound **VII** at $>$ 315 nm resulted in a rapid loss of estradiol-binding sites. This process was inhibited strongly by inclusion of estradiol, implying site specificity for the inactivation reaction. However, when the amount of covalent attachment of Compound **VII** was investigated using [$^3$H]-**VII,** little reduction of covalent coupling occurred when estradiol was present during photolysis. The investigators suggest that most of the covalent coupling was of a nonspecific nature and that the process leading to site inactivation did not involve covalent attachment of the steroid.

Qualitatively different results were obtained with Compound **VI.** When $\alpha$-fetoprotein was irradiated at 300 nm or $>$ 315 nm in the presence of Compound **VI,** site inactivation was observed. The site was protected by estradiol. When the kinetics of reagent covalent attachment were measured, attachment correlated with inactivation kinetically and that attachment was suppressed by estradiol. This shows that the inactivation reaction involves attachment of the reagent, in contrast to the case of Compound **VII.** However, attachment continued long after all of Compound **VI** had been photolyzed. In addition, prephotolysis of Compound **VI** followed by addition of $\alpha$-fetoprotein and dark incubation resulted in attachment which was suppressed by estradiol. This indicated that photolysis of Compound **VI** generated a long-lived photoproduct that reacted with the protein like a classical affinity reagent. This photoproduct had a half-life in the buffer of 17 h. The photoproduct could be inactivated rapidly by 2-mercaptoethanol. Not all of the covalent attachment of Compound **VI** proceeded via the long-lived photoproduct, however. When $\alpha$-fetoprotein was irradiated in the presence of Compound **VI** plus 2-mercaptoethanol, attachment still occurred, though in a lesser amount than in the absence of the thiol. The kinetics of this attachment process closely followed the photolysis kinetics of Compound **VI,** indicating that it proceeded directly from Compound **VI** or a very short-lived photoproduct (such as the ketocarbene).

Thus, the photoinactivation of $\alpha$-fetoprotein by photolyzing Compound **VI** proceeds via two divergent mechanisms, a chromophore-dependent attachment reaction proceeding via a short-lived reactive species and a chromophore-independent attachment mechanism involving the intermediacy of a long-lived reactive species. The short-lived reactive species couples to the site faster than it can dissociate from it (mercaptoethanol does not scavenge it). The long-lived species may be generated either in free solution or at the target site, but since it is scavenged by mercaptoethanol, it must dissociate from the site before it couples to it. The chemical nature of the long-lived species is currently under investigation (*59*).

Electronically Excited States. Another kind of light-activated reagent that has been used increasingly widely in steroid-binding-site labeling experiments is one which, upon absorption of light, is raised to an excited state which, then, either reacts directly with residues in the binding site or decays back to the ground state. Unsaturated ketones are intended to be reagents of this type, though evidence for their reaction directly from the excited state rather than from a short-lived reactive species derived from the excited state has not been obtained yet. Operationally, this class of reagent can be distinguished from the carbene/nitrene proreagents by the kinetics of the site modifications they promote in the light. With carbene/nitrene generating reagents, the kinetics of site labeling parallels the kinetics of proreagent photolysis. With excited-state reagents, first-order inactivation reactions are expected (if the excited state reverts to proreagent ground state when it fails to react with another species such as the target protein). This is the situation with $\alpha,\beta$ unsaturated cyclohexenones for which the quantum efficiency for photorearrangement is 0.01 or less (*60*). Some of the photoreactions of ketone-excited states have been summarized by Benisek (*13*).

Most steroid hormones possess 4,5 unsaturated 3-keto groups in their structures (e.g. tesosterone, progesterone, and corticosteroids). Consequently, the underivatized hormones might function as photoaffinity reagents. This approach was tested first by Martyr and Benisek (*61, 62*) using the *P. testosteroni* $\Delta^5$-3-ketosteroid isomerase as the target. This early work showed that the enzyme was subjected to a $\Delta^4$-3-ketosteroid-dependent photoinactivation ($\lambda > 300$ nm) that satisfied criteria for active-site specificity in that nonchromophoric competitive inhibitors protected the enzyme activity, that the noninhibitor, cyclohexeneone, did not stimulate photoinactivation, and that the rates of photoinactivation produced by a series of $\Delta^4$-3-one steroids of varying affinity for the active site correlated with their binding affinities when the photoreactions were conducted at subsaturating concentrations of steroids. Comparing gross amino acid analyses of native enzyme and those of enzyme photoinactivated in the presence of the competitive inhibitor, 19-nortestosterone acetate, showed that destruction of a residue aspartic acid or asparagine correlated quantitatively with the fraction of sites inactivated. Later studies by Ogez and co-workers (*44*) identified the modified residue as aspartic acid-38 of the polypeptide chain, implicating a role for this residue in enzyme catalysis. The importance of this residue to isomerase function has been substantiated further by its specific amidation with various amines and a carbodiimide (*63*). A surprising discovery of Ogez and co-workers (*44*) was the identification of the photoproduct of aspartic acid-38 as alanine. Little covalent attachment of the steroid was obtained (*63*). Neither the mechanism of this unique photodecarboxylation nor the fate of the lost aspartate $\beta$-carboxylate has been elucidated.

Extension of the original approach of Benisek and his colleagues using monounsaturated eneones to other systems usually has not been successful. For example, Katzenellenbogen has reported (*2*) that preliminary attempts to obtain site-specific covalent photoattachment of several $\Delta^4$-3-one steroids to rat liver glucocorticoid receptor resulted in little labeling of protein when irradiation at either 254 nm or 315 nm was performed and even that was nonspecific. Recently Dure et al. (*70*) found that crude chick oviduct cytosol progesterone receptor was not labeled by progesterone when irradiated at $\lambda > 300$ nm. Whether or not binding-site inactivation occurred in either system was not reported. However, success was obtained with another steroid ketone (*see* below). Recently several laboratories have obtained encouraging results using ketonic steroids possessing more than one conjugated double bond.

During the period of time when the nature of the 19-nortestosterone acetate-dependent photoinactivation was under investigation, a new bacterial steroid isomerase was obtained from extracts of *Pseudomonas putida* (Biotype B) in nearly homogeneous form and some of its physical and enzymatic properties were characterized (*64, 65, 66*). The *putida* isomerase is similar in its molecular weight and quaternary structure to the *testosteroni* isomerase. Chemically, the most striking difference between the two isomerases is the presence of four residues of cysteine per polypeptide chain of the *putida* isomerase whereas no cysteine or cystine is present in the *testosteroni* isomerase. *N*-Terminal sequence analysis of the *putida* isomerase demonstrated substantial sequence homology between the two enzymes.

When attempts were made to specifically modify the active site of the *putida* isomerase using $\Delta^4$-3-ketosteroids as photoactivatable affinity reagents, severe enzyme photoinactivation occurred in the absence of steroid. This was in marked contrast to the *testosteroni* enzyme which was much more stable to the irradiation conditions. In order to subvert the inherent photoinstability of the *putida* isomerase, the intensity of the irradiating light was reduced greatly. In order to compensate for the reduced-light flux a photoactivatable steroid of higher absorbancy than a $\Delta^4$-3-one at $\lambda > 300$ nm was required. This requirement suggested to us the use of polyunsaturated 3-ketosteroids, which have much greater absorbancy than do monounsaturated 3-ketosteroids above 300 nm. Thus low-light levels could be used under which conditions the *putida* isomerase activity was stable for many minutes. When Compounds **VIII** and **IX** were tested as photoreagents we found that both supported rapid photoinactivation of the *putida* enzyme. Both compounds are competitive inhibitors of the enzyme.

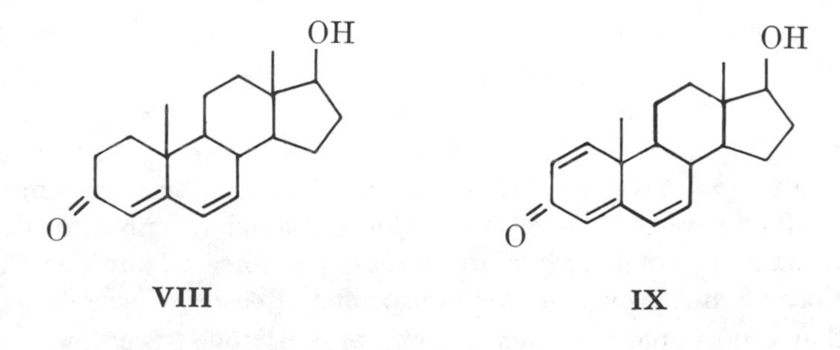


The bile acid, deoxycholic acid, Compound **X,** is also a good competitive inhibitor, whereas cholic acid, Compound **XI,** is not.

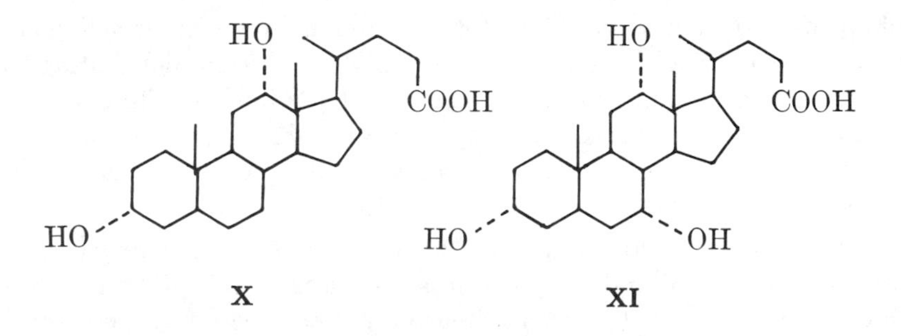


Accordingly, Compound **X** protected the enzyme against Compound **IX**-sensitized photoinactivation whereas Compound **XI** was without effect on the inactivation kinetics. Preliminary experiments utilizing 4-[$^{14}$C]-**IX** indicate that no covalent attachment of Compound **IX** accompanies photoinactivation. Amino acid composition studies and thiol group analyses demonstrate that the extent of Compound **IX**-sensitized photoinactivation correlates quantitatively with the loss of one of the four cysteine thiols. Statistically significant changes in other amino acids were not observed. The chemical nature of the thiol-destroying reaction is still unknown, but it is under active investigation. The results summarized above were reported in preliminary form in 1978 (*67*) and have been described more fully by Smith and Benisek (*68*).

Other groups independently have used polyunsaturated ketosteroids as photoaffinity reagents directed towards mammalian steroid-binding proteins. Katzenellenbogen attempted to obtain site-specific photoinactivation and photoattachment of 6-oxoestradiol, Compound **XII,** to uterine estrogen receptor

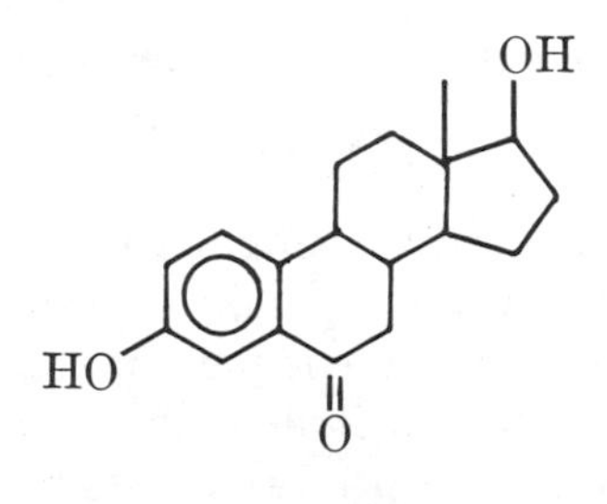


(*2, 56*) with irradiation at 315 nm. A rapid first-order loss of binding sites occurred ($\tau_{1/2} \approx 7$ min) when crude rat uterine estrogen receptor was irradiated in the presence of 144n*M* Compound **XII.** Site specificity for this inactivation process was suggested by the excellent protective effect of 30n*M* estradiol. However, when the amount of covalent attachment of Compound **XII** was measured using [$^{3}$H]-**XII,** much less was photoattached than could account for the site inactivation. Thus, some unknown-site-localized reaction not resulting in reagent attachment is involved here. Such a reaction may prove useful in delineating the structure–function relationships of the estrogenic binding site once sufficient quantities of pure receptor become available, but it is useless for most studies involving crude systems in which the purpose of the labeling is to specifically attach a radioactive tag to the receptor.

Danzo and his co-workers (*72*) have investigated extensively the testicular androgen-binding protein in rats and rabbits. This protein, which is distinct from the intracellular androgen receptor, is secreted by the Sertoli cells of the seminiferous tubules and is transported to the epididymis via the testicular fluid. The physiological role of this protein is not understood. One suggestion has been that it transports androgens from their production sites in the testes to the androgen target tissue, the epididymis. This target tissue requires a continuous supply of androgen in order to maintain its structural and functional integrity. Essentially pure ($> 98\%$) rat epididymal androgen-binding protein has a molecular weight of 85,000. On SDS gel electrophoresis two bands migrating with apparent molecular weights of 41,000 and 45,000 and present in a mass ratio of 1:3 are observed (*71*). Thus, androgen-binding protein may be a mixed-hybrid dimer. The protein binds androgens with high affinity with $K_d$ for 5α-dihydrotestosterone being $5.8 \times 10^{-9}M$ (*71*).

Danzo and co-workers (*72*), in equilibrium binding studies using crude epididymal cytosol, have shown that 5α-dihydrotestosterone and $\Delta^6$-testosterone (Compound **VIII**) bind to the same high-affinity site and that $\Delta^6$-testosterone binds with as nearly high affinity ($K_d$, the dissociation constant of the $\Delta^6$-testosterone complex, equals $1.15 \times 10^{-8}M$) as 5α-dihydrotestosterone ($K_d = 6.1 \times 10^{-9}M$) in good agreement with the value for $K_d$ determined by Musto and co-workers (*71*) for 5α-dihydrotestosterone binding to pure androgen-binding protein. Thus, $\Delta^6$-tesosterone was a potential photoactivatable site-specific reagent for androgen-binding protein. Another important reason for selecting $\Delta^6$-testosterone rather than testosterone or dihydrotestosterone was the favorable absorption spectrum of the dieneone chromophore. This has substantial absorbance at wavelengths longer than the absorbance of crude cytosol, whereas the absorption spectra of testosterone and dihydrotestosterone are masked by the endogenous cytosolic component's spectra. Recently Taylor and colleagues (*69*) have found that Compound **VIII** does, indeed, photolabel site specifically androgen-binding protein. Irradiation of epididymal cytosol by light of $\lambda > 300$ nm in the presence of Compound **VIII** resulted in inactivation of androgen-binding sites. Inactivation was suppressed by the presence of 5α-dihydrotestosterone in the reaction mixture. SDS gel electrophoretic analysis of cytosol inactivated in the presence of [$^3$H]-**VIII** showed that covalent labeling occurred predominantly on a polypeptide migrating with an apparent molecular weight of 47,000. Inclusion of 5α-dihydrotestosterone in the photolysis reaction mixture blocked the labeling of the 47,000-dalton species. In more recent experiments using gel electrophoresis techniques of greater resolving power, Taylor and co-workers (*73*) have found that the labeled 47,000-dalton species seen on lower-resolution gels was resolved into a 3.3:1 mixture of a 48,000-dalton species and a 43,000-dalton species. Both species were labeled with tritium and labeling of both was suppressed by 5α-dihydrotestosterone. The strong similarity of the SDS-gel-labeling pattern of the crude cytosol preparation and the subunit molecular weights and relative proportions of pure androgen-binding protein subunits and the fact that 5α-dihydrotestosterone suppresses the labeling provide strong evidence that the androgen-binding protein is the major cytosolic species labeled. This result attests to a considerable labeling selectivity by $\Delta^6$-testosterone plus light since the androgen-binding protein is a very minor protein constituent of epididymal cytosol. Danzo and his colleagues plan to use this labeling technique to help elucidate the physiological role and mechanism of action of androgen-binding protein in the epididymis.

Conceptually similar photoaffinity labeling experiments have been reported by Schrader's laboratory quite recently (*70*). These workers have sought to label the cytosolic progesterone receptor of chick oviduct using the polyunsaturated synthetic progestin, R5020, Compound **XIII.**

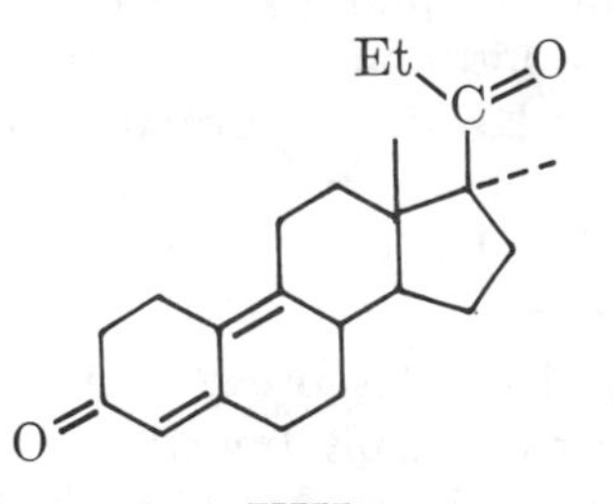


**XIII**

This steroid exhibits a high affinity for the receptor protein which exceeds that of progesterone by several fold (*74*). The oviduct progesterone receptor has been the subject intensive physiochemical investigation in spite of the minute quantities that are available in pure form. This protein exists as a noncovalent complex of two nonidentical subunits, A and B, having molecular weights of 79,000 and 115,000, respectively. Both subunits possess progesterone-binding sites. Efforts to affinity label this protein in the past with other steroidal reagents have not been successful. Dure and co-workers (*70*) using [$^3$H]-**XIII** and crude oviduct cytosol performed irradiation at $\lambda > 300$ nm. Incorporation of tritium into cytosolic proteins was assayed by SDS gel electrophoresis and radioactivity analyses of the sliced gels. Three major radioactive bands were observed to accumulate during the course of irradiation. These had electrophoretic mobilities corresponding to apparent molecular weights of 39,000, 78,000, and 106,000. The two larger species were labeled more heavily and rapidly than the 39,000-dalton species. Progesterone added to the reaction mixtures prior to photolysis protected the 78,000- and 106,000-dalton species but not the 39,000 mol wt material. According to the authors, the efficiency of labeling of the larger species was about 10% and both were labeled at similar rates. On the basis of the above data, it was suggested that the species undergoing labeling were the A and B subunits of the progesterone receptor. Additional evidence for this conclusion came from the chromatographic properties of the photolabeled 78,000 and 106,000 species on DEAE–cellulose which were found to be similar to those of unlabeled A and B subunits. The nature of the labeled 39,000-dalton species is unknown; it may be derived from a major 39,000 species known to be present in the crude cytosol. It should be pointed out that the receptor subunits are present at levels that are too low for them to be detected visually on Coomassie Blue-stained electrophoresis gels of cytosol, whereas a prominent 39,000-dalton band is apparent. In subsequent unpublished studies the photolabeled receptor subunits have been purified and peptide mapping and sequence studies are in progress in order to locate the R5020 site (*75*).

Following a very different photochemical strategy, Katzenellenbogen and colleagues (*76*) have introduced a new type of photoreactive chromophore for photoaffinity labeling. Cornelisse and Havinga (*77*) have observed that *m*-nitrophenyl ethers undergo a light dependent nucleophilic displacement of the alkoxy substituent by aliphatic amines. The reaction proceeds via a very short-lived excited state ($\tau_{1/2} = 10^{-7}$–$10^{-9}$ s). Reactions with other nucleophiles present in proteins (e.g. thiol, thioether, and indole) also can occur, but the reaction products are not known. The *m*-nitrophenyl ethers would appear to be much more selective in their reactivity than are carbenes and nitrenes. Jelenc and coworkers (*78*) have used this type of photoreagent to cross-link fetal hemoglobin with a heterobifunctional reagent.

The photoreagent used by Katzenellenbogen (*76*) is Compound **XIV**, an analog of hexestrol, but lacking a complete hexane backbone. Irradiation of lamb uterine cytosol at $\lambda > 315$ nm in the presence of Compound **XIV** resulted in the loss of 67% of the estrogen receptor binding sites. Most of this photodestruction was abolished by inclusion of 30n*M* estradiol in the photoreaction mixture, indicating that the inactivation process is site located. Whether site-specific covalent coupling takes place in this interesting reaction is not determined yet. Also, the

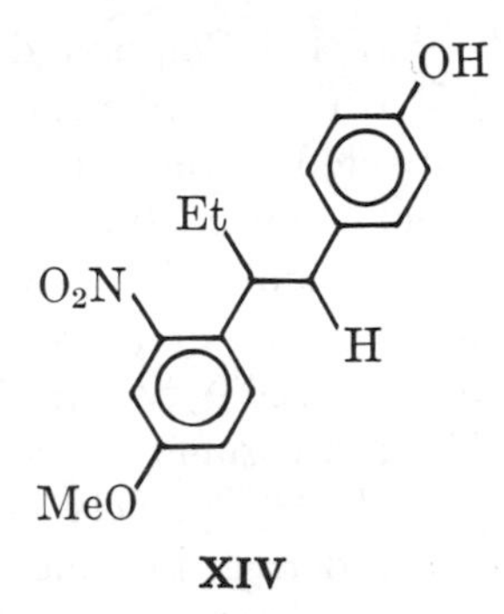


**XIV**

amount of nonspecific photoreaction is unknown. These questions will be addressable when radiolabeled Compound **XIV** is obtained.

**Enzyme-Generated Reagents: $k_{cat}$ Reagents.** An increasing number of steroidal $k_{cat}$ reagents have been inflicted on various steroid-metabolizing enzymes. All of these reagents have been designed as substrates of their target enzymes which give rise to reactive intermediates or products that can undergo coupling reactions with protein nucleophiles. If it is demonstrated that Pollack's epoxysteroids (*49, 52*) react with steroid isomerase via enzyme-mediated general acid catalysis (as is postulated to occur in the regular enzymatic reaction) then these reagents also should be regarded as $k_{cat}$ types.

The first steroidal $k_{cat}$ reagent was that of Batzold and Robinson (*79, 80*) who synthesized the 5,10-secosteroids **XV** and **XVI** as reagents directed towards *P. testosteroni* steroid isomerase.

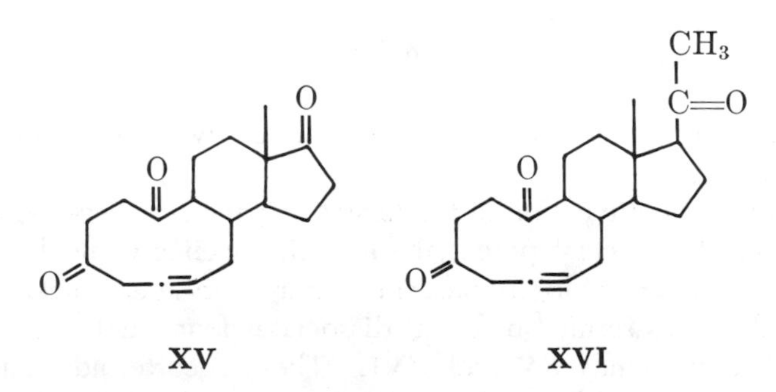


As isomerase substrates, **XV** and **XVI** would be enzymatically isomerized to $\Delta^{4,5}$-allenic ketones. Since allenic ketones react with nucleophiles by addition across C=C, it was hoped that such a process would occur at the isomerase catalytic site. The desired reaction sequence is given in Figure 11. A crystallographic determination of the structures of Compound **XV**, its allenic isomer, and the analogous normal substrate androst-5-ene-3,17-dione has been conducted by Carrell and co-workers

(*51*). This study demonstrated that Compound **XV** and androst-5-ene-3,17-dione possess marked structural similarities, particularly in the C and D rings, as would be expected. Even the ten-membered ring of the secosteroid has a similarity of conformation to the A and B rings of androst-5-ene-3,17-dione. For example, the C-10 carbonyl oxygen of the secosteroid occupies the same position in space as does the 19-methyl of the normal substrate. The distances between C-4 and C-6 of the two compounds are similar. There are some differences though. The C-3 carbonyl oxygen of Compound **XV** projects almost axially above the mean plane of the ten-membered ring, but the C-3 carbonyl oxygen of androst-5-ene-3,17-dione radiates equatorially and lies slightly below the ring plane. Also the enzymatically transferred 4$\beta$ proton of androst-5-ene-3,17-dione is axial and lies above the ring plane whereas the 4$\beta$ proton of the secosteroid is closer to equatorial and lies only slightly above the corresponding plane. In spite of these structural differences between the molecules in the crystalline state, Compound **XV** definitely serves as a substrate of the isomerase and is converted rapidly by it to Compound **XVII** (*81*).

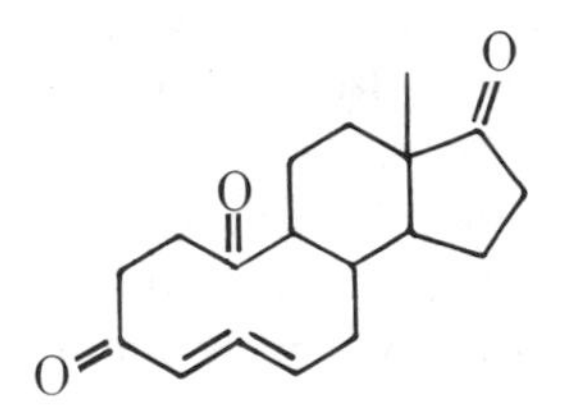


**XVII**

Compound **XVII** is then available for reaction with the catalytic site which created it.

Although catalytic action of the target enzyme is a necessary step in $k_{cat}$ inhibition, the special potential for high specificity by this type of reagent is realized only if the nascent reactive product reacts with the site at which it was formed before it dissociates from that site. Is this a property of Compounds **XV** and **XVI**? These secosteroids inactivated isomerase according to first-order kinetics and the dependence of inactivation rate on reagent concentration obeyed Kitz–Wilson kinetics (*79*). The parameters $K_R$ and $k_2$ (Equation 9) for Compounds **XV** and **XVI** are given in Figure 12. In a subsequent report Covey and Robinson (*81*) reported the syntheses of the four allenic secosteroids derived from Compounds **XV** and **XVI** by isomerizations catalyzed by triethylamine. All four compounds inactivated the isomerase. Analysis of the products of enzymatic isomerization showed that Compounds **XVII** and **XIX** (*see*

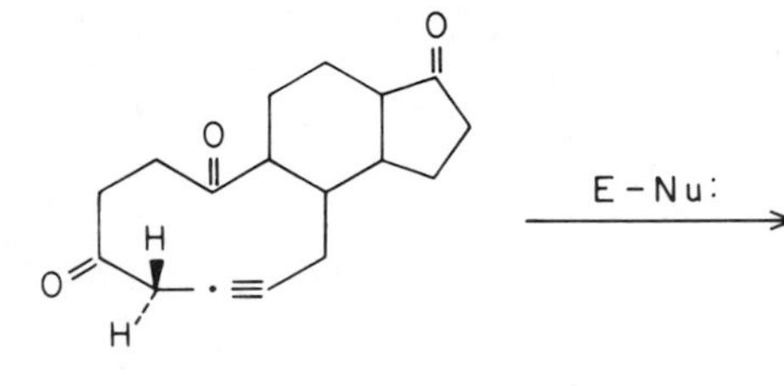


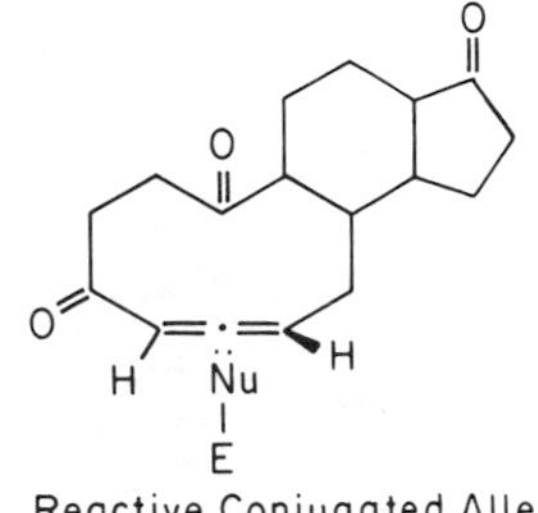


Non Reactive

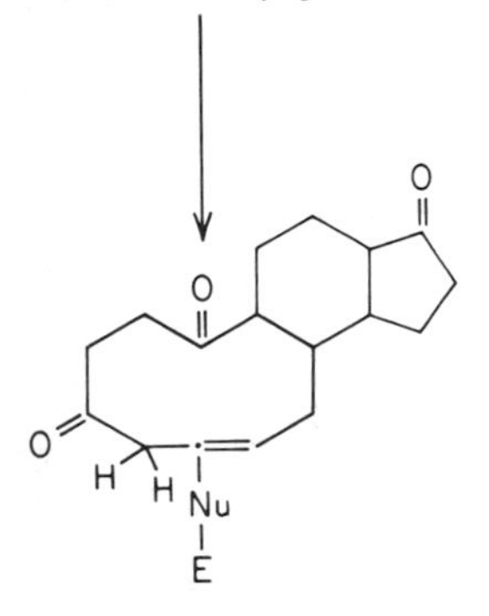


*Figure 11. Acetylenic secosteroids as* $k_{cat}$ *inhibitors of* P. testosteroni *steroid isomerase.*

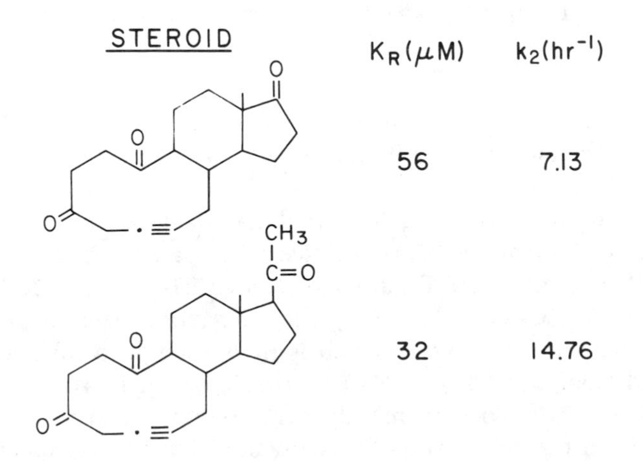


*Figure 12. Inactivation kinetics of acetylenic secosteroids.*

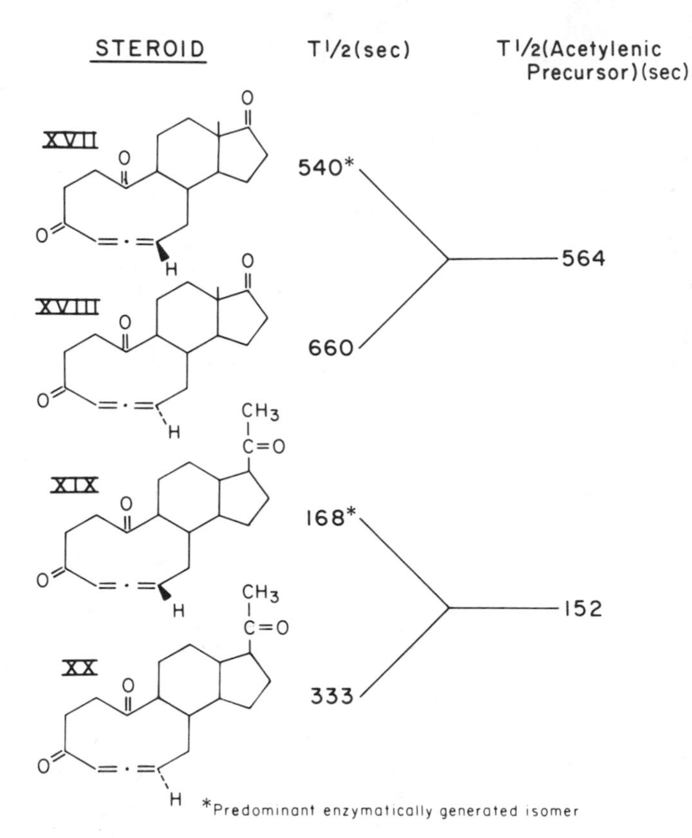


*Figure 13. Inactivation kinetics of allenic secosteroids: [all secosteroids] = 200μM.*

Figure 13) were the predominant product isomers, as would be predicted from the known $4\beta \rightarrow 6\beta$ isomerization stereochemistry for normal $\Delta^5$-3-one substrates.

When the inactivation half-lives determined at $200\mu M$ steroid concentration of Compounds **XVII–XX** were compared with those for Compounds **XV** and **XVI** (*see* Figure 13), $\tau_{1/2}$ (**XVII**) $\approx$ $\tau_{1/2}$ (**XV**) and $\tau_{1/2}$ (**XIX**) $\approx$ $\tau_{1/2}$ (**XVI**). Moreover, the enzyme concentration used in these incubations was $4.8\mu M$, which is high enough to instantly convert any added Compound **XV** or **XVI** to allenic product. We conclude that enzyme inactivation occurs mainly after release of the allenic steroid products from the active site. Consequently, the special specificity that characterizes true $k_{cat}$ inhibitors (which covalently react before product release) is not expected for Compound **XV** or **XVI.** Nonetheless, specificity typical of classical affinity reagents might be obtained, and this can be quite high in favorable cases. In the present instance Penning and colleagues (*46*) have shown using 7-[$^3$H]-**XV** that inactivation correlates well with the covalent coupling of only one **XV** moiety per enzyme subunit. Since Covey and Robinson (*81*) reported that 19-nortestosterone significantly slows the inactivation induced by Compound **XVII,** one may presume that the covalent coupling of Compound **XV** is at the steroid-binding site.

The chemistry of the coupling reaction of the allenic secosteroids to isomerase has been under investigation by Penning and Talalay (*82, 83, 84*) who report that the site of attachment is in the tetrapeptide sequence 55–58. Identification of the actual site has been hampered by a lability of the bound steroid to both low and high pH.

In the meantime, Covey and colleagues (*85*) have identified the products of model reactions between various nucleophiles and Compound **XVII.** They obtained two types of product, depending on the nature of the nucleophile. Adducts of the type of Compound **XXI** were formed when nucleophile = imidazole, thiol, phenol, and carboxylate; adducts of the structural type **XXII** were obtained when nucleophile = pyrrolidine, water, or methanol.

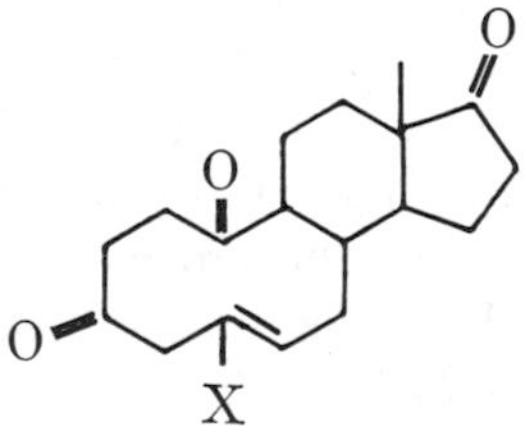

**XXI** 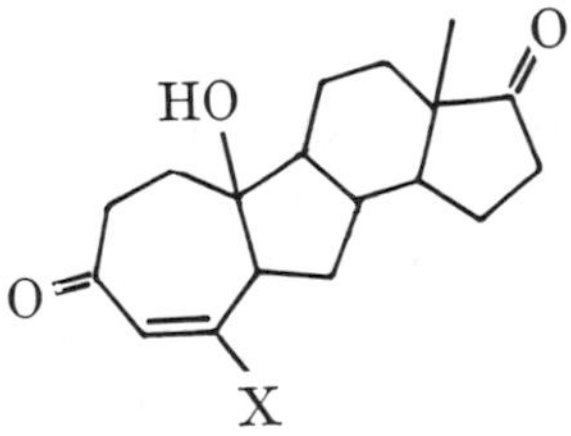 **XXII**

Robinson's group has identified some in vivo effects of acetylenic and allenic secosteroids. Thus, Compounds **XV, XVI,** and **XIX,** when administered to male rats intraperitoneally (*86*) produced 25–42% reductions in the weight of the ventral and dorsal lateral prostate. It was suggested that these compounds acted by interfering with a step in androgen biosynthesis, perhaps the rat steroid isomerase. Other studies (*87, 88*) by this group have shown that the allenic secosteroids **XVII** and **XIX** are potent noncompetitive, irreversible inhibitors of rat epididymal steroid $\Delta^4$-5$\alpha$-reductase and that they exhibit antiandrogenic activity when evaluated by the hamster flank organ test.

From a therapeutic and pharmacologic standpoint the choice of target protein is at least as important as the choice of reagent. A particularly appropriate target for the blocking action of an affinity reagent is the enzyme estrogen synthetase. This enzyme, unique to the estrogen biosynthetic pathway, catalyzes the conversion of androgens such as testosterone and androst-4-ene-3,17-dione to estrogens. Selective, irreversible inhibition of this enzyme activity could be a highly efficient way of blocking estrogen synthesis without affecting the pathways for the other steroid hormones. The resulting reduction in circulating estrogen levels could be helpful in a nonsurgical (e.g. ovariectomy, adrenalectomy, and hypophysectomy) chemotherapy of estrogen-responsive tumors of the breast. Approximately 30% of human female breast tumors are estrogen responsive in the sense that they require estrogens for growth (*89*).

Covey and others recently have reported the development of potential $k_{cat}$ affinity reagents designed to inhibit estrogen synthetase (*90, 91*). The rationale behind Covey's approach is based on what is known about the mechanism of the synthetase reaction. Figure 14 (*left side*) shows some of the steps of the mechanism. The substrate, androst-4-ene-3,17-dione, is subjected to two hydroxylations on C-19 forming Compound **XXIV,** and then Compound **XXV,** which dehydrates to give Compound **XXVI,** which, in a subsequent multistep oxidation, loses the aldehydic carbon (as formate) and Ring A becomes aromatized. The 19-hydroxy derivatives such as Compound **XIV** are substrates of the enzyme in addition to Compound **XIII** and testosterone. Covey has suggested that the 19-acetylenic and 19-hydroxy-19-acetylenic analogues of Compounds **XXIII** and **XXIV** might serve as substrates of the enzyme. Following the normal reaction sequence (*see* Figure 14, *right side*) these would be converted to the intermediate Compound **XXVI** A. This compound then might participate in a Michael addition reaction with a nucleophilic group on the synthetase inactivating it. Covey and co-workers (*90*) synthesized both 19-epimers of Compound **XXIV** A. In subsequent tests with synthetase from placental microsomes the S isomer inactivated the synthetase in an NADPH-dependent reaction. Omission of NADPH did not result in inactivation, showing that S-**XXIV** A itself is not the reactive species. Significantly, Compound R-**XXIV** A, though it was a competitive inhibitor of the enzyme, did not irreversibly inactivate it. More recently Covey has evaluated the nonhydroxylated substrate analogue Compound **XXIII** A as an estrogen-synthetase inactivator and has found (*92*) that it, too, behaves as a $k_{cat}$ inhibitor.

A chemically similar approach to $k_{cat}$ inhibition has been used by Strickler and others (*93*). These workers have used 17$\beta$-[(1S)-1-hydroxy-2-propnyl]androst-4-en-3-one, Compound **XVIII,** as an inactivator of the S. *hydrogenans* 3$\alpha$,20$\beta$ steroid dehydrogenase. This derivative is a substrate of the enzyme, and is converted by it to Compound **XIX,** the actual reactive species, which presumably inactivates by an addition reaction across the triple bond.

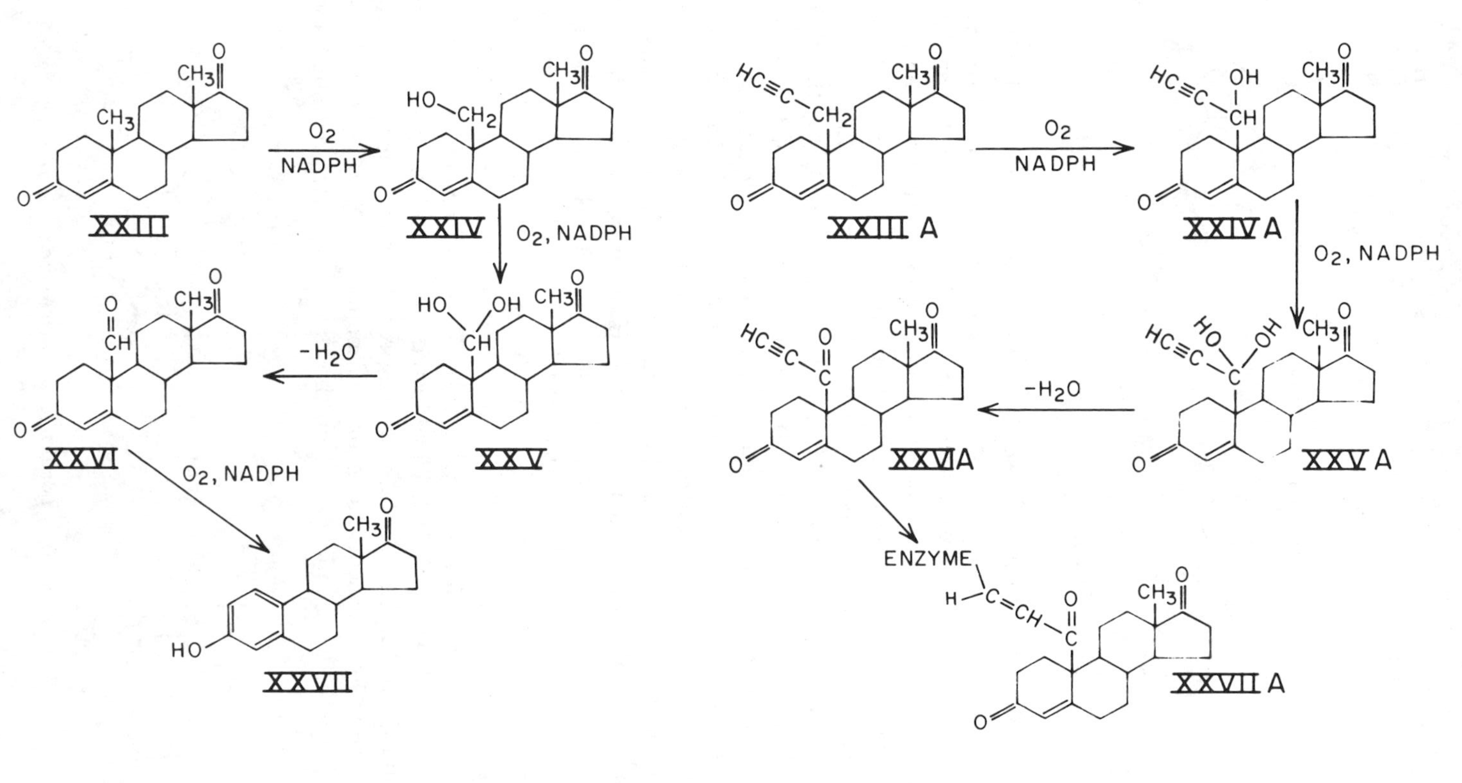


*Figure 14. Compounds* **XXIII** *and* **XXIIIA** *through* **XXIIIA** *and* **XXVIIA**, *respectively*

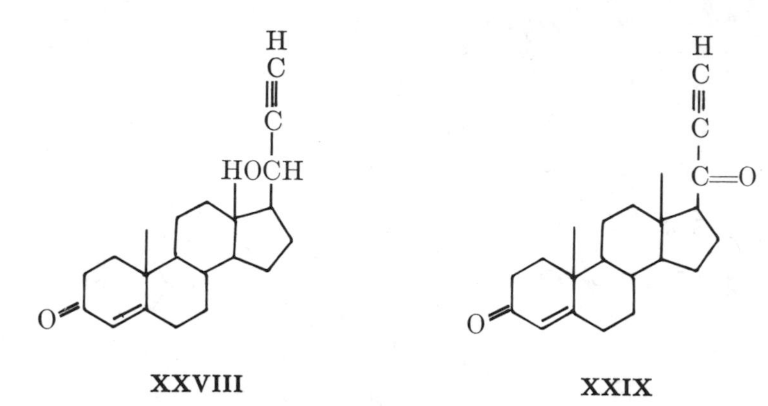


Another physiologically strategic target in steroid biochemistry is the $\Delta^4$-5$\alpha$-steroid reductase present in many androgen target tissues. Inhibition of this enzyme would have effects selective for androgen-dependent processes in animals since for many target tissues the most potent androgen is 5$\alpha$-dihydrotestosterone rather than testosterone. The reaction catalyzed by the 5$\alpha$-reductase is shown in Figure 15. In essence, testosterone is reduced by adding $H_{(b)}$ as a hydride donated by the NADPH and $H_{(a)}$ as a proton donated by some acidic group of the reductase. Blohm and colleagues (*94*) have proposed the diazo derivative, Compound **XXX,** as a $k_{cat}$ inhibitor of the reductase. Its structure and intended mechanism of reductase inactivation are presented in Figure 16. Compound **XXX** is designed to be analogous to the intermediate in the reduction that is obtained when NADPH donates a hydride to the substrate's C-5. When the reductase donates a proton to C-4 of Compound **XXX,** as it would in the normal reaction, an aliphatic diazonium ion, Compound **XXXI,** is created. The $N_2$ group is an excellent leaving group in nucleophilic displacement reactions. Compound **XXXI** is postulated to react with some nucleophile, *X*, on the enzyme, thereby inactivating it. Blohm et al. (*94*) have tested Compound **XXX** for its ability to inactivate the 5$\alpha$-reductase of rat prostate microsomes. They found that very low concentrations of Compound **XXX** ($1 \times 10^{-8}M$ $-3 \times 10^{-7}M$) affected a rapid ($\tau_{1/2}$ s on the order of tens of minutes) loss of

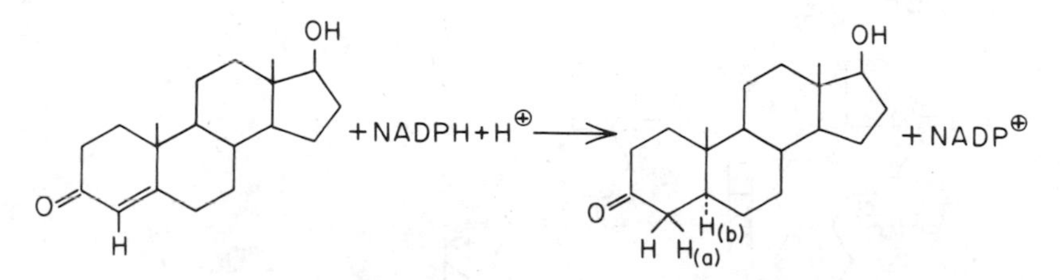


*Figure 15. Testosterone 5α-reductase reaction.*

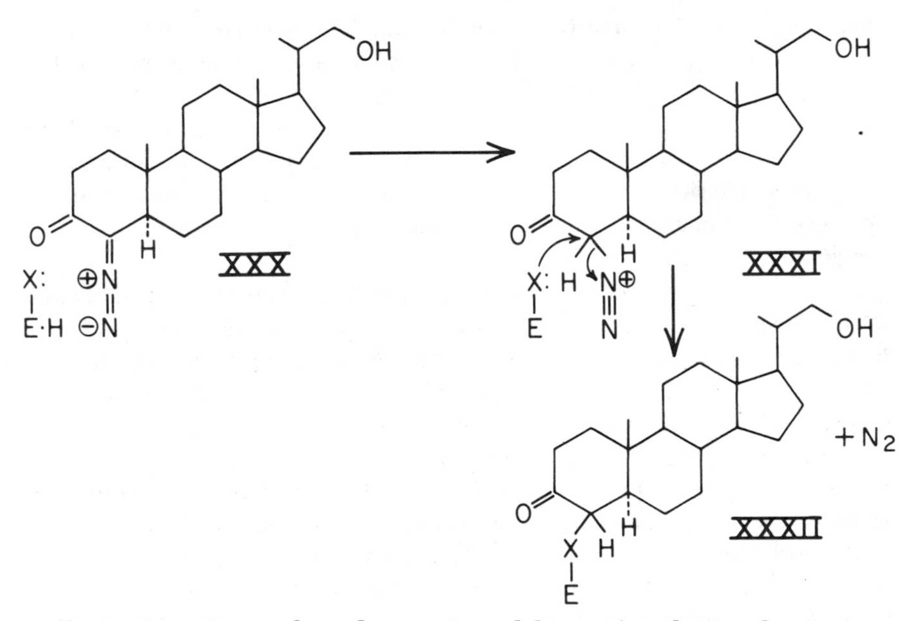


*Figure 16. Proposed mechanism for inhibiting 5α-reductase by Compound XXX.*

reductase activity which followed pseudo-first-order kinetics, though instability of the enzyme interfered with the precision of the data. Nonetheless, replots of the kinetics on a reciprocal plot demonstrated that Kitz–Wilson kinetics were followed, implying that Compound **XXX** forms a complex with the reductase. From the Kitz–Wilson plot a $K_R$ of $3.5 \times 10^{-8}M$ was estimated for Compound **XXX.** Testosterone concentrations greater than $1.25 \times 10^{-6}M$ significantly reduced the rate of inactivation by Compound **XXX.** Interestingly, inactivation by Compound **XXX** was obtained in the intact rat as well, though a rather high dose (100 mg/kg) was used. Unfortunately, the degree of irreversibility of the inactivation by Compound **XXX** is difficult to address due to the instability of the enzyme preparation.

### Conclusions and Future Prospects

In reviewing the field of affinity labeling with an eye to recognizing some unifying principles that can guide workers in developing affinity-reagent-class drugs, one is struck by the strong bias of the literature toward reagents that have been successful. Little published information is available about reagents that have failed to achieve their intended specificity or have been unreactive toward their target proteins. Because of this bias one cannot base guidelines for future efforts on empirical experience in a meaningful way. All approaches to site-specific labeling look more promising than they actually are. The best that one can do at present is to apply to drug design those theoretical concepts of the affinity-labeling approach that significantly affect reagent reactivity and specificity. Published work can help to direct one's attention to those reagent types that have been successful in other systems; but since each application has its unique requirements and potential pitfalls, prior experience is no guarantee that a candidate reagent will hit its target with the requisite specificity.

In in vivo applications, specificity is probably the greatest challenge to the protein chemist. Thus, efforts to maximize reagent specificity would seem to be the direction in which future research should proceed. The surest route to maximize specificity is to use the lowest feasible concentration of reagent and to select those reagents that have the highest value for $k_2/K_R$. In the case of enzyme targets, particularly low $K_R$ values are expected for ligands that mimic the structure of the transition state of the enzyme-catalyzed reaction. For multisubstrate enzymes a multisubstrate inhibitor can bind with high affinity. Thus, reagents based on the structures of transition states and multisubstrate inhibitors should prove to be more specific than reagents based on the structures of single substrates or conventional competitive inhibitors. Certainly, more worthwhile effort will be spent in developing $k_{cat}$ inhibitors for steroid-metabolizing enzymes, since this strategy offers the potential for particularly high specificity. Only the ingenuity of protein chemists limits this approach. The recent report of Blohm et al. (*94*) constitutes a partcularly notable example of a combination of these approaches in which a reagent is designed to be both a transition-state analogue and a $k_{cat}$ proreagent.

Similar strategies for the enhancement of specificity would not seem to be available to those workers attempting to develop reagents directed towards nonenzymatic binding proteins. Though photoaffinity reagents have proved useful in in vitro systems, the application of photoaffinity reagents in vivo would be limited necessarily to those tissues to which light has access (e.g. skin and internal sites that can be reached by fiber optical devices).

Even so, the binding activity of nonenzymatic targets might be exploited in new ways. If enough information were known about the structural determinants of ligand affinity for a binding site, it might be possible to design a ligand that can undergo a reaction (it need not have any physiological counterpart) for which the transition-state structure is optimized more nearly to fit the target-binding site than that of the ligand itself. This reaction will be catalyzed necessarily by the binding protein (*95*). Consequently, the opportunity for developing a $k_{cat}$ proreagent for the binding protein would exist if such a reacting ligand were in hand. The authors are not aware of any attempt to exploit this idea, but note that a few examples of catalytic activities that have been associated with nonenzymatic protein-binding sites have been reported recently (*96, 97*).

*Acknowledgment*

The authors wish to extend their thanks to the investigators who generously made available to them unpublished findings of significance to the subject matter of this review prior to their publication. In addition, we especially wish to thank Bryce V. Plapp for very helpful guidance in the derivations of the kinetic equations. We also wish to acknowledge the National Institute of Arthritis and Metabolic Disease for continuing research support (AM-14729).

*Literature Cited*

1. Warren, J. C.; Arias, F.; Sweet, F. *Methods Enzymol.* **1975,** *36,* 374.
2. Katzenellenbogen, J. A. In "Biochemical Actions of Hormones"; Litwack, G., Ed.; Academic: New York, 1977; Vol. 4, p. 1.
3. Wofsy, L.; Metzger, H.; Singer, S. J. *Biochemistry* **1962,** *1,* 1031.
4. Baker, B. R.; Lee, W. W.; Tong, E.; Ross, L. O. *J. Am. Chem. Soc.* **1961,** *83,* 3713.
5. Knowles, J. R. *Acc. Chem. Res.* **1972,** *5,* 155.
6. Bayley, H.; Knowles, J. R. *Methods Enzymol.* **1977,** *46,* 69.
7. Rando, R. R. *Science* **1974,** *185,* 320.
8. Singh, A.; Thorton, E. A.; Westheimer, F. H. *J. Biol. Chem.* **1962,** *237,* PC 3006.
9. Reiser, A.; Willets, F. W.; Terry, G. C.; Williams, V.; Marley, R. *J. Chem. Soc., Faraday Trans.* **1968,** *64,* 3265.
10. Cornelisse, J.; DeGurst, G. P.; Havinga, E. *Adv. Phys. Org. Chem.* **1975,** *11,* 225.
11. Turro, N, J.; Schuster, G. *Science* **1975,** *187,* 303.
12. March, J. "Advanced Organic Chemistry: Reaction, Mechanisms, and Structure"; McGraw-Hill: New York, 1968; p. 164.
13. Benisek, W. F. *Methods Enzymol.* **1977,** *46,* 469.
14. Kitz, R.; Wilson, I. B. *J. Biol. Chem.* **1962,** *237,* 3245.
15. Meloche, H. P. *Biochemistry* **1967,** *6,* 2273.
16. Baker, B. R. "Design of Active-Site-Directed Irreversible Enzyme Inhibitors"; John Wiley & Sons: New York, 1967.
17. Farney, D. E.; Gold, A. M. *J. Am. Chem. Soc.* **1963,** *85,* 997.
18. Fersht, A. "Enzyme Structure and Mechanism"; Freeman: San Francisco, 1977, p. 130.
19. Segal, I. H. "Enzyme Kinetics"; John Wiley & Sons: New York, 1975.
20. Cornish-Bowden, A. "Fundamentals of Enzyme Kinetics"; Butterworths: London, 1979.
21. Eisenthal, R.; Cornish-Bowden, A. *Biochem. J.* **1974,** *139,* 715.
22. Pons, M.; Nicolas, J-C.; Boussioux, A.; Descomps, B.; Crastes de Paulet, A. *Eur. J. Biochem.* **1976,** *68,* 385.
23. Schoellmann, G.; Shaw, E. *Biochemistry* **1963,** *2,* 252.
24. Naider, F.; Becker, J. M.; Wilchek, M. *Isr. J. Chem.* **1974,** *12,* 441.
25. Chin, C. C.; Warren, J. C. *Biochemistry* **1972,** *11,* 2720.
26. Edwards, C. A. F.; Orr, J. C. *Biochemistry* **1978,** *17,* 4370.

27. Sweet, F.; Samant, B. R. *Biochemistry* **1980,** *19,* 978.
28. Szymanski, E. S.; Furfine, C. S. *J. Biol. Chem.* **1977,** *252,* 205.
29. Fersht, A. "Enzyme Structure and Mechanism"; Freeman: San Francisco, 1977; p. 99.
30. Blomquist, C. H. *Arch. Biochem. Biophys.* **1973,** *159,* 590.
31. Ganguly, M.; Warren, J. C. *J. Biol. Chem.* **1971,** *246,* 3646.
32. Gurd, F. R. N. *Methods Enzymol.* **1967,** *11,* 532.
33. Sweet, F.; Arias, F.; Warren, J. C. *J. Biol. Chem.* **1972,** *247,* 3424.
34. Ibid., **1973,** *248,* 5641.
35. Strickler, R. C.; Sweet, F.; Warren, J. C. *J. Biol. Chem.* **1975,** *250,* 7656.
36. Ibid., **1978,** *253,* 1385.
37. Clark, S. W.; Sweet, F.; Warren, J. C. *Biol. Reprod.* **1974,** *11,* 519.
38. Sweet, F.; Judd, R. M.; Samant, B. R., personal communication, 1980.
39. Samant, B. R.; Sweet, F. *J. Med. Chem.* **1977,** *20,* 833.
40. Kenyon, G. L.; Hegeman, G. D. *Methods Enzymol.* **1977,** *46,* 541.
41. O'Connell, E. L.; Rose, I. A. *Methods Enzymol.* **1977,** *46,* 381.
42. Batzold, F. H.; Benson, A. M.; Covey, D. F.; Robinson, C. H.; Talalay, P. *Adv. Enzyme Regul.* Weber, G., Ed.; **1976,** *14,* 243.
43. Benson, A. M.; Jarabak, R.; Talalay, P. *J. Biol. Chem.* **1971,** *246,* 7514.
44. Ogez, J. R.; Tivol, W. F.; Benisek, W. F. *J. Biol. Chem.* **1977,** *252,* 6151.
45. Ogez, J. R.; Benisek, W. F. *Biochem. Biophys. Res. Commun.* **1978,** *85,* 1082.
46. Penning, T. M.; Westbrook, E. M.; Talalay, P. *Eur. J. Biochem.* **1980,** *105,* 461.
47. Weintraub, H.; Alfsen, A.; Baulieu, E. E. *Eur. J. Biochem.* **1970,** *12,* 217.
48. Viger, A.; Marquet, A. *Biochim. Biophys. Acta* **1977,** *485,* 482.
49. Pollack, R. M.; Kayser, R. H.; Bevins, C. L. *Biochem. Biophys. Res. Commun.* **1979,** *91,* 783.
50. Cook, C. E.; Corley, R. C.; Wall, M. W. *J. Org. Chem.* **1968,** *33,* 2789.
51. Carrell, H. L.; Glusker, J. P.; Covey, D. F.; Batzold, F. H.; Robinson, C. H. *J. Am. Chem. Soc.* **1978,** *100,* 4282.
52. Bevins, C. L.; Kayser, R. H.; Pollack, R. M.; Ekiko, D. B.; Sadoff, S. *Biochem. Biophys. Res. Commun.* **1980,** *95,* 1131.
53. Addy, J. K.; Parker, R. E. *J. Chem. Soc.* **1963,** 915.
54. Chowdry, V.; Westheimer, F. H. *Annu. Rev. Biochem.* **1979,** *48,* 293.
55. Katzenellenbogen, J. A.; Carlson, K. E.; Johnson, H. J.; Myers, H. N. *Biochemistry* **1977,** *16,* 1970.
56. Katzenellenbogen, J. A.; Johnson, H. J.; Carlson, K. E.; Myers, H. N. *Biochemistry* **1974,** *13,* 2986.
57. Marver, D.; Chin, W. H.; Wolff, M. E.; Edelman, I. S. *Proc. Natl. Acad. Sci., U.S.A.* **1976,** *73,* 4462.
58. Payne, D. W.; Katzenellenbogen, J. A.; Carlson, K. E. *J. Biol. Chem.* **1980,** *255,* 10359.
59. Katzenellenbogen, J. A., personal communication, 1980.
60. Schuster, D. I.; Brown, R. H.; Resnick, B. M. *J. Am. Chem. Soc.* **1978,** *100,* 4504.
61. Martyr, R. J.; Benisek, W. F. *Biochemistry* **1973,** *12,* 2172.
62. Martyr, R. J.; Benisek, W. F. *J. Biol. Chem.* **1975,** *250,* 1218.
63. Benisek, W. F.; Ogez, J. R.; Smith, S. B. *Ann. N.Y. Acad. Sci.* **1980,** *346,* 115.
64. Richards, J. W.; Smith, S. B. *Fed. Proc. Fed. Am. Soc. Exp. Biol.* **1978,** *37,* 1342.
65. Smith, S. B.; Richards, J. W.; Benisek, W. F. *J. Biol. Chem.* **1980,** *255,* 2678.
66. Ibid., 2685.
67. Smith, S. B.; Benisek, W. F. *Fed. Proc. Fed. Am. Soc. Exp. Biol.* **1978,** *37,* 1805.

68. Smith, S. B.; Benisek, W. F. *J. Biol. Chem.* **1980**, *255*, 2690.
69. Taylor, C. A.; Smith, H. E.; Danzo, F. J. *Proc. Natl. Acad. Sci. U.S.A.* **1980**, *77*, 234.
70. Dure, L. S.; Schrader, W. T.; O'Malley, B. *Nature* **1980**, *283*, 784.
71. Musto, N. A.; Gunsalus, G. L.; Bardin, C. W. *Biochemistry* **1980**, *19*, 2853.
72. Danzo, B. J.; Taylor, C. A.; Schmidt, W. N., submitted for publication.
73. Taylor, C. A.; Smith, H. A.; Danzo, B. J., submitted for publication.
74. Philbert, D.; Ragnoud, J. P. *Steroids* **1973**, *22*, 89.
75. Schrader, W. T., personal communication, 1980.
76. Katzenellenbogen, J. A.; Kilbourn, M. R.; Carlson, K. E. *Ann. N.Y. Acad. Sci.* **1980**, *346*, 18.
77. Cornelisse, J.; Havinga, E. *Chem. Rev.* **1975**, *75*, 353.
78. Jelenc, P. C.; Cantor, C. R.; Simon, S. R. *Proc. Natl. Acad. Sci. U.S.A.* **1978**, *75*, 3564.
79. Batzold, F. H.; Robinson, C. H. *J. Am. Chem. Soc.* **1975**, *97*, 2576.
80. Batzold, F. H.; Robinson, C. H. *J. Org. Chem.* **1976**, *41*, 313.
81. Covey, D. F.; Robinson, C. H. *J. Am. Chem. Soc.* **1976**, *98*, 5038.
82. Penning, T. M.; Covey, D. F.; Talalay, P. *Fed. Proc. Fed. Am. Soc. Exp. Biol.* **1978**, *37*, 1297.
83. Penning, T. M.; Talalay, P. *Fed. Proc. Fed. Am. Soc. Exp. Biol.* **1979**, *38*, 512.
84. Ibid., **1980**, *39*, 2000.
85. Covey, D. F.; Albert, K. A.; Robinson, C. H. *J. Chem. Soc., Chem. Commun.* **1979**, 795.
86. Batzold, F. H.; Covey, D. F.; Robinson, C. H. *Cancer Treat. Rep.* **1977**, *61*, 255.
87. Robaire, B.; Covey, D. F.; Robinson, C. H.; Ewing, L. L. *J. Steroid Biochem.* **1977**, *8*, 307.
88. Voigt, W.; Castro, A.; Covey, D. F.; Robinson, C. H. *Acta Endocrinol.* **1978**, *87*, 668.
89. McGuire, W. L.; Chamness, G. C.; Costlow, M. E.; Horwitz, K. B. In "Hormone-Receptor Interaction: Molecular Aspects"; Levey, G. S., Ed.; Dekker: 1976; 265.
90. Covey, D. F.; Parikh, V. D.; Chien, W. W. *Tetrahedron Lett.* **1979**, 2105.
91. Covey, D. F.; Hood, W. F.; Parikh, V. D.; Chien, W. W. "Abstracts of Papers," *62nd Annu. Meet., Endocrine Society, 1980;* p. 254.
92. Covey, D. F.; Hood, W. F.; Parikh, V. D. *J. Biol. Chem.* **1981**, *256*, 1076.
93. Strickler, R. C.; Covey, D. F.; Tobias, B. *Biochemistry* **1980**, *19*, 4950.
94. Blohm, T. R.; Metcalf, B. W.; Laughlin, M. E.; Sjoerdsma, A.; Schatzman, G. L. *Biochem. Biophys. Res. Commun.* **1980**, *95*, 273.
95. Wolfenden, R. *Acc. Chem. Res.* **1972**, *5*, 10.
96. Kohen, F.; Kim, J. B.; Lindner, H. R.; Eshhar, Z. *FEBS Lett.* **1980**, *111*, 427.
97. Kohen, F.; Kim, J. B.; Barnard, G.; Lindner, H. *Biochim. Biophys. Acta* **1980**, *629*, 328.

RECEIVED November 17, 1980.

# 11

# Characterization of a Major Drug Binding Site in Human Serum Albumin

GARY E. MEANS, NICHOLAS P. SOLLENNE,
and ABUAGLA MOHAMED

Department of Biochemistry, The Ohio State University,
484 West 12th Avenue, Columbus, OH 43210

*The interactions of many important drugs, metabolites, and structurally related compounds with HSA have been characterized on the basis of their inhibition of its acetylation by* p-*nitrophenyl acetate. Dissociation constants of* $10^{-5}$M *or lower have been determined for tryptophan, phenylpyruvate, several pharmacologically important benzodiazepines, several types of small, apolar, anionic drugs and related compounds that interact specifically with the inhibitory site. Two other classes of inhibitory compounds have been identified and similarly characterized, the members of which interact either to a comparable extent or preferentially with one other site. The inhibitory site appears to be elongated, apolar, and has a monoanion binding site near one end.*

Albumin, the most abundant protein in human serum, has been studied widely (i.e., *see* Refs. *1, 2, 3, 4, 5*) but its physiological functions are not clearly known. Its strong binding of many different apolar compounds, however, suggests that it may play an important role in the solubilization and transport of many relatively insoluble compounds in the blood stream. Thus blood often contains substances that are quite insoluble in water but are soluble in serum as a result of their binding to human serum albumin (HSA).

Large fatty acid anions, particularly oleate and palmitate, are normally present in human serum at concentrations greatly exceeding their solubility in water (*6, 7, 8*). With HSA being normally at about 0.6m*M*, the total serum content of nonesterified fatty acids usually ranges from about the same level to a normal high of about 2m*M* or to 4m*M*

0065-2393/82/0198-0325$05.50/0

during severe lipemia (*7, 8, 9*). Throughout this range most of the fatty acid is bound to HSA. Under physiological conditions HSA usually is associated with one or more equivalents of fatty acid.

Bilirubin is another physiologically important compound that interacts strongly with HSA and would not be soluble in serum otherwise. Bilirubin normally is transported by the blood stream to the liver where it is metabolized but may accumulate in the blood when that metabolism is disrupted as often happens during certain types of liver disease (*10, 11, 12*). HSA has only one high-affinity binding site for bilirubin, however, and excessive amounts of bilirubin also may be deposited in body tissues. At high bilirubin concentrations irreversible brain damage sometimes results, particularly in newborn children, due to its deposition in their central nervous systems (*13, 14, 15*). Binding to HSA thus facilitates the transport of bilirubin and affords important protection from its adverse effects on the brain.

Other normal and abnormal serum constituents also interact strongly with HSA. Many drugs, for example, are not very soluble in water and would not be soluble in serum were it not for their binding to HSA. Again, however, the blood stream is more than just a transport system and binding to HSA has several important consequences. Thus the blood stream also serves as a central pool from which many drugs must equilibrate with other tissues and body fluids (*see* Figure 1). Binding

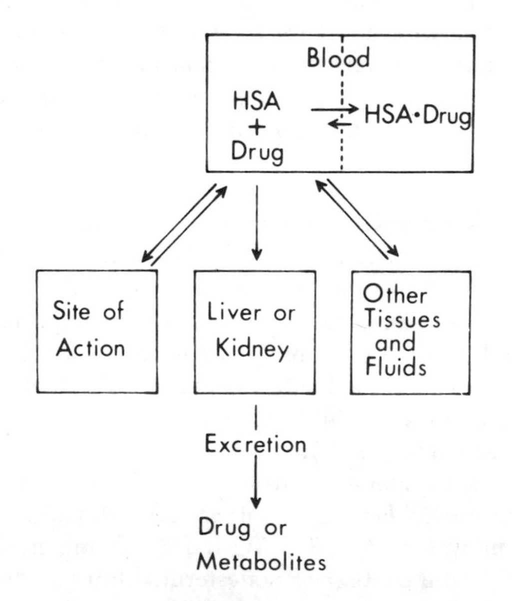


*Figure 1. Schematic illustrating the equilibration of a free drug in blood with body tissues and organs.*

to HSA competes with binding to tissue and organ binding sites (i.e., receptors, transport systems, catabolic enzymes, etc.); strong binding presumably necessitates larger dosages to achieve a desired therapeutic level of free drug than would otherwise be the case. Since only unbound drugs are able to pass freely from one body compartment to another, binding also tends to slow transport between the different body parts (e.g., *see* Refs. *16, 17, 18, 19*).

Larger dosages can be administered usually to compensate for the amount of a drug expected to bind to HSA but may entail serious problems. In other cases, however, important advantages may result. Thus, for many drugs, strong binding to HSA allows for both a large pool of readily available drug and a much lower therapeutically desired level of the free drug. Strong interactions with HSA thus tend to moderate the effects of such a drug and to prolong its action.

Because the blood normally contains a variety of substances that interact with HSA, competition for a limited number of sites sometimes may result or one ligand may influence interactions with another (e.g., *see* Refs. *20–25*). The variable sensitivities of patients to some drugs and different sensitivities of an individual patient to a drug at different times may result from differences in the kind or amount of other ligands bound to HSA. Abnormal drug effects during uremia, jaundice, and other disease states, and upon concurrent administration of more than one drug, sometimes may result from similar competitive interactions with HSA (i.e. *see* Refs. *22, 26, 27,* and *28*).

To help characterize the interactions of some drugs and metabolites with HSA and their competition for its limited number of binding sites, we and others have studied the influence of those substrates on its acetylation by *p*-nitrophenyl acetate (*29, 30, 31, 32, 33*). This reaction is very fast and results in the irreversible acetylation of a particular tyrosine residue (i.e. #411) which is located in a major apolar binding site of that protein (*34*). Because this reaction is so fast under most conditions, competing reactions are not significant and formation of *p*-nitrophenolate is a convenient reflection of the reaction rate (*35*).

As illustrated in Equation 1a, the reaction involves prior rapidly reversible binding of *p*-nitrophenyl acetate which, as indicated by Equation 1b, is competitive with some other ligands known to interact strongly with HSA.

$$+ \text{NphOAc} \overset{K_s}{\rightleftarrows} [\text{HSA} \cdot \text{NphOAc}] \overset{k_2}{\rightarrow} \text{AcHSA} + \text{NphO}^- \quad (1a)$$

HSA

$$+ \text{Ligand} \overset{K_d}{\rightleftarrows} [\text{HSA} \cdot \text{Ligand}] \quad (1b)$$

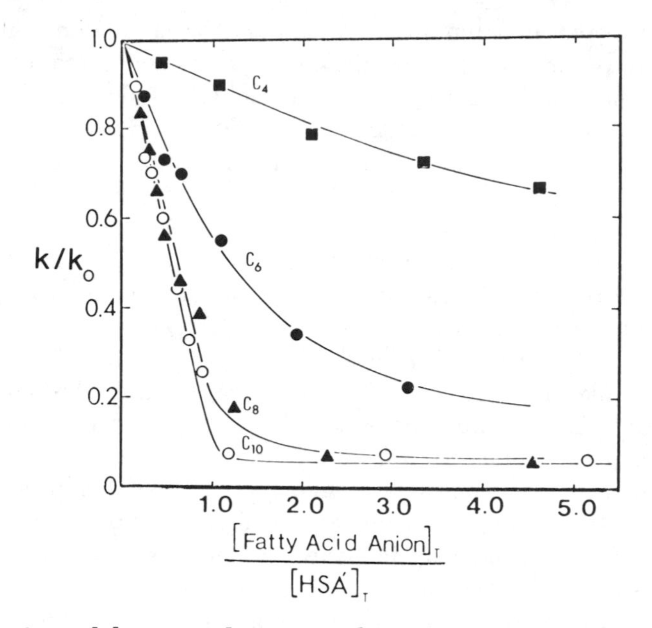


*Figure 2. Inhibition of the reaction between* p-*nitrophenyl acetate and HSA by carboxylate ions. Relative reactivities are shown as a function of the mole ratio of decanoate (○), octanoate (▲), hexanoate (●), and butyrate (■) as determined at 25°C in pH 7.5, 0.02*M *triethanolamine/HCl (29).*

Small fatty acids, for example, bind strongly to HSA and, as shown in Figure 2, strongly inhibit its reaction with *p*-nitrophenyl acetate (*29*). In each case, and as is particularly obvious with decanoate ion, complete inhibition appears to require a single carboxylate ion, each, however, having a different affinity in proportion to its hydrophobicity.

Such data can be converted into a more convenient linear form from which dissociation constants and binding-site concentrations can be obtained if plotted according to the relationship

$$\frac{[\text{Ligand}]_{\text{total}}}{1 - \dfrac{k - k_b}{k_0 - k_b}} = \frac{K_d}{\dfrac{k - k_b}{k_0 - k_b}} + [\text{HSA}]_{\text{total}} \tag{2}$$

where $k_o$, $k$, and $k_b$ are rate constants for the reaction with *p*-nitrophenyl acetate in the absence and presence of an inhibitory ligand, and in the presence of a large excess of an inhibitory ligand, respectively.

According to such plots the intercept should correspond to the concentration of binding sites, and the slopes should give the respective dissociation constants. Data obtained with three common drugs are plotted in this form in Figure 3. Note that although they appear to be

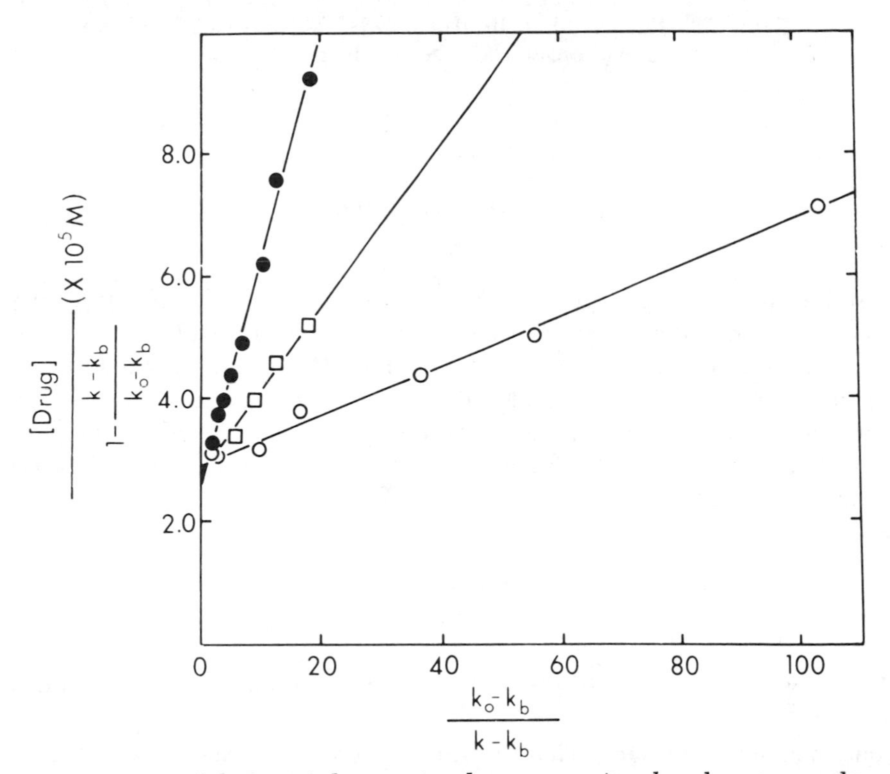


*Figure 3. Inhibition of the reaction between* p-nitrophenyl acetate *and HSA by drugs. Data for oxazepam (●), flufenamic acid (□), and naproxen (○) obtained at pH 7.4 in 0.03*M *triethanolamine/HCl at 25°C with 3.5 × 10*$^{-5}$M *HSA. Dissociation constants are given in Table II.*

structurally dissimilar (*see* Scheme I), each interacts strongly and, as indicated by their common intercept, selectively with the inhibitory site.

However HSA has several binding sites and some drugs interact with those other sites. For those that interact only with other sites no information can be obtained by this method. Table I lists five widely

*Scheme I.*

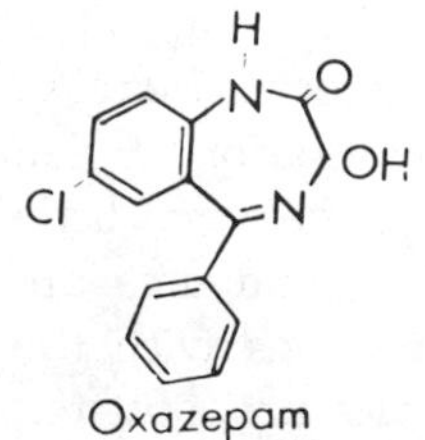


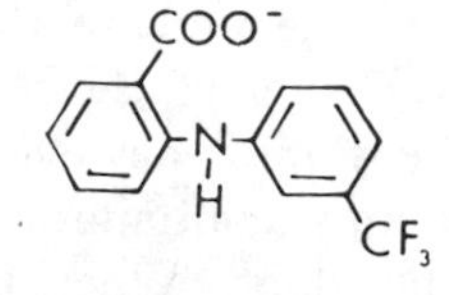


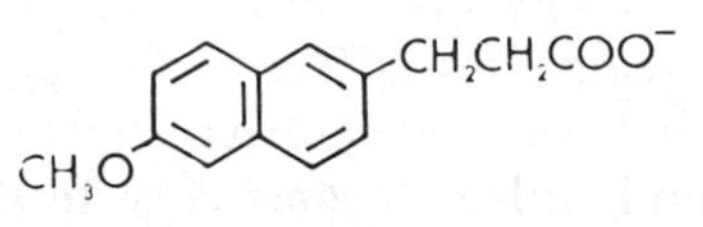

**Table I. Drugs That Bind to HSA Without Affecting Its Reaction with *p*-Nitrophenyl Acetate**

| |
|---|
| Penicillin G |
| Chloropromazine |
| Imipramine |
| Propranolol |
| Diphenylhydantoin |

used drugs, each of which is known to bind strongly to HSA but none of which alters its reactivity with *p*-nitrophenyl acetate.

Many drugs, however, interact with both the inhibitory site and one or more noninhibitory sites and while in those cases the resulting inhibition may be quite complex, it still can be very useful. The effects of such additional interactions can be imagined readily according to the simple scheme illustrated in Equations 3a and 3b involving one inhibitory and

$$\text{Ligand} + \text{Site \# 1} \overset{K_{d_1}}{\rightleftarrows} [\text{Ligand Site \# 1}] \text{ (inhibitory)} \quad (3a)$$

$$\text{Ligand} + \text{Site \# 2} \overset{K_{d_2}}{\rightleftarrows} [\text{Ligand Site \# 2}] \text{ (noninhibitory)} \quad (3b)$$

one noninhibitory site. Thus Figure 4 shows a series of theoretical inhibition curves calculated at a concentration of $10^{-5}M$ HSA interacting with an inhibitory substance with a dissociation constant of $10^{-5}M$ which also interacts with one noninhibitory site with dissociation constants from $10^{-3}M$ to $10^{-7}M$. With a value of $10^{-3}M$ for the noninhibitory site, the curve is, for practical purposes, indistinguishable from that of a specific inhibitor. With a dissociation constant of $10^{-7}M$, however, a pronounced initial plateau is observed and at intermediate values a series of intermediate inhibition curves is obtained. The shapes of such curves should provide a qualitative indication of interactions with other sites.

When such data are plotted according to Equation 2, a series of lines is obtained, each of which has a nearly linear portion with a slope approximating the inhibitory-site dissociation constant (*see* Figure 5). When interactions with the noninhibitory site are weak, the extrapolated intercept corresponds to the concentration of competent inhibitory sites. As interactions with the noninhibitory site increase, the intercepts increase giving a value corresponding to the sum of their concentrations when the two sites interact equally. The slopes, however, change very little and, although some curvature develops, continue to be a good measure of the inhibitory site dissociation constant. When interactions with the noninhibitory site predominate, upward curvature becomes pronounced,

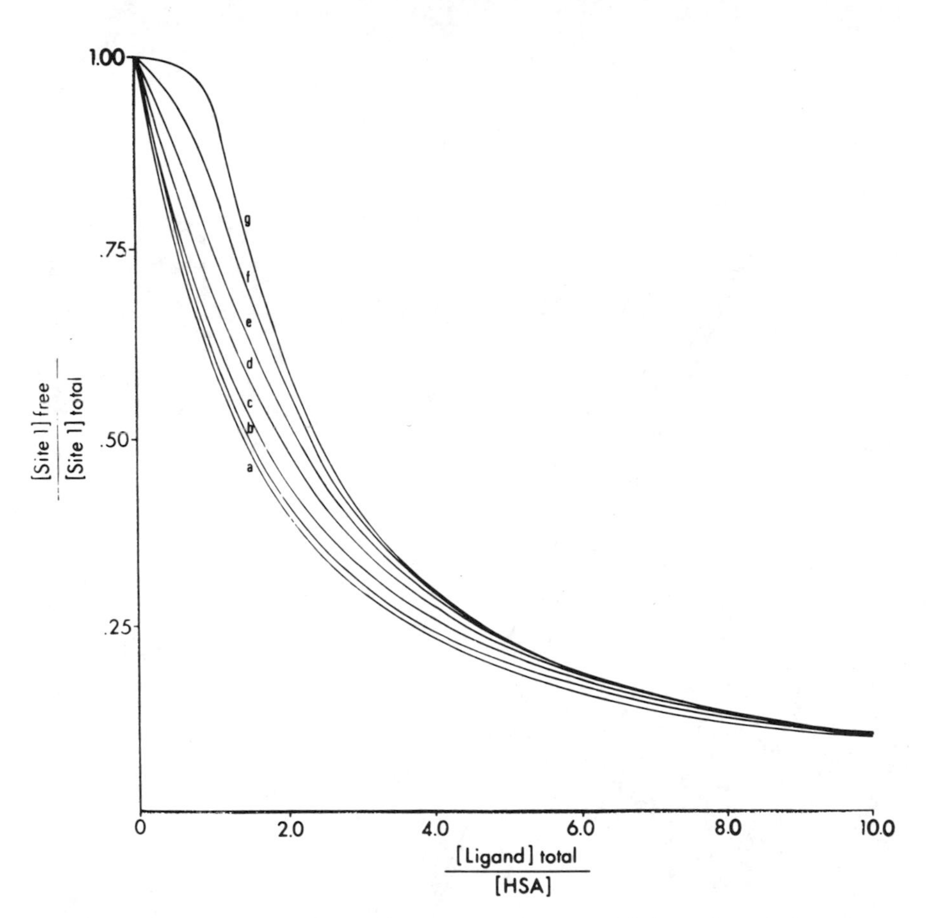


*Figure 4. Theoretical inhibition curves for an inhibitory ligand that also interacts with one noninhibitory site. Calculated for $10^{-5}$M HSA for an inhibitor with $K_i$ of $10^{-5}$M interacting also with a noninhibitory site with $K_d$ of (a) $10^{-3}$M, (b) $10^{-4}$M, (c) $3 \times 10^{-5}$M, (d) $10^{-5}$M, (e) $3.3 \times 10^{-6}$M, (f) $10^{-6}$M, and (g) $10^{-7}$M. The inhibitory site concentration was assumed to be 0.80 of the total HSA concentration.*

but a major portion of the line still retains a slope approximating the inhibitory-site dissociation constant. At different HSA concentrations and different assumed dissociation constants or with additional non-inhibitory sites, similar curves are obtained from which inhibitory-site dissociation constants still can be estimated. Thus extensive binding at other sites does not preclude the characterization of interactions with that site.

Figure 6 shows the inhibitory curves obtained with four drugs, each representing a major inhibitory type. Thus diphenylhydantoin binds very strongly to HSA but has no appreciable inhibitory effect. Diazepam

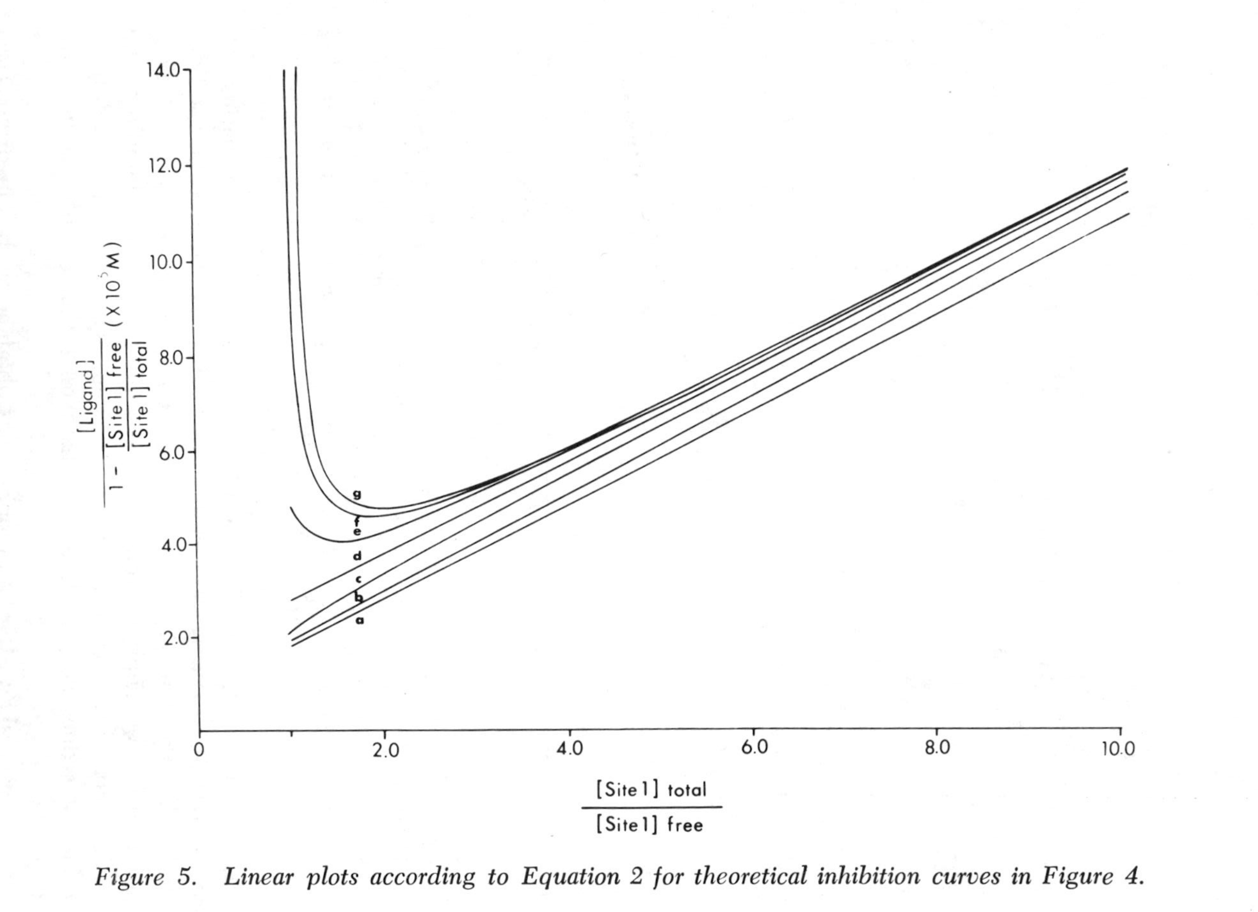


*Figure 5. Linear plots according to Equation 2 for theoretical inhibition curves in Figure 4.*

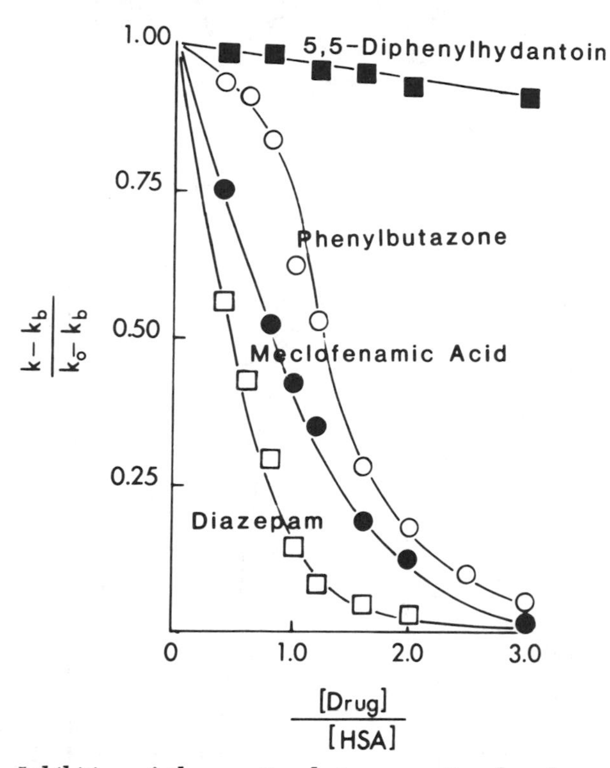


*Figure 6. Inhibition of the reaction between* p-nitrophenyl acetate *and HSA by diphenylhydantoin (■), phenylbutazone (○), meclofenamic acid (●), and diazepam (□) at 25°C and pH 7.4 in 0.03*M *triethanolamine/HCl and 3.5 × $10^{-5}$*M *HSA.*

gives strong stoichiometric inhibition, meclofenamic acid interacts to about the same extent with another site, and phenylbutazone interacts preferentially with another site. Linear plots of the data for the last three drugs are shown in Figure 7.

Data similar to that presented in Figures 6 and 7 have been obtained for a large number of drugs, other normal and abnormal blood constituents, and a hundred or so structurally related compounds. Some drugs and metabolically important compounds that are specific inhibitors of the reaction between HSA and *p*-nitrophenyl acetate and which therefore appear to interact only with that site are listed in Table II. These compounds thus appear to share the same high-affinity binding site on HSA and when they are present together they may compete. Those that bind very strongly and achieve high serum concentrations or any combination of two or more that bind strongly and achieve a high total concentration are likely to affect the binding and, to some extent, the therapeutic action of each of the others.

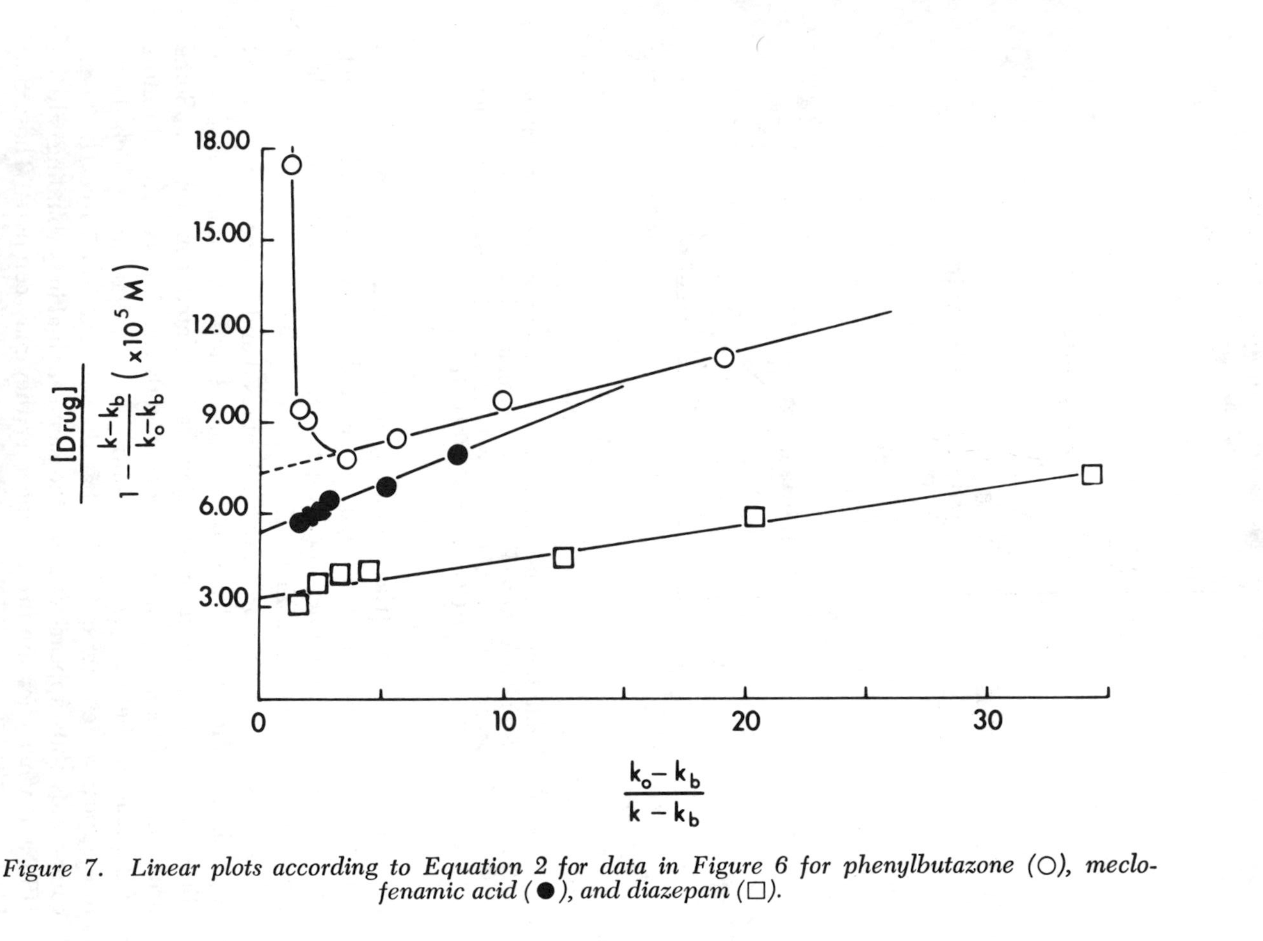


*Figure 7. Linear plots according to Equation 2 for data in Figure 6 for phenylbutazone* (○), *meclofenamic acid* (●), *and diazepam* (□).

**Table II. Stoichiometric Inhibitors of the Reaction Between HSA and *p*-Nitrophenyl Acetate**[a]

| *Compound* | $K_d$ (*μM*)[b] |
|---|---|
| Decanoate | 0.01 |
| Octanoate | 1.6 |
| Tryptophan | 17.8 |
| *N*-Acetyltryptophan | 16.0 |
| Phenylpyruvate | 9.6 |
| *Anti-inflammatory Agents* | |
| Clofibric acid | 1.1 |
| Fenoprofen | 1.0 |
| Flufenamic acid | 1.3 |
| Ibuprofen | 1.4 |
| Naproxen | 0.40 |
| *Tranquilizers* | |
| Chlordiazepoxide | 6.2 |
| Diazepam | 1.6 |
| Desmethyldiazepam | 2.0 |
| 3-Hydroxydiazepam | 1.3 |
| Oxazepam | 3.7 |
| *Diuretic* | |
| Ethacrynic acid | 0.62 |
| *Uricosuric* | |
| Probenecid | 4.5 |

[a] A partial list.
[b] At pH 7.4 in 0.03*M* triethanolamine–0.02*M* HCl and 25°C.

A list of drugs that bind more or less equally to the *p*-nitrophenylacetate binding site and one noninhibitory site, with dissociation constants differing by no more than a factor of two is given in Table III. At present very little is known about these other sites, e.g. even the total number of such sites is not known.

Among the many compounds we have examined by this method, some of the most interesting are those that are strong inhibitors but appear to interact more strongly with a noninhibitory site. The inhibition curves of four such compounds are shown in Figure 8. Each gives an initial plateau followed by a relatively rapid decline in reactivity. Such behavior appears to reflect a strong interaction with the nitrophenyl acetate binding site but even stronger interactions with one other site.

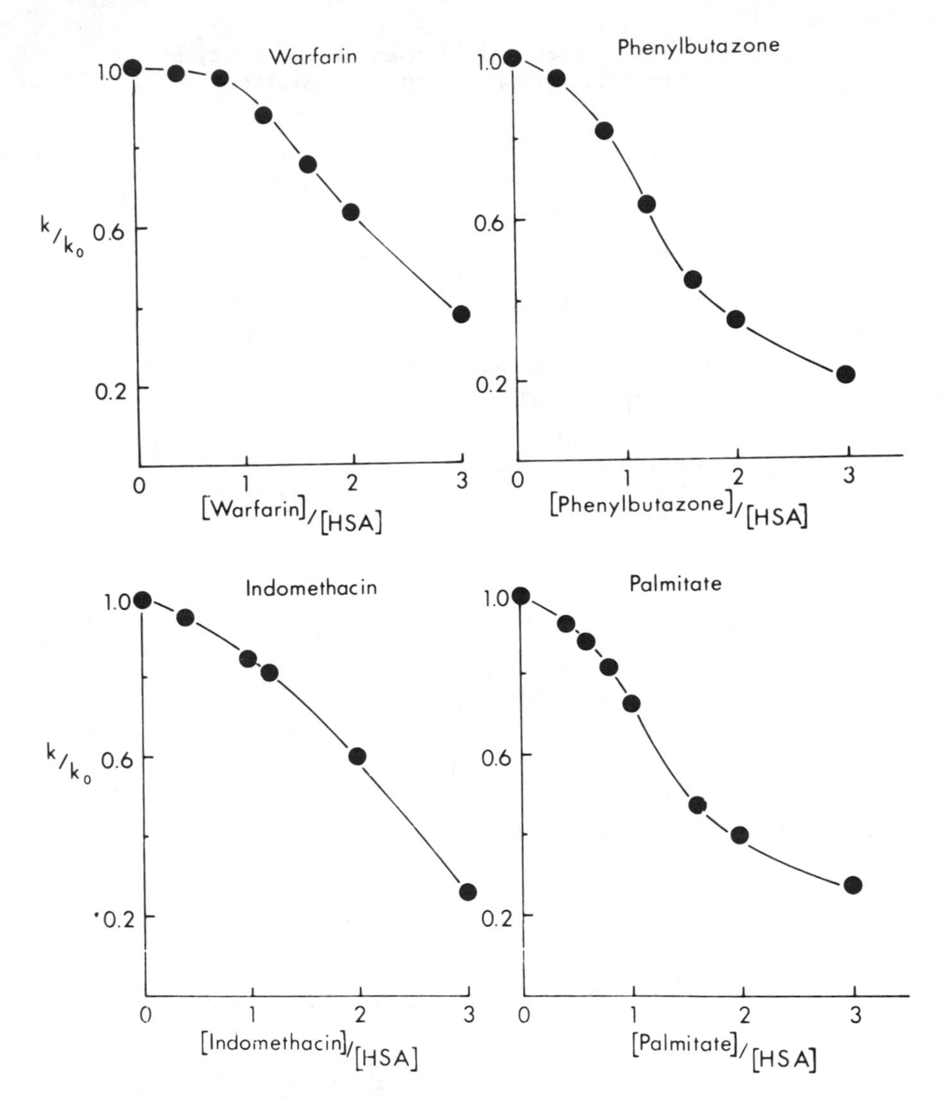


*Figure 8. Inhibition curves for warfarin, phenylbutazone, indomethacin, and palmitate at 25°C and pH 7.4 in 0.03*M *triethanolamine/HCl and* $10^{-4}$M *HSA.*

As mentioned earlier, the binding sites of HSA have not been characterized well and it is not clear what these compounds have in common that would cause each of them to interact with the nitrophenyl acetate binding site (*see* Scheme II). Furthermore it is not clear whether the other site with which each of them interacts is the same site or a series of four different sites. To help shed light on this latter question we have determined the effect of each on the inhibition of the others.

**Table III. Drugs That Bind Simultaneously at Two HSA Binding Sites and Inhibit Its Reaction with *p*-Nitrophenyl Acetate**

| | n | $K_d$ ($\mu$M)[a] |
|---|---|---|
| Meclofenamic acid | 2 | 3.2 |
| Salicylate | 2 | 3.4 |
| Tolmetin | 2 | 5.6 |
| Valproate | 2 | 2.6 |

[a] At pH 7.4 in 0.03*M* triethanolamine and 0.02*M* HCl and at 25°C.

*Scheme II.*

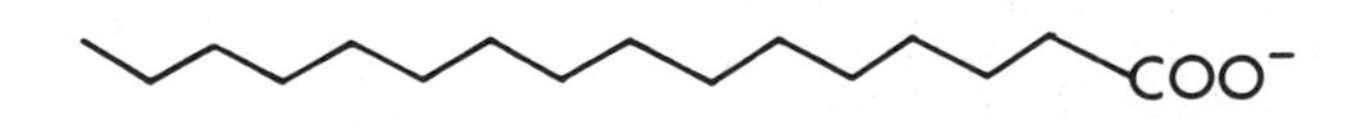


Palmitate

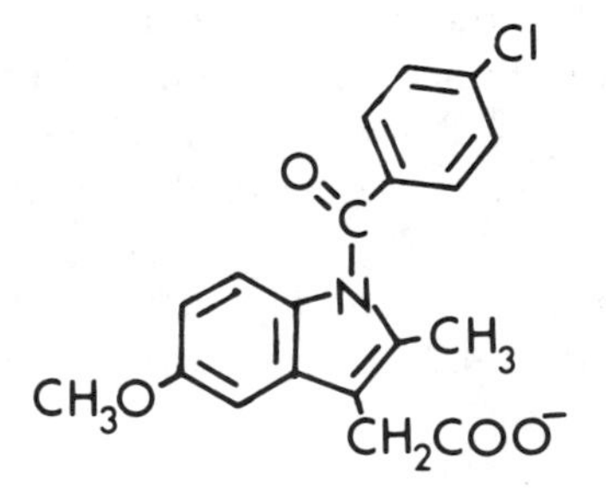


Indomethacin

Phenylbutazone

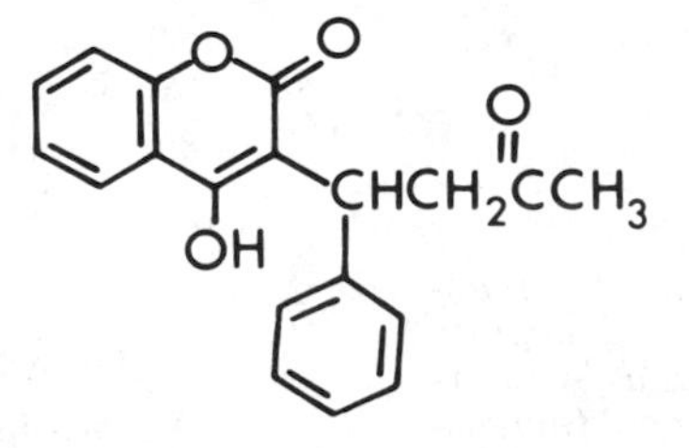


Warfarin

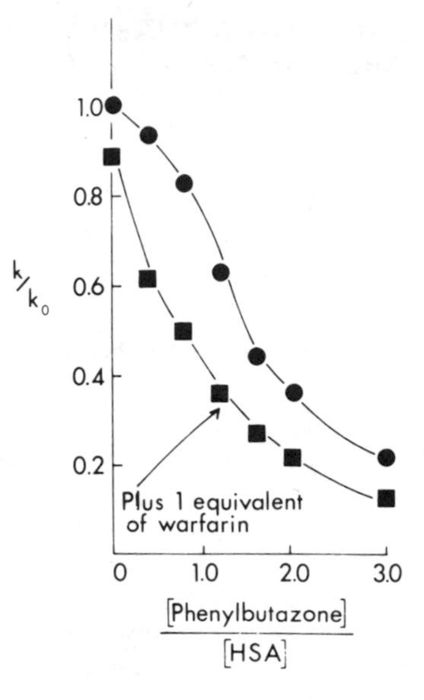


*Figure 9. Inhibition of the reaction between* p-*nitrophenyl acetate and HSA by phenylbutazone and by phenylbutazone plus warfarin. Increasing amounts of phenylbutazone were added to a solution of HSA as described in Figure 8 (●) and after the addition of $10^{-4}$M warfarin (■).*

Figure 9 shows the effects of warfarin on the inhibition by phenylbutazone and Figure 10 shows the effects of phenylbutazone on the inhibition by warfarin. The presence of either at equal concentration with HSA eliminates the plateau otherwise observed in the inhibition curve of the other. These results clearly seem to indicate a common noninhibitory site for these compounds. Thus the high-affinity binding site (the noninhibitory site) and the secondary binding site (the inhibitory site) for these compounds are the same.

Bilirubin which, as we mentioned earlier, has one high-affinity HSA binding site, has relatively little effect on the inhibition by either phenyl-

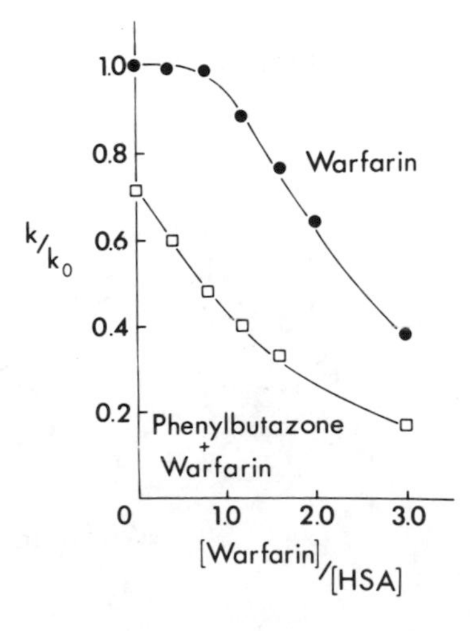


*Figure 10. Inhibition of the reaction between* p-*nitrophenyl acetate and HSA by warfarin and by warfarin plus phenylbutazone. Increasing amounts of warfarin were added to a solution of HSA as described in Figure 8 (●) and after adding $10^{-4}$M phenylbutazone (■).*

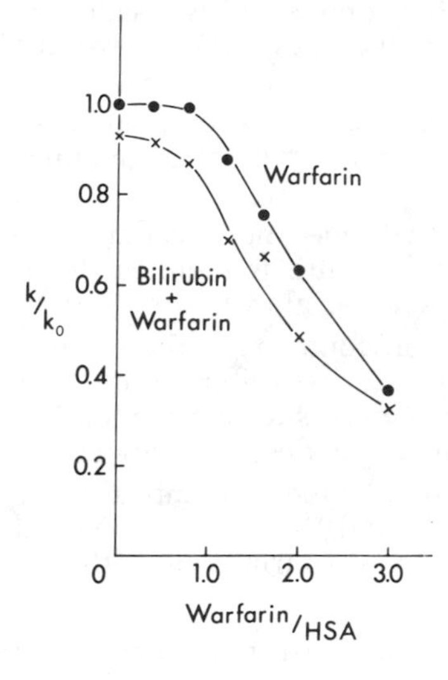


*Figure 11. Inhibition of the reaction between* p-*nitrophenyl acetate and HSA by warfarin and by warfarin plus bilirubin. Increasing amounts of warfarin were added to a solution of HSA as described in Figure 8 (●) and after adding* $10^{-4}$M *bilirubin (×).*

butazone or warfarin (*see* Figures 11 and 12). Thus the binding site for bilirubin appears to be independent of that which binds phenylbutazone and warfarin in keeping with earlier reports (e.g. *see* Ref. 33). Similarly we have determined the effects of other substances known to associate with HSA on the inhibition curves of each other and have arrived at a tentative identification of their respective binding sites as follows:

> Tryptophan—a single strong binding site that coincides with or overlaps the binding site for NphOAc, small fatty acid anions, benzodiazepines, fenamic acids, naproxen, ibuprofen,

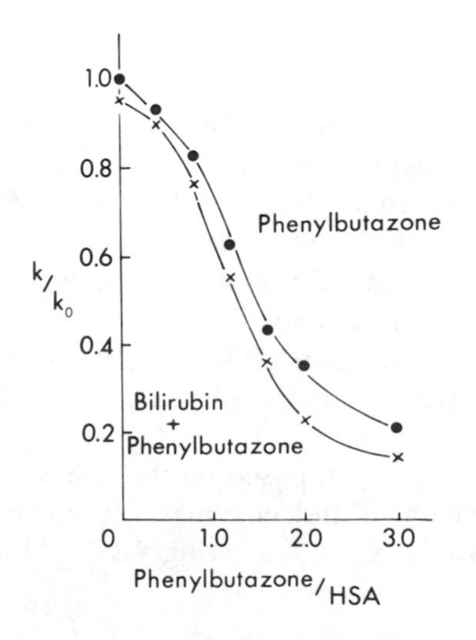


*Figure 12. Inhibition of the reaction between* p-*nitrophenyl acetate and HSA by phenylbutazone and phenylbutazone plus bilirubin. Increasing amounts of phenylbutazone were added to a solution of HSA as described in Figure 8 (●) and after adding* $10^{-4}$M *bilirubin (×).*

and many other small hydrophobic anions. It is located in the carboxyl terminal domain and includes tyrosine residue #411.

Bilirubin—one high-affinity site that does not overlap that for tryptophan described above nor that for any other compounds so far examined.

Palmitate—one high-affinity site that does not overlap with either of the above sites. A secondary site is identical to or overlaps that for tryptophan as described above.

Warfarin, Indomethacin, and Phenylbutazone—these compounds share a single high-affinity site that differs from those described above. They share a secondary site that is identical to or overlaps with that for tryptophan as described above.

Salicylate—two primary binding sites. One is identical to or overlaps the tryptophan binding site and the other is identical to or overlaps the high-affinity site for warfarin, indomethacin, and phenylbutazone as described above.

These studies provide evidence for at least four binding sites in HSA; one for tryptophan, that also interacts strongly with many other compounds; one that appears to be very specific for bilirubin; one high-affinity site for palmitate and one high-affinity site for warfarin, indomethacin, phenylbutazone, and a number of other drugs (i.e. *see* Ref. *36*). Each of these sites appears to be separate and distinct from each of the others.

Additional studies of the same kind are continuing with other drugs to determine whether or not we may be able to identify further distinct binding sites. It is our expectation, however, that the total number of discrete sites probably will be quite small, probably not much greater than the four so far indicated. The evidence at hand certainly rules out the large numbers of sites that sometimes have been suggested in the past.

### *Characterization of the Inhibitory Site*

Based on the amino acid sequence determined by Behrens et al. (*37*) and Meloun et al. (*38*) and the identity of the diisopropylfluorophosphate reactive residue (*39*), we tentatively have located the tryptophan binding site in the carboxyl terminal third of HSA including or adjacent to tyrosine #411 (*34*). However Gambhir et al. (*40*) previously have identified a histidine residue near the amino terminal end of the sequence as being in that site. The conformation of HSA may be such that both residues are in that site; however other explanations are also possible.

Figure 13 shows a simplified model of HSA based on its amino acid sequence (*37, 38*) and results from chemical and enzymatic fragmentation studies (*41, 42, 43, 44*) and low-angle X-ray scattering (*45*). Much

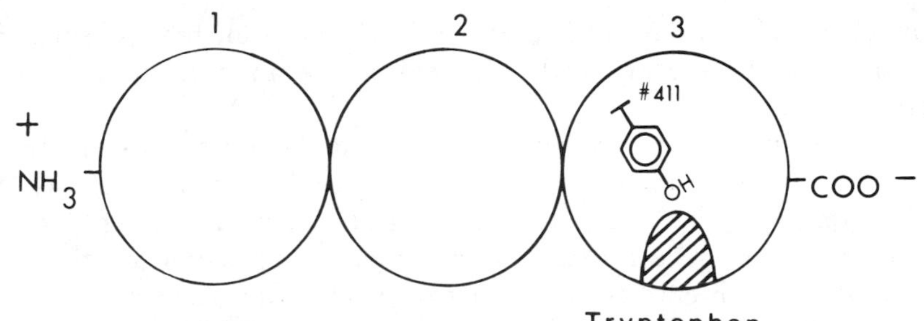


Tryptophan
& other Indoles
Decanoate
& other small fatty
acid anions
Benzodiazepines
Clofibric acid
Ethacrynic acid
Fenoprofen
Flufenamic acid
Ibuprofen
Naproxen
Probenecid

*Figure 13. A simplified, three-domain model of HSA showing the primary binding site for tryptophan and many other small hydrophobic substances in a carboxyl terminal domain adjacent to tyrosine residue # 411.*

more detailed three-dimensional models also have been proposed which are consistent with the same model (*46*). In the present case we show the tryptophan binding site in domain number three where we believe it is located and list a few of the compounds that also bind quite strongly to that site. From the structures of those compounds and others with which it interacts, some ideas are emerging as to the nature of that site. Most of the compounds that bind strongly are obviously anions; however, there are also some compounds like diazepam and *p*-nitrophenyl acetate, that although uncharged still bind quite strongly. So far no cations have been found to bind and dianions generally bind much more weakly than monoanions.

To help clarify the nature of the site we have determined systematically its affinity for many different structurally related compounds. Thus, in the case of simple aliphatic carboxylate ions, affinities vary in proportion to their nonpolar surface areas as shown in Figure 14 (*29, 47*). The straight line is drawn for the unbranched carboxylates, butyrate through decanoate, and has a slope of 25.2 cal/mol/Å$^2$, approximately as expected for transfering an aliphatic hydrocarbon from an aqueous to an apolar medium (*48, 49, 50, 51*).

All branched-chain carboxylates appear to bind less strongly than those of the same approximate nonpolar surface area without branches. The difference may indicate something about the structure of the binding site but might also reflect systematic error in the surface area estimates. We can see from Figure 14 that undecanoate and dodecanoate ions are bound much weaker than their shorter homologs. We believe that this is consistent with a site which, although able to bind decanoate and smaller anions, is too small to accommodate those slightly larger anions (*29*).

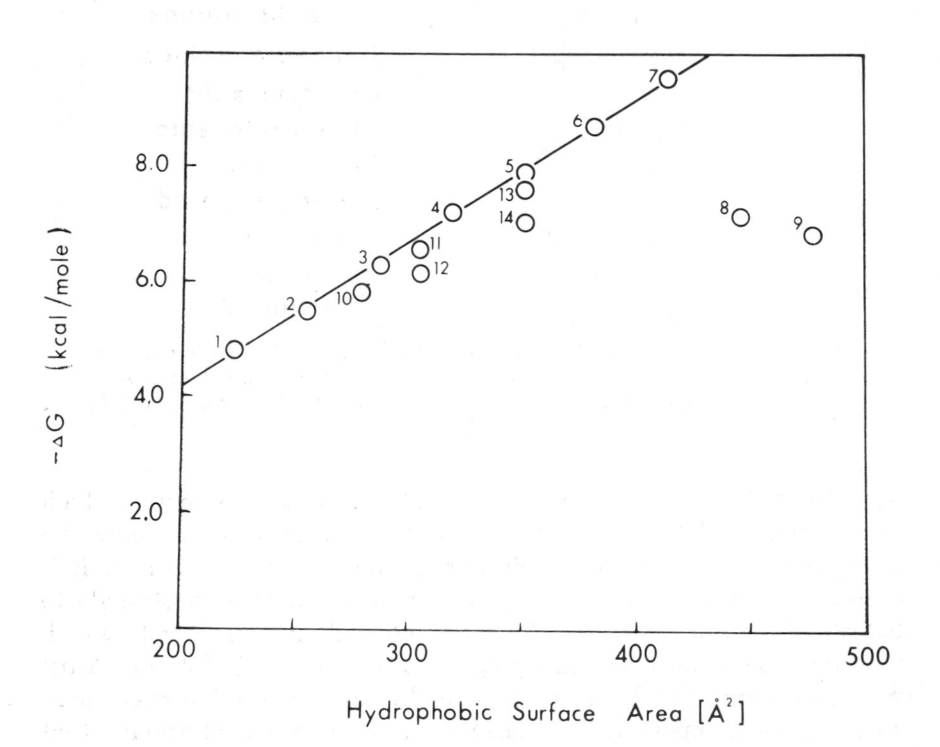


*Figure 14. Free energy of binding vs. nonpolar surface areas for a series of aliphatic carboxylate ions interacting with HSA.*

*Dissociation constants were determined from their inhibition of the reaction between HSA and* p*-nitrophenyl acetate as determined in pH 7.4, 0.03*M *triethanolamine/HCl at 25°C. Nonpolar surface areas were calculated according to Hermann (48). The carboxylate ions are as follows: (1) butyrate; (2) valerate; (3) hexanoate; (4) heptanoate; (5) octanoate; (6) nonanoate; (7) decanoate; (8) undecanoate; (9) dodecanoate; (10) cyclohexanecarboxylate; (11) 1-methyl-1-cyclohexanecarboxylate; (12) cyclohexylacetate; (13) 2-n-propylvalerate; and (14) 2-ethylhexanoate.*

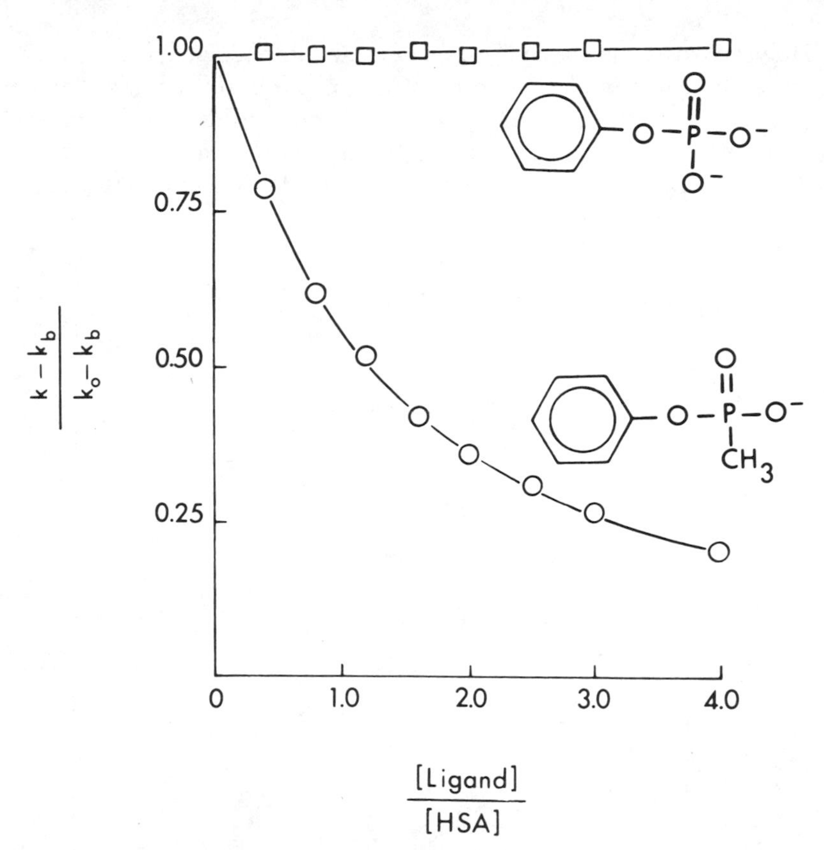


*Figure 15. Inhibition of the reaction between* p-*nitrophenyl acetate and HSA by a mono- and dianion. Increasing amounts of the monoanion, phenylmethylphosphonate (○), and the dianion, phenylphosphate (□), were added to a solution of 3.5 × $10^{-5}$*M *HSA in pH 8.0, 0.06*M *triethanolamine/HCl at 25°C.*

We also have looked at a large number of other anions, particularly aromatic carboxylates, including many containing heteroatoms. On that basis, we generally have concluded that the site is very hydrophobic and relatively long and narrow such that the decanoate ion, e.g., probably binds in a more or less extended conformation.

The affinity of this site for anions presumably reflects its content of one or more cationic groups. To characterize this cationic center we have compared the inhibitory effects of a number of different kinds of anions. Figure 15 shows the effect of two closely related anions, phenylphosphate, a dianion at pH 8.0, and phenylmethylphosphonate, a monoanion under the same conditions, on the reaction between HSA and *p*-nitrophenyl acetate. Only the latter compound is inhibitory which along with other results suggests that the site is specific for monoanions.

However even monoanions differ in their inhibitory effects, as shown in Figure 16. Thus in Figure 16 data obtained for six inorganic anions are plotted according to the Debye–Hückel relationship which, incidentally, does not seem to pertain except perhaps in the cases of phosphate and fluoride. However reactivities decline with higher concentrations of

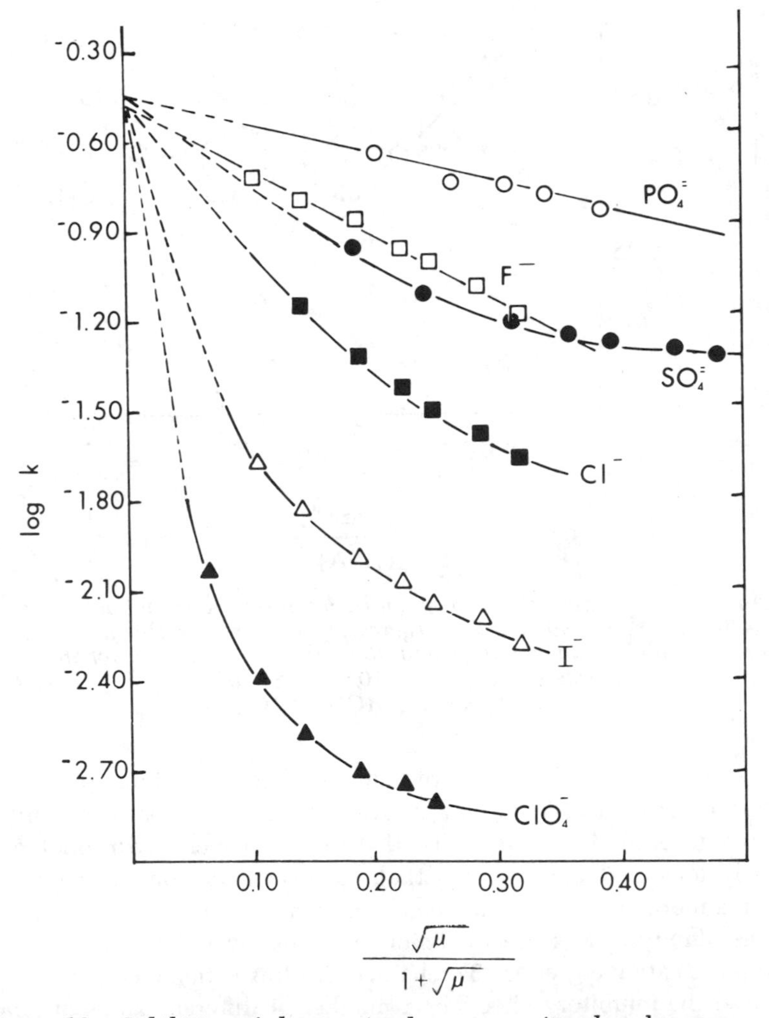


*Figure 16. Inhibition of the reaction between* p-*nitrophenyl acetate and HSA by inorganic anions. Solutions of each anion, triethanolamine, and 3.5 × $10^{-5}$M HSA were prepared at the appropriate concentrations and pH 8.1. Logarithms of the observed pseudo-first-order rate constants are plotted vs. a function of ionic strength, μ, for triethylammonium phosphate (○), fluoride (□), sulfate (●), chloride (■), iodide (△), and perchlorate (▲).*

each anion, particularly in the case of perchlorate. Although a much weaker inhibitor than perchlorate, even chloride gives considerable inhibition such that, at 20m*M*, as used for all of the experiments described above, reaction rates are only about one-fourth as fast as they apparently would be in the absence of any anions.

Many low-molecular-weight blood constituents interact strongly with HSA. Thus the metabolism of tryptophan, bilirubin, fatty acids, many drugs, and other compounds may be influenced by those interactions. In the present study we have tried to characterize one of the major ligand binding sites of HSA and to identify those substances with which it may interact. Studies designed to further characterize this and other HSA binding sites are currently in progress.

## *Literature Cited*


1. Peters, T. *Adv. Clin. Chem.* **1970,** *13,* 37–111.
2. Franglin, G. In "Structure and Function of Plasma Proteins"; Allison, A. C., Ed.; Plenum: New York, 1974; Vol. 1, pp. 265–281.
3. Peters, T. In *"The Plasma Proteins,"* 2nd ed.. Putnam, F. W., Ed.; Academic Press: New York, 1975; pp. 133–181.
4. Rosenoer, V. M.; Oratz, M.; Rothschild, M. A. In "Albumin Structure, Function and Uses"; Pergamon: Oxford, 1977.
5. Peters, T.; Reed, R. G. In "Albumin: Structure, Biosynthesis, Function"; Peters, T.; Sjoholm, I., Eds.; Pergamon: Oxford, 1978; pp. 11–20.
6. Chlouverakis, C.; Harris, P. *Nature* **1960,** *188,* 1111–1112.
7. Scow, R. O.; Chernick, S. S. In "Comprehensive Biochemistry"; Florkin, M.; Stotz, E., Eds.; Elsevier: Amsterdam, 1970; Vol. 18, pp. 19–49.
8. Spector, A. A.; Fletcher, J. E. "Disturbances in Lipid and Lipoprotein Metabolism"; Dietschy, J. M.; Gotto, A. M.; Ontko, J. A., Eds.; Am. Phys. Soc.: Bethesda, 1978; pp. 229–249.
9. Gallin, J. I.; Kaye, D.; O'Leary, W. M. *N. Engl. J. Med.* **1969,** *281,* 1081–1086.
10. Routh, J. I. "Fundamentals of Clinical Chemistry"; 2nd ed.; Tietz, N. W., Ed.; Saunders: Philadelphia, 1976; pp. 1025–1062.
11. Odell, G. B. *J. Clin. Invest.* **1959,** *38,* 823–833.
12. Cooke, J. R.; Roberts, L. B. *Clin. Chim. Acta* **1969,** *26,* 425–436.
13. Odell, G. B. *J. Pediatr.* **1959,** *55,* 268–279.
14. Odell, G. B. *Ann. N. Y. Acad. Sci.* **1973,** *226,* 225–237.
15. Yeary, R. A.; Davis, D. R. *Toxical. Appl. Pharmacol.* **1974,** *28,* 269–283.
16. Gillette, J. R. *Ann. N. Y. Acad. Sci.* **1973,** *226,* 6–17.
17. Schoeneman, P. T.; Yesair, D. W.; Coffey, J. J.; Bullock, F. J. *Ann. N. Y. Acad. Sci.* **1973,** *226,* 162–171.
18. Dayton, P. G.; Israili, Z. H.; Perel, J. M. *Ann. N. Y. Acad. Sci.* **1973,** *226,* 172–194.
19. Tillement, J. P. *Proc. Int. Congr. Pharmacol., 7th* **1978,** *7,* 143–152.
20. Rudman, D.; Bixler, T. J.; Del Rio, A. E. *J. Pharmacol. Exp. Ther.* **1971,** *176,* 261–271.
21. Birkett, D. J.; Myers, S. P.; Sudlow, G. *Mol. Pharmacol.* **1977,** *13,* 987–992.
22. Rasmussen, L. F.; Ahlfors, C. E.; Wennberg, R. P. *J. Clin. Pharmacol.* **1978,** *18,* 477–481.
23. Oie, S.; Levy, G. *J. Pharm. Sci.* **1979,** *68,* 1–6.

24. Wiegand, U. W.; Slattery, J. T.; Hintze, K. L.; Levy, G. *Life Sci.* **1979,** *25,* 471–478.
25. Sellers, E. M. *Proc. Int. Congr. Pharmacol., 7th* **1978,** *7,* 153–162.
26. Reidenberg, M. M.; Affrime, M. *Ann. N. Y. Acad. Sci.* **1973,** *226,* 115–126.
27. Shoeman, D. W.; Benjamin, D. M.; Azarnoff, D. L. *Ann. N. Y. Acad. Sci.* **1973,** *226,* 127–130.
28. Borga, O. *Proc. Int. Congr. Pharmacol., 7th* **1978,** *7,* 143–152.
29. Koh, S.-W. M.; Means, G. E. *Arch. Biochem. Biophys.* **1979,** *192,* 73–79.
30. Sollenne, N. P.; Means, G. E. *Mol. Pharmacol.* **1979,** *15,* 754–757.
31. Ikeda, K.; Kurono, Y.; Ozeki, Y.; Yotsuyanagi, T. *Chem. Pharm. Bull.* **1979,** *27,* 80–87.
32. Kurono, Y.; Maki, T.; Yotsuyanagi, T.; Ikeda, K. *Chem. Pharm. Bull.* **1979,** *27,* 2781–2786.
33. Ozeki, Y.; Kurono, Y.; Yotsuyanagi, T.; Ikeda, K. *Chem. Pharm. Bull.* **1980,** *28,* 535–540.
34. Means, G. E.; Wu, H.-L. *Arch. Biochem. Biophys.* **1979,** *194,* 526–530.
35. Means, G. E.; Bender, M. L. *Biochemistry* **1975,** *14,* 4989–4994.
36. Sudlow, G. *Proc. Int. Congr. Pharmacol., 7th* **1974,** *7,* 113–123.
37. Behrens, P. Q.; Spiekerman, A. M.; Brown, J. R. *Fed. Proc.* **1975,** *34,* 591.
38. Meloun, B.; Moravek, L.; Kosta, V. *FEBS Lett.* **1975,** *58,* 134–137.
39. Sanger, F. *Proc. Chem. Soc. (London)* **1963,** 76–86.
40. Gambhir, K. K.; McMenamy, R. H.; Watson, F. *J. Biol. Chem.* **1975,** *250,* 6711–6719.
41. Weber, G.; Young, L. B. *J. Biol. Chem.* **1964,** *239,* 1424–1431.
42. Pederson, D. M.; Foster, J. F. *Biochemistry* **1969,** *8,* 2357–2365.
43. Markus, G.; McClintock, D. K.; Castellani, B. A. *J. Biol. Chem.* **1967,** *242,* 4395–4401.
44. King, T. P.; Spencer, M. *J. Biol. Chem.* **1970,** *245,* 6134–6148.
45. Bloomfield, V. *Biochemistry* **1966,** *5,* 684–689.
46. Squire, P. G.; Moser, P.; O'Konski, C. T. *Biochemistry* **1968,** *7,* 4261–4272.
47. Sollenne, N. P., Ph.D. Thesis, The Ohio State University, 1980.
48. Hermann, R. B. *J. Phys. Chem.* **1972,** *76,* 2754–2759.
49. Harris, M. J.; Higuchi, T.; Rytting, J. H. *J. Phys. Chem.* **1972,** *77,* 2694–2703.
50. Reynolds, J. A.; Gilbert, D. B.; Tanford, C. *Proc. Natl. Acad. Sci.* **1974,** *72,* 2925–2927.
51. Gelles, J.; Klapper, M. H. *Biochim. Biophys. Acta* **1978,** *533,* 465–477.

RECEIVED December 1, 1980.

# 12

# Proteolytic Enzymes and Their Active-Site-Specific Inhibitors: Role in the Treatment of Disease

JAMES C. POWERS

School of Chemistry, Georgia Institute of Technology, Atlanta, GA 30332

*Proteolytic enzymes (proteases) are involved in a wide variety of physiological processes including digestion, fertilization, coagulation, and the immune response. Outside of their normal environment, proteases can be extremely destructive and natural human plasma inhibitors inhibit most proteases that escape. Imbalance in protease–protease inhibitor systems can lead to a number of diseases of which pulmonary emphysema is one well-characterized example. This disease results when the protease elastase attacks elastin, the major elastic protein in the lung. Considerable effort has been devoted to the synthesis of inhibitors of proteolytic enzymes such as elastase for possible therapeutic use. In the future, specific and selective synthetic protease inhibitors should be useful for treating specific diseases that range from the common cold to chronic disorders such as emphysema.*

Proteases (protein-hydrolyzing enzymes or proteolytic enzymes) have been known for over 100 years since trypsin was first isolated from pancreatic juice by Kühne (*1*). Because several pancreatic proteases could be isolated and crystallized easily, these enzymes became some of the best characterized and studied. For most of the period since their discovery, proteases were thought to be involved only in digestion. However, in the last decade, proteases have been shown to be involved in many other important physiological processes in addition to digestion.

Many enzymes, hormones, and other physiologically active proteins are synthesized as inactive precursors or zymogens (*2*). The zymogens

0065-2393/82/0198-0347$05.25/0

are converted into physiologically active forms by proteases that selectively cleave one or a limited number of peptide bonds. This is basically an irreversible process since there are no biochemical pathways for resynthesizing a cleaved peptide bond in a protein. Thus, proteolysis is a prompt and irreversible method of turning on any particular physiological process. Likewise, cleavage of a peptide bond or bonds in an active enzyme or protein can lead to an inactive molecule. This in turn provides a quick method terminating some processes.

Proteases are involved in the regulation of many important biological processes. Some of the better known examples are blood coagulation, fibrinolysis, complement activation, fertilization, protein processing, digestion, and phagocytosis. Some of these processes involve cascades of proteolytic events. For example, in the resting state, the coagulation and complement systems are composed of a number of plasma zymogens of proteases. Upon receiving an appropriate signal some of these zymogens are converted into active proteases which in turn activate other zymogens producing a cascade of enzymatic reactions. The initial signal in the coagulation system is the formation of new surfaces at the site of a wound and the end result is formation of the fibrin clot. In the complement system, initial formation of antigen–antibody complexes leads to the lysis of foreign cells. In both cases, the cascade of proteolytic events gives a system for prompt amplification and control of the initial signal.

### *Proteases and Disease*

Since a major component of all cells is protein, proteases could be very destructive if they were not carefully controlled or compartmentalized. The potential seriousness of uncontrolled proteolysis can be recognized by the fact that ca. 10% of the proteins by weight found in human plasma are protease inhibitors. The currently recognized plasma protease inhibitors are listed in Table I (*3, 4*). In addition to the plasma inhibitors, there are other inhibitors that are more localized and have not been as well characterized.

We now believe that many diseases result when there is an imbalance between specific proteases and their natural inhibitors. Some examples of proteases that have been linked to specific diseases are listed in Table II.

**Emphysema.** Emphysema is one disease where the linkage between proteolysis and the disease is fairly well understood. Pulmonary emphysema is a disease characterized by a progressive loss of lung elasticity due to the destruction of lung elastin and alveoli. Respiration becomes increasingly difficult and death often results.

The first clue to the involvement of proteases in emphysema was

**Table I. Major Protease Inhibitors of Human Plasma**

| *Inhibitor* | *Molecular Weight* | *Concentration (mg/ 100 mL)* | *Representative Proteases Inhibited* |
|---|---|---|---|
| $\alpha_1$-Protease inhibitor | 52,000 | 290 | elastase (leukocyte)<br>cathepsin G |
| $\alpha_1$-Antichymotrypsin | 69,000 | 49 | cathepsin G |
| Inter-$\alpha$-trypsin inhibitor | 160,000 | 50 | trypsin |
| $\alpha_2$-Antiplasmin | 70,000 | 7 | plasmin |
| Antithrombin III | 65,000 | 24 | coagulation serine proteases<br>thrombin |
| $C_1$-Inactivator | 70,000 | 24 | $C1\overline{r}$, $C1\overline{s}$<br>Factor $XII_a$ |
| $\alpha_2$-Macroglobulin | 720,000 | 260 | most proteases |

**Table II. Diseases and Possible Target Proteases**

| *Disease* | *Enzyme* |
|---|---|
| Emphysema | elastase (leukocyte)<br>cathepsin G |
| Arthritis | collagenase |
| Pancreatitis | pancreatic serine proteases |
| Cancer | collagenase<br>plasminogen activator |
| Hypertension | renin<br>angiotensin converting enzyme |
| Inflammation | mast cell chymotrypsin-like proteases |
| Amyloidosis | elastase |

the observation by Laurell and Ericksson that individuals who were homozygotes (PiZ) in an $\alpha_1$-protease inhibitor ($\alpha_1$-antitrypsin) deficiency were predisposed to the disease (5). This serum protease inhibitor inhibits a spectrum of proteases including elastase and cathepsin G which are found in the granules of human leukocytes. Leukocyte proteases are involved normally in phagocytosis in the lung and contribute to the turnover of damaged lung cells and the digestion of invading bacteria. In normal individuals $\alpha_1$-PI protects the lung from being digested by any of the proteases that may leak from leukocytes. In individuals not so protected, proteolysis of lung elastin leads to emphysema. The $\alpha_1$-PI of PiZ individuals has a one amino acid substitution (Lys for Glu) somewhere in the sequence (*6*). As a result, the $\alpha_1$-PI of PiZ individuals accumulates in the liver and is not exported to the plasma.

A major predisposing factor for developing chronic obstructive pulmonary disease is cigarette smoking. It now is believed that $\alpha_1$-PI is inactivated either directly by oxidants in the smoke or by myeloperoxidase which is released from leukocytes (*7*). The interrelationship between $\alpha_1$-PI, elastase, and emphysema is shown schematically in Figure 1.

The reactive site of $\alpha_1$-PI has a Ala–Ile–Pro–Met*Ser–Ile–Pro–Pro sequence where the asterisk indicates the bond cleaved when $\alpha_1$-PI · protase complexes are dissociated at high pH or by using nucleophiles. We have synthesized a number of peptides with the amino acid sequence at the $\alpha_1$-PI reactive site and have shown them to be perfectly adequate substrates for human leukocyte elastase. However, oxidation of the methionine residue of the substrates to a methionine sulfoxide residue (*see* Table III) almost completely destroys their reactivity toward human leukocyte elastase and other proteases (*8, 9*). Oxidation of $\alpha_1$-PI itself destroys its ability to inhibit most proteases (*10*).

Cigarette smoke decreases the $\alpha_1$-PI activity in rat lung and produces a functional deficiency of protease inhibitors in the lower respiratory tract of humans (*11, 12*). It is quite reasonable to assume that this is due to the oxidation of the essential methionine residue in $\alpha_1$-PI. Thus, chemical modification of $\alpha_1$-PI by oxidation of a methionine residue to a methionine sulfoxide residue by some component of cigarette smoke or by myeloperoxidase which is released by cigarette smoke, results in inactivation of this essential protease inhibitor. The resulting imbalance of proteases and protease inhibitors in the lung then results in the development of emphysema.

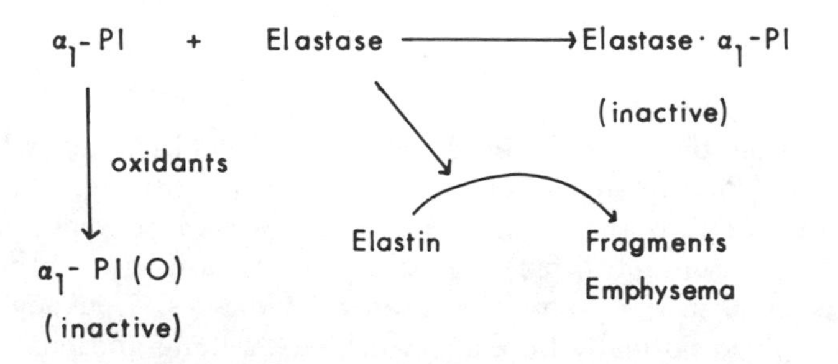


*Figure 1. Interrelationship between human leukocyte elastase, its natural inhibitor, $\alpha_1$-protease inhibitor ($\alpha_1$-PI), and emphysema.*

**Table III. Hydrolysis of Peptide Substrates by Human Leukocyte Elastase**

| Substrate[a] | $k_{cat}/K_m$ ($M^{-1}$ $s^{-1}$) Methionine | Oxidized Methionine[b] |
|---|---|---|
| MeO–Suc–Ala–Ala–Pro–Met*NA[c] | 300 | 0 |
| Ac–Ala–Ala–Pro–Met*Ser–Ala–$NH_2$[d] | 850 | 157 |
| Ac–Ala–Ile–Pro–Met*Ser–Ile–Pro–Pro–$NH_2$[d] | 10,000 | 27 |

[a] The asterisk indicates the bond cleaved in the substrate.
[b] The methionine residue has been oxidized to the sulfoxide.
[c] The pH 7.5, 0.1*M* Hepes buffer, 0.5*M* NaCl, 9.8% dimethyl sulfoxide at 25°C.
[d] The pH 7.5, 0.040*M* phosphate, 0.50*M* NaCl, 37°C.

**Arthritis, Inflammation, and Related Diseases.** The synovial fluid of arthritic joints is rich in collagenase derived from leukocytes and other phagocytic cells that massively infiltrate into the joint spaces during the acute inflammatory episodes of rheumatoid arthritis (*13*). This enzyme is the most important protease that is able to degrade collagen, the major connective tissue protein in the human body. In arthritis significant degradation of articular cartilage, which is composed of collagen and other connective tissue proteins, takes place. It now appears that a protease–protease inhibitor imbalance in the joint is responsible for this tissue proteolysis. Collagenase and other leukocyte proteases such as cathepsin G and elastase have been implicated also. Leukocyte elastase, in fact, can degrade basement membrane collagen (*14*). At present, the mechanism(s) by which the disease is initiated is not known.

Another disease involving collagenase is idiopathic pulmonary fibrosis. This chronic disease results in fibrosis in the lung. Recently degradation of collagen in the lung by a collagenase active against Type I collagen has been observed in the disease and this leads to the suspicion that a protease–protease inhibitor imbalance is occurring in this disease (*15*).

Leukocytes, macrophages, and mast cells are observed at one stage or another of most inflammatory diseases. All of these cells are rich in proteases and are important mediators of allergic, infective, and anaphylactic reactions of tissues in most animals. It is likely that proteases released from these cells contribute to the inflammatory process by degrading connective tissue components and by releasing kinins.

**Muscular Dystrophy.** A striking feature of muscular dystrophy is the extensive loss of contractile proteins and their replacement by fat and connective tissue. The involvement of proteases in the degradation of muscle tissue in muscular dystrophy and a variety of other physiological and pathological conditions of the muscle are now well documented. Protease inhibitors indeed decrease muscle protein turnover in both normal and diseased muscle (*16, 17*). At present, the protease or proteases which cause the disease have not been well characterized although lysosomal cysteine proteases such as cathepsin B may be involved (*18*).

The breakdown and dissolution of myelin in multiple sclerosis also may be related to muscular dystrophy and emphysema. Multiple sclerosis has been attributed to the action of proteases, some of which are highly elevated in and around the plaques of multiple sclerosis (*19*). Again, the proteases responsible and their natural inhibitors are not known.

**Tumors.** Neoplastic cells display a number of biochemical changes compared with normal cells. For example, a number of invasive tumors have been shown to secrete collagenase. This enzyme gives the malignant cells the ability to invade adjacent tissues and to spread as the collagenase dissolves connective tissue collagen.

Plasminogen activator is another protease that increases in many cell types after neoplastic transformation (*20*). This enzyme activates the zymogen plasminogen to the protease plasmin. Plasmin is a protease of broad specificity and high local proteolytic activity can be achieved in the vicinity of the transformed cells.

Proteases are involved also in the expression of mutations in bacteria, in tumor promotion in mice, and in the X-ray-induced neoplastic transformation in hamster and mouse cells (*21, 22*). Protease inhibitors will block tumor promotion in mice (*23*) and plant protease inhibitors in the diet have been proposed to account for the low incidence of breast and bladder cancer in countries where diet is high in plant proteins.

**Proteases in Bacterial and Viral Infections.** A number of bacteria produce proteases that may participate in their infectivity. For example, *Pseudomonas aeruginosa* is an antibiotic-resistant pathogen that causes hemorrhagic pneumonia and corneal ulcers. The major cause of morbidity and mortality in cystic fibrosis is severe chronic pulmonary infections with *P. aeruginosa* and other bacterial pathogens. Most strains of *P. aeruginosa* produce elastase and another neutral protease. Those strains with the elastase are much more pathogenic than those without and the elastase is likely the factor responsible for destruction of corneal tissue (melting eye syndrome) and hemorrhages of the lung in *P. aeruginosa* infections. In addition, *P. aeruginosa* elastase is capable of inactivating $\alpha_1$-protease inhibitor (*24*) and thus is able to partially overcome the body's protease inhibitor screen.

*Neisseria gonorrhoeae* and *Neisseria meningitides* are pathogenic for humans in which they cause gonorrhea and meningitis. Both of these organisms, which infect human mucosal surfaces, elaborate a highly specific protease (*25*) that cleaves immunoglobulin A (IgA). IgA is the principal mucosal antibody and these two organisms thus are able to partially overcome the body's antibody screen. This may explain why individuals who have been infected with gonorrhea and have built up antibodies can become reinfected.

Proteases also are involved in viral maturation. For example, in picornavirus replication the virus RNA is translated into large virus precursor polypeptides. These then are cleaved by a viral protease(s) into the viral proteins (*26, 27*). This opens the possibility that specific inhibitors for the viral processing protease could be used as antiviral agents.

### *Protease Inhibitors*

It now is evident that selective protease inhibitors could be used for treating a large number of diseases. Other possible target proteases for inhibitors are listed in Table IV. The inhibitors could be natural protein protease inhibitors isolated from animal or plant sources, small-molecular-weight inhibitors isolated from fermentation broths, or synthetic protease inhibitors. Small-molecular-weight inhibitors have a number of advantages over the larger protein protease inhibitors for therapeutic uses. The protein protease inhibitors usually must be given by injection and there is always the possibility of immunogenicity. In addition, they are often more difficult to obtain in large quantities. For example, if $\alpha_1$-protease inhibitor was isolated in good yield from all of the human blood processed in this country, there would be at most enough inhibitor to treat 25% of the patients with emphysema in the United States.

Low-molecular-weight inhibitors, on the other hand, probably could be obtained in large amounts either by synthesis or fermentation. In most cases they would be likely to be orally active. In addition, it would be feasible to carry out structural modifications on the inhibitor to increase reactivity and selectivity or to reduce toxicity.

Synthetic inhibitors can be designed to bind either reversibly or irreversibly to the target protease. In this review I will concentrate on

**Table IV. Other Target Proteases**

| *Proteases* | *Reason for Inhibition* |
|---|---|
| Acrosin | antifertility drug |
| Thrombin and other blood coagulation factors | anticoagulant |
| Complement proteases | suppression of the immune response |
| Enkephalinase and other hormone degrading proteases | extend the in vivo life of hormones |
| Pepsin | treat ulcers |

irreversible active-site-directed protease inhibitors. They have the advantage that the target protease usually cannot become reactivated once the inhibition reaction takes place and the organism or cell must resynthesize the protease at a considerable expense in cellular energy. Reversible inhibitors in contrast always must be present in sufficient concentration and as the inhibitor is degraded or excreted, the target protease regains its activity. A disadvantage of irreversible inhibitors is their possible toxicity due to reaction with other proteins in the cell. Irreversible inhibitors often contain reactive functional groups that can react with many side-chain functional groups of proteins.

**Protease Classification.** In order to rationally design an inhibitor for a protease it is first necessary to place it into one of four families of proteases (*see* Table V). For a new enzyme, a study of its inhibition profile with a series of general protease inhibitors is sufficient to classify it into one of the four families. The inhibitors usually used are diisopropylphosphofluoridate (DFP) or phenylmethane sulfonyl fluoride (PMSF) for serine proteases, 1,10-phenanthroline for metalloproteases, thiol reagents such as iodoacetate or *N*-ethylmaleimide for thiol proteases, and pepstatin or diazo compounds such as diazoacetyl–norleucine methyl ester for carboxyl proteases.

Within each family of proteases, the catalytic residues in the active site of each member of the family are likely to be identical or very similar both in structure and in geometry. X-ray crystallographic studies of several members each of the serine protease and the metalloprotease families have shown this to be true even where the individual proteases such as chymotrypsin and subtilisin have evolved from different precursors. Thus, once a new enzyme is classified into one of the four protease families, it is possible to derive a fairly complete working model of the catalytic site of the new enzyme from studies that have been carried out with the better known members of the family. This allows the investigator to direct inhibitors against specific functional groups in the enzyme's active site before he has obtained sufficient enzyme or has enough time to actually prove those functional groups are present. For example, we have been able to design inhibitors for the *Pseudomonas aeruginosa* elastase using information derived from studies on thermolysin (*28*) and inhibitors for the angiotensin-converting enzyme have been developed based on earlier work with carboxypeptidase (*29*).

Within each protease family, individual members will differ in their substrate specificity. Most proteases have extended substrate binding sites and will bind to and recognize several amino acid residues of a polypeptide substrate (*see* Figure 2). Usually one of these will be the primary binding site. For example, in the serine proteases chymotrypsin, trypsin, and elastase, the primary substrate binding site is the $S_1$ subsite

**Table V. Mechanistic Families of Proteases**

| *Class* | *Protease* | *Substrate Specificity* |
|---|---|---|
| Serine | acrosin | trypsin-like |
| | mast cell proteases | chymotrypsin-like |
| | elastase (human leukocyte) | cleaves elastin, cleaves peptide bonds following Val > Ala |
| | elastase (porcine pancreatic) | cleaves elastin, cleaves peptide bonds following Ala > Val |
| Metallo-protease | collagenase (mammalian) | cleaves a Gly–Leu or Gly–Ile bond in native triple helical collagen |
| | gelatinase | cleaves gelatin (denatured collagen) |
| | *Psuedomonas aeruginosa* elastase | cleaves elastin |
| | angiotensin converting enzyme | cleaves C-terminal His–Leu from angiotensin I forming the hypertensive peptide angiotensin II |
| | carboxypeptidase A | cleaves C-terminal aromatic residues |
| | carboxypeptidase B | cleaves C-terminal basic residues |
| | thermolysin | cleaves peptide bonds preceding Ile, Phe, and Leu residues |
| Thiol | cathepsin B | broad specificity |
| | papain | broad specificity |
| | cathepsin H | endoaminopeptidase |
| Carboxyl (aspartate, acid) | pepsin | cleaves peptide bonds following hydrophobic or aromatic residues |
| | penicillopepsin | similar to pepsin |
| | cathepsin D | similar to pepsin |

and they prefer respectively aromatic residues, basic residues, and residues with small aliphatic side chains. In addition, recognition at other subsites contributes to binding and catalysis and often quite similar enzymes (e.g., human leukocyte elastase and porcine pancreatic elastase) can be distinguished by their secondary specificity.

Older methods of classifying proteases by their substrate specificity are of no assistance with regard to the design of inhibitors. There are elastases in both the serine protease and metalloprotease family (*see*

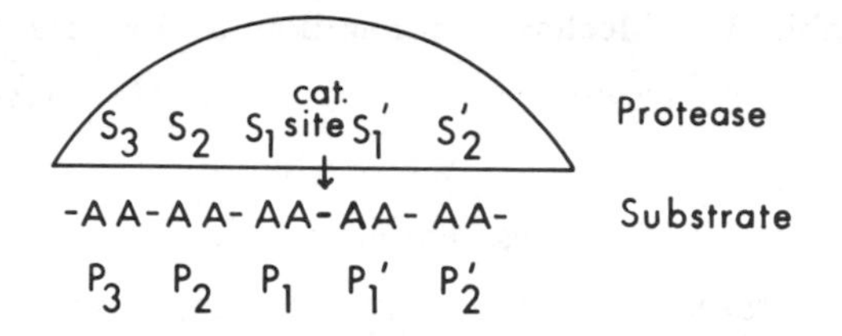


*Figure 2. A representation of the interaction of a polypeptide substrate with the extended substrate binding site of a protease. The individual amino acid (AA) residues are designated $P_1$, $P_2$, $P_1'$ etc. and the corresponding subsites of the enzyme are $S_1$, $S_2$, $S_1'$ etc. The bond cleaved by the enzyme is the peptide bond between the $P_1$ and $P_1'$ residues.*

Table V) and there are carboxypeptidases that also belong to both families. Substrate specificity alone is not sufficient to rationally design an inhibitor for a new protease since it tells nothing about the active-site functional groups.

Two things must be in the proper juxtaposition in order to achieve specific inhibition of a targeted protease. The first requirement is a functional group that will react with (affinity label) or be activated by (suicide substrate) the catalytic groups in one particular protease family. Ideally the functional group should exhibit low reactivity toward other functional groups found in proteins of physiological systems. Secondly, the functional group must be attached to a peptide, peptide-like, or some other structural element that is recognized specifically by the targeted protease. Ideally this structural feature should exclude reaction (or at least slow down) reaction with other similar proteases. Obviously, the ideal specific protease inhibitor is realized rarely in practice. In the remainder of this review, I will discuss specific examples of inhibitors designed for various classes of proteases.

**Serine Protease Inhibitors.** The two major catalytic residues of serine proteases are a serine residue and a histidine residue. Affinity labels have been developed that modify both of these residues.

**Peptide Chloromethyl Ketones.** Peptide chloromethyl ketone inhibitors have been studied extensively and a fairly detailed picture of the inhibition reaction (*see* Figure 3) has emerged from numerous chemical and crystallographic studies (*30, 31*). The inhibitor resembles a serine protease substrate with the exception that the scissile peptide bond of the substrate is replaced with a chloromethyl ketone functional group in the inhibitor. The inhibitor binds to the serine protease in the extended substrate binding site and the reactive chloromethyl ketone functional group is placed then in the proper position to alkylate the active-site histidine residue. In addition, the serine OH reacts with the inhibitor carbonyl group to form a hemiketal.

Peptide chloromethyl ketone inhibitors have been developed for almost every serine protease that has been characterized adequately (*30*). For example, human leukocyte elastase, due to its involvement in emphysema, has been studied extensively with this class of inhibitor (*32*). The rate at which peptide chloromethyl ketones inhibit elastase is influenced by their interaction with the primary substrate binding site ($S_1$) of the enzyme and by interactions at other subsites. The most effective chloromethyl ketone elastase inhibitor found thus far is MeO–Suc–Ala–Ala–Pro–Val$CH_2Cl$ (MeO–Suc– = $CH_3OCOCH_2CH_2CO$–). This will not inhibit the other major leukocyte protease, cathepsin G (*see* Table VI). In contrast, Z–Gly–Leu–Phe–$CH_2Cl$ (Z = $C_6H_5CH_2OCO$–) inhibits cathepsin G, but not elastase. Both enzymes can be inhibited with Ac–Ala–Ala–Pro–Val$CH_2Cl$.

One concern that is expressed often about chloromethyl ketones is their ability to alkylate a variety of nucleophiles in proteins. This is thought to be a major limitation to the use of chloromethyl ketones in physiological situations. In order to evaluate the significance of side reactions with other nucleophiles, the rate of reaction of the thiol glutathione with MeO–Suc–Ala–Ala–Pro–Val$CH_2Cl$ was measured (*see* Table VI). The rate is quite slow and the inhibitor would discriminate in favor of leukocyte elastase over glutathione by a factor of 1770 if the concentrations were equivalent. Thus, MeO–Suc–Ala–Ala–Pro–Val$CH_2Cl$ appears to be a highly reactive and selective elastase inhibitor for use in physiological situations.

One goal in the design of an inhibitor is specificity. Often this is a considerable challenge since physiological systems contain a number of closely related proteases. For example, there are at least four chymotrypsin-like enzymes in humans. These include pancreatic chymotrypsin, cathepsin G, and two mast cell proteases (the human enzymes have not been characterized yet, but two are found in rats). All of these enzymes

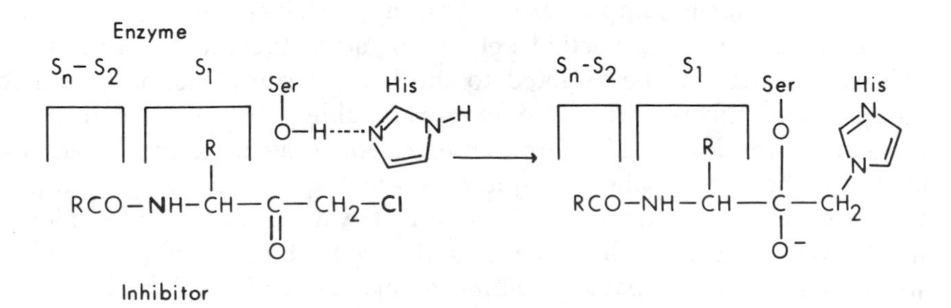


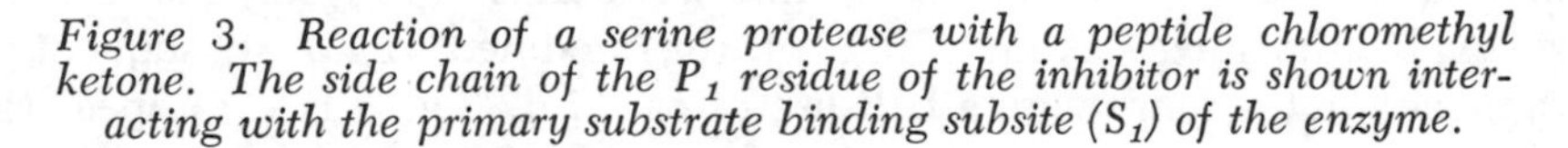

*Figure 3. Reaction of a serine protease with a peptide chloromethyl ketone. The side chain of the $P_1$ residue of the inhibitor is shown interacting with the primary substrate binding subsite ($S_1$) of the enzyme.*

**Table VI. Specificity of Peptide Chloromethyl Ketone Inhibitors at pH 7.5**

| *Inhibitor* | $k_{obs}/[I]$ $(M^{-1} s^{-1})$ |
|---|---|
| Z–Gly–Leu–PheCH$_2$Cl | |
| elastase (human leukocyte) | 0 |
| cathepsin G | 51 |
| MeO–Suc–Ala–Ala–Pro–ValCH$_2$Cl | |
| elastase (human leukocyte) | 1560 |
| cathepsin G | 0 |
| glutathione | 0.88 |
| Ac–Ala–Ala–Pro–ValCH$_2$Cl | |
| elastase (human leukocyte) | 219 |
| cathepsin G | 3.7 |
| Suc–Pro–Leu–PheCH$_2$Cl | |
| mast cell protease I (rat) [a] | 37 |
| mast cell protease II (rat) [a] | 0.4 |
| cathepsin G | 0.7 |

[a] pH 8.0.

will react with chloromethyl ketones of the general formula RCO–Phe-CH$_2$Cl. We recently have been able to design a specific inhibitor for one of the mast cell enzymes by utilizing its secondary specificity (*33*). Most serine proteases don't like $P_3$ prolyl residues in substrates or inhibitors. However, rat mast cell protease I is not affected adversely by a $P_3$ prolyl residue and Suc–Pro–Leu–Phe–CH$_2$Cl is a very selective inhibitor for this enzyme. This inhibitor reacts with rat mast cell protease I 93 and 53 times faster than with rat mast cell protease II and cathepsin G, respectively. Inhibitors of this type should be useful for elucidating the function of mast cell proteases and for uncovering their role in the inflammatory process.

Another possible approach to obtaining specificity in vivo involves attaching peptide chloromethyl ketones or other therapeutics to a suitable carrier that can be targeted to the desired site of action. Human albumin microspheres (HAM) offer considerable potential for delivering reagents to the lung. HAM are nontoxic, nonantigenic, and biodegradable, and because of their unique size are trapped in the pulmonary capillary bed after intravenous injection. HAM with a peptide chloromethyl ketone elastase inhibitor attached have been prepared (*see* Figure 4) and shown to be capable of inhibiting elastase (*34*). When the modified HAM were injected into rats, they were taken up rapidly and exclusively by the lungs. About 50% of the modified HAM subsequently remained in the lungs with a half-life of ca. 17 d. These HAM appear to offer promise as a therapeutic agent for emphysema.

**Azapeptides.** Azaamino acid residues are analogs of amino acids in which the α-CH has been replaced by a nitrogen. This substitution has a profound effect on the reactivity of azapeptides (peptides containing an azaamino acid residue) with serine proteases. In particular azapeptide esters are inhibitors and active-site titrants of several serine proteases (*35, 36*). Inhibition of these enzymes is believed to arise from the acylation of the active-site serine yielding an acylated enzyme (*see* Figure 5). The acyl enzyme formed is a carbazic acid derivative and is sterically quite similar to the acyl enzyme formed when a serine protease reacts with a normal peptide substrate. However, in contrast to a normal acyl enzyme (an ester), the carbazyl enzyme (*see* Figure 5) is much more stable toward deacylation and the enzyme remains inactivated in favorable cases.

The reaction of human leukocyte elastase has been studied with a number of azapeptide *p*-nitrophenyl esters and some of the results are listed in Table VII. All of the azapeptides acylate elastase except Ac–Ala–Aphe–ONp, which reacts very slowly. However this azapeptide will react with cathepsin G as expected from the differing substrate specificity of the two enzymes. The kinetics of the reaction are described in detail elsewhere (*36*), but with most of the inhibitors, $k_{cat}$ is equal to the deacylation (or reactivation) rate of the acylated enzyme. Azapep-

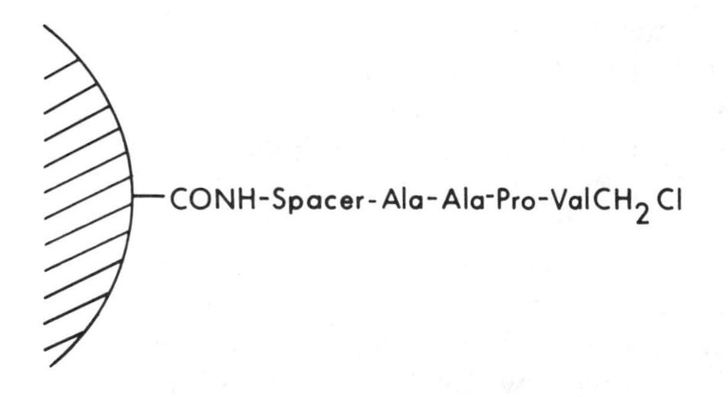


*Figure 4. Chloromethyl ketone elastase inhibitor attached to a HAM.*

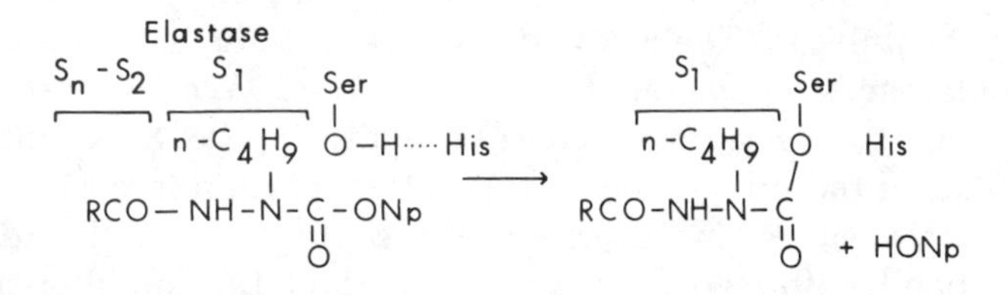


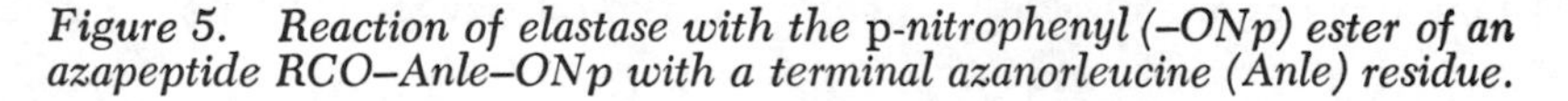

*Figure 5. Reaction of elastase with the* p-*nitrophenyl (–ONp) ester of an azapeptide RCO–Anle–ONp with a terminal azanorleucine (Anle) residue.*

**Table VII. Reaction of Human Leukocyte Elastase with Azapeptide *p*-Nitrophenyl Esters at pH 6.0**

| *Azapeptide* | $k_{cat}(s^{-1}) \times 10^4$ |
|---|---|
| Ac–Ala–Ala–Aval–ONp[a] | 3500 |
| Ac–Ala–Ala–Aile–ONp | 59 |
| Ac–Ala–Ala–Aala–ONp | 20 |
| Ac–Ala–Ala–Anva–ONp | 14 |
| Ac–Ala–Ala–Aleu–ONp | < 1.9 |
| Ac–Ala–Ala–Anle–ONp | < 1.9 |
| Ac–Ala–Aphe–ONp | no reaction |

[a] Aval, Aile, Aala, Anva, Aleu, Anle, and Aphe are valine, isoleucine, alanine, norvaline, leucine, norleucine, and phenylalanine residues where the α-CH has been replaced by a nitrogen atom.

tides with low $k_{cat}$ values make the best inhibitors since the acylated enzymes formed will regain enzyme activity more slowly. The best inhibitor in the series is Ac–Ala–Ala–Anle–ONp, which acylates elastase in less than 10 s and has no measurable $k_{cat}$ at either pH 6.0 or 7.0.

Other active esters of the azapeptides can be used in place of the *p*-nitrophenyl esters. The trifluoroethyl esters will inhibit elastase just as effectively as the *p*-nitrophenyl esters, while the ethyl esters of azapeptides will not acylate elastase (*37*).

**Sulfonyl Fluorides.** Sulfonyl fluorides inhibit serine proteases by reacting with the active-site serine residue. Previously we investigated the rates of inhibition of human leukocyte elastase and cathepsin G by a variety of sulfonyl fluorides and found relatively little selectivity or reactivity (*38*). However, we have discovered recently that the introduction of fluoroacyl groups into the sulfonyl fluoride structure gives considerable reactivity and selectivity for elastase (*39*).

Kinetic data with two inhibitors is given in Table VIII. The 2-trifluoroacetylaminobenzene sulfonyl fluoride reacts effectively with both porcine pancreatic and human leukocyte elastase at rates that are 10–200 times faster than rates observed with cathepsin G and chymotrypsin. This illustrates the considerable selectivity that can be obtained with this class of inhibitor. The acetyl derivative and the 3- and 4-isomers of the 2-trifluoroacetyl compound are unreactive. We propose that the benzene ring of the sulfonyl fluoride binds in the $S_1$ subsite of the enzyme placing the sulfonyl fluoride functional group in position to react with the active-site serine of elastase (*see* Figure 6). In addition, a hydrogen bond is formed between the N–H of the inhibitor and some H-bonding donor (possibly a peptide carbonyl oxygen) of the enzyme.

The electronegative trifluoroacetyl group would strengthen this interaction accounting for the difference in reactivity between the acetyl and trifluoroacetyl derivatives.

Sulfonyl fluorides also have been investigated as inhibitors for numerous other serine proteases including some of the enzymes involved in complement activation (*40*).

**Other Serine Protease Inhibitors.** An interesting approach to the design of serine protease inhibitors is found in furylsaccharin (*41*). This was designed as an agent to acylate serine proteases. It was thought that specificity could be obtained by altering the acyl group. The saccharin was chosen as a leaving group since it is known to be relatively nontoxic. Furylsaccharin (*see* Figure 7) does indeed inhibit several serine proteases including elastase; however, reaction takes place not on the acyl group, but probably at the carbonyl group in the heterocyclic ring.

Suicide substrates have not been applied yet to proteases as extensively as they have to other classes of enzymes such as pyridoxal dependent enzymes (*42, 43*). The potential high selectivity obtainable with suicide substrates certainly makes pursuit of such inhibitors a worthy goal.

**Table VIII. Reaction of Sulfonyl Fluorides with Serine Proteases at pH 7.5**

| | $k_{obs}/[I]$ ($M^{-1}$ $s^{-1}$) | |
|---|---|---|
| *Enzyme* | *2-$CF_3CONH$–$C_6H_4$–$SO_2F$* | *2-$CH_3CONH$–$C_6H_4$–$SO_2F$* |
| Elastase (human leukocyte) | 590 | 6.6 |
| Elastase (porcine pancreatic) | 2300 | <1 |
| Cathepsin G (human leukocyte) | 12 | <1 |
| Chymotrypsin (bovine) | 48 | <1 |

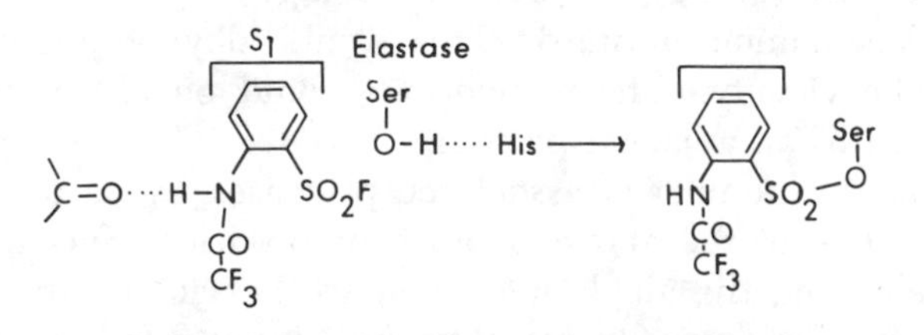


*Figure 6. Proposed model for the reaction of elastase with 2-trifluoroacetylaminobenzene sulfonyl fluoride.*

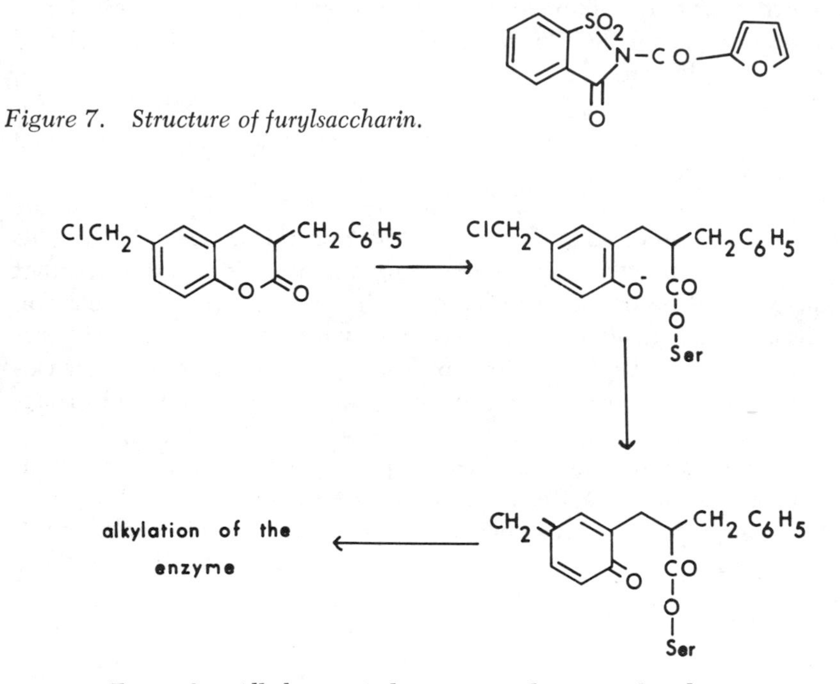


*Figure 7. Structure of furylsaccharin.*

*Figure 8. Alkylation of chymotrypsin by a suicide substrate.*

One successful approach to suicide inhibitors for serine proteases is outlined in Figure 8. The dihydrocoumarin reacts with chymotrypsin to form an acyl enzyme and uncover a *p*-hydroxybenzyl chloride functional group. This is an extremely reactive alkylating agent due to formation of the quinone methide and the enzyme is rapidly inactivated (*44*). It is likely that suicide substrates will be applied to other proteases in the future.

A novel approach to inhibition of serine proteases (or another class of proteases) involves the release of a relatively nondiscriminating agent only in the vicinity of the protease (*45*). Plasmin will cleave D–Val–Leu–Lys–NH–$C_6H_4$–N($CH_2CH_2Cl$)$_2$ releasing a powerful alkylating agent, phenylenediamine mustard. This would alkylate almost any nucleophile in the vicinity. The nucleophile could be either part of the protease or part of an adjacent protein.

Tumors have a number of associated proteases, especially plasminogen activator. Use of the above peptide or prodrug offers a means of selectively delivering the alkylating agent to the vicinity of the tumor. The prodrug inhibited virally transformed chicken embryo fibroblasts and showed much greater selectivity than the nitrogen mustard itself.

**Metalloprotease Inhibitors.** Relatively few irreversible inhibitors have been developed for metalloproteases (*46*). Some representative examples are given in Table IX. Most of the inhibitors are alkylating agents.

The *N*-hydroxy amino acid derivatives are likely to be applicable to other metalloproteases. Thermolysin is inhibited irreversibly at pH 7.2 by $ClCH_2CO$–DL–HOLeu–$OCH_3$ where HOLeu is *N*-hydroxyleucine (*47*). The inhibition reaction involves coordination of the hydroxamic acid functional group to the active-site zinc atom of the enzyme. This then places the chloroacetyl group adjacent to Glu-143, an essential catalytic residue of thermolysin (*see* Figure 9). An ester linkage is formed and the enzyme is inactivated irreversibly. This reagent also inactivated two neutral metalloproteases from *B. subtilis*, but reacted only very slowly with carboxypeptidase A ($t_{1/2} > 3$ d).

This class of inhibitors also has been extended to the *P. aeruginosa* elastase (*28*). $ClCH_2CO$–DL–HOLeu–$OCH_3$ would not inhibit this elastase. However, extending the peptide chain gave a moderate inhibitor. Elastase was inhibited by the tripeptide $ClCH_2CO$–HOLeu–Ala–Gly–$NH_2$ with a $k_3/K_I = 0.092M^{-1}$ $s^{-1}$. As expected, the tripeptide also inhibited thermolysin with $k_3/K_I = 40M^{-1}$ $s^{-1}$. This is another example where extension of the peptide chain of an inhibitor profoundly influ-

**Table IX. Representative Irreversible Inhibitors of Metalloproteases**

| *Enzyme* | *Inhibitor* |
|---|---|
| Thermolysin | $ClCH_2CO$–HOLeu–$OCH_3$ |
| *P. aeruginosa* elastase | $ClCH_2CO$–HOLeu–Ala–Gly–$NH_2$ |
| Carboxypeptidase A | $BrCH_2CO$–MePhe–OH |
| Carboxypeptidase B | $BrCH_2CO$–D–Arg–OH |

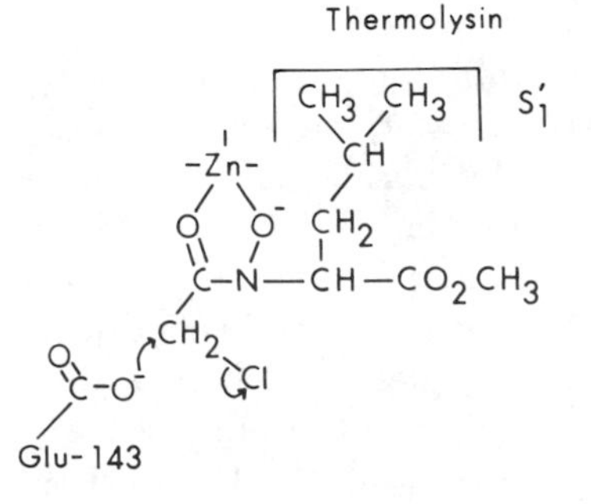


*Figure 9. Reaction of $ClCH_2CO$–HOLeu–OMe with the active site of thermolysin (HOLeu is* N*-hydroxyleucine).*

ences the rate of reaction. Both enzymes must have extended substrate binding sites, but with *P. aeruginosa* elastase interaction with a long peptide chain is essential in order to get inhibition.

Numerous reversible inhibitors have been developed for metalloproteases and some have therapeutic applications.

**Inhibitors of Thiol and Carboxyl Proteases.** Thiol proteases are inactivated by peptide chloromethyl ketones (*30*) and other alkylating agents. Peptide diazomethyl ketones are much more selective reagents since they do not react with serine proteases as do chloromethyl ketones. Diazoketones have been applied to papain and cathepsin B (*48*) thus far and it appears that they should be applicable to most thiol proteases. Specificity should be obtainable by changing the peptide sequence of the inhibitor to match that of the enzyme being studied.

No selective irreversible inhibitors for carboxyl proteases have been developed yet. Reagents such as diazoacetyl–DL–norleucine methyl ester and 1,2-epoxy-3-(-*p*-nitrophenoxyl)propane will inhibit many of the carboxyl proteases that have been examined, but little specificity is likely to be observed.

**Reviews.** A number of reviews are available on the general topic of affinity labeling or on specific types of inhibitors (*30, 42, 43, 46, 49, 50, 51*). These should be consulted for a more complete discussion of irreversible inhibitors for proteases.

### *Protease Inhibitors in Therapy*

A number of protease inhibitors are being used currently for treating disease. Trasylol [Bayer's trademark for pancreatic trypsin inhibitor (Kunitz)] is being used currently in Europe for treating pancreatitis. This disease, which is often fatal in young alcoholic men and older women, results in the leakage of pancreatic proteases into the plasma.

A reversible inhibitor of the angiotensin converting enzyme, Captopril [developed by a group at Squibb (*29*)] is being tested for treating hypertension.

Synthetic elastase inhibitors have considerable potential for treating emphysema. A chloromethyl ketone elastase inhibitor, Ac–Ala–Ala–Pro–Ala$CH_2Cl$, significantly diminishes the extent of experimental elastase-induced emphysema in hamsters (*52*). MeO–Suc–Ala–Ala–Pro–Val$CH_2Cl$ is orally active in providing protection against induced emphysema in rats (*53*). Although the chloromethyl ketones showed no toxic effects during the period of the studies, there is considerable question whether such reactive reagents could be used for treating emphysema in humans.

For use in treating human disease, an enzyme inhibitor has to show a high degree of selectivity and must have minimal side reactions or toxic side effects. Inhibitors that modify proteins by their nature have reactive functional groups (either masked or unmasked) and it is difficult to get the high degree of selectivity required. As a result, most drugs are molecules that reversibly bind to specific enzymes or receptor sites. Irreversible reagents have the advantage that once inhibition of the enzyme has occurred, clearance of the inhibitor does not lead to reactivation of the enzyme as it does with reversible inhibitors. At present, only a limited number of drugs are used that are irreversible enzyme inhibitors. One good example is aspirin, which probably acylates the dioxygenase involved in prostaglandin biosynthesis. It is likely, however, that many future drugs will be irreversible enzyme inhibitors.

It is now evident that protease inhibitors could be used to treat diseases that range from the common cold (inhibitors for the viral processing protease) to chronic diseases such as cancer (tumor protease inhibitors) and emphysema (elastase inhibitors). At present we have just reached the stage where the role of proteases in a number of diseases is being appreciated. The full potential of protease inhibitors has not been reached and the future will see the development of more specific and selective protease inhibitors that will be applied to treating specific diseases.

### *Acknowledgments*

The author's research was supported by grants from the National Institutes of Health (HL 18679) and the Council for Tobacco Research. The human leukocyte enzymes used in this research were a generous gift from James Travis and his group at the University of Georgia. The *Pseudomonas aeruginosa* elastase was the generous gift of Kazuyuki Morihara, Shionogi Research Laboratory, Osaka, Japan.

### *Literature Cited*

1. Neurath, H. *Trends Biochem. Sci.* **1976,** *1,* N27.
2. Neurath, H.; Walsh, K. A. *Proc. Natl. Acad. Sci. USA* **1976,** *73,* 3825.
3. Laurell, C.-B.; Jeppson, J.-O. In "The Plasma Proteins"; Putnam, F. W., Ed.; Academic: New York, 1975; Vol. 1, p. 229.
4. Laskowski, M., Jr.; Kato, I. *Ann. Rev. Biochem.* **1980,** *49,* 593.
5. Mittman, C., Ed. "Pulmonary Emphysema and Proteolysis"; Academic: New York, 1972.
6. Jeppsson, J. O.; Laurell, C. B.; Fagerhol, M. *Eur. J. Biochem.* **1978,** *83,* 143.
7. Matheson, N. R.; Wong, P. S.; Travis, J. *Biochem. Biophys. Res. Commun.* **1979,** *88,* 402.
8. McRae, B.; Nakajima, K.; Travis, J.; Powers, J. C. *Biochemistry* **1980,** *19,* 3973.
9. Nakajima, K.; Powers, J. C.; Ashe, B. M.; Zimmerman, M. *J. Biol. Chem.* **1979,** *254,* 4027.

10. Johnson, D.; Travis, J. *J. Biol. Chem.* **1979,** *254,* 4022.
11. Janoff, A.; Carp, H.; Lee, D. K.; Drew, R. T. *Science* **1979,** *206,* 1313.
12. Gadek, J. E.; Fells, G. A.; Crystal, R. G. *Science* **1979,** *206,* 1315.
13. Oronsky, A. L.; Buermann, C. W. *Ann. Rep. Med. Chem.* **1979,** *14,* 219.
14. Mainardi, C. L.; Dixit, S. N.; Kang, A. H. *J. Biol. Chem* **1980,** *255,* 5435.
15. Gadek, J. E.; Kelman, J. A.; Fells, G. A.; Weinberger, S. E.; Horwitz, A. L.; Reynolds, H. Y.; Fulmer, J. P.; Crystal, R. G. *New Eng. J. Med.* **1979,** *301,* 737.
16. Stracher, A.; McGowan, E. B.; Shafiq, S. A. *Science* **1978,** *200,* 50.
17. Libby, P.; Goldberg, A. L. *Science* **1978,** *199,* 534.
18. Schwartz, W. N.; Bird, J. W. C. *Biochem. J.* **1977,** *167,* 811.
19. Hirsch, H. E.; Parks, M. E. *J. Neurochem.* **1979,** *32,* 505.
20. Rohrlich, S. T.; Rifkin, D. B. *Ann. Rep. Med. Chem.* **1979,** *14,* 229.
21. Borek, C.; Miller, R.; Pain, C.; Troll, W. *Proc. Natl. Acad. Sci. USA* **1979,** *76,* 1800.
22. Meyn, M. S.; Rossman, T.; Troll, W. *Proc. Natl. Acad. Sci. USA* **1977,** *74,* 1152.
23. Troll, W.; Klassen, A.; Janoff, A. *Science* **1970,** *169,* 1211.
24. Morihara, K.; Tsuzuki, H.; Oda, K. *Infect. Immun.* **1979,** *24,* 188.
25. Plaut, A. G.; Gilbert, J. V.; Artenstein, M. S.; Capra, J. D. *Science* **1975,** *190,* 1103.
26. Korant, B.; Chow, N.; Lively, M.; Powers, J. C. *Proc. Natl. Acad. Sci. USA* **1979,** *76,* 2992.
27. Lockart, R. Z., Jr.; Colonno, R. J.; Korant, B. D. *Ann. Rep. Med. Chem.* **1979,** *14,* 240.
28. Nishino, N.; Powers, J. C. *J. Biol. Chem.* **1980,** *255,* 3482.
29. Cushman, D. W.; Cheung, H. S.; Sabo, E. F.; Ondetti, M. A. *Biochemistry* **1977,** *16,* 5484.
30. Powers, J. C. In "Chemistry and Biochemistry of Amino Acids, Peptides and Proteins"; Weinstein, B., Ed.; Marcel Dekker: New York, 1977; Vol. 4, 65.
31. Powers, J. C. *Methods Enzymol.* **1977,** *46,* 197.
32. Powers, J. C.; Gupton, B. F.; Harley, A. D.; Nishino, N.; Whitley, R. J. *Biochem. Biophys. Acta* **1977,** *485,* 156.
33. Yoshida, N.; Everitt, M. T.; Neurath, H.; Woodbury, R. G.; Powers, J. C. *Biochemistry* **1980,** *19,* 5799.
34. Martodam, R. R.; Twumasi, D. Y.; Liener, I. E.; Powers, J. C.; Nishino, N.; Krejcarek, G. *Proc. Nat. Acad. Sci. USA* **1979,** *76,* 2128.
35. Powers, J. C.; Carroll, D. L. *Biochem. Biophys. Res. Comm.* **1975,** *67,* 639.
36. Powers, J. C.; Gupton, B. F. *Methods Enzymol.* **1977,** *46,* 208.
37. Kam, C. M.; Powers, J. C., unpublished data.
38. Lively, M. O.; Powers, J. C. *Biochim. Biophys. Acta* **1978,** *525,* 171.
39. Yoshimura, T.; Powers, J. C., unpublished data.
40. Bing, D. H.; Laura, R.; Andrews, J. M.; Cory, M. *Biochemistry* **1978,** *17,* 5713.
41. Zimmerman, M.; Morman, H.; Mulvey, D.; Jones, H.; Frankshun, R.; Ashe, B. *J. Biol. Chem.* **1980,** *255,* 9848.
42. Abeles, R. H.; Maycock, A. L. *Acct. Chem. Res.* **1976,** *9,* 313.
43. Jung, M. J. *Ann. Rep. Med. Chem.* **1978,** *13,* 249.
44. Bechet, J. J.; Dupaix, A.; Blagoeva, I. *Biochemie* **1977,** *59,* 231.
45. Carl, P. L.; Chakravarty, P. K.; Katzenellenbogen, J. A.; Weber, M. J. *Proc. Nat. Acad. Sci. USA* **1980,** *77,* 2224.
46. Glazer, A. N. In "The Proteins", 3rd ed.; Neurath, H.; Hill, R. L., Eds.; Academic: New York, 1976; Vol. 2, p. 1.
47. Rasnick, D.; Powers, J. C. *Biochemistry* **1978,** *17,* 4363.

48. Leary, R.; Larsen, D.; Watanabe, H.; Shaw, E. *Biochemistry* **1977**, *16*, 5857.
49. Lawson, W. B. *Ann. Rep. Med. Chem.* **1978**, *13*, 261.
50. Shaw, E. *Physiol. Rev.* **1970**, *50*, 244.
51. Baker, B. R. "Design of Active Site Directed Irreversible Enzyme Inhibitors"; John Wiley & Sons: New York, 1967; pp. 122–153.
52. Kleinerman, J.; Ranga, V.; Rynbrandt, D.; Ip, M. P. C.; Sorensen, J.; Powers, J. C. *Am. Rev. Respir. Dis.* **1980**, *121*, 381.
53. Janoff, A.; Dearing, R. *Am. Rev. Respir. Dis.* **1980**, *121*, 1025.

RECEIVED October 3, 1980.

# 13

# Properties In Vivo of Chelate-Tagged Proteins and Polypeptides

CLAUDE F. MEARES, LESLIE H. DeRIEMER,
CHARLES S.-H. LEUNG, SIMON M. YEH,
MICHIKO MIURA, and DAVID G. SHERMAN

Department of Chemistry, University of California, Davis, CA 95616

DAVID A. GOODWIN and CAROL I. DIAMANTI

Department of Nuclear Medicine, Palo Alto V.A. Medical Center and Stanford University, Palo Alto, CA 94304

*Attaching chelating groups to biological molecules and adding radioactive metal ions into the chelating groups provide short-lived radiotracers for diagnostic studies. Choosing the attachment site is restricted when a small polypeptide such as bleomycin (mol wt ≈ 1400) is to be conjugated to a derivative of 1-benzyl-EDTA (mol wt ≈ 400), but the terminal amine region of bleomycin may be modified without destroying its in vivo cancer-localizing properties. A larger number of acceptable labeling sites is expected for a macromolecule such as albumin (mol wt ≈ 68,000). However, the in vivo properties of chelate-tagged albumin depend exquisitely on the chemical-labeling procedure that is used. For example, diazonium coupling yields chelate–protein conjugates which, upon being added to the circulation, are broken down quickly in the liver. On the other hand, alkylating the albumin sulfhydryl group with a relatively selective bromoacetamide reagent yields a product which resembles the native protein in vivo.*

Special chelating agents that may be used to link metal ions stably to biological molecules are a new aspect of chemical modifications of proteins. Because they may be attached to biological molecules and also may bind metal ions, these reagents are known colloquially as bifunctional chelating agents. Since the original work of Sundberg et al. (*1,2*), we

0065-2393/82/0198-0369$05.00/0

have investigated the synthesis of this type of reagent in order to develop convenient, versatile methods for preparing compounds with a variety of possible chelating groups and reactive side chains. Recently we have used the readily available $\alpha$-amino acids as starting materials in a procedure that has the desired features (*3, 4, 5*).

To date, the primary application of bifunctional chelating agents has been for preparing radiopharmaceuticals; these are compounds radiolabeled with $\gamma$-emitters, which now are being used for in vivo diagnosis of disease. For example, BLEDTA, a bleomycin analog bearing an EDTA group, accumulates in some types of cancer tissue and thus can make that tissue detectable by an external radiation counter (gamma camera). There are many other uses for bifunctional chelating agents, such as the attachment of very heavy elements to biological molecules for electron microscopy, the attachment of luminescent metals for energy-transfer studies, the attachment of redox centers incorporating chelated metals for localized chemical reactions, or (perhaps) the use of bifunctional chelating agents to remove toxic metal ions from the body.

The rational application of these molecules to in vivo studies with human subjects requires a detailed knowledge of their behavior in the body, including modes of metabolism and elimination, sites of accumulation, and time dependence. Developing a radiopharmaceutical having desirable biological transport properties is often a matter of much experimentation (e.g., to find conditions for chemical modification of a protein that do not alter its biological activity). The studies with albumin that are discussed below provide some insight into this process.

### *Bifunctional Chelating Agents*

The basis of the present technology for preparing these molecules is the reduction of $\alpha$-amino amides. These amides are prepared easily from $\alpha$-amino acids by converting them to the methyl esters and reacting them with ammonia or amines (the latter provide a versatile route to many different chelating groups). After reducing the amides, all primary and secondary amino groups are carboxymethylated using bromoacetate to yield a polyaminocarboxylate chelating group such as ethylenediaminetetraacetic acid (EDTA). A lengthy but straightforward example is provided by synthesizing (S)-1-(*p*-bromoacetamidobenzyl)-EDTA from L-nitrophenylalanine, as outlined in Figure 1.

If ethanolamine rather than ammonia had been used for preparing the amide, the final chelating group formed would have been *N*-hydroxyethylethylenediaminetriacetate (HED3A); if ethylenediamine had been used for preparing the amide, the chelating group would have been diethylenetriaminepentaacetate (DTPA). The range of further options

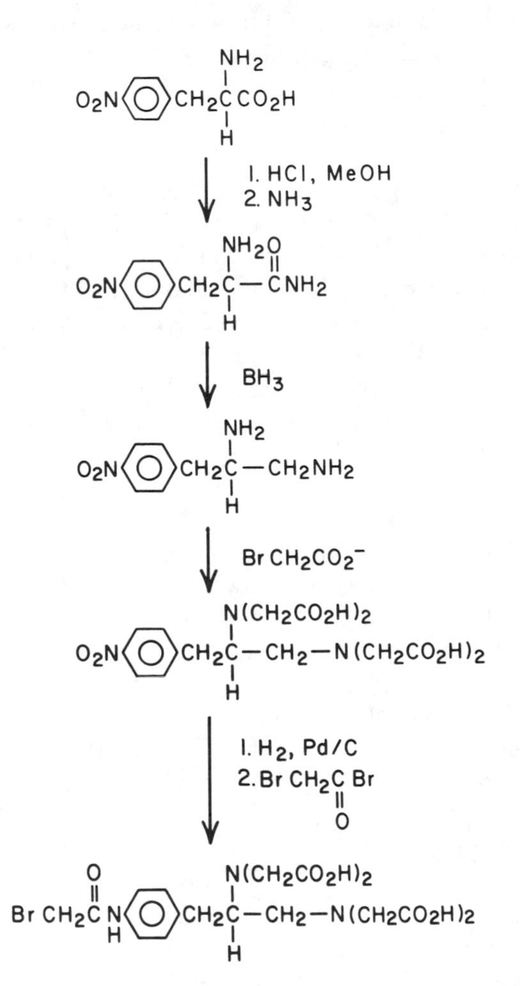


Journal of Labelled Compounds and Radiopharmaceuticals

*Figure 1. Synthesizing a bifunctional chelating agent from an amino acid. In this case, L-nitrophenylalanine was prepared by nitrating the aromatic ring of L-phenylalanine. The illustrated steps then lead to the alkylating agent (S)-1-(p-bromoacetamidobenzyl)-EDTA, which can be attached to bleomycin as shown in Figure 2 (4).*

is enormous. Also, the aromatic amino group of the intermediate *p*-aminobenzyl–EDTA in Figure 1 may be modified in many other ways to provide reactive or biologically active side chains. For example, reaction with thiophosgene yields an isothiocyanate (amine-group reagent), reaction with nitrous acid yields the reactive diazonium compound, acylation with a fatty acid chloride yields an amphiphilic product which binds to cell membranes, and so on. Using other amino acids, or even oligopeptides, as starting materials further extends the number of possible side chains and configurations.

### *Bleomycin and BLEDTA*

The anticancer drug bleomycin has been investigated extensively as an in vivo carrier of $\gamma$-emitting metal ions (*7–12*). Because bleomycin is accumulated selectively in some cancer cells (*13*), radiolabeled bleomycin is potentially useful in locating tumors. Bleomycin is able to bind a number of metal ions, but most bleomycin–metal ion complexes are unstable in vivo and unsuitable for diagnostic use (*7–12*). Cobalt–bleomycin is stable in vivo, and the excellent clinical results that were obtained with this compound have shown that a stable radiolabeled bleomycin can be a very useful diagnostic tool (*12*). However, the long half-life of $^{57}$Co (270 d) poses contamination problems which preclude its widespread clinical use.

The resistance to decomposition exhibited by $^{57}$Co–bleomycin is characteristic of cobalt(III) compounds. $^{57}$Co–bleomycin is prepared by adding cobalt(II) salts to aqueous solutions of bleomycin; in the presence of nitrogen-containing ligands such as bleomycin, complexed cobalt(II) is air-oxidized easily to the cobalt(III) complex (*14*). Due to the $d^6$ electronic configuration of cobalt(III), most of its complexes undergo ligand exchange extremely slowly (*15*). Cobalt(III)–bleomycin is kinetically inert in vivo and for prolonged periods in vitro in the presence of EDTA (*16, 17*).

The uptake of bleomycin by mouse Ehrlich solid ascites tumor is enhanced when bleomycin is bound to cobalt (*16*). The kinetic inertness and higher tumor uptake of cobalt–bleomycin suggested that it would be a suitable starting material for chemical-modification studies. The goal of these studies was the addition of a powerful metal-chelating group at a nonessential site on bleomycin, such that its tumor-localizing properties would be retained. Such a compound could be used as a carrier of radionuclides which are more useful clinically than $^{57}$Co—in particular those that emit only $\gamma$ radiation, with energy between 100 and 400 keV and with half-lives ranging from several hours to a few days (*12*).

The overall strategy is shown in Figure 2. In the first step, nonradioactive cobalt(II) is bound to bleomycin $A_2$ and air-oxidized to form the stable cobalt(III)–bleomycin $A_2$ complex. The dimethyl($\gamma$-aminopropyl)-sulfonium group at the right side of the structure is unique to bleomycin $A_2$; the other bleomycins have quite different terminal amine residues. Since the biological transport properties of all of the bleomycins are similar (*17*), the structure of the terminal amine residue appears to have relatively little influence on transport into cancer cells; therefore this residue is a promising site for chemical modification.

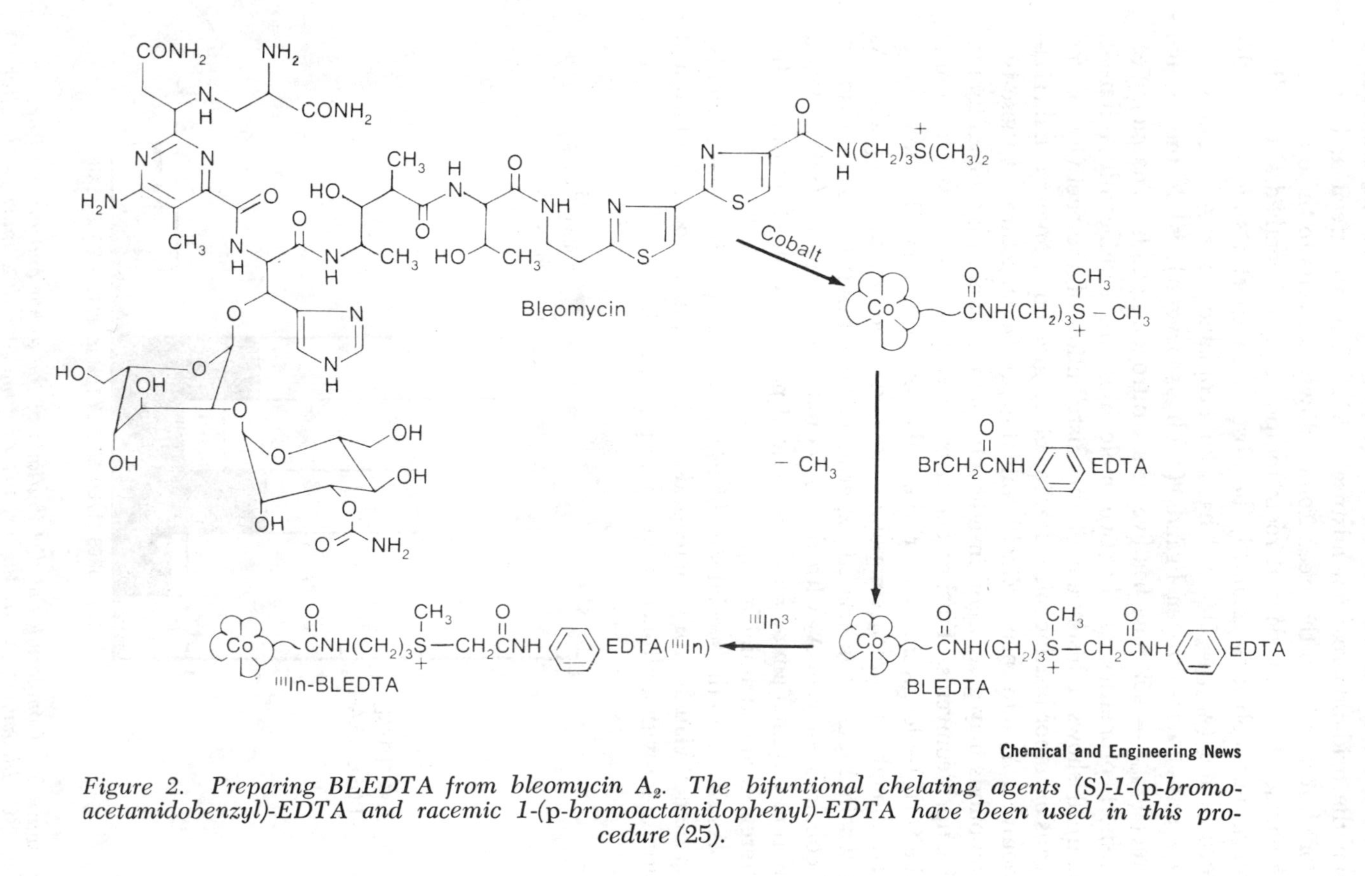


*Figure 2. Preparing BLEDTA from bleomycin $A_2$. The bifuntional chelating agents (S)-1-(p-bromoacetamidobenzyl)-EDTA and racemic 1-(p-bromoactamidophenyl)-EDTA have been used in this procedure (25).*

Two coordination isomers of cobalt(III)–bleomycin $A_2$ are isolated from the oxidation reaction mixture (*18*). They are green and orange colored, respectively; the green isomer slowly converts to the orange over the course of several days at room temperature. If octahedral coordination around cobalt is assumed, the absorption spectra suggest that the green isomer ($\lambda_{max}$ = 590 nm) has four (coplanar) nitrogen ligands and two (trans) non-nitrogen ligands (such as oxygen), while the orange isomer ($\lambda_{max}$ = 452 nm) has five or six nitrogen ligands. The biological transport properties of the two molecules are significantly different. Figure 3 shows a histogram of the distribution of radioactivity in the organs of tumor-bearing mice after injecting green or orange $^{57}$cobalt(III)–bleomycin $A_2$ into them. Elucidating the precise structures of these two compounds may give useful insight into the mechanism by which tumors take up bleomycins. Further chemistry was done on the stable orange cobalt(III)–bleomycin $A_2$ for the sake of experimental convenience, even though the green compound showed superior tumor uptake. As shown in Figure 4, a related experiment comparing the orange isomer of cobalt–bleomycin $A_2$ with that of cobalt–bleomycin $B_2$ (which has a different terminal amine residue) showed no significant effect due to the different terminal amine group.

As shown in the second step of Figure 2 orange cobalt–bleomycin $A_2$ was demethylated with a mercaptide and then alkylated at the terminal sulfur atom with a bifunctional chelating agent (*4, 18*).

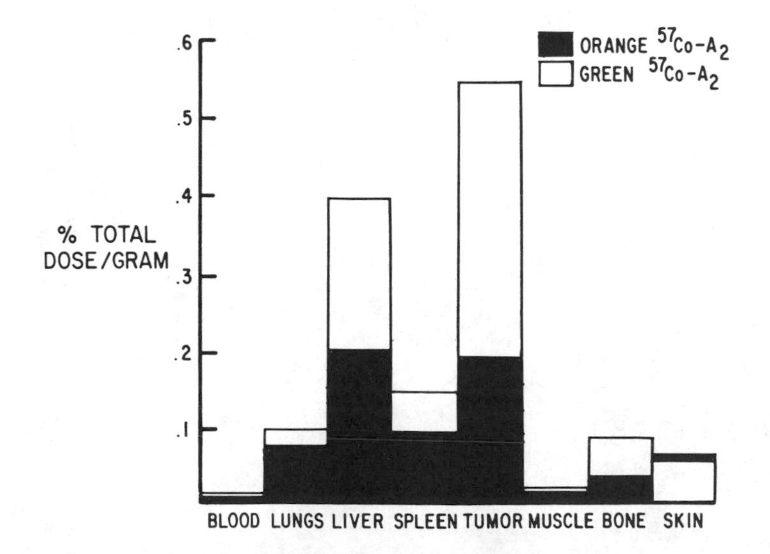


*Figure 3. Comparing the distributions of the green and orange isomers of $^{57}$Co-bleomycin $A_2$ in the organs of tumor-bearing mice 24 h after intravenous injection* (18).

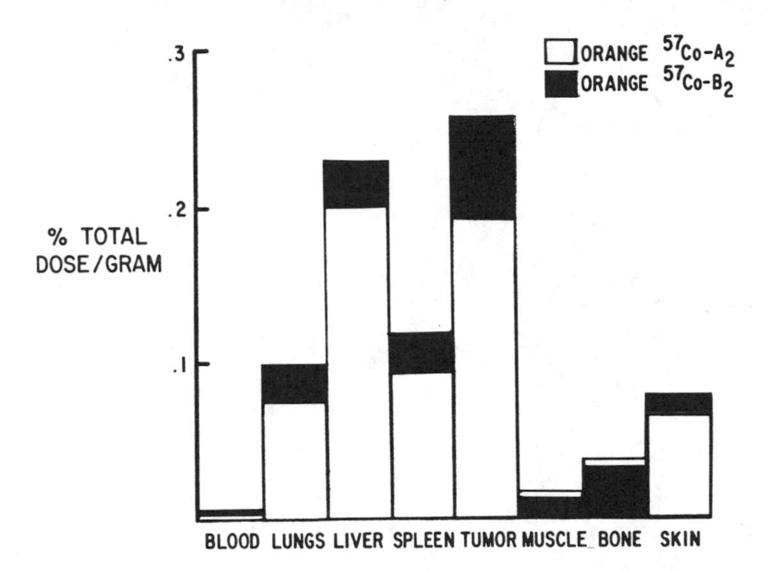


*Figure 4. Comparing the distributions of the orange isomers of $^{57}Co$–bleomycin $A_2$ and $B_2$ in the organs of tumor-bearing mice 24 h after intravenous injection.*

### *Biological Properties of BLEDTA*

BLEDTA can be labeled quickly with a metal ion such as the clinically useful $^{111}In^{2+}$. The cancer-localizing property of $^{111}$In–BLEDTA is illustrated by comparing it with phenyl-EDTA–$^{111}$In and with the complex prepared by mixing $^{111}In^{3+}$ with unmodified bleomycin. The structures of these compounds are compared in Figure 5 and their distributions in tumor-bearing mice are compared in Figure 6. The simple phenyl-EDTA–$^{111}$In chelate is excreted rapidly and shows little uptake in any organs except liver and spleen. The unmodified bleomycin–$^{111}$In complex dissociates in vivo; the serum protein transferrin binds the $^{111}In^{3+}$ and deposits it in several organs. But $^{111}$In–BLEDTA has a higher uptake in the mouse tumor than in any other organ that was assayed (except the kidneys).

Clinical trials with $^{111}$In–BLEDTA have begun and the results to date are summarized in Table I. This radiopharmaceutical is particularly useful for imaging squamous cancers (*19*), and the results obtained with it are comparable with those reported by the French group with $^{57}$Co–bleomycin (*12*). A gamma-camera scan of a patient with thyroid cancer is shown in Figure 7; several metastases are indicated by arrows (*18*). Figure 8 compares whole-body scans of the same patient using $^{131}I^-$ and $^{111}$In–BLEDTA; this thyroid cancer does not take up iodide (because it is anaplastic), but it does accumulate $^{111}$In–BLEDTA.

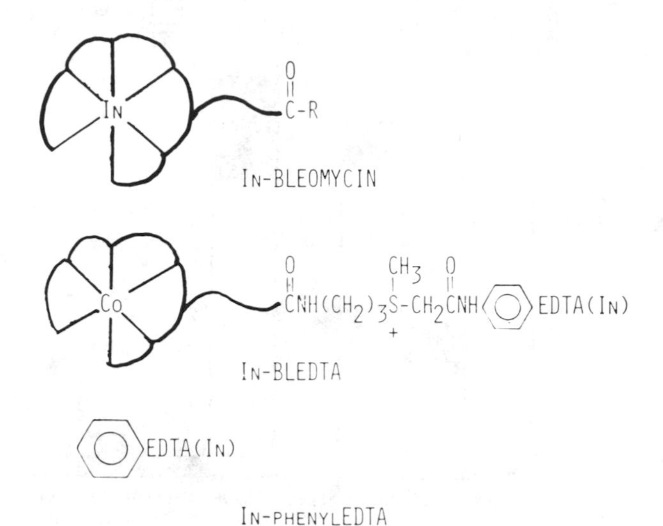


*Figure 5. Comparing the structures of the $^{111}In^{3+}$ complex of unmodified bleomycin, $^{111}In$–BLEDTA, and $^{111}In$–phenyl-EDTA.*

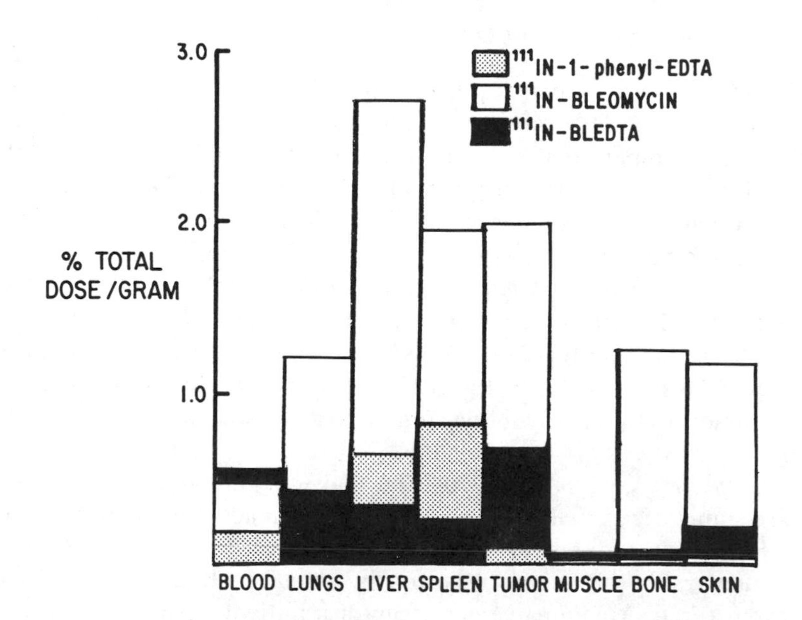


*Figure 6. The organ distributions in tumor-bearing mice of the compounds shown in Figure 5.*

**Table I. Overall Clinical Results Relating $^{111}$In–BLEDTA Scans and Biopsy Results in 95 Patients[a]**

| | *Biopsy Positive* | *Biopsy Negative* | *Row Totals* |
|---|---|---|---|
| *Scan Positive* | 71 | 3 | 74 |
| *Scan Negative* | 17 | 4 | 21 |
| | 88 | 7 | 95 |

[a] A variety of sites and cancer types were represented in this group (*19*). Sensitivity = 71/88 = 81%; positive predictive value = 71/74 = 96%; accuracy = 75/95 = 79%.

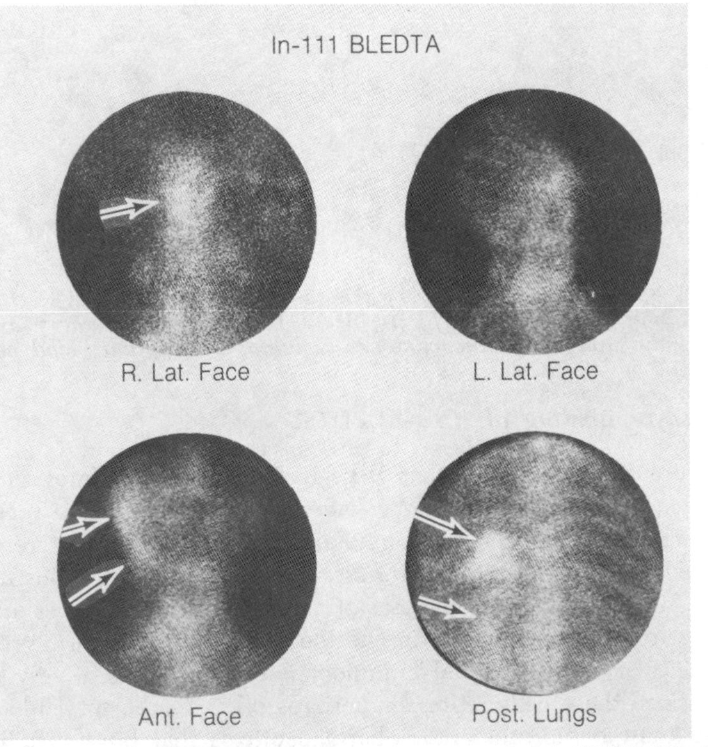


Journal of Medicinal Chemistry

*Figure 7. Gamma-camera views of a human subject with anaplastic thyroid cancer taken 24 h after intravenous injection of 25 nanomoles (2 mCi) of $^{111}$In–BLEDTA. Arrows indicate metastases to the soft tissue of the face and lungs* (18).

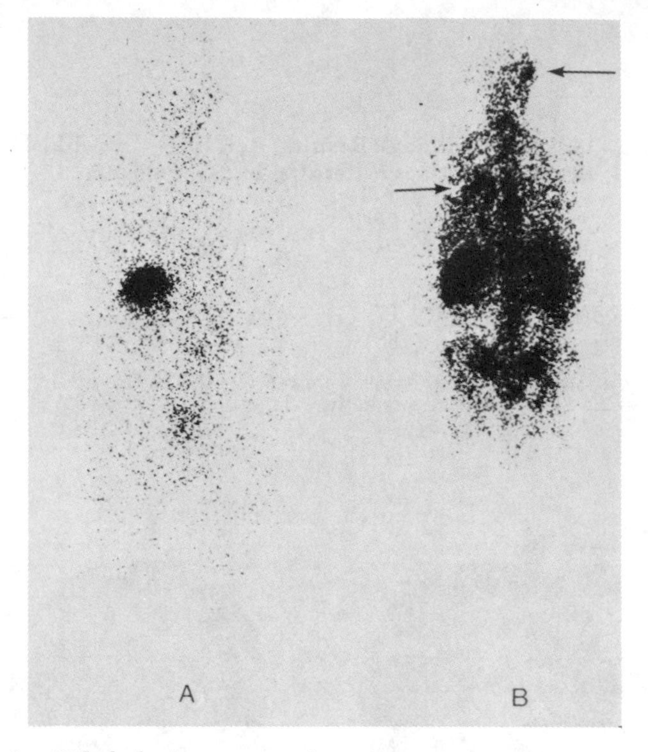


*Figure 8. Whole-body scans of the subject in Figure 7: (A) $^{131}I^{-}$ scan shows only the stomach; (B) $^{111}$In–BLEDTA scan shows tumors (arrows) and also uptake of radioactivity in skeleton, kidney, and blood pool.*

### *Granulocyte Binding of $^{111}$In–BLEDTA*

Figure 8B also shows that $^{111}$In–BLEDTA accumulates in several other tissues besides cancers. We have begun to investigate these processes, first by tracing the small amount of radioactivity that remains in the bloodstream 24 h after injecting $^{111}$In–BLEDTA. As illustrated by typical data in Table II, analysis of patients' blood samples indicated that most of the radioactivity in the blood was attached firmly to white blood cells of the neutrophil granulocyte type. This is a very unusual finding since these cells represent perhaps 0.1% of the total blood cells, and we know of no other radiopharmaceuticals that bind specifically to these cells in vivo.

A related experiment was performed with doubly labeled [$^{57}$Co] $^{111}$In–BLEDTA in mice bearing both a tumor and a sterile abscess. Granulocytes are part of the immune system and accumulate in abscesses. Clear evidence for cleavage of the radiopharmaceutical in vivo was provided by observing that the $^{111}$In/$^{57}$Co ratio was very high in the abscess, blood, and tumor, as shown in Table III. The sulfonium linkage

**Table II. Distribution of $^{111}$In–BLEDTA in Blood 23 H after Injection**

| *Blood Component* | *$^{111}$In Bound (%)* |
|---|---|
| Plasma | 20.9 |
| Red cells | 0.8 |
| Platelets | 0.2 |
| Lymphocytes | 0.6 |
| Granulocytes (PMN) | 77.5 |

**Table III. Comparison of the Uptake of $^{57}$Co–Bleomycin, $^{111}$In–BLEDTA, and a Doubly Labeled Compound with $^{111}$In in the EDTA Moiety and $^{57}$Co Bound to the Bleomycin Moiety[a]**

| | | | *$^{111}$In/$^{57}$Co–BLEDTA* | |
|---|---|---|---|---|
| | *$^{57}$Co–BLM* | *$^{111}$In–BLEDTA* | *$^{111}$In* | *$^{57}$Co* |
| *Tumor* | 0.18 | 0.55 | 0.81 | 0.20 |
| *Abscess* | 0.07 | 0.49 | 0.61 | 0.14 |
| *Blood* | 0.01 | 0.29 | 0.57 | 0.07 |

[a] See Figure 2. BALB/c mice with a tumor in one flank and an abscess in the other were assayed 24 h after injection; the data are reported as percent of injected radioactivity per gram of tissue. Data in the third and fourth columns are very different, implying that the EDTA moiety becomes separated from the bleomycin moiety in vivo. It is interesting to note that the third column roughly parallels the second column (they would be identical if the experiments were perfect) and that the fourth column parallels the first.

of BLEDTA is subject to nucleophilic attack, and cleavage of the S–$CH_2$ bond closest to the chelator is favored because of the adjacent carbonyl group. Thus it appears that mouse granulocytes (and perhaps tumor cells) possess nucleophiles that can be alkylated by $^{111}$In–BLEDTA.

Human granulocytes have been labeled in vitro with $^{111}$In–BLEDTA and then fractionated. As shown in Figure 9, examination of the cell-surface proteins by polyacrylamide gel electrophoresis (PAGE) and scintillation counting showed that a single protein of molecular weight $1.3 \times 10^5$ had been radiolabeled. These studies are continuing and we feel that they ultimately may reveal how bleomycin is taken up selectively by tumor cells.

### *Human Serum Albumin*

Radiolabeled albumin is a convenient radiopharmaceutical for several nuclear medicine procedures (*20*). In the past, we have attached bifunctional chelating agents to albumin by diazonium coupling (*21, 22*). Techniques were developed to specifically radiolabel the product chelat-

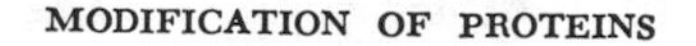


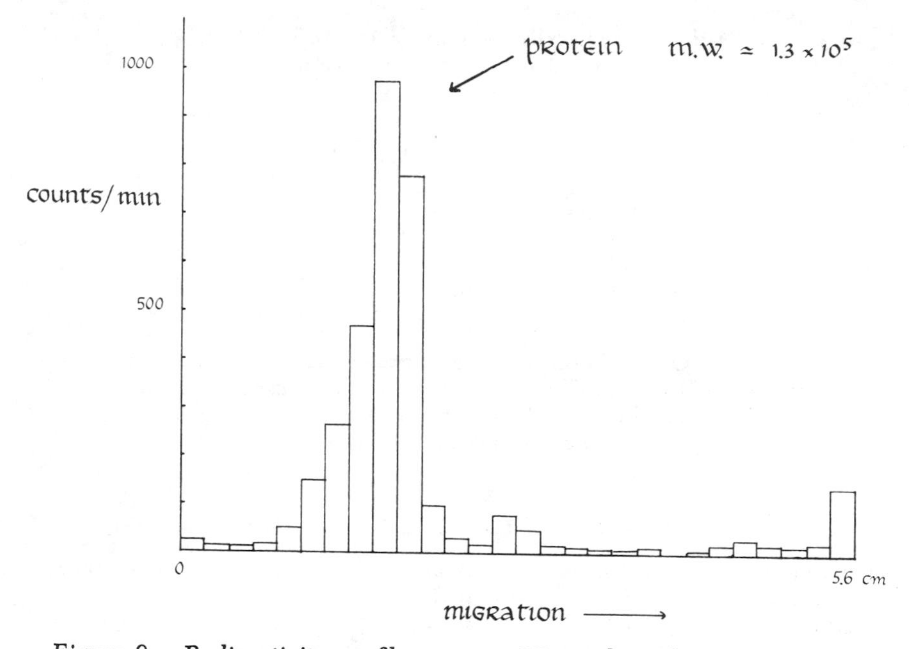


*Figure 9. Radioactivity profile across a SDS–polyacrylamide (5%) gel in which surface–membrane proteins of human neutrophil granulocytes have been separated. Before this experiment, isolated granulocytes had been incubated 2 h at 37°C with $^{111}In$–BLEDTA in plasma. As indicated by the arrow, a single protein with molecular weight $1.3 \times 10^5$ was labeled with $^{111}In$.*

ing groups with metal ions, and the properties of the tagged protein as a tracer for albumin were studied. As shown in Figure 10, most of the product was cleared rapidly from the circulation by the liver, as a foreign protein would be. This behavior was partially the result of the diazonium-coupling conditions which involve low temperature (4°C) with fluctuating pH (6 to 10). By titrating the protein-bound chelating groups with luminescent terbium(III), we found that only about 40% of them were available to bind metals (*22*). As shown in Figure 11, proteolysis of the tagged albumin led to practically complete availability of the chelating groups to bind metals. The presence of an EDTA group buried within the protein surely would have a significant effect on the conformation of alubumin.

### *Nonselective Protein Labeling*

Another problem was the number of chelating groups attached to each protein. Diazonium reagents react with many different amino acid side chains including those of lysine, tyrosine, and histidine (*24*). Therefore labeling a protein by diazonium coupling generally will not produce a homogeneous product.

A simple calculation is instructive. Consider reaction with one type of side chain of which there are, say, dozens on each protein molecule (in identical environments) and suppose we wish to have an average of one chelator per protein molecule. Poisson statistics are appropriate to describe a situation like this (*23*), where the number of labeled side chains will be a small fraction of the number of available side chains.

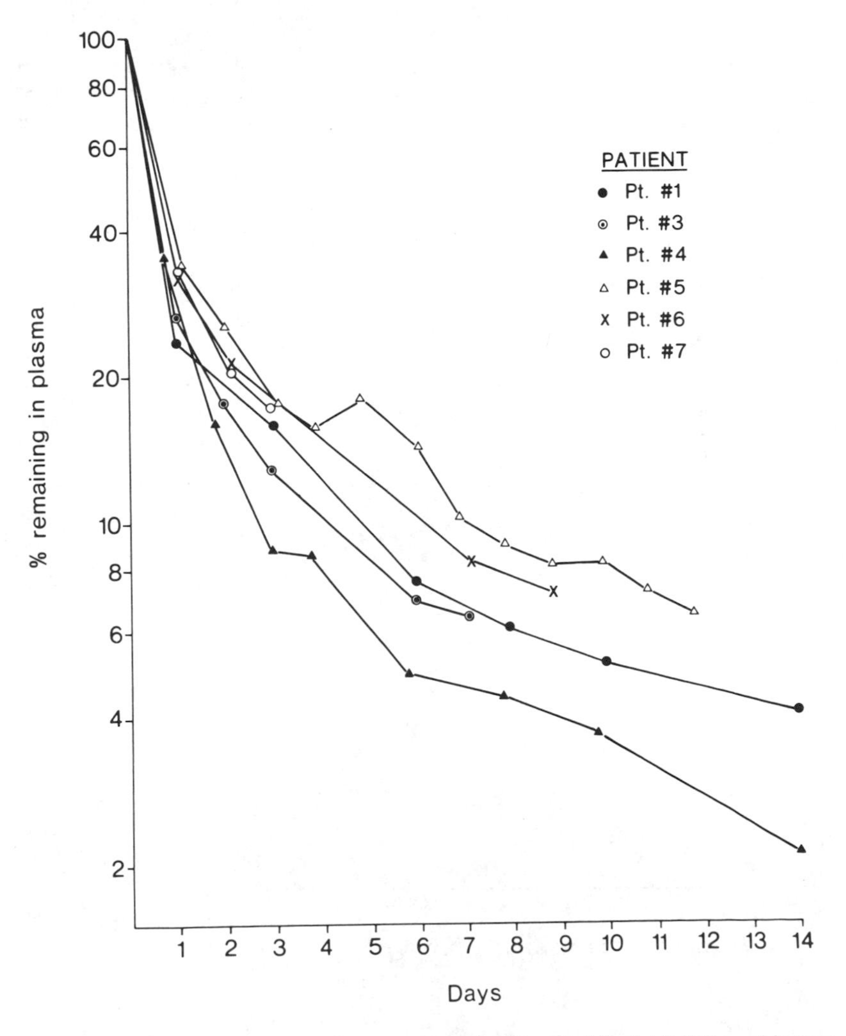


International Journal of Applied Radiation and Isotopes

*Figure 10. Semilogarithmic plot of radioactivity remaining in plasma as a function of time for seven patients injected intravenously with* $^{111}$*In–azoalbumin* (*22*).

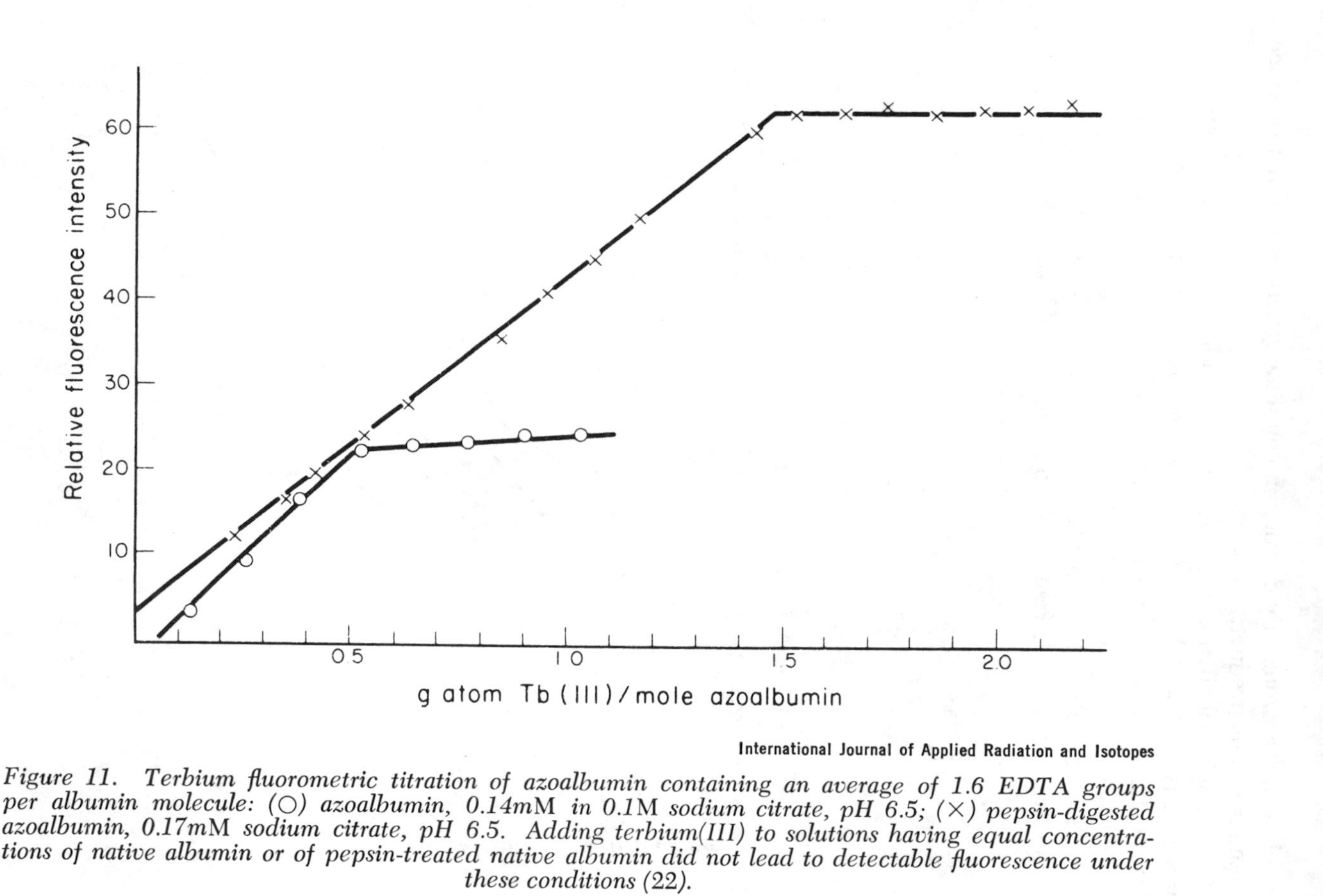


International Journal of Applied Radiation and Isotopes

*Figure 11. Terbium fluorometric titration of azoalbumin containing an average of 1.6 EDTA groups per albumin molecule: (○) azoalbumin, 0.14mM in 0.1M sodium citrate, pH 6.5; (×) pepsin-digested azoalbumin, 0.17mM sodium citrate, pH 6.5. Adding terbium(III) to solutions having equal concentrations of native albumin or of pepsin-treated native albumin did not lead to detectable fluorescence under these conditions (22).*

If we let $X$ be the number of chelators on a particular protein molecule ($X = 0, 1, 2, \ldots$) and $\mu$ is the average number of chelators per protein (we chose $\mu = 1.0$ for this calculation), then the number of chelators per protein molecule is distributed according to:

$$P(X) = \frac{\mu^X}{X!} e^{-\mu}$$

Here $P(X)$ is the probability that an individual protein molecule will have $X$ chelating groups and $e = 2.71828$.

Therefore, the probability that a protein molecule is not labeled at all is $P(0) = e^{-1.0} = 0.37$; the probability that a protein has one chelator is $P(1) = 0.37$ (because of our particular choice of $\mu$); the probability of two chelators per protein is $P(2) = 0.37/2 = 0.18$ and the probabilities of more chelators per protein decline as $0.37/X!$.

Of course, the signal that we observe (e.g., counts/minute) is directly proportional to the number of chelators present. That is, the fraction of radiolabel that is bound to a protein molecule containing $X$ chelating groups is:

$$\left(\frac{X}{\mu}\right) P(X) = \frac{\mu^{X-1}}{(X-1)!} e^{-\mu}$$

Therefore, the fraction of counts/minute coming from singly labeled proteins in our example is $(1/1)P(1) = 0.37$, for doubly labeled proteins it is $(2/1)P(2) = 0.37$, and for triply labeled it is $(3/1)P(3) = 0.18$ and so on.

This simple example shows that even though only 26% of the protein molecules contain more than one chelating group, 63% of the signal will come from these multiply labeled molecules! Refinements in the mathematical model will not change this conclusion qualitatively. Since the more chelators the product contains, the less it resembles the unmodified protein, it is evident that with a nonselective reagent the average number of chelators per protein must be kept well below 1.0 if the labeled product is to retain its native properties.

### *Single-Site Protein Labeling*

However, if a reagent would react only with a single site on a macromolecule, then a chemically homogeneous product having one chelator per protein molecule could be formed. Alkylation of the single sulfhydryl group of albumin with a haloacetyl reagent probably comes reasonably close to this situation.

After extensive dialysis against 0.15*M* NaCl, samples of 1.3m*M* HSA were reacted with 1-(*p*-bromoacetamidophenyl)-EDTA (*18*) at 37°C, pH 8, for 9 h. A sample reaction mixture initially containing 2 mol of chelator per mole of albumin yielded an alkylalbumin product that contained an average of 1.5 chelators per protein molecule; a reaction mixture with one fourth as much chelator gave a product containing 0.3 chelators per protein. In each case, more than 95% of the protein-bound chelating groups were available for metal-binding. These compounds were radiolabeled with $^{111}In^{3+}$ and their disappearance from the blood after injection was studied in a human volunteer. As shown in Figure 12, the lightly labeled protein sample exhibited properties in vivo that were only slightly inferior to commercial $^{125}I$–albumin sold for radiopharmaceutical use.

One further refinement produced interesting results; when the albumin was alkylated without prior dialysis, the product had properties that were indistinguishable from $^{125}I$–albumin, as shown in Figure 13. This

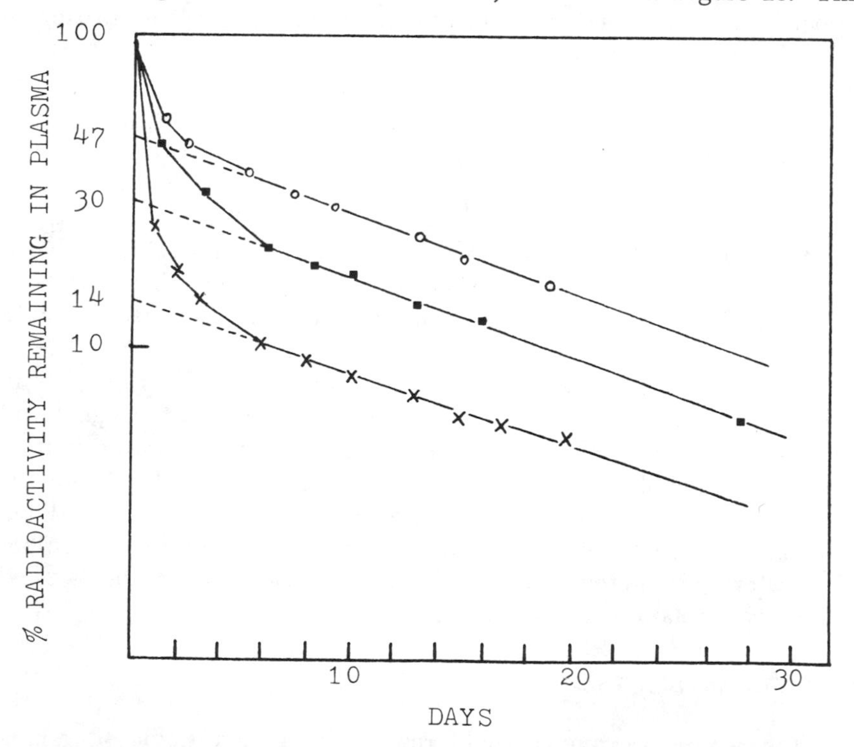


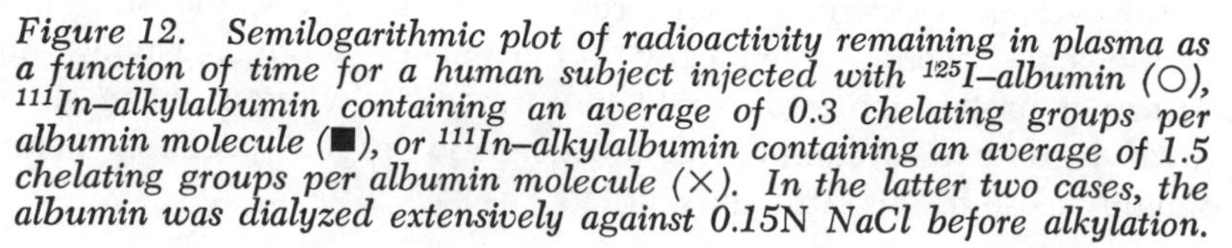

*Figure 12. Semilogarithmic plot of radioactivity remaining in plasma as a function of time for a human subject injected with $^{125}I$–albumin (○), $^{111}In$–alkylalbumin containing an average of 0.3 chelating groups per albumin molecule (■), or $^{111}In$–alkylalbumin containing an average of 1.5 chelating groups per albumin molecule (×). In the latter two cases, the albumin was dialyzed extensively against 0.15*N *NaCl before alkylation.*

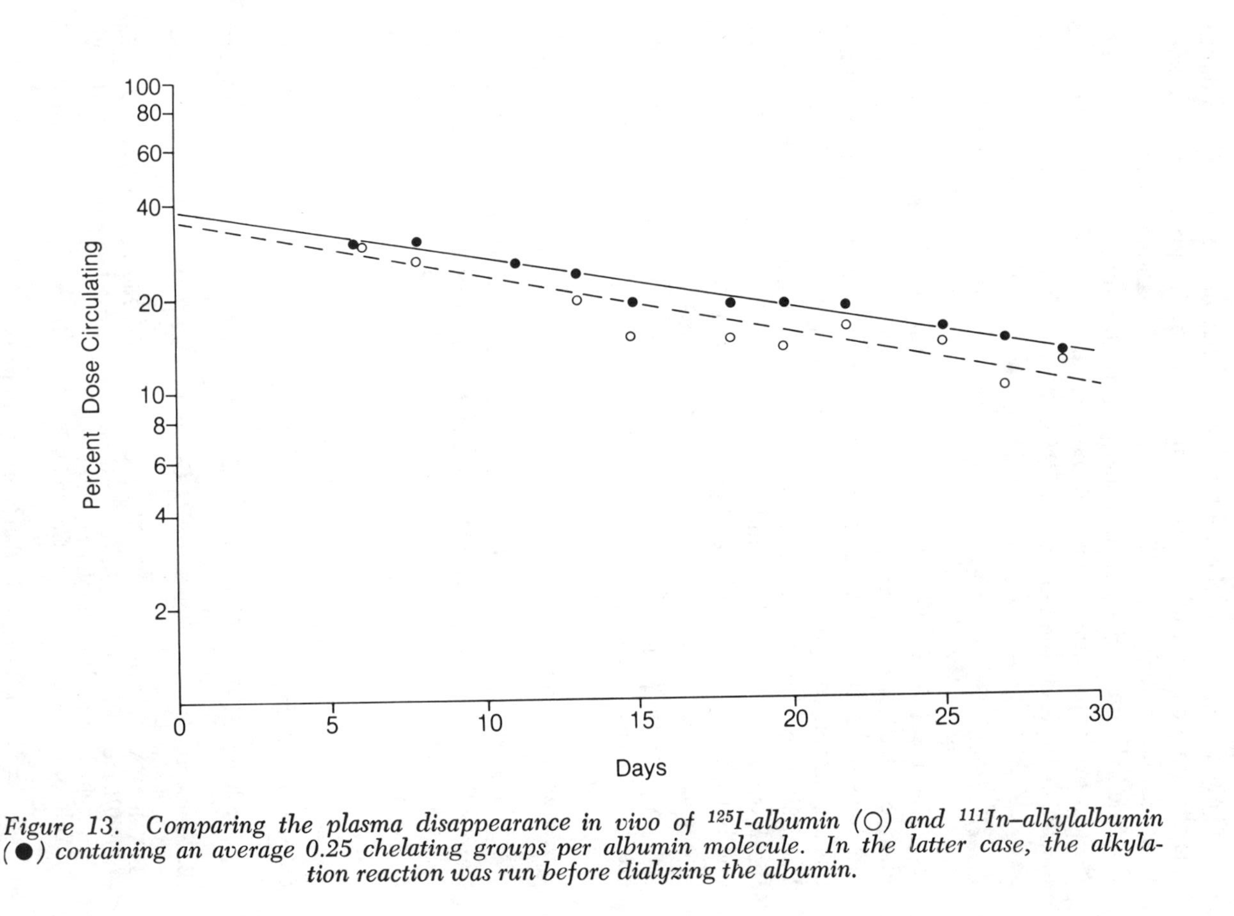


*Figure 13. Comparing the plasma disappearance in vivo of* $^{125}$*I-albumin* (○) *and* $^{111}$*In–alkylalbumin* (●) *containing an average 0.25 chelating groups per albumin molecule. In the latter case, the alkylation reaction was run before dialyzing the albumin.*

suggests that the native conformation of albumin requires binding of some small molecules or ions to the protein.

The clinical use of chelate-tagged albumin in place of $^{125}$I–albumin has the fundamental advantage of not having the protein radiolabeled until just before it is used. This permits a choice among radionuclides so that the half-life, radiations, and energies can be suited to each particular clinical procedure, and it removes the radioactive half-life as a limitation on the shelf-life of the modified protein.

### *Conclusions*

The results of this work underscore the importance of specific chemical reactions at particular sites on biological molecules. It has been our experience that great care is needed in designing and preparing a chelate-tagged compound if it is to retain its normal properties in vivo. Using bifunctional chelating agents effectively separates chemical synthesis from radiochemistry and puts the choice of a radionuclide (or another metal probe) at the end of the labeling procedure. The practical applications of this approach are still at an early stage of development, but its versatility and fundamental advantages over prior technology are now becoming evident.

### *Acknowledgments*

The work was supported by Research Grant CA 16861 and Research Career Development Award CA 00462 to CFM from the National Cancer Institute, NIH, and by a Veterans Administration Research Grant to DAG. We thank R. E. Feeney for his advice on chemical modifications and numerous other topics.

### *Literature Cited*

1. Sundberg, M. W.; Meares, C. F.; Goodwin, D. A.; Diamanti, C. I. *Nature* **1974,** *250,* 587.
2. Sundberg, M. W.; Meares, C. F.; Goodwin, D. A.; Diamanti, C. I. *J. Med. Chem.* **1974,** *17,* 1304.
3. Yeh, S. M.; Sherman, D. G.; Meares, C. F. *Anal. Biochem.* **1979,** *100,* 152.
4. DeRiemer, L. H.; Meares, C. F.; Goodwin, D. A.; Diamanti, C. I. J. Labelled Compd. Radiopharm., in press.
5. Yeh, S. M.; Meares, C. F.; Goodwin, D. A. *J. Radioanal. Chem.* **1979,** *53,* 327.
6. Meares, C. F.; Rice, L. S. *Biochemistry,* in press.
7. Lilien, D. L.; Jones, S. E.; O'Mara, R. E.; Salmon, S. E.; Durie, B. G. M. *Cancer* **1975,** *35,* 1936.
8. Ryo, U. Y.; Ice, R. D.; Jones, J. D.; Beirwaltes, W. H. *J. Nucl. Med.* **1975,** *16,* 127.
9. Hall, J. N.; O'Mara, R. E.; Cruz, P. *J. Nucl. Med.* **1974,** *15,* 498.

10. Lin, M. S.; Goodwin, D. A.; Kruse, S. L. *J. Nucl. Med.* **1974,** *15,* 338.
11. Goodwin, D. A.; Sundberg, M. W.; Diamanti, C. I.; Meares, C. F. In "18th Annual Clinical Conference Monograph, Radiologic and Other Biophysical Methods in Tumor Diagnosis"; Yearb. Med.: Chicago, 1975; pp. 55–88.
12. Nouel, J. P. *GANN Monogr. Cancer Res.* **1975,** *19,* 301.
13. Fujii, A.; Takita, T.; Maeda, K.; Umezawa, H. *J. Antibiot.* **1973,** *26,* 398.
14. Nunn, A. *Int. J. Nucl. Med. Biol.* **1977,** *4,* 204.
15. Taube, H. *Chem. Rev.* **1952,** *50,* 69.
16. Kono, A.; Matsushima, Y.; Kojima, M.; Maeda, T. *Chem. Pharm. Bull.* **1977,** *25,* 1725.
17. Eckelman, W. C.; Rzeszotarski, W. J.; Siegel, B. A.; Kubota, H.; Chelliah, M.; Stevenson, J.; Reba, R. C. *J. Nucl. Med.* **1975,** *16,* 1033.
18. DeRiemer, L. H.; Meares, C. F.; Goodwin, D. A.; Diamanti, C. I. *J. Med. Chem.* **1979,** *22,* 1019.
19. Goodwin, D. A.; Meares, C. F.; DeRiemer, L. H.; Diamanti, C. I.; Goode, R. L.; Baumert, J. E., Jr.; Sartoris, D. J.; Lantieri, R. L.; Fawcett, H. D. *J. Nucl. Med.*, in press.
20. Rhodes, B. A. *Sem. Nucl. Med.* **1974,** *4,* 281.
21. Meares, C. F.; Goodwin, D. A.; Leung, C. S.-H.; Girgis, A. Y.; Silvester, D. J.; Nunn, A. D.; Lavender, P. J. *Proc. Natl. Acad. Sci. USA* **1976,** *73,* 3803.
22. Leung, C. S–H.;Meares, C. F.; Goodwin, D. A. *Int. J. Appl. Radiat. Isot.* **1978,** *29,* 687.
23. Bevington, P. R. "Data Reduction and Error Analysis for the Physical Sciences"; McGraw-Hill: San Francisco, 1969; p. 36.
24. Means, G. E.; Feeney, R. E. "Chemical Modification of Proteins"; Holden-Day: San Francisco, 1971; p. 12.
25. Worthy, W. *Chem. Eng. News* **Sept. 15, 1980,** *58,* 43–44.

RECEIVED October 17, 1980.

# INDEX

# INDEX

## O



## P

## R

*Jacket design by Kathleen Schaner.*
*Production by Robin Giroux and Karen Gray.*

*Composed by Service Composition Co., Baltimore, MD.*
*Printed and bound by Braun-Brumfield, Inc., Ann Arbor, MI.*